W9-CFB-405

# FACILITIES PLANNING

# FACILITIES PLANNING

## SECOND EDITION

JAMES A. TOMPKINS
Tompkins Associates, Inc.

JOHN A. WHITE
Georgia Institute of Technology

YAVUZ A. BOZER
University of Michigan

EDWARD H. FRAZELLE
Georgia Institute of Technology

J. M. A. TANCHOCO
Purdue University

JAIME TREVINO
ABB, Inc.

JOHN WILEY & SONS, INC.

New York   Chichester   Brisbane   Toronto   Singapore

ACQUISITIONS EDITOR   Charity Robey
MARKETING MANAGER   Debra Riegert
PRODUCTION EDITOR   Cathy Ronda
COVER DESIGNER   Madelyn Lesure
COVER IMAGE   Lois and Bob Schlowsky/Tony Stone Images
MANUFACTURING MANAGER   Susan Stetzer
ILLUSTRATION COORDINATOR   Rosa Bryant
OUTSIDE PRODUCTION MANAGEMENT   Ingrao Associates

This book was set in Garamond by University Graphics, Inc., and printed and bound by Hamilton Printing. The cover was printed by Phoenix Color.

Recognizing the importance of preserving what has been written, it is a policy of John Wiley & Sons, Inc. to have books of enduring value published in the United States printed on acid-free paper, and we exert our best efforts to that end.

The paper on this book was manufactured by a mill whose forest management programs include sustained yield harvesting of its timberlands. Sustained yield harvesting principles ensure that the number of trees cut each year does not exceed the amount of new growth.

***Library of Congress Cataloging in Publication Data:***
Facilities planning / James A. Tompkins . . . [et al.]. — 2nd ed.
         p.     cm.
     Includes index.
     ISBN 0-471-00252-6 (alk. paper)
        1. Factories—Design and construction.   2. Facility management.
     I. Tompkins, James A.
     TS177.F324   1996
725′.4—dc20                                                   95-25388
                                                                CIP

Printed in the United States of America

10 9 8 7 6 5 4 3 2 1

# PREFACE

This edition of *Facilities Planning* is quite different from the first edition. For one thing, four co-authors have been added, bringing new perspectives and expertise. Secondly, the revision incorporates the significant advances that have occurred in the past decade, including advances in facilities planning, material handling, and computing technologies, as well as engineering and management philosophies embodied in Just-in-Time, Time Based Competition, Total Quality Management, and Concurrent Engineering.

However, in our attempt to capture contemporary thought, we have not departed from what we considered to be the strength of the first edition—an insistence on the content being relevant to real world facilities planning. We continue to take a pragmatic approach to the subject, and continue to blend theory with practice. We continue to provide a balanced exposure to both the "tried-and-true" and the "state-of-the-art".

Although facilities planning has been taught for more than 50 years, it continues to be challenging and very much alive. While the gap between professional practice and academic preparation has closed during the past decade, professional practice is still ahead of many colleges and universities in terms of what is taught on the subject. (Many engineers and architects perform facilities planning using more advanced computer technology than what is available to engineering and architecture students.)

Over the past decade, we used the first edition in our teaching and consulting and identified its weaknesses; based on our practice and research, we accumulated new information which we taught our students; we directed graduate student research on various aspects of the subject; we presented papers on the subject at national and international conferences. Not only is the subject one that merits "lifetime learning" for our students, but also for ourselves. We continue to learn new things about facilities planning. Each new consulting opportunity provides new insights; each thesis or dissertation provides new understanding and approaches to planning facilities.

In this edition, we have resisted taking a cookbook approach to planning facilities. Instead, we have attempted to provide a window through which you can catch a glimpse of our most recent experiences in planning new and existing facilities. As with the previous edition, we realize there are few "givens" in facilities planning; we continue to resist using "rules-of-thumb" and checklists as the foundations for design; we continue to advocate the use of analytical models and fact-based decision making.

The text is divided into five parts. Part One focuses on the determination of the requirements for people, equipment, space, and material in the facility. Part Two presents concepts and techniques to facilitate the generation of alternative facilities plans. Part Three continues the focus on generating alternative facilities plans, but concentrates on the functions of the organization. Part Four presents a variety of quantitative approaches that can be used to model specific aspects of facilities planning problems. Part Five concludes the treatment of facilities planning and includes chapters dealing with evaluating, selecting, preparing, presenting, implementing, and maintaining the facilities plan.

Depending on the background and interests of the reader, one might eliminate consideration of certain parts of the text. We have endeavored to organize the text in such a way that considerable flexibility is provided to the instructor. Further, we have prepared an Instructor's Manual to accompany the text.

Many people have influenced the development of the text. Deserving special mention are Marvin H. Agee, James M. Apple, Paul T. Eaton, and Ruddell Reed, whose memories we cherish. Additionally, James M. Apple, Jr., Thomas P. Cullinane, Robert P. Davis, Richard L. Francis, Robert Graves, Marc Goetschalckx, Hugh D. Kinney, Leon F. McGinnis, Benoit Montreaul, Colin L. Moodie, James M. Moore, Gunter P. Sharp, H. Donald Ratliff, James Solberg, Jerry D. Smith, John C. Spain, Richard E. Ward, and Richard A. Wysk, among others, have influenced our thinking on the subject. The contributions of numerous graduate students and administrative assistants in our organizations are significant and appreciated.

Finally, we thank our wives and children for their patience, understanding, and support. Their encouragement for us to be more efficient in preparing the revision than we were in preparing the first edition reduced the cycle time significantly.

*James A. Tompkins*
*John A. White*
*Yavuz A. Bozer*
*Edward H. Frazelle*
*J. M. A. Tanchoco*
*Jaime Trevino*

# CONTENTS

# FACILITIES PLANNING

# *1*

# *INTRODUCTION*

## *1.1* **FACILITIES PLANNING DEFINED**

*Facilities planning* is a complex and broad subject that cuts across several specialized disciplines. For example, within the engineering profession, civil, electrical, industrial, and mechanical engineers are all involved with facilities planning. Additionally, architects, consultants, general contractors, managers, real estate brokers, and urban planners also participate in facilities planning. In this text, we do not attempt to address the subject of facilities planning from all these viewpoints; rather, we have approached the subject from an industrial engineer's viewpoint. We have, however, adopted a very broad perspective on the industrial engineer's role in developing effective and efficient facilities plans. Even though we intend for the book to be used by those who prepare facility layouts, select material handling systems, and specify process equipment, it can also be used by those who are responsible for the overall planning of facilities.

The subject of facilities planning has been a popular topic for many years. In spite of its deep heritage, it is one of the most popular subjects of current publications, conferences, and research. The treatment of facilities planning as a subject has ranged from checklist, cookbook-type approaches to highly sophisticated mathematical modeling approaches. In this text, we intend to employ a practical approach to facilities planning, taking advantage of empirical and analytical approaches using both traditional and contemporary concepts.

It should be noted that facilities planning, as addressed in this text, has broad applications. For example, the contents of this book can be applied equally to the planning of a new hospital, an assembly department, an existing warehouse, or the baggage department in an airport. Whether the activities in question occur in the

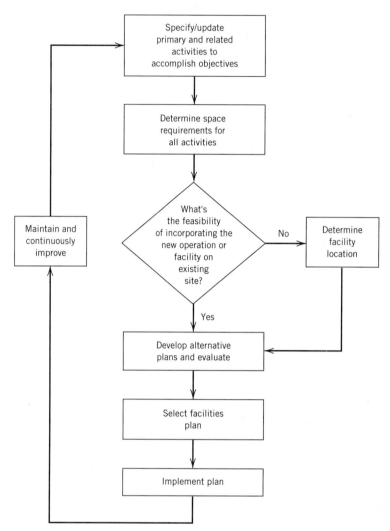

**Figure 1.1**   Continuous improvement facilities planning cycle.

context of a hospital, production plant, warehouse, airport, retail store, school, bank, office, or any portion of these activities, the material presented in this text should be useful in planning facilities. It is important to recognize that contemporary facilities planning considers the facility as a dynamic entity and that a key requirement to a facilities plan is its adaptability, that is, the facility's ability to become suitable for new use. In this regard as a facilities planner, the notion of continuous improvement must be an integral element of the facilities planning cycle. The continuous improvement facilities planning cycle shown in Figure 1.1, details this concept. Whether you are involved in planning a new facility or planning to update an existing facility, the subject matter should be of considerable interest and benefit.

   *Facilities planning determines how an activity's tangible fixed assets best support achieving the activity's objective.* For a manufacturing firm, facilities planning involves the determination of how the manufacturing facility best supports production. In the case of an airport, facilities planning involves determining how the airport facility is to support the passenger–airplane interface. Similarly, facilities planning for a hospital determines how the hospital facility supports providing medical care to patients.

It is important to recognize that we do not use the term facilities planning as a synonym for such related terms as facilities location, facilities design, facilities layout, or plant layout. As depicted in Figure 1.2, it is convenient to divide a facility into its location and its design components.

The *location* of the facility refers to its placement with respect to customer, suppliers, and other facilities with which it interfaces. Also, the location includes its placement and orientation on a specific plot of land.

The design *components* of a facility consist of the facility systems, the layout, and the handling system. The facility systems consist of the structural systems, the atmospheric systems, the enclosure systems, the lighting/electrical/communication systems, the life safety systems, and the sanitation systems; the layout consists of all equipment, machinery, and furnishings within the building envelope; and the handling system consists of the mechanisms needed to satisfy the required facility interactions. The facility systems for a manufacturing facility may include the envelope (structure and enclosure elements), power, light, gas, heat, ventilation, air conditioning, water, and sewage needs. The layout consists of the production areas, production-related or support areas, and personnel areas within the building. The handling system consists of the materials, personnel, information, and equipment handling systems required to support production.

Determining how the *location* of a facility supports meeting the facility's objective is referred to as *facilities location*. The determination of how the design components of a facility support achieving the facility's objectives is referred to as *facilities design*. Therefore, facilities planning may be subdivided into the subject of facilities

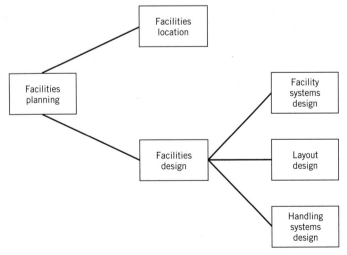

**Figure 1.2** Facilities planning hierarchy.

location and facilities design. Facilities location addresses the macro issues whereas facilities design looks at the micro elements.

The general terms facilities planning, facilities location, facilities design, facility systems design, layout design, and handling system design are utilized to indicate the breadth of the applicability of this text. In Figure 1.3, the facilities planning hierarchy is applied to a number of different types of facilities. It is because of its breadth of application that we employ a unified approach to facilities planning.

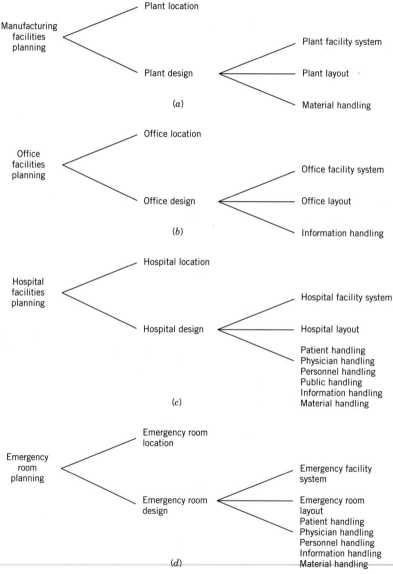

**Figure 1.3**  Facilities planning for specific types of facilities: (*a*) Manufacturing plant. (*b*) Office. (*c*) Hospital. (*d*) Emergency room.

# *1.2*  SIGNIFICANCE OF FACILITIES PLANNING

Since 1955, approximately 8% of the gross national product (GNP) has been spent annually on new facilities in the United States. Table 1.1 indicates the typical expenditures, in percentage of GNP, for major industry groupings. The size of the investment in new facilities each year makes the field of facilities planning important. As stated previously, contemporary facilities planning must include the notion of continuous improvement in the design approach. Adaptability, as a key design criteria, is evidence by the ever increasing performance of previously purchased facilities, which are modified each year and require replanning. For these reasons, it seems reasonable to suggest that over 250 billion will be spent annually in the United States alone on facilities that will require planning or replanning.

Although the scope of facilities planning is indicated by the annual dollar volume of the facilities planned or replanned, it does not appear that adequate planning is being performed. Based on our collective experience, it appears that there exists a significant opportunity to improve the facilities planning process as practiced in industry today.

To stimulate your thoughts on the breath of the facilities planning opportunities, consider the following questions:

1.  What impact does facilities planning have on handling and maintenance costs?
2.  What impact does facilities planning have on employee morale, and how does employee morale impact operating costs?
3.  In what do organizations invest the majority of their capital, and how convertible is their capital once invested?
4.  What impact does facilities planning have on the management of a facility?
5.  What impact does facilities planning have on a facility's capability to adapt to change and satisfy future requirements?

Although these questions are not easily answered, they tend to highlight the importance of effective facilities planning. As an example, consider the first question. It has been estimated that between 20 and 50% of the total operating expenses within

Table 1.1  *Percentage of the Gross National Product (GNP) Typically Expended on New Facilities between 1955 and Today by Industry Grouping*

| Industry | GNP Percentage |
|---|---|
| Manufacturing | 3.2 |
| Mining | 0.2 |
| Railroad | 0.2 |
| Air and other transportation | 0.3 |
| Public utilities | 1.6 |
| Communication | 1.0 |
| Commercial and other | 1.5 |
| All industry | 8.0 |

*Source:* U.S. Bureau of Census.

manufacturing are attributed to materials handling. Furthermore, it is generally agreed that effective facilities planning can reduce these costs by at least 10 to 30%. *Hence, if effective facilities planning were applied, the annual manufacturing productivity in the United States would increase approximately three times more than it has in any year in the last 15 years.*

It is difficult to make similar projections for the other sectors of our economy. However, there is reason to believe that facilities planning will continue to be one of the most significant fields of the future. It represents one of the most promising areas for increasing our rate of productivity improvement. The "reindustrialization of the United States" and the relative performances of Asian, European, and other nations' industries in competition with the United States are strongly related to a need for improved facilities planning.

Economic considerations force a constant reevaluation and recognition of existing systems, personnel, and equipment. New machines and processes render older models and methods obsolete. Facilities planning must be a continuing activity in any organization that plans to keep abreast of developments in its field.

With the rapid changes in production techniques and equipment that have taken place in the recent past and those that are expected in the future, very few companies will be able to retain their old facilities or layouts without severely damaging their competitive position in the marketplace. Productivity improvements must be realized as quickly as they become available for implementation.

One of the most effective methods for increasing plant productivity and reducing costs is to reduce or eliminate all activities that are unnecessary or wasteful. A facilities design should accomplish this goal in terms of material handling, personnel and equipment utilization, reduced inventories, and increased quality.

If an organization continually updates its production operations to be as efficient and effective as possible, then there must be continuous relayout and rearrangement activity in progress. Only in very rare situations can a new process or piece of equipment be introduced into a system without disrupting ongoing activities. A single change may have a significant impact on integrated technological, management, and personnel systems, resulting in suboptimization problems that can only be avoided or resolved through the redesign of the facility.

Employee health and safety is an area that has become a major source of motivation behind many facilities planning studies. In 1970, the Occupational Safety and Health Act was voted into law and brought with it a far-reaching mandate: "to assure as far as possible every working man and woman in the nation safe and healthful working conditions and to preserve our human resources."

Because the act covers nearly every employer in a business affecting commerce who has one or more employees, it has had and will continue to have a significant impact on the structure, layout, and material handling systems of any facility within its scope. Under the provisions of the law an employer is required to provide a place of employment free from recognized hazards and to comply with those occupational safety and health standards set forth in the act.

Because of these stringent requirements and attendant penalties, it is imperative during the initial design phases of a new facility or the redesign and revamping of an existing facility that adequate consideration be given to established health and safety norms and that possible hazardous conditions within the work environment be eliminated or minimized.

Equipment and/or processes that may create hazards to workers' health and safety must be located in areas where the potential for employee contact is minimal. By incorporating vital health and safety measures into the initial design phase, the employer may avoid fines for unsafe conditions and losses in both money and manpower resulting from industrial accidents.

Energy conservation is another major motivation for the redesign of a facility. Energy has become an important and expensive raw material. Equipment, procedures, and materials for conserving energy are introduced to the industrial marketplace as fast as they can be developed. As these energy-conserving measures are introduced, companies should incorporate them into their facilities and their manufacturing process.

These changes often necessitate changes in other aspects of the facility design. For example, in some of the energy-intensive industries, companies have found it economically feasible to modify their facilities to include provisions for using the energy discharged from the manufacturing processes to heat water and office areas. In some cases, the addition of ducting and service lines has forced changes in material flows and the relocation of in-process inventories. In one large office complex, the basic facility design was altered so that many rooms could be heated by the excess energy created by the computing equipment.

If a company is going to retain a competitive edge today, it must reduce its consumption of energy. One method of doing this is to modify and relayout facilities or redesign material handling systems and manufacturing processes to accommodate new energy-saving measures.

Other factors that motivate investment in new facilities or the alteration of existing facilities are community considerations, fire protection, security, and the American Disabilities Act (ADA) of 1989. Community rules and regulations regarding noise, air pollution, and liquid and solid waste disposal are frequently cited as reasons for the installation of new equipment that requires modification of facilities and systems operating policies.

On nearly a daily basis in any large city newspaper, reports of fires that completely destroy a whole facility can be found. In many instances, these fires can be attributed to poor housekeeping or poor facilities design. Companies are now carefully seeking modifications to existing material handling systems, storage systems, and manufacturing processes that will lower the risk of fire.

Pilferage is yet another major and growing problem in many industries today. Several billion dollars' worth of merchandise is stolen annually from manufacturing companies in the United States. The amount of control devoted to material handling, flow of materials, and design of the physical facility can help reduce losses to a firm.

One of the most significant challenges to facilities planners today is how to make the facility "barrier free" in compliance with the ADA of 1989. The enactment of this legislature has resulted in a significant increase in the alternation of existing facilities and has radically shaped the way facilities planners approach the planning and design of facilities. This new act impacts all elements of the facility, from parking space allocation and space design, ingress and egress ramp requirements, and restroom layout to drinking fountain recommended rim heights. Companies are aggressively spending billions of dollars to comply with the law, and those involved with facilities planning must be the leaders in pursuing the required changes.

# *1.3* OBJECTIVES OF FACILITIES PLANNING

Facilities planning may be likened to painting a picture or playing a musical instrument. There are general guidelines, principles, and techniques that, if followed, may lead to an effective facilities plan, a beautiful piece of art, or a splendid performance. However, one must go beyond an intellectual understanding of guidelines, principles, and techniques and develop a feel for the interrelated, often conflicting, objectives that affect the overall results. Facilities planning then, although becoming more scientific, continues to rely greatly on the experience of planners.

A meaningful objective for facilities planning is difficult to state succinctly. However, by subdividing facilities planning into facilities location and facilities design, a list of workable objectives may be established. In 1930, W. G. Holmes stated the following widely accepted objective of industrial facilities location: ". . . to determine the location which, in consideration of all factors affecting deliver-to-customers cost of the product(s) to be manufactured, will afford the enterprise the greatest advantage to be obtained by virtue of location."[2] This objective may be easily generalized to include both industrial and nonindustrial facilities by considering not only products that are manufactured but also services that are offered.

Although the task of designing or determining the arrangement of equipment and related work areas is only one facet of the total facility planning process, the development of the best possible facility design must be central to the facility planning activity. Some typical facilities design objectives are to:

1.  Support the organization's vision through improved material handling, material control, and good housekeeping.
2.  Effectively utilize people, equipment, space, and energy.
3.  Minimize capital investment.
4.  Be adaptable and promote ease of maintenance.
5.  Provide for employee safety and job satisfaction.

It is not reasonable to expect that a facility design will be superior to all others for every objective listed. Some of the objectives are conflicting. Hence, it is important to evaluate carefully the performance of each alternative, using each of the appropriate criteria.

Facilities planning is the composite of facilities location and facilities design. Hence, the objective of facilities planning is to plan a facility that achieves facilities location and design objectives.

# *1.4* FACILITIES PLANNING PROCESS

The facilities planning process is best understood by placing it in the context of a facility life cycle. Although a facility is planned only once, it is frequently replanned to synchronize the facility and its constantly changing objectives. The facilities planning and facilities replanning process are linked by the continuous improvement facilities planning cycle shown in Figure 1.1. This process continues until a facility is torn down as the facility is continuously improved to satisfy its constantly changing objectives.

Even though facilities planning is not an exact science, it can be approached using an organized, systematic approach. Traditionally, the engineering design process can be applied; it consists of the following six steps:

1. Define the problem.
2. Analyze the problem.
3. Generate alternative designs
4. Evaluate the alternatives.
5. Select the preferred design.
6. Implement the design.

Applying the engineering design process to facilities planning results in the following process:

1. *Define (or redefine) the objective of the facility.* Whether planning a new facility or planning the improvement of an existing facility, it is essential that the product(s) to be produced and/or service(s) to be provided be specified quantitatively. Volumes or levels of activity are to be identified when possible.

2. *Specify the primary and support activities to be performed in accomplishing the objective.* The primary and support activities to be performed and requirements to be met should be specified in terms of the operations, equipment, personnel, and material flows involved. Support activities allow primary activities to function with minimal interruption and delay. As an example, the maintenance function is a support activity for manufacturing.

3. *Determine the interrelationships among all activities.* Establish if and how activities interact or support one another within the boundaries of the facility and how this is to be undertaken. Both quantitative and qualitative relationships should be defined.

4. *Determine the space requirements for all activities.* All equipment, material, and personnel requirements must be considered when calculating space requirements for each activity.

5. *Generate alternative facilities plans.* The alternative facilities plans will include both alternative facilities locations and alternative designs for the facility. The facilities design alternatives will include alternative layout designs, structural designs, and material handling system designs. Depending on the particular situation, the facility location decision and the facility design decision can be decoupled.

6. *Evaluate alternative facilities plans.* On the basis of accepted criteria, rank the plans specified. For each, determine the subjective factors involved and evaluate if and how these factors will affect the facility or its operation.

7. *Select a facilities plan.* The problem is to determine which plan, if any, will be the most acceptable in satisfying the goals and objectives of the organization. Most often cost is not the only major consideration when evaluating a facilities plan. The information generated in step 6 should be utilized to arrive at final selection of a plan.

8. *Implement the facilities plan.* Once the plan has been selected, a considerable amount of planning must precede the actual construction of a facility or the layout of an area. Supervising installation of a layout, getting ready to start up, actually starting up, running, and debugging are all part of the implementation phase of facilities planning.

9. *Maintain and adapt the facilities plan.* As new requirements are placed on the facility, the overall facilities plan must be modified accordingly. It should reflect any energy-saving measures or improved material handling equipment that become available. Changes in product design or mix may require changes in handling equipment or flow patterns that, in turn, require an updated facilities plan.

10. *Redefine the objective of the facility.* As indicated previously in Step 1, it is necessary to identify the products to be produced or services to be provided in specific quantifiable terms. In the case of potential modifications, expansions, and so on for existing facilities, all recognized changes must be considered and integrated into the layout plan.

A novel contemporary facilities planning approach is the winning facilities planning process, as shown in Figure 1.4. More detailed explanation of the winning facilities planning process steps is shown in Table 1.2.

The Model of Success referred to in Figure 1.4 presents a clear direction for where a business is headed. Experience has shown that in order for the facilities plan to be successful, not only a clear understanding of the vision is needed but also, the mission, the requirements of success, the guiding principles and the evidence of success. It is the total of these five elements (Vision, Mission, Requirements of Success, Guiding Principles, and Evidence of Success) that form an organization's Model of Success.

The definitions of these five elements are:

• Vision: A description of where you are headed

• Mission: How to accomplish the vision

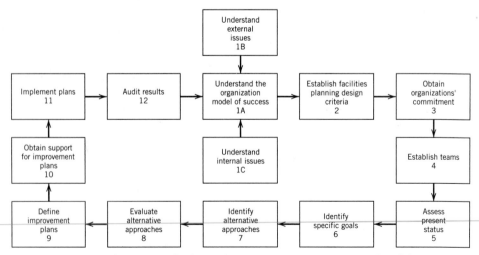

**Figure 1.4** Winning facilities planning process. *Source:* Tompkins [3].

Table 1.2 *Explanation of the Winning Facilities Planning Process*

| Step | Function | Comment |
|------|----------|---------|
| 1A | Understand the organization model of success | This requires an education program for all levels of an organization. Understanding an organization's model of success is a prerequisite for successful facilities planning. |
| 1B | Understand external issues | This requires an external outreach via professional society involvement, participation in trade shows, seminars and conferences and reading of magazines and books. A coordinated effort is required if external issues are to be well understood. |
| 1C | Understand internal issues | A winning organization must understand not only the model of success, but also an organization's business plan, resources, constraints and objectives of the overall organization. A prerequisite of winning is understanding a company's future. |
| 2 | Establish facilities planning design criteria | To implement improvements, an organization must have focus. This step requires that management determine the facilities planning design criteria. |
| 3 | Obtain organizational commitment | Management must make a clear commitment to implement the justified improvements consistent with the facilities planning design criteria. This commitment must be uncompromised. |
| 4 | Establish teams | Teams having a broad-based representation and the ability to make decisions should be established for each design requirement. These teams must be uncompromised. |
| 5 | Assess present status | This assessment will result in the baseline against which improvements will be measured. Both quantitative and qualitative factors should be assessed. |
| 6 | Identify specific goals | The identification of clear, measurable, time-related goals for each design criteria. For example, reduce raw material inventory to $300,000 by June 1. |
| 7 | Identify alternative approaches | The creative process of identifying alternative systems, procedures, equipment or methods to achieve the specified goals. The investigation of all feasible alternatives. |
| 8 | Evaluate alternative approaches | The economic and qualitative evaluation of the identified alternatives. The economic evaluation should adhere to corporate guidelines while estimating the full economic benefit of pursuing each alternative. |
| 9 | Define improvement plans | Based upon the evaluation done in Step 8, select the best approach. Define a detailed implementation and cash flow schedule. |
| 10 | Obtain support for improvement plans | Sell the improvement plans to management. Document the alternatives, the evaluation, and the justification. Help management visualize the improved operation. |

**Table 1.2** *Continued*

| Step | Function | Comment |
|------|----------|---------|
| 11 | Implement plans | Oversee development, installation, soft load, start-up and debug. Train operators and assure proper systems utilization. Stay with effort until results are achieved. |
| 12 | Audit results | Document actual systems operation. Compare results with the specified goal and anticipated performance. Identify and document discrepancies. Provide appropriate feedback. |

*Source:* Tompkins [3].

- Requirements of Success: The science of your business
- Guiding Principles: The values to be used while pursuing the vision.
- Evidence of Success: Measurable results that will demonstrate when an organization is moving towards their vision.

To help people understand where their organization is headed, it is often useful to illustrate the first four elements of the Model of Success in graphical form as shown in Figure 1.5. This graphical representation is often called the winner's circle and is viewed as the organization's bullseye.

In Table 1.3, the traditional engineering design process and the winning facilities planning process are compared. The first phases of the facilities planning process involve either the initial definition of the objectives of a new facility or the updating of an existing facility. These first phases are undertaken by the people charged with overall responsibility for facilities planning and management of the facility.

The second phase of the facilities planning process is assessing the present status, identifying specific goals, identifying alternative approaches, evaluating alternative approaches, defining improvement plans, and obtaining support for improvement. The final phase consists of implementing the plans and auditing the results. In

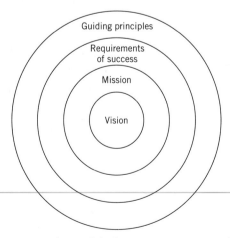

**Figure 1.5** The Model of Success "Winning Circle."

**Table 1.3** *Comparison of the Engineering Design Process, the Winning Facilities Planning Process, and the Facilities Planning Process*

| Phase | The Engineering Design Process | The Facilities Planning Process | Facilities Planning Steps | |
|---|---|---|---|---|
| I | Define problem | 1. Define or redefine objective of the facility<br>2. Specify primary and support activities | 1A | Understand the organization model of success |
| | | | 1B | Understand external issues |
| | | | 1C | Understand internal issues |
| | | | 2 | Establish facilities planning design criteria |
| | | | 3 | Obtain organizational commitment |
| II | Analyze the problem<br>Generate alternatives<br>Evaluate the alternatives<br>Select the preferred design | 3. Determine the interrelationships<br>4. Determine space requirements<br>5. Generate alternative facilities plan<br>6. Evaluate alternative facilities plan<br>7. Select a facilities plan | 4 | Establish teams |
| | | | 5 | Assess present status |
| | | | 6 | Identify specific goals |
| | | | 7 | Identify alternative approaches |
| | | | 8 | Evaluate alternative approach |
| | | | 9 | Define improvement plans |
| | | | 10 | Obtain support for improvement plans |
| III | Implement the design | 8. Implement the plan<br>9. Maintain and adopt the facilities plan<br>10. Redefine the objective of the facility | 11 | Implement plans |
| | | | 12 | Audit results |

applying the facilities planning concepts, an iterative process is often required to develop satisfactory facilities plans. The iterative process might involve considerable overlap, backtracking, and cycling through the analysis, generation, evaluation, and selection steps of the engineering design process.

At this point, a word of caution seems in order. Namely, you should not infer from our emphasis on a unified approach to facilities planning that the process of replanning a pantry in a cafeteria is identical to that involved in planning a new manufacturing facility. The scope of a project does affect the intensity, magnitude, and thoroughness of the study. However, the facility planning process described above and depicted in Figure 1.6 should be followed.

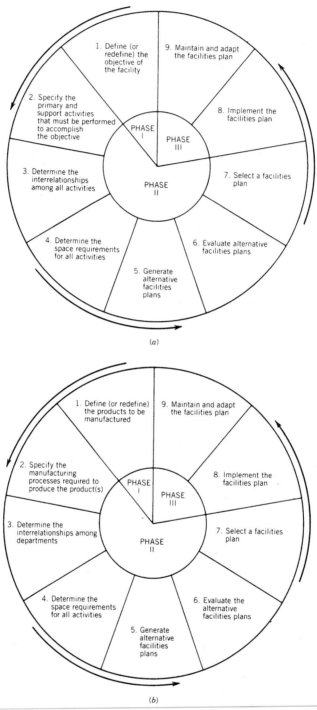

**Figure 1.6** The facilities planning process. (*a*) General. (*b*) Manufacturing facilities. (*c*) Hospital facilities.

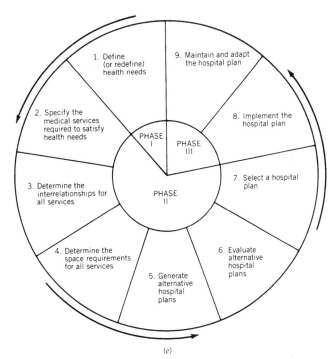

*(c)*

**Figure 1.6** Continued.

## 1.5 ORGANIZATION OF THE BOOK

This book is aimed primarily at undergraduate students majoring either in engineering or engineering technology. A secondary audience is practicing facilities planners. Finally, some of the material presented will be of interest to first-year graduate students interested in quantitative facilities planning techniques.

The text is organized on the basis of the facilities planning process. Part 1 focuses on defining requirements and encompasses the first four steps in the facilities planning process. The importance of carefully defining (or redefining) the objectives of the facility leads to a separate treatment of strategic facilities planning in Chapter 2. The relationship between facilities planning and product, process, and schedule design is explored in Chapter 3. The specification of primary and support activities to be performed in the facility, coupled with the determination of interrelationships and space requirements for all activities, results in consideration of flow, space, and activity relationships in Chapter 4, and the presentation of personnel requirements in Chapter 5.

The generation of alternative facilities plans serves as the basis for Parts 2, 3, and 4. Part 2 is devoted to basic concepts involved in developing alternatives; individual chapters are devoted to a consideration of material handling system design (Chapter 6) and facilities layout design (Chapter 7 and 8).

Part 3 utilizes a function organization in addressing the generation of alternatives. Individual chapters treat such functions as receiving, shipping, storage, and warehousing (Chapter 9), manufacturing (Chapter 10), and facility systems (Chapter 11).

Quantitative approaches that have proven to be useful in developing alternative facilities plans are the subject of Part 4. Chapter 12 presents deterministic and probabilistic models. The models presented are intended to aid the facilities planner in determining the locations of facilities, the layout of a warehouse, the type of storage system to be used, the design of a recirculating conveyor, the design of belt and roller conveyors, the design of automated storage systems and order picking systems, and the analysis of waiting-line problems.

The process of evaluating, selecting, preparing, presenting, implementing, and maintaining the facilities plan is the subject of Part 5. Chapter 13 addresses the evaluation and selection process; in addition to presenting an economic justification procedure, we illustrate the use of spreadsheets in performing economic justifications. Preparation, presentation, implementation, and maintenance of the facilities plan is the subject of Chapter 14.

## *1.6* SUMMARY

Facilities planning is treated in this book from an industrial engineering viewpoint, but every attempt has been made to provide a comprehensive approach that can benefit anyone responsible for the overall planning of facilities. We have endeavored to use a practical approach in presenting the subject. A balance of empirical and analytical approaches to facilities planning is presented. Facilities planning

1. Determines how an activity's tangible fixed assets should contribute to meeting the activity's objectives.
2. Consists of facilities location and facilities design.
3. Is part art and part science.
4. Can be approached using the engineering design process.
5. Is a continuous process and should be viewed from a life cycle perspective.
6. Represents one of the most significant opportunities for cost reduction and productivity improvement.

## BIBLIOGRAPHY

1. Cullinane, T. P., and Tompkins, J. A., "Facility Layout in the 80's: The Changing Conditions," *Industrial Engineering,* vol. 12, no. 9, pp. 34–42, September 1980.
2. Holmes, W. G., *Plant Location,* McGraw-Hill, New York, 1930.
3. Tompkins, J. A., *Winning Manufacturing: The How To Book Of Successful Manufacturing,* IIE, Norcross, GA, 1989.
4. Material Handling Industry of America, *ProMat 93 Proceedings,* Charlotte, NC, vol. 6, pp. 75–83.

# PROBLEMS

1.1 Describe both the planning and operating activities required to conduct a professional football game from the point of view of the
   a. visiting team's football coach.
   b. home team quarterback.
   c. manager of refreshment vending.
   d. ground crew manager.
   e. stadium maintenance manager.

1.2 List 10 components of a football stadium facility.

1.3 Describe the activities that would be involved in the (a) facilities location, (b) facilities design, and (c) facilities planning of an athletic stadium. Consider baseball, football, soccer, and track and field.

1.4 Assume you are on a job interview and you have listed on your resume a career interest in facilities planning. The firm where you are interviewing is a consulting firm that specializes in problem solving for transportation, communication, and service industries. React to the following statement directed to you by the firm's personnel director: "Facilities planning is possibly of interest to a firm involved in manufacturing, but it is not clear that our customers have needs in this area sufficient for you to pursue your field of interest."

1.5 What criterion should be utilized to determine the optimal facilities plan?

1.6 Evaluate the facilities plan for your campus and list potential changes you would consider if you were asked to replan the campus. Why would you consider these?

1.7 Chart the facilities planning process for
   a. a bank.
   b. a university campus.
   c. a warehouse.
   d. a consulting and engineering office.

1.8 Describe the procedure you would follow to determine the facilities plan for a new library on your campus.

1.9 Is facilities planning ever completed for an enterprise? Why or why not?

1.10 With the aid of at least three references, write a paper on the industrial engineer's and architect's roles in the planning of a facility.

1.11 Consider the definition of industrial engineering approved by the Institute of Industrial Engineers. Discuss the extent to which the definition applies to facilities planning.

# Part One

# DEFINING
# REQUIREMENTS

# 2

# STRATEGIC FACILITIES PLANNING[1]

## 2.1 INTRODUCTION

*Facilities planning* was defined in the previous chapter to be the composite of facilities location and facilities design. While it is true that the concerns of facilities planning are the location and the design of the facility, there exists another primary responsibility—*planning!* The importance of planning in facilities planning cannot be overemphasized, for it is the planning emphasis that distinguishes the activities of the facilities planner from the facilities designer and the facilities "locator."

Dwight D. Eisenhower said, "The plan is nothing, but planning is everything." As an indication of its importance in facilities planning, consider the process of planning and designing a manufacturing facility, building it, and installing and using the equipment. As shown in Figure 2.1, the costs of design changes increase exponentially as a project moves beyond the planning and designing phases.

In this chapter, we focus on a special type of planning, **strategic planning.** The term appears to have originated in the military. Webster defines strategy as "the science and art of employing the armed strength of a belligerent to secure the objects of a war." Today, the term is frequently used in politics, sports, investments, and business. Our concern is with the latter usage.

Based on Webster's definition of strategy, *business* strategies can be defined as *the art and science of employing the resources of a firm to achieve its business objectives.* Among the resources that are available are marketing resources, manufacturing resources, and distribution resources. Hence, marketing strategies, manufacturing

---

[1]A condensed version of the material presented in this chapter was published in *Industrial Engineering.* See reference 15, at the end of this chapter.

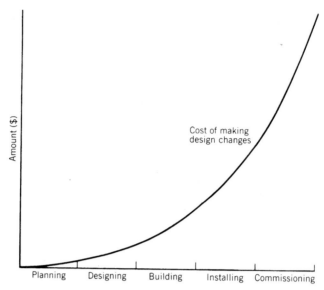

**Figure 2.1** Cost of design changes during a project.

strategies, and distribution strategies can be developed to support the achievement of the business objectives.

Recall, facilities planning was defined as determining how a firm's resources (tangible fixed assets) best support achieving the business objectives. In a real sense, facilities planning is itself a strategic process and must be an integral part of overall corporate strategy.

Historically, the development of corporate strategies has been restricted to the presidential level in many companies. Furthermore, business strategies tended to be limited to a consideration of such issues as acquisition, finance, and marketing. Consequently, decisions were often made without a clear understanding of the impact on manufacturing and distribution, as well as such support functions as facilities, material handling, information systems, and purchasing.

As an illustration, suppose an aggressive market plan is approved without the realization that manufacturing capacity is inadequate to meet the plan; furthermore, suppose the lead times required to achieve the required capacity are excessive. As a result, the market plan will fail because the impact of the plan on people, equipment, and space was not adequately comprehended. A winning facilities plan must consider integrating all elements that will impact the plan. An example of the accumulating benefits that can result from integrating operations is shown in Figure 2.2.

*Business Week, Industry Week, Time, Fortune,* and other business publications have focused on the competitiveness of America. This attention reflects the growing awareness in the business community of the importance of improved manufacturing productivity and technology. General Electric, Tektronix, Texas Instruments, and Westinghouse, among others, have expanded the strategic planning process to include the development of manufacturing strategies. Top management is recognizing the need to become involved in manufacturing decision making.

As noted by Skinner [10, p. 25], "When companies fail to recognize the relationship between manufacturing decisions and corporate strategy, they may become

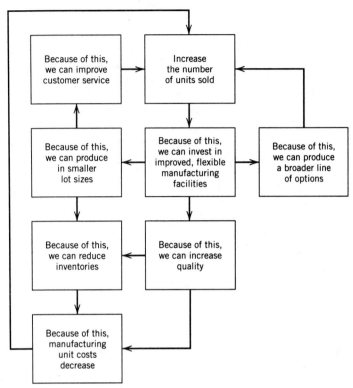

**Figure 2.2** Synergistic benefit of winning manufacturing on an integrated manufacturing-marketing team. *Source:* Tompkins [1].

saddled with seriously noncompetitive production systems that are expensive and time-consuming to change." He goes on to state, "Manufacturing affects corporate strategy, and corporate strategy affects manufacturing. Even in an apparently routine operating area such as a production scheduling system, strategic considerations should outweigh technical and conventional industrial engineering factors invoked in the name of 'productivity'." [10, p. 27]

Manufacturing strategies must be integrated with the other elements of overall business strategy. Each element of the overall organization should have a strategic plan that supports the objectives of the firm. Within the manufacturing and distribution strategies, the storage, movement, protection, and control of material must be addressed.

## *2.2* DEVELOPING FACILITIES PLANNING STRATEGIES

The process of effectively translating objectives into actions can only take place if the power of the individuals inside an organization is unleashed. Team-based implementation of company objectives will ensure that all members of the organization are involved in its achievement.

As noted in the previous section, strategies are needed for such functions as marketing, manufacturing, distribution, purchasing, facilities, material handling, and data processing/information systems, among others. It is important to recognize that each functional strategy is multidimensional. Namely, each must support or contribute to the strategic plan for the entire organization. Furthermore, each must have its own set of objectives, strategies, and tactics.

As previously stated, one method used to ensure that the objectives are effectively translated into action is the model of success. The model of success is effective because it is lateral as opposed to hierarchical in its approach. With the traditional top-down approach, only a handful of people are actively involved in ensuring that the objectives are met by driving these goals and plans into action. The lateral structure of the model of success communicates to everyone in an organization where the organization is headed.

The facilities planning process can be improved in a number of ways. Three potential dimensions for improvement are illustrated in Figure 2.3. Suppose the objective is to increase the size of the box shown. One approach is to make it taller by focusing on the physical aspects of facilities planning for example, buildings, equipment, and people. Another approach is to make the box wider by focusing on control aspects of facilities planning, for example, space standards, materials control, stock locator systems, and productivity measures. While it is possible to make the box taller and wider, we must not overlook the benefits provided by the third dimension—time. To make the box deeper requires time for planning. The old adage of "There's never time to do it right, but there's always time to do it over" has been demonstrated repeatedly with respect to facilities planning. Sufficient lead time is needed to do it right!

A number of internal functional areas tend to have a significant impact on facilities planning, including marketing, product development, manufacturing, production and inventory control, human resources, and finance. Marketing decisions affect the location of facilities and the handling system design. For example, material handling will be affected by decisions related to unit volume, product mix, packaging, service levels for spares, and delivery times.

Product development and design decisions affect processing and materials requirements, which in turn affect layout and material handling. Changes in materials,

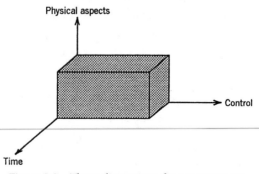

**Figure 2.3**  Three dimensions for improvement

component shapes, product complexity, number of new part numbers and package sizes introduced (due to a lack of standardization in design), stability of product design, and the number of products introduced will affect the handling, storage, and control of materials. Manufacturing decisions will have an impact on both facilities location and facilities design. Decisions concerning the degree of vertical integration, types and levels of automation, types and levels of control over tooling and work-in-process, plant sizes, and general-purpose versus special-purpose equipment can affect the location and design of manufacturing and support facilities.

Production planning and inventory control decisions affect the layout and handling system. Lot size decisions, production scheduling, in-process inventory requirements, inventory turnover goals, and approaches used to deal with seasonal demand affect the facilities plan.

Human resources and finance decisions related to capital availability, labor skills and stability, staffing levels, inventory investment levels, organization design, and employee services and benefits will impact the size and design of facilities, as well as their number and location. Space and flow requirements will be affected by financial and human resources decisions; in turn, they have an impact on the storage, movement, protection, and control of material.

For the facilities plan to support the overall strategic plan, it is necessary for facilities planners to participate in the development of the plan. Typically, facilities planners tend to *react* to the needs defined by others, rather than participate in the decision making that creates the needs. A proactive rather than a reactive role for facilities planning is recommended. The model of success approach will ensure that facilities planners are on board focusing on the overall direction of the company.

Close coordination is required in developing facilities plans to support manufacturing and distribution. Manufacturing/facilities planning and distribution/facilities planning interfaces are especially important. As the manufacturing plan addresses automatic load/unload of machines, robotics, group technology, transfer lines, flexible manufacturing systems, numerically controlled machines, just-in-time and computer-integrated manufacturing, alternative storage systems for tooling and work-in-process, real-time inventory control, shop floor control, and waste handling/removal systems, the facilities plan must be supportive of changes in manufacturing technology. Likewise, the facilities plan must be supportive of a distribution plan that addresses automatic palletizers, shrinkwrap/stretchwrap, automatic identification, automatic load/unload of vehicles and trailers, automated storage and retrieval of unit loads and small parts, and automated guided vehicle systems.

It is important for the level of manufacturing/distribution technology in use currently to be assessed objectivity and compared with the state-of-the-art. Five- and 10-year technology targets should be identified and an implementation plan developed to facilitate the required evolution.

To illustrate the level of detail required in planning, the following kinds of questions need to be resolved in developing a material handling plan that will support manufacturing and distribution.

1.   Should automated storage/retrieval systems (AS/RS), computer-controlled narrow-aisle trucks, manually operated trucks, or some combination be used for palletized storage/retrieval?

2.   Should miniloads, automated carousels, manually operated carousels, operator

aboard storage/retrieval machines, or some combination be used for storage/retrieval of small parts?

3. Should automated guided vehicles, tow lines, pallet conveyor, tractor-trailer trains, pallet trucks, or some combination be used to deliver loads to/from palletized storage?

4. Should fixed path, variable path, or some combination be used for material handling to/from/within manufacturing?

5. Should centralized or distributed storage of work-in-process be used? How should it be stored, moved, protected, and controlled?

6. Should transporter-conveyors, light duty roller conveyors, or carts be used to transport kits and parts to/from assembly stations? Should kitting be performed at all? If so, what issue quantities should be used?

7. Should modular workstations, modular handling systems, and/or modular storage units be used in manufacturing and assembly?

8. Should real-time inventory control be used for shop floor control and storage of raw material/work-in-process/finished goods? What data entry technology is appropriate?

9. Should block stacking, deep-lane storage, mobile rack, double-deep rack, drive-in/drive-through rack, selective rack, or some combination be used for pallet storage?

10. Should automatic loading/unloading of trailers be planned for receiving and shipping? If so, when, where, and for what materials?

The strategic planning process continues to be heavily influenced by its military roots. For example, developing contingency plans by asking numerous "what if" questions is an important element of the process. By asking such questions, an uncertainty envelope can be developed for facility requirements. Also, in translating market projections to requirements for facilities, it is important to consider learning curve effects, productivity improvements, technological forecasts, and site capacity limits.

The following 10 issues may have a long-range impact on the strategic facilities plan:

1. Number, location, and sizes of warehouses and/or distribution centers.

2. Centralized versus decentralized storage of supplies, raw materials, work-in-process, and finished goods for single- and multi-building sites, as well as single- and multi-site companies.

3. Acquisition of existing facilities versus design of modern factories and distribution centers of the future.

4. Flexibility required because of market and technological uncertainties.

5. Interface between storage and manufacturing.

6. Level of vertical integration, including "subcontract versus manufacture" decisions.

7. Control systems, including material control and equipment control, as well as level of distributed processing.

8. Movement of material between buildings, between sites, both inbound and outbound.

9. Changes in customers' and suppliers' technology as well as a firm's own manufacturing technology and material movement, protection, storage, and control technology.

10. Design-to-cost goals for facilities.

## *2.3*  EXAMPLES OF INADEQUATE PLANNING

Numerous examples exist of situations where inadequate planning was being performed. The following actual situations are presented to illustrate the need for improved planning.

- A major manufacturer in the midwest made a significant investment in storage equipment for a parts distribution center. The selection decision was based on the need for a "quick fix" to a pressing requirement for increased space utilization. The company soon learned that the "solution" would not provide the required throughput and was not compatible with long-term needs.

- An electronics manufacturer was faced with rapid growth. Management received proposals that required approximately equivalent funding for large warehouses at two sites having essentially the same storage and throughput requirements. Management questioned the rationale for one "solution" being a high-rise AS/RS and the other being a low-rise warehouse with computer-controlled industrial trucks.

- Another firm installed miniload systems at two sites. One system was designed for random storage; the other system was designed for dedicated storage. The storage and throughout requirements were approximately the same for the two systems; however, different suppliers had provided the equipment and software. Management raised the questions: Why are they different? and Which is best?

- A textile firm installed a large high-rise AS/RS for one of its divisions. Requirements subsequently changed in terms of the amount and size of the product to be stored. Other changes in technology were projected. The system became obsolete before it was operational.

- An engine manufacturer was planning to develop a new site. The facilities planners and architects were designing facilities for the site. Decisions had not been made concerning which products would be off-loaded to the new site, nor what effect the off-load would have on requirements for moving, protecting, storing, and controlling material.

- An electronics manufacturer was planning to develop a new site. The facilities planners and architects were designing the first building for the site. No projections of space and throughput had been developed since decisions had not been made concerning the occupant of the building.

- The mission for a major military supply center was changed in order that additional bases could be serviced. The throughput, storage, and control requirements for the new "customers" were significantly different from those for which the system was originally designed. However, no modifications to the system were funded.

- A manufacturer of automotive equipment acquired the land for a new manufacturing plant. The manufacturing team designed the layout, and the architect began designing the facility *before the movement, protection, storage, and control system was designed.*

- An aerospace-related manufacturer implemented group technology in its process planning and converted to manufacturing cells in a machining department. No analyses had been performed to determine queue or flow requirements. Subsequent analyses showed the manufacturing cells were substantially less efficient as a result of their impact on movement, protection, storage, and control of work-in-process [14].

In practically every case the projects were interrupted and significant delays were incurred because proper facilities planning had not been performed. These examples emphasize once more the importance of providing adequate lead times for planning.

The previous list of examples of inadequate facilities planning could possibly create a false impression that no one is doing an adequate planning job. Such is not the case; several firms have recognized the need for strategic facilities planning and are doing it.

A major U.S. airline developed 10-year and 20-year facilities plans to facilitate decision making regarding fleet size and mix. Maintenance and support facilities requirements were analyzed for wide-body and midsized aircraft. The impact of route planning, mergers and acquisitions, and changes in market regions to include international flights were considered in developing the plan.

The airline industry operates in a dynamic environment. Governmental regulations and attitudes toward business are changeable, energy costs and inflationary effects are significant, and long lead times are required for aircraft procurement. As an example of the latter, for new generation aircraft an airline company might negotiate procurement conditions, including options, eight years before taking delivery of the airplane.

## *2.4*  SUMMARY

In summary, a strategic facilities plan is needed to support management decision making regarding facilities acquisition, new construction, lease terms, new and existing facilities integration, and improvement of existing operations. The model of success can be used to ensure that the overall strategic plan supports manufacturing, distribution, and marketing plans. The lateral emphasis of the model encourages a team-based approach in achieving the objectives of the facilities plan. Additionally, the facilities plan must include the following functions: receiving, inbound inspection, storage, retrieval, dispatching to production and assembly, production and assembly, movement within production and assembly, work-in-progress storage, materials control, packaging, dispatching to and from storage, order picking, dispatching to shipping, order accumulation, shipping, maintenance, and personnel service. Finally, input is required from inventory control, production control, quality control, purchasing, marketing, product design, packaging, product development, manufacturing, information systems, and management.

# BIBLIOGRAPHY

1. Apple, J. M., Jr., "Strategic Planning for Material Handling Improvements," presentation to the AIIE/MHI Seminar, San Francisco, March 1980.
2. Collins, J. D., "Material Handling System Concept," presentation to the 26th Annual Material Handling Management Course, American Institute of Industrial Engineers, Mt. Pocono, PA., June 1979.
3. Collins, J. D., "Strategic Planning for Material Handling and Storage," presentation to the Computer Integrated Manufacturing: Productivity for the 1980's, A Joint Industry—DoD Manufacturing Technology Workshop, Society of Manufacturing Engineers, Detroit, MI., September 1979.
4. Geoffrion, A. M., Graves, G. W., and Lees, S. J., "Strategic Distribution System Planning: A Status Report," Chapter 7 in A. Hax (ed.), *Studies in Operations Management,* North-Holland, Amsterdam, N.Y. 1978.
5. Kinney, H. D., "A Totally Integrated Material Management System," presentation to the 1980 Annual Conference of the International Material Management Society, Cincinnati, OH, June 1980.
6. McCord, A. R., "TI's Preparations for the Challenges of the Eighties," a presentation to members of the Institute de L'Enterprise given at the University of Texas at Austin, March 1979.
7. Phipps, C. H., "Strategic Development in a Technology Intensive Company: Texas Instruments OST System," presented to U.S. Army Corporate planners, May 1979.
8. Radford, K. J., *Strategic Planning: An Analytical Approach,* Reston Publishing, Reston, VA., 1980.
9. Rothschild, W. E., "How to Ensure the Continued Growth of Strategic Planning," *The Journal of Business Strategy,* vol. 1, no. 1, pp. 11–18, Summer 1980.
10. Skinner, W., *Manufacturing in the Corporate Strategy,* Wiley, New York, 1978.
11. Tompkins, J. A., *Winning Manufacturing: The How To Book Of Successful Manufacturing,* IIE, Norcross, GA 1989.
12. White, J. A., "The Planning Process Need Not Be Painful," presentation to the National Material Handling Show and Seminar, The Material Handling Institute, Chicago, IL., June 1978.
13. White, J. A., "SFP—Strategic Facilities Planning," *Modern Materials Handling,* pp. 23–25, November 1978.
14. White, J. A., *Yale Management Guide to Productivity,* Eaton Corp., Philadelphia, PA., September 1979.
15. White, J. A., "Strategic Facility Planning," *Industrial Engineering,* vol 2, no. 9, pp. 94–100, September 1980.
16. *Measuring Productivity in Physical Distribution: A 40 Billion Dollar Goldmine,* National Council of Physical Distribution Management, Chicago, IL., 1978.

# PROBLEMS

**2.1**   Read three papers on strategic planning published in the *Harvard Business Review,* summarize the material, and relate it to strategic facilities planning.

**2.2**   Develop a list of strategic issues that must be addressed in performing facilities planning for:
   **a.**  an airport.
   **b.**  a community college.
   **c.**  a bank.
   **d.**  a grocery store chain.
   **e.**  a soft drink bottler and distributor.
   **f.**  a library.
   **g.**  an automobile dealership.
   **h.**  a shopping center developer.
   **i.**  a public warehousing firm.
   **j.**  a professional sports franchise.

**2.3**   Develop a set of responses to the following "reasons" for not doing strategic facilities planning:
   **a.**  There are more critical short-term problems to be solved.
   **b.**  The right people internally are too busy to be involved in the project.
   **c.**  The future is too hard to predict, and it will probably change anyway.
   **d.**  Nobody really knows what alternatives are available and which ones might apply.
   **e.**  Technology is developing very rapidly; any decisions we make will be obsolete before they can be implemented.
   **f.**  The return on investment in strategic planning is hard to measure.

**2.4**   What is the impact of facilities planning on the competitiveness of manufacturing facilities?

**2.5**   What are the implications of strategic planning on your personal career planning?

**2.6**   What are the impacts of automation on facilities planning?

**2.7**   What are the issues to be addressed in strategic planning for warehousing/distribution? What are the cost and customer services implications?

**2.8**   Based on [11], what are the impacts of the 20 requirements of manufacturing success on facilities planning?

**2.9**   What are the differences between strategic planning and contingency planning?

**2.10**  How does the issue of time impact the process of facilities planning?

# 3

# PRODUCT, PROCESS, AND SCHEDULE DESIGN

## 3.1 INTRODUCTION

Based on the presentation given in Chapter 1, the facilities planning process for manufacturing facilities can be listed as follows:

1. Define the products to be manufactured.
2. Specify the manufacturing processes and related activities required to produce the products.
3. Determine the interrelationships among all activities.
4. Determine the space requirements for all activities.
5. Generate alternative facilities plans.
6. Evaluate the alternative facilities plans.
7. Select the preferred facilities plan.
8. Implement the facilities plan.
9. Maintain and adapt the facilities plan.
10. Update the products to be manufactured and redefine the objective of the facility.

As indicated in Chapter 2, the facilities planning process will be greatly impacted by the business strategic plan and the concepts, techniques, and technologies to be considered in the manufacturing strategy.

Before the 1980s, most American companies were not considering small lot purchasing and production, multiple-receiving docks, deliveries to the points of use, decentralized storage areas, cellular manufacturing, flat organizational structures, "pull" approaches with kanbans, team decision making, electronic data interchange, process quality, self-rework, multifunctional "associates," and many other concepts that are changing drastically the facilities planning process and the final facilities plan (see Chapter 10).

Among the questions to be answered before alternative facility plans can be generated are the following:

1.   What is to be produced?
2.   How are the products to be produced?
3.   When are the products to be produced?
4.   How much of each product will be produced?
5.   For how long will the products be produced?
6.   Where are the products to be produced?

The answers to the first five questions are obtained from product design, process design, and schedule design, respectively. The sixth question might be answered by facilities location determination or it might be answered by schedule design when production is to be allocated among several existing facilities.

Product designers specify what the end product is to be in terms of dimensions, material composition, and perhaps, packaging. The process planner determines how the product will be produced. The production planner specifies the production quantities and schedules the production equipment. The facilities planner is dependent on timely and accurate input from product, process, and schedule designers to carry out his task effectively.

In this chapter we focus on the product, process, and schedule (PP&S) design functions as they relate to facilities planning. The success of a firm is dependent on having an efficient production system. Hence, it is essential that product designs, process selections, production schedules, and facilities plans be mutually supportive. Figure 3.1 illustrates the need for close coordination among the four functions.

Recently, many organizations have created teams with product, process, scheduling, and facilities design planners and with personnel from marketing, purchasing, and accounting to address the design process in an integrated, simultaneous, or concurrent way. Customer and supplier representatives are also involved in this process.

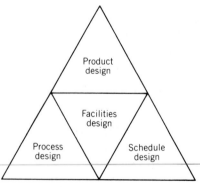

**Figure 3.1**   Relationship between product, process, and schedule (PP&S) design and facilities planning.

These teams are often referred to as *concurrent* or *simultaneous engineering* teams and are reducing the design cycle time, improving the design process, and eliminating engineering changes. Companies implementing this integrated approach have reported impressive results in cost, quality, productivity, sales, customer satisfaction, delivery time, amount of inventories, space and handling requirements, and building size, among others [14].

We can no longer say that product, process, schedule, and facilities design decisions can be done independently and sequentially. Furthermore, management needs to provide a clear vision of what to do and how to do it (including concepts, techniques, and technologies to consider). Management commitment to the use of multiple-receiving docks, smaller lot sizes, decentralized storage areas, open offices, decentralized cafeterias, self-managing teams, and focused factories will guide the design team in the generation of the best alternatives to satisfy business objectives and goals and make the organization more competitive.

In the case of an existing facility with ongoing production operations, a change in the design of a product, the introduction of a new product, changes in the processing of products, and modifications to the production schedule can occur without influencing the location or design of the facility.

In many cases, however, changes in product, process, and schedule design of current manufacturing facilities will require layout, handling, and/or storage modifications (especially if the design change requires more or less inventories, space, people, offices, and machines).

In this chapter, we are concerned with the interactions between facilities planning and PP&S design. Hence, we assume either a new facility or a major expansion/modification to an existing facility is being planned.

The Seven Management and Planning Tools methodology presented in Section 3.5 illustrates the integration of the facilities design process for a given scenario.

## *3.2* PRODUCT DESIGN

Product design involves both the determination of which products are to be produced and the detailed design of individual products. Decisions regarding the products to be produced are generally made by top management based on input from marketing, manufacturing, and finance concerning projected economic performance. In other instances, as illustrated in Chapter 2, the lead times to plan and build facilities, in the face of a dynamic product environment, might create a situation in which it is not possible to accurately specify the products to be produced in a given facility.

The facilities planner must be aware of the degree of uncertainty that exists concerning the mission of the facility being planned, the specific activities to be performed and the direction of those activities. As an illustration, a major electronics firm initially designed a facility for semiconductor manufacture. Before the facility was occupied, changes occurred in space requirements and another division of the company was assigned to the facility; the new occupant of the site used the space for manufacturing and assembling consumer electronic products. As the division grew in size, many of the manufacturing and assembly operations were off-loaded to newly developed sites and the original facility was converted to predominantly an administrative and engineering site.

Depending on the type of products being produced, the business philosophy concerning facilities, and such external factors as the economy, labor availability and attitudes, and competition, the occupants of a facility might change frequently or never change at all. Decisions must be made very early in the facilities planning process regarding the assumptions concerning the objectives of the facility.

If it is decided that the facility is to be designed to accommodate changes in occupants and mission, then a highly flexible design is required and very general space will be planned. On the other hand, if it is determined that the products to be produced can be stated with a high degree of confidence, then the facility can be designed to optimize the production of those particular products.[1]

The design of a product is influenced by aesthetics, function, materials, and manufacturing considerations. Marketing, purchasing, industrial engineering, manufacturing engineering, product engineering, and quality control, among others, will influence the design of the product. In the final analysis, the product must meet the needs of the customer.

This challenge can be accomplished through the use of Quality Function Deployment (QFD) [1]. QFD is an organized planning approach to identify customer needs and to translate the needs to product characteristics, process design, and tolerances required.

Benchmarking can also be used to identify what the competition is doing to satisfy the needs of your customer or to exceed the expectations of your customer [8]. It can also be used to identify the best of the best organizations.

Through QFD and Benchmarking, product designers can focus their work on customer needs being provided marginally or not provided at all (compared to the competition and to the best organizations).

Detailed operational specifications, pictorial representations, and prototypes of the product are important inputs for the facilities planner. Exploded assembly drawings, such as that given in Figure 3.2, are quite useful in designing the layout and handling system. These drawings generally omit specifications and dimensions, although they are drawn to scale.

As an alternative to the exploded assembly drawing, a photograph can be used to show the parts properly oriented. Such a photograph is given in Figure 3.3. Photographs and drawings allow the planner to visualize how the product is assembled, provide a reference for part numbers, and promote clearer communications during oral presentations.

Detailed component part drawings are needed for each component part. The drawings should provide part specifications and dimensions in sufficient detail to allow part fabrication. Examples of component part drawings are given in Figures 3.4 and 3.5. The combination of exploded assembly drawings and component part drawings fully documents the design of the products.

The drawings can be prepared and analyzed with *computer aided design* (CAD) systems. CAD is the creation and manipulation of design prototypes on a computer to assist the design process of the product. A CAD system consists of a collection

---

[1]Minor changes in product design and the addition of similar products to the product family would be included in this scenario.

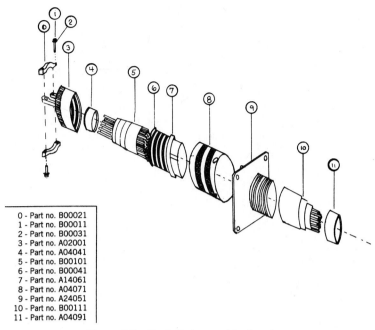

0 - Part no. B00021
1 - Part no. B00011
2 - Part no. B00031
3 - Part no. A02001
4 - Part no. A04041
5 - Part no. B00101
6 - Part no. B00041
7 - Part no. A14061
8 - Part no. A04071
9 - Part no. A24051
10 - Part no. B00111
11 - Part no. A04091

**Figure 3.2** Exploded assembly drawing.

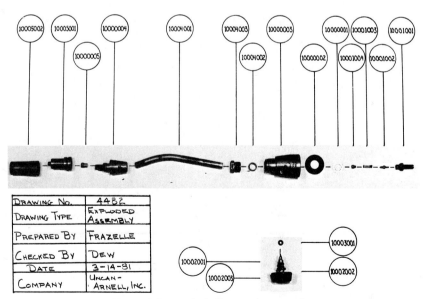

| DRAWING No. | 4432 |
| DRAWING TYPE | EXPLODED ASSEMBLY |
| PREPARED BY | FRAZELLE |
| CHECKED BY | DEW |
| DATE | 3-14-81 |
| COMPANY | UNCAN-ARNELL, INC. |

**Figure 3.3** Exploded parts photograph.

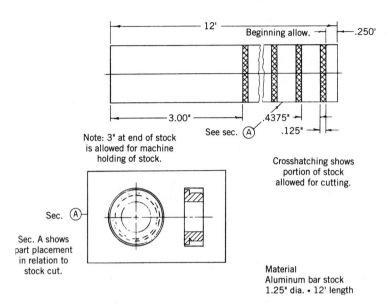

**Figure 3.4**   Component part drawing of a plunger.

of many application modules under a common database and graphics editor. The blending of computers and the human ability to make decisions enable us to use CAD systems in design, analysis, and manufacturing [9].

During the facilities design process, the computer's graphics capability and computing power allow the planner to visualize and test ideas in a flexible manner. The CAD system also can be used for area measurement, building and interior design, layout of furniture and equipment, relationship diagramming, generation of block and detailed layouts, and interference checks for process oriented plants [2].

In addition to CAD, *concurrent engineering* (CE) can be used to improve the relationship between the function of a component or product and its cost. Concurrent engineering provides a simultaneous consideration in the design phase of life cycle factors such as product, function, design, materials, manufacturing processes, testa-

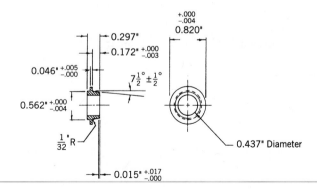

**Figure 3.5**   Component part drawing of a seat.

bility, serviceability, quality, and reliability. As a result of this analysis, a less expensive but functionally equivalent product design might be identified. Concurrent engineering is important because it is at the design stage that many of the costs of a product are specified. It has been estimated that more than 70% of a product's manufacturing cost is dictated by design decisions [14].

In addition to the design tools we have described, many other tools can be used to help the product-process designer in the evaluation and selection of the best product and process design combination [10, 21, 23].

# 3.3   PROCESS DESIGN

The process designer or process planner is responsible for determining *how* the product is to be produced. As a part of that determination, the process planner addresses *who* should do the processing; namely, should a particular product, subassembly, or part be produced in-house or subcontracted to an outside supplier or contractor? The "make-or-buy" decision is part of the process planning function.

In addition to determining whether a part will be purchased or produced, the process designer must determine how the part will be produced, which equipment will be used, and how long it will take to perform the operation. The final process design is quite dependent on input from both the product and schedule designs.

## Identifying Required Processes

Determining the scope of a facility is a basic decision and must be made early in the facilities planning process. For a hospital whose objective is to serve the health needs of a community, it may be necessary to limit the scope of the facility by not including in the facility a burn-care clinic, specific types of x-ray equipment, and/or a psychiatric ward. The excluded services, although needed by the community, may not be feasible for a particular hospital. Patients requiring care provided elsewhere would be referred to other hospitals. Similarly, the scope of a manufacturing facility must be established by determining the processes that are to be included within the facility. The extremes for a manufacturing facility may range from a vertically integrated firm that purchases raw materials and proceeds through a multitude of refining, processing, and assembly steps to obtain a finished product, to another firm that purchases components and assembles finished products. Therefore, it is obvious that the scope and magnitude of activities within a manufacturing facility are dependent on the decisions concerning the level of vertical integration. Such decisions are often referred to as "make-or-buy" decisions.

Large corporations are downsizing and breaking down large facilities into small business units that keep only processes that are economically feasible. Small business units operate with low overhead, low management levels, and frequently with self-managing operator teams. Buildings for this type of organizations are smaller and management functions and offices are decentralized.

Make-or-buy decisions are typically managerial decisions requiring input from finance, industrial engineering, marketing, process engineering, purchasing, and perhaps human resources, among others. A brief overview of the succession of questions

| Secondary Questions | Primary Questions | Decisions |
| --- | --- | --- |

1. Is the item available?
2. Will our union allow us to purchase the item?
3. Is the quality satisfactory?
4. Are the available sources reliable?

1. Is the manufacturing of this item consistent with our firm's objectives?
2. Do we possess the technical expertise?
3. Is the labor and manufacturing capacity available?
4. Is the manufacturing of this item required to utilize existing labor and production capacities?

1. What are the alternative methods of manufacturing this item?
2. What quantities of this item will be demanded in the future?
3. What are the fixed, variable, and investment costs of the alternative methods and of purchasing the item?
4. What are the product liability issues which impact the purchase or manufacture of this item?

1. What are the other opportunities for the utilization of our capital?
2. What are the future investment implications if this item is manufactured?
3. What are the costs of receiving external financing?

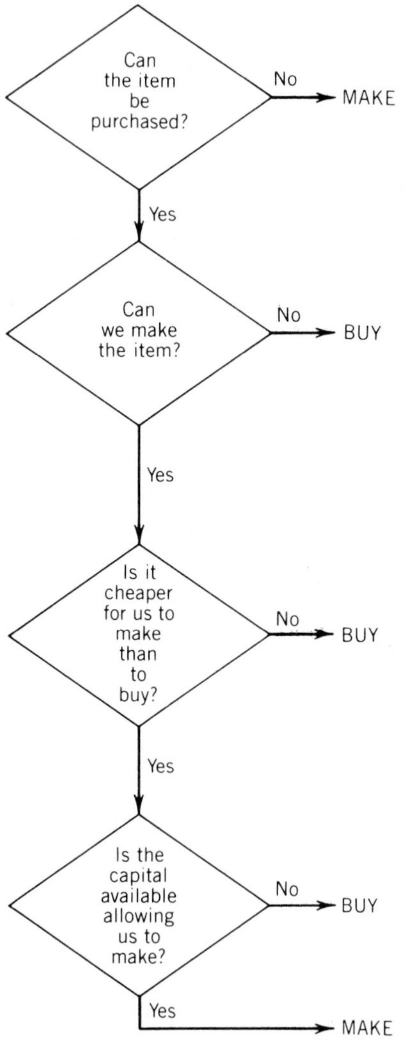

**Figure 3.6**  The make-or-buy decision process.

leading to make-or-buy decisions is given in Figure 3.6. The input to the facilities planner is a listing of the items to be made and the items to be purchased. The listing often takes the form of a parts list or a bill of materials.

The parts list provides a listing of the component parts of a product. In addition to make-or-buy decisions, a parts list includes at least the following:

1. Part numbers.
2. Part name.
3. Number of parts per product.
4. Drawing references.

A typical parts list is given in Figure 3.7.

**PARTS LIST**

Company ___T. W., Inc.___                    Prepared by ___J. A.___
Product ___Air Flow Regulator___           Date _____

| Part No. | Part Name | Drwg. No. | Quant./ Unit | Material | Size | Make or Buy |
|---|---|---|---|---|---|---|
| 1050 | Pipe plug | 4006 | 1 | Steel | .50" × 1.00" | Buy |
| 2200 | Body | 1003 | 1 | Aluminum | 2.75" × 2.50" × 1.50" | Make |
| 3250 | Seat ring | 1005 | 1 | Stainless steel | 2.97" × .87" | Make |
| 3251 | O-ring | — | 1 | Rubber | .75" dia. | Buy |
| 3252 | Plunger | 1007 | 1 | Brass | .812" × .715" | Make |
| 3253 | Spring | — | 1 | Steel | 1.40" × .225" | Buy |
| 3254 | Plunger housing | 1009 | 1 | Aluminum | 1.60" × .225" | Make |
| 3255 | O-ring | — | 1 | Rubber | .925" dia. | Buy |
| 4150 | Plunger retainer | 1011 | 1 | Aluminum | .42" × 1.20" | Make |
| 4250 | Lock nut | 4007 | 1 | Aluminum | .21" × 1.00" | Buy |

**Figure** 3.7    Parts list for an air flow regulator.

A bill of materials is often referred to as a structured parts list, as it contains the same information as a parts list plus information on the structure of the product. Typically, the product structure is a hierarchy referring to the level of product assembly. Level 0 usually indicates the final product; level 1 applies to subassemblies and components that feed directly into the final product; level 2 refers to the subassemblies and components that feed directly into the first level, and so on. A bill of materials is given in Figure 3.8 and is illustrated in Figure 3.9 for the product described in the parts list in Figure 3.7.

## Selecting the Required Processes

Once a determination has been made concerning the products to be made "in-house," decisions are needed as to how the products will be made. Such decisions are based on previous experiences, related requirements, available equipment, production rates, and future expectations. Therefore, it is not uncommon for different processes to be selected in different facilities to perform identical operations. However, the selection procedure used should be the same. In Figure 3.10, the procedure for process selection is given.

Input into the process selection procedure is called **process identification.** Process identification consists of a description of what is to be accomplished. For a manufactured product, process identification consists of a parts list indicating what is to be manufactured, component part drawings describing each component, and the quantities to be produced.

## BILL OF MATERIALS

Company   T. W., Inc.              Prepared by  J. A.

Product   Air Flow Regulator          Date

| Level | Part No. | Part Name | Drwg. No. | Quant./ Unit | Make or Buy | Comments |
|---|---|---|---|---|---|---|
| 0 | 0021 | Air flow regulator | 0999 | 1 | Make | |
| 1 | 1050 | Pipe plug | 4006 | 1 | Buy | |
| 1 | 6023 | Main assembly | — | 1 | Make | |
| 2 | 4250 | Lock nut | 4007 | 1 | Buy | |
| 2 | 6022 | Body assembly | — | 1 | Make | |
| 3 | 2200 | Body | 1003 | 1 | Make | |
| 3 | 6021 | Plunger assembly | — | 1 | Make | |
| 4 | 3250 | Seat ring | 1005 | 1 | Make | |
| 4 | 3251 | O-ring | — | 1 | Buy | |
| 4 | 3252 | Plunger | 1007 | 1 | Make | |
| 4 | 3253 | Spring | — | 1 | Buy | |
| 4 | 3254 | Plunger housing | 1009 | 1 | Make | |
| 4 | 3255 | O-ring | — | 1 | Buy | |
| 4 | 4150 | Plunger retainer | 1011 | 1 | Make | |

**Figure 3.8**   Bill of materials for an air flow regulator.

*Computer aided process planning* (CAPP) can be used to automate the manual planning process. There are two types of CAPP Systems: *variant* and *generative*. In a variant CAPP, standard process plans for each part family are stored within the computer and called up whenever required. In generative process planning, process plans are generated automatically for new components without requiring the existing plans. Selection of these systems basically depends on product structure and cost considerations. Typically, variant process planning is less expensive and easier to implement.

Since process planning is a critical bridge between design and manufacturing, CAPP systems can be used to test the different alternative routes and interact with the facility design process. The input of a CAPP system is commonly a three dimensional model from a CAD database including information related to tolerances and special features. Based on these inputs, manufacturing leadtime and resource requirements can also be determined [2].

The facilities planner will not typically perform process selection. However, an understanding of the overall procedure provides the foundation for the facilities plan. Step 1 of the procedure involves the determination of the operations required to

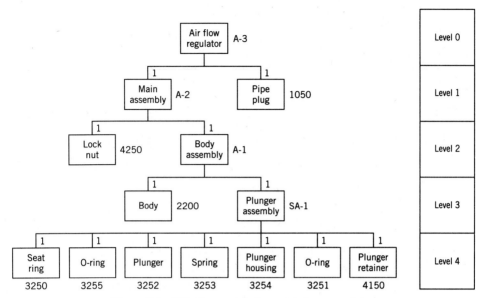

**Figure 3.9**  Bill of materials for an air flow regulator.

produce each component. In order to make this determination, alternative forms of raw materials and types of elemental operations must be considered. Step 2 involves the identification of various equipment types capable of performing elemental operations. Manual, mechanized, and automated alternatives should be considered. Step 3 includes the determination of unit production times and equipment utilizations for various elemental operations and alternative equipment types. The utilizations are the input into step 4 of the procedure. Step 5 involves an economic evaluation of alternative equipment types. The results of the economic evaluation along with intangible factors such as flexibility, versatility, reliability, maintainability, and safety serve as the basis for step 6.

Computer programs that perform time and cost evaluations of different process combinations have also been developed to aid process planners [10, 21].

The outputs from the process selection procedure are the processes, equipment, and raw materials required for the in-house production of products. Output is generally given in the form of a **route sheet.** A route sheet should contain at least the data given in Table 3.1. Figure 3.11 is a route sheet for the production given in part in Table 3.1.

### PROCESS IDENTIFICATION

| | |
|---|---|
| Define elemental operations | Step 1 |
| Identify alternative process for each operation | Step 2 |
| Analyze alternative processes | Step 3 |
| Standardize processes | Step 4 |
| Evaluate alternative processes | Step 5 |
| Select processes | Step 6 |

**Figure 3.10**  Process selection procedure.

**Table 3.1**  *Route Sheet Data Requirements*

| Data | Production Example |
|------|--------------------|
| Component name and number | Plunger housing—3254 |
| Operation description and number | Shape, drill, and cut off—0104 |
| Equipment requirements | Automatic screw machine and appropriate tooling |
| Unit times | Setup time—5 hr Operation time—.0057 hr per component |
| Raw material requirements | 1 in. dia. × 12 ft. aluminum bar per 80 components |

## Sequencing the Required Processes

The only process selection information not yet documented is the method of assembling the product. An **assembly chart,** given in Figure 3.12, provides such documentation. The easiest method of constructing an assembly chart is to begin with the completed product and to trace the product disassembly back to its basic components. For example, the assembly chart given in Figure 3.12 would be constructed by beginning in the lower right-hand corner of the chart with a finished air flow regulator. The first disassembly operation would be to unpackage the air flow regulator (operation A-4). The operation that precedes packaging is the inspection of the air flow regulator. Circles denote assembly operation; inspections are indicated on assembly charts as squares. Therefore, in Figure 3.12, a square labeled I-1 immediately precedes operation A-4. The first component to be disassembled from the air flow regulator is part number 1050, the pipe plug, indicated by operation A-3. The lock nut is then disassembled, followed by the disassembly of the body assembly (the subassembly made during subassembly operation SA-1) and the body. The only remaining steps required to complete the assembly chart are the labeling of the circles and lines of the seven components flowing into SA-1.

Although route sheets provide information on production methods and assembly charts indicate how components are combined, neither provides an overall understanding of the flow within the facility. However, by superimposing the route sheets on the assembly chart, a chart results that does give an overview of the flow within the facility. This chart is an **operation process chart.** An example of an operation process chart is given in Figure 3.13.

To construct an operation process chart, begin at the upper right side of the chart with the components included in the first assembly operation. If the components are purchased, they should be shown as feeding horizontally into the appropriate assembly operation. If the components are manufactured, the production methods should be extracted from the route sheets and shown as feeding vertically into the appropriate assembly operation. The operation process chart may be completed by continuing in this manner through all required steps until the product is ready for release to the warehouse.

The operation process chart can also include materials needed for the fabricated components. This information can be placed below the name of the component. Additionally, operation times can be included in this chart and placed to the left of

**Route Sheet**

| Company: A.R.C., Inc. | Part Name: Plunger Housing | Prepared by: J. A. |
|---|---|---|
| Produce: Air Flow Regulator | Part No.: 3254 | Date: |

| Oper. No. | Operation Description | Machine Type | Tooling | Dept. | Set-up Time (hr.) | Operation Time (hr.) | Materials or Parts Description |
|---|---|---|---|---|---|---|---|
| 0104 | Shape, drill, cut off | Automatic screw machine | .50 in. dia. collet, feed fingers, cir. form tool, .45 in. dia. center drill, .129 in. twist drill, finish spiral drill, cut off blade | | 5 | .0057 | Aluminum 1.0 in. dia. × 12 ft. |
| 0204 | Machine slot and thread | Chucker | .045 in. slot saw, turret slot attach. 3/8–32 thread chaser | | 2.25 | .0067 | |
| 0304 | Drill 8 holes | Auto. dr. unit (chucker) | .078 in. dia. twist drill | | 1.25 | .0038 | |
| 0404 | Deburr and blow out | Drill press | Deburring tool with pilot | | .5 | .0031 | |
| SA1 | Enclose subassembly | Dennison hyd. press | None | | .25 | .0100 | |

**Figure 3.11**  Route sheet for one component of the air flow regulator.

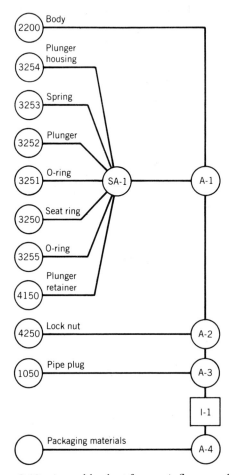

**Figure 3.12**  Assembly chart for an air flow regulator.

operations and inspections. A summary of the number of operations and inspections and operation times can be provided below the chart.

The operation process chart can be complemented with transportations, storages, and delays (including distances and times) when the information is available (referred to as the process chart by Barnes [3]).

Assembly charts and operation process charts may be viewed as analog models of the assembly process and the overall production process, respectively. As noted previously, circles and squares represent time and horizontal connections represent sequential steps in the assembly of the product.

Notice in the assembly chart that components have been identified with a four-digit code starting with 1, 2, 3, and 4. Furthermore, subassemblies (SA) and assemblies (A) have been identified with letters and numbers. The same identification approach has been used in the operation process chart. Additionally, fabrication operations have been represented with a four-digit code starting with 0.

A second viewpoint (from graph and network theory) is to interpret the charts as network representations, or more accurately, tree representations of a production

**OPERATION PROCESS CHART**

Company    A.R.C., Inc.       Prepared by    J. A.

Product    Air Flow Regulator       Date

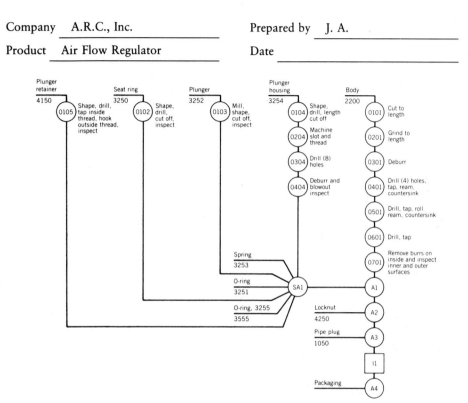

**Figure 3.13**   Operation process chart for the air flow regulator.

process. A variation of the network viewpoint is to treat the assembly chart and the operation process chart as special cases of a more general graphical model, the **precedence diagram.** The precedence diagram is a directed network and is used in project planning; critical path diagrams and PERT charts are examples of precedence diagrams.

A precedence diagram for the air flow regulator is given in Figure 3.14. The precedence diagram shows part numbers on the arcs and denotes operations and inspections by circles and squares, respectively. A procurement operation, 0100, is used in Figure 3.14 to initiate the process.

Notice in the precedence diagram that purchased parts and materials that do not require modifications are placed on the top and bottom part of the diagram so that they can be inserted in the center part of the diagram when needed (packaging materials, pipe plug, lock nut, spring, and O-rings). The center part of the diagram is then used to include purchased materials and/or components that require some work before being assembled (body, plunger housing, plunger, seat ring, and plunger retainer). Fabrication and assembly operations are placed in the center part of the diagram.

The precedence diagram representation of the operations and inspections involved in a process can be of significant benefit to the facilities planner. It establishes

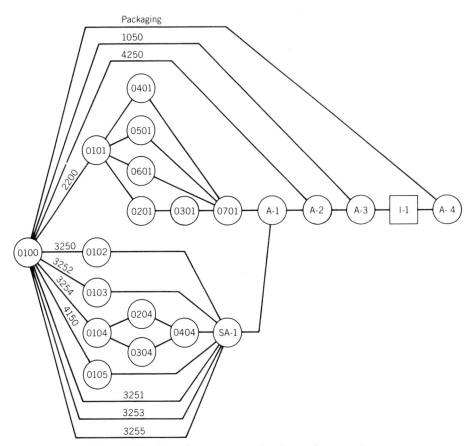

**Figure 3.14**  Precedence diagram for the air flow regulator.

the precedence relationships that must be maintained in manufacturing and assembling a product. No additional constraints are implicitly imposed; no assumptions are made concerning which parts move to which parts; no material handling or layout decisions are implicit in the way the precedence diagram is constructed. Unfortunately, the same claims cannot be made for the assembly chart and operation process chart.

Recall the instructions we gave for constructing the assembly chart: "The easiest method of constructing an assembly chart is to begin with the completed product and to trace the product disassembly back to its basic components." Implicit in such an instruction is that *the sequence used to disassemble the product should be reversed to obtain the sequence to be used to assemble the product.* Just as there are alternative disassembly sequences that can be used, there are also alternative assembly sequences. *The assembly chart and the operation process chart depict a single sequence.* The particular sequence used can have a major impact on space and handling system requirements.

Notice operations 0101, 0201, 0301, 0401, 0501, 0601, and 0701 are not shown in series in Figure 3.14, even though they were so represented on the operation process chart. Hence, there exists some latitude in how the product is assembled.

Unfortunately, there doesn't exist a mechanism to show the possibility of alternative processing sequences on the operation process chart.

In order to motivate further our concerns regarding the misuse of the operation process chart in layout planning, consider the processes involved in manufacturing an axle for an over-the-road tractor. Using the advice typically provided in texts that describe the construction of operation process charts, the axle itself should be shown at the extreme right side of the chart. Subassemblies, components, and purchased parts would be shown sequentially feeding into the axle until a finished assembly was produced.

By observing the operation process chart one might be tempted to develop an assembly line for the axle (assuming sufficient quantities are to be produced). The axle would be moved along the assembly line and subassemblies, components, and purchased parts would be attached to it. Using such an approach, *space and handling equipment requirements for the line would be based on the largest component part in the assembly.*

As an alternative, after performing the last processing operation on the axle it could remain stationary and all subassemblies could flow to it. No further movement of the axle would have to occur.

Because of the limitations of the assembly chart and the operation process chart, we recommend a precedence diagram be constructed first. Based on the precedence diagram, alternative assembly charts and operation process charts should then be constructed.

Other techniques to generate and evaluate assembly sequences have been explored. These techniques consider the assembly according to the relationship among parts instead of the order in which parts will be assembled. With these approaches, liaison sequences are generated instead of assembly sequences.

The original *Liaison Sequence Analysis* (LSA) method for assembly sequencing was proposed by Bourjault [4]. Bourjault's method is an algorithmic approach that uses the relationships between parts to generate all the possible assembly sequences.

The second LSA method is a modification of Bourjault's technique developed by Whitney and DeFazio [22]. This second method reduces the number of questions while keeping the algorithmic nature of the original approach.

Improvements to LSA are currently being addressed. For example, Silva [20] developed the assembly sequences generator system (ASGS) to produce all the physically possible assembly sequences for a given functional product design. ASGS also considers the factors affecting the decision making process to select the best assembly sequence.

Group technology is having an impact on product and process design. Group technology refers to the grouping of parts into families and then making design decisions based on family characteristics. Groupings are typically based on part shapes, part sizes, material types, and processing requirements. In those cases where there are thousands of individual parts, the number of families might be less than 100. Group technology is an aggregation process that has been found to be quite useful in achieving standardized part numbers, standard specifications of purchased parts, for example, fasteners and standardized process selection.

The importance of the process design or process plan in developing the facilities plan cannot be overemphasized. Furthermore, it is necessary that the process planner understands the impact of process design decisions on the facilities plan. Our experience indicates that process planning decisions are frequently made without such an understanding. As an example, it is often the case that alternatives exist in both the selection of the processes to be used and their sequence of usage. The final choice should be based on interaction between schedule design and facilities planning. Unfortunately, the choice is often based on "We've always done it this way" or "The computer indicated this was the best way."

In many companies the process selection procedure has become sufficiently routine that computer-aided process planning systems (CAPPS) have been developed. The resulting standardization of process selection has yielded considerable labor savings and reductions in production lead times. At the same time, standardization in process selection might create disadvantages for schedule and facility design. If such a situation occurs, a mechanism should exist to allow exceptions.

The facilities planner must take the initiative and participate in the process selection decision process to ensure that "route decisions" are not unduly constraining the facilities plan. Many of the degrees of freedom available to the facilities planner can be affected by process selection decisions.

## *3.4*  SCHEDULE DESIGN

Schedule design decisions provide answers to questions involving how much to produce and when to produce. Production quantity decisions are referred to as lot size decisions; determining when to produce is referred to as production scheduling. In addition to how much and when, it is important to know how long production will continue; such a determination is obtained from market forecasts.

Schedule design decisions impact machine selection, number of machines, number of shifts, number of employees, space requirements, storage equipment, material handling equipment, personnel requirements, storage policies, unit load design, building size, and so on. Consequently, schedule planners need to interface continuously with marketing and sales personnel and with the largest customers to provide the best information possible to facilities design planners.

To plan a facility, information is needed concerning production volumes, trends, and the predictability of future demands for the products to be produced. The less specificity provided regarding product, process, and schedule designs, the more general purpose will be the facility plan. The more specific the inputs from product, process, and schedule designs, the greater the likelihood of optimizing the facility and meeting the needs of manufacturing.

### Marketing Information

A facility that produces 10,000 television sets per month should differ from a facility that produces 1000 television sets per month. Likewise, a facility that produces 10,000 television sets for the first month and increases production 10% per month thereafter should not be considered the same as a facility that produces 10,000 television sets

Table 3.2  *Minimum Market Information Required for Facilities Planning*

| Product or Service | First Year Volume | Second Year Volume | Fifth Year Volume | Tenth Year Volume |
|---|---|---|---|---|
| A | 5000 | 5000 | 8000 | 10,000 |
| B | 8000 | 7500 | 3000 | 0 |
| C | 3500 | 3500 | 3500 | 4,000 |
| D | 0 | 2000 | 3000 | 8,000 |

per month for the foreseeable future. Lastly, consider a facility that produces 10,000 television sets per month for the next 10 years versus one that produces 10,000 television sets per month for 3 months and is unable to predict what product or volume will be produced thereafter; they too should differ.

As a minimum, the market information given in Table 3.2 is needed. Preferably, information regarding the dynamic value of demands to be placed on the facility is desired. Ideally, information of the type shown in Table 3.3 would be provided. If such information is available, a facilities plan can be developed for each demand state and a facility designed with sufficient flexibility to meet the yearly fluctuations in product mix. By developing facilities plans annually and noting the alterations to the plan, a facilities master plan may be established [16]. Unfortunately, information of the type given in Table 3.3 is generally unavailable. Therefore, facilities typically are planned using deterministic data. The assumptions of deterministic data and known demands must be dealt with when evaluating alternative facilities plans.

In addition to the volume, trend, and predictability of future demands for various products, the qualitative information listed in Table 3.4 should be obtained. Additionally, the facilities planner should solicit input from marketing as to why market trends are occuring. Such information may provide valuable insight to the facilities planner.

An Italian economist, Pareto, observed that 85% of the wealth of the world is held by 15% of the people. Surprisingly, his observations apply to several aspects of facilities planning. For example, Pareto's law often applies to the product mix of a facility. That is, 85% of the production volume is attributed to 15% of the product line. Such a situation is depicted by the volume-variety chart given in Figure 3.15. This chart suggests that the facilities plan should consist of a mass production area for the 15% of high-volume items and a job shop arrangement for the remaining 85% of the product mix. By knowing this at the outset, development of the facilities plan may be significantly simplified.

This volume-variety information is very important in determining the layout type to use. As described in Chapter 4, identifying the layout type will impact among others the type of material handling alternatives, storage policies, unit loads, building configuration, and the location of receiving/shipping docks.

Pareto's law may not always describe the product mix of a facility. Figure 3.16 represents the product mix for a facility where 15% of the product line represents approximately 25% of the production volume; Pareto's law clearly does not apply. However, knowing that Pareto's law does not apply is valuable information; because if no product dominates the production flow, a general job shop facility is suggested.

Table 3.3  *Market Analysis Indicating the Stochastic Nature of Future Requirements for Facilities Planning*

| Product or Service | Demand State | First Year | | Second Year | | Fifth Year | | Tenth Year | |
|---|---|---|---|---|---|---|---|---|---|
| | | Probability | Volume | Probability | Volume | Probability | Volume | Probability | Volume |
| A | Pessimistic | .3 | 3000 | .2 | 3500 | .1 | 5500 | .1 | 7,000 |
| | Most likely | .5 | 5000 | .6 | 5500 | .8 | 8000 | .9 | 10,000 |
| | Optimistic | .2 | 6000 | .2 | 6500 | .1 | 9500 | | |
| B | Pessimistic | .1 | 7000 | | | | | | |
| | Most likely | .6 | 8000 | .7 | 7000 | .9 | 3000 | 1.0 | 0 |
| | Optimistic | .3 | 8500 | .3 | 8000 | .1 | 3500 | | |
| C | Pessimistic | .2 | 2000 | .2 | 2000 | .2 | 2000 | .2 | 2,000 |
| | Most likely | .7 | 4000 | .7 | 4000 | .7 | 4000 | .6 | 4,000 |
| | Optimistic | .1 | 4500 | .1 | 4500 | .1 | 4500 | .2 | 5,000 |
| D | Pessimistic | | | .1 | 1500 | .1 | 2500 | .2 | 7,000 |
| | Most likely | 1.0 | 0 | .9 | 2000 | .8 | 3000 | .6 | 8,000 |
| | Optimistic | | | | | .1 | 3500 | .2 | 9,000 |
| Confidence level or degree of certainty | | 90% | | 85% | | 70% | | 59% | |

**Table 3.4**   *Valuable Information that Should Be Obtained from Marketing and Used by a Facilities Planner*

| Information to Be Obtained from Marketing | Facilities Planning Issues Impacted by the Information |
|---|---|
| Who are the consumers of the product? | 1. Packaging<br>2. Susceptibility to product changes<br>3. Susceptibility to changes in marketing strategies |
| Where are the consumers located? | 1. Facilities location<br>2. Method of shipping<br>3. Warehousing systems design |
| Why will the consumer purchase the product? | 1. Seasonability<br>2. Variability in sales<br>3. Packaging |
| Where will the consumer purchase the product? | 1. Unit load sizes<br>2. Order processing<br>3. Packaging |
| What percentage of the market does the product attract and who is the competition? | 1. Future trends<br>2. Growth potential<br>3. Need for flexibility |
| What is the trend in product changes? | 1. Space allocations<br>2. Materials handling methods<br>3. Need for flexibility |

## Process Requirements

Process design determines the specific equipment types required to produce the product. Schedule design determines the number of each equipment type required to meet the production schedule.

Specification of process requirements typically occurs in three phases. The first phase determines the quantity of components that must be produced, including scrap

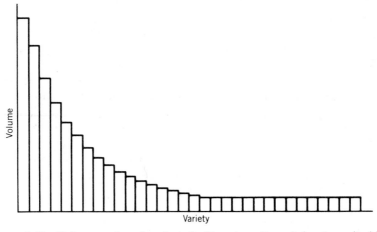

**Figure 3.15**   Volume-variety chart for a facility where Pareto's law is applicable.

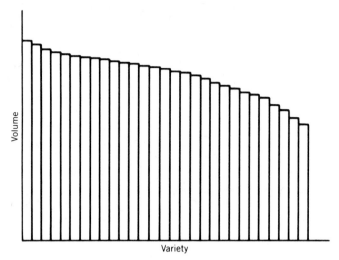

**Figure 3.16**   Volume-variety chart for a facility where Pareto's law is not applicable.

allowance, in order to meet the market estimate. The second phase determines the equipment requirements for each operation, and the third phase combines the operation requirements to obtain overall equipment requirements.

### Scrap Estimates

The market estimate specifies the annual volume to be produced for each product. To produce the required amount of product, the number of units scheduled through production must equal the market estimate plus a scrap estimate. Hence, production capacity must be planned for the production of scrap. Otherwise, when scrap is produced the market estimate will not be met. Scrap is the material waste generated in the manufacturing process due to geometric or quality considerations. For example, scrap due to geometry is generated when a rectangular steel plate is used to create circular components or when rolls of fabric are used to make shirts. Scrap due to bad quality might be caused by mistakes in machining or assembly operations. Ideally, a company should strive for continuous improvement to achieve zero scrap. An estimate of the percentage of scrap to be incurred from each operation must be made. This may be based on historical data or estimated from similar operations. In general, the more automated a process, the less scrap is produced; the looser the part tolerance, the less scrap is produced; the bigger the number of certified suppliers, the less scrap is generated; the more quality at the source and prevention techniques are applied, the less scrap is generated; and the higher the grade of material, the less scrap is produced.

Let $P_k$ represent the percentage of scrap produced on the $k^{\text{th}}$ operation, $O_k$ the desired output of nondefective product from operation $k$, and $I_k$ the production input to operation $k$. It follows that, on the average

$$O_k = I_k - P_k I_k \tag{3.1}$$

or

$$O_k = I_k (1 - P_k)$$

Hence,

$$I_k = \frac{O_k}{1 - P_k} \tag{3.2}$$

Thus, the expected number of units to start into production for a part having $n$ operations is

$$I_1 = \frac{O_n}{(1 - P_1)(1 - P_2) \cdots (1 - P_n)} \tag{3.3}$$

where in this case $O_n$ is the market estimate.

## Example 3.1

A product has a market estimate of 97,000 components and requires three processing steps (turning, milling, and drilling) having scrap estimates of $P_1 = 0.04$, $P_2 = 0.01$, and $P_3 = 0.03$. The market estimate is the output required from step 3. Therefore,

$$I_3 = \frac{97,000}{1 - 0.03} = 100,000$$

Assuming no damage between operations 2 and 3 and an inspection operation to remove all rejects, the output of good components from operation 2 ($O_2$) may be equated to the input to operation 3 ($I_3$). Therefore, the number of components to start into operation 2 ($I_2$) is

$$I_2 = \frac{100,000}{1 - 0.01} = 101,010$$

Likewise, for operation 1:

$$I_1 = \frac{101,010}{1 - 0.04} = 105,219$$

The calculations are identical to:

$$I_1 = \frac{97,000}{(1 - 0.03)(1 - 0.01)(1 - 0.04)} = 105,219$$

The amount of raw material and processing on operation 1 is not to be based on the market estimate of 97,000 components, but on 105,219 components, as summarized in Table 3.5.

Table 3.5  *Summary of Production Requirements for Example 3.1*

| Operation | Production Quantity Scheduled (units) | Expected Number of Good Units Produced |
|---|---|---|
| Turning | 105,219 | 101,010 |
| Milling | 101,010 | 100,000 |
| Drilling | 100,000 | 97,000 |

### Equipment Fractions

The quantity of equipment required for an operation is referred to as the equipment fraction. The equipment fraction may be determined for an operation by dividing the total time required to perform the operation by the time available to complete the operation. The total time required to perform an operation is the product of the standard time for the operation and the number of times the operation is to be performed. For example, if it takes ½ hr to duplicate a paper and if six papers are to be duplicated in 2 hr, it clearly follows that 1.5 duplicating machines are needed to complete the operation. Whether or not 1.5 duplicating machines are actually adequate to duplicate the required papers depends on the following questions.

1.  Are the papers actually being duplicated according to the ½ hr per paper standard?
2.  Is the duplicating machine available when needed during the 2-hr period?
3.  Are the standard time, the number of papers, and the time the equipment is actually available known with certainty and fixed over time?

The first question may be handled by dividing the standard time by the historical efficiency of performing the operation. The second question may be handled by multiplying the time the equipment is available to complete an operation by the historical reliability factor for the equipment. The reliability factor is the percentage of time the equipment is actually producing and is not down because of malfunctioning or planned maintenance.

The third question dealing with the uncertainty and time-varying nature of machine fraction variables can be an important factor in determining process requirements. If considerable uncertainty and variation over time exists, it may be useful to consider using probability distributions, instead of point estimates for the parameters, and utilize a probabilistic machine fraction model. Typically, such models are not utilized and the approach taken is to use a deterministic model and plan the facility to provide sufficient flexibility to handle changes in machine fraction variables.

The following deterministic model can be used to estimate the equipment fraction required:

$$F = \frac{SQ}{EHR} \qquad (3.4)$$

where

$F$ = number of machines required per shift
$S$ = standard time (minutes) per unit produced
$Q$ = number of units to be produced per shift
$E$ = actual performance, expressed as a percentage of standard time
$H$ = amount of time (minutes) available per machine
$R$ = reliability of machine, expressed as percent "up time"

Additionally, equipment requirements are a function of the following factors:

•  Number of shifts (the same machine can work in more than one shift).
•  Set-up times (if machines are not dedicated, the longer the setup, the more machines needed).

- Degree of flexibility (customers may require small lot sizes of different products delivered frequently—extra machine capacity will be required to handle these requests).
- Layout type (dedicating manufacturing cells or focused factories to the production of product families may require more machines).
- Total productive maintenance (will increase machine up time and improve quality, thus fewer machines will be needed).

## Example 3.2

A machined part has a standard machinery time of 2.8 min per part on a milling machine. During an 8-hr shift 200 units are to be produced. Of the 480 min available for production, the milling machine will be operational 80% of the time. During the time the machine is operational, parts are produced at a rate equal to 95% of the standard rate. How many milling machines are required?

For the example, S = 2.8 min per part, Q = 200 units per shift, H = 480 min per shift, E = 0.95, and R = 0.80. Thus,

$$F = \frac{2.8(200)}{0.95(480)(0.80)} = 1.535 \text{ machines per shift}$$

### Specifying Total Equipment Requirements

The next step in determining process requirements is to combine the equipment fractions for identical equipment types. Such a determination is not necessarily straightforward. Even if only one operation is to be performed on a particular equipment type, overtime and subcontracting must be considered. If more than one operation is to be run on a particular equipment type, several alternatives must be considered.

## Example 3.3

The machine fractions for an ABC drill press are given in Table 3.6. No drill press operator, overtime, or subcontracting is available for any operations on the ABC drill press. It may be seen that a minimum of four and a maximum of six machines are required. How many should be purchased? The answer is either four, five, or six. With no further information, a specific recommendation cannot be made. Information on the cost of the equipment, the length of machine setups, the cost of in-process inventories, the cost and feasibility of overtime, production, and/or setups, the expected future growth of demand, and several other qualitative factors must be analyzed to reach a decision.

Table 3.6   *Total Equipment Requirement Specification Example*

| Operation Number | Equipment Fraction | Next Highest Whole Number |
|---|---|---|
| 109 | 1.1 | 2 |
| 206 | 2.3 | 3 |
| 274 | 0.6 | 1 |

# 3.5  FACILITIES DESIGN

Once the product, process, and schedule design decisions have been made, the facilities planner needs to organize the information and generate and evaluate layout, handling, storage, and unit load design alternatives. As explained in Chapter 2 and Section 3.1, the facilities planner must be aware of top management's objectives and goals to maximize the impact of the facilities design effort on such objectives and goals. Some typical business objectives include breakthroughs in production cost, on-time delivery, quality, and lead time.

Some tools frequently used by quality practitioners (e.g., Pareto chart) can be very useful in facilities planning efforts. More recently, the seven management and planning tools have gained acceptance as a methodology for improving planning and implementation efforts in general [5]. The tools have their roots in post-WWII operations research work and the total quality control (TQC) movement in Japan.

In the mid 1970s a committee of engineers and scientists in Japan refined and tested the tools as an aid for process improvement, as proposed by the Deming cycle. In 1950 Dr. W. E. Deming proposed a model for continuous process improvement that involves four steps: planning and goal setting, doing or execution, checking or analysis, and specifying corrective actions (Plan–Do–Check–Act).

The seven management and planning tools are the affinity diagram, the interrelationship digraph, the tree diagram, the matrix diagram, the contingency diagram, the activity network diagram, and the prioritization matrix. Each is described below and illustrated with examples related to facilities design.

## Affinity Diagram

The affinity diagram is used to gather language data, such as ideas and issues, and organize it into groupings. Suppose we are interested in generating ideas for reducing

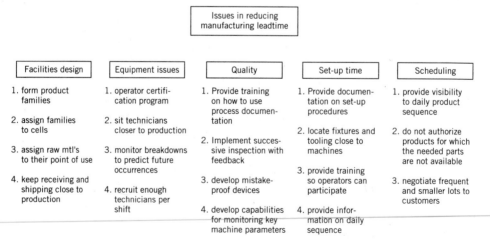

**Figure 3.17**  Affinity diagram example for reducing manufacturing leadtime.

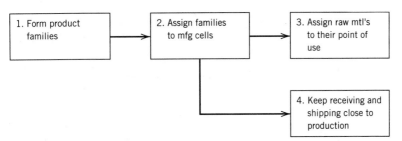

**Figure 3.18**   Interrelationship diagraph for facilities design.

manufacturing leadtime. In a brainstorming session, the issues are written down on "post-it" notes and grouped on a board or wall. Each group then receives a heading. An affinity diagram for reducing manufacturing leadtime is presented in Figure 3.17. The headings selected were: facilities design, equipment issues, quality, set-up time, and scheduling.

## Interrelationship Digraph

The interrelationship digraph is used to map the logical links among related items, trying to identify which items impact others the most. The term *digraph* is employed because the graph uses directed arcs. Suppose we want to study the relationship between the items in Figure 3.17 under facilities design. The interrelationships are presented in Figure 3.18. Note that this graph helps us understand the logical sequence of steps for the facilities design. The efforts must be initiated with the formation of product families.

## Tree Diagram

The tree diagram is used to map in increasing detail the actions that need to be accomplished in order to achieve a general objective. Assuming that we want to construct a tree diagram for the formation of product families, the tree is presented in Figure 3.19. Note that the same exercise can be performed for each item in the interrelationship digraph.

## Matrix Diagram

The matrix diagram organizes information such as characteristics, functions, and tasks into sets of items to be compared. A simple application of this tool is the design of a table in which the participants and their role within the small teams are defined. This tool provides visibility to key contacts on specific issues and helps identify individuals who are assigned to too many teams. Table 3.7 assumes that three teams are formed in response to the actions listed in the tree diagram (Figure 3.19). The teams focus

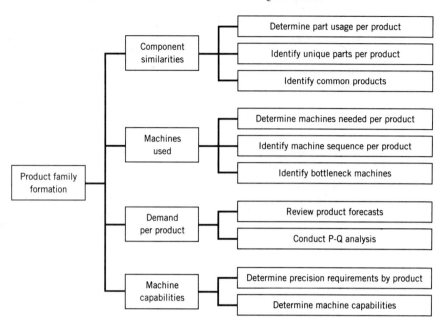

**Figure 3.19**   Tree diagram for the formation of product families.

on (1) part usage, (2) machine usage and capability, and (3) demand forecasts. Note that additional teams will be formed for the other activities identified in the interrelationship digraph (Figure 3.18). In the table, team leaders and coordinators have been identified since they might carry a heavier workload and it is desirable for them to have less involvement in other efforts.

## Contingency Diagram

The contingency diagram, formally known as process decision program chart, maps conceivable events and contingencies that might occur during implementation. It is particularly useful when the project being planned consists of unfamiliar tasks. The benefit of preventing or responding effectively to contingencies makes it worthwhile to look at these possibilities during the planning phase.

Table 3.7   *Matrix Diagram for Team Participation*

| Team\Participants | Joe | Mary | Jerry | Lou | Linda | Daisy | Jack |
|---|---|---|---|---|---|---|---|
| Part usage team | P | C | P | L | | | P |
| Machine use & cap team | L | | C | | | | P |
| Demand forecast team | | | | P | C | L | |

*Note.*   L: Team Leader
  C: Team Coordinator
  P: Team Participant

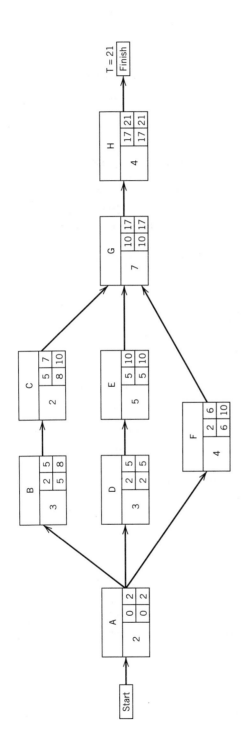

*Note.* A: Schedule shutdown periods for equipment movement and installation

B: Interview, evaluate, select, and hire new employees

C: Train new employees

D: Interview, evaluate, and select equipment vendors

E: Order equipment

F: Interview, evaluate, select, and hire construction contractors

G: Meet to review installation plan (Facilities Design Team, contractors, new employees, vendor representatives, and management

H: Execute installation plan and test system

D: Activity duration
ES: Early Start
EF: Early Finish
LS: Late Start
LF: Late Finish

**Figure 3.20** Activity network diagram example for a production line expansion facilities design project.

**Table 3.8**   *Weekly Timetable for Team Work Sessions*

| Time\Day | Mon | Tue | Wed | Thu | Fri |
|---|---|---|---|---|---|
| 8–10 A.M. | | | | | |
| 10–12 P.M. | Parts | | Parts | | |
| 1–3 P.M. | | Mach's | | Demand | Mach's |
| 3–5 P.M. | | | | Demand | |

## Activity Network Diagram

The activity network diagram is used to develop a work schedule for the facilities design effort. This diagram is synonymous to the critical path method (CPM) graph. It can also be replaced by a Gantt chart and if a range is defined for the duration of each activity, the Program Evaluation and Review Technique (PERT) chart can also be used. The important message is that a well thought out time table is needed to understand the length of the facilities design project. This timetable can be developed after the actions on the tree diagram (Figure 3.19) have been evaluated with the prioritization matrix. An example of an activity network diagram for a production line expansion is illustrated in Figure 3.20.

After understanding the magnitude of the facilities design effort, it might be necessary to form several small teams to work concurrently on the various tasks. In a participative environment, production, material handling, and representatives from support functions are included in the facilities design project. They have the best understanding on how the plant processes are being executed. With their participation, more ideas are generated, there is significantly less resistance to change in the organization, and the end result is improved. If a team has varying levels of understanding about facilities design, the team interaction should be started with education and training.

Team activities can also be planned, and a weekly schedule can be prepared, assigning one or two work sessions per team. Scheduling the work sessions help team members in allocating time for the effort. A typical weekly work schedule for the teams defined above is presented in Table 3.8.

## Prioritization Matrix

In developing facilities design alternatives it is important to consider:

(a)   Layout characteristics
- total distance travelled
- manufacturing floor visibility
- overall aesthetics of the layout
- ease of adding future business

(b)   Material handling requirements
- use of current material handling equipment
- investment requirements on new equipment
- space and people requirements

(c)  Unit load implied
- impact on WIP levels
- space requirements
- impact on material handling equipment

(d)  Storage strategies
- space and people requirements
- impact on material handling equipment
- human factor risks

(e)  Overall building impact
- estimated cost of the alternative
- opportunities for new business.

The prioritization matrix can be used to judge the relative importance of each criterion as compared to each other. Table 3.9 presents the prioritization of the criteria for the facilities design example. The criteria are labeled to help in building a table with weights.

| | |
|---|---|
| A.  Total distance travelled | G.  Space requirements |
| B.  Manufacturing floor visibility | H.  People requirements |
| C.  Overall aesthetics of the layout | I.  Impact on WIP levels |
| D.  Ease of adding future business | J.  Human factor risks |
| E.  Use of current MH equipment | K.  Estimated cost of alternative |
| F.  Investment in new MH equipment | |

Table 3.9  *Prioritization Matrix for the Evaluation of Facilities Design Alternatives*

| | A | B | C | D | E | F | G | H | I | J | K | Row totals (%) |
|---|---|---|---|---|---|---|---|---|---|---|---|---|
| A | 1 | 5 | 10 | 5 | 1 | 1 | 1 | 1 | 1 | 5 | 1 | 32. (9.9) |
| B | 1/5 | 1 | 5 | 1/5 | 1/5 | 1/10 | 1/5 | 1/5 | 1/10 | 1/5 | 1/5 | 7.6 (2.4) |
| C | 1/10 | 1/5 | 1 | 1/10 | 1/10 | 1/10 | 1/5 | 1/5 | 1/10 | 1/10 | 1/10 | 2.3 (0.7) |
| D | 1/5 | 5 | 10 | 1 | 1/5 | 1/5 | 1/5 | 1/5 | 1/10 | 1/5 | 1/10 | 17.4 (5.4) |
| E | 1 | 5 | 10 | 5 | 1 | 1 | 5 | 5 | 1/5 | 1 | 1/5 | 34.4 (10.7) |
| F | 1 | 10 | 10 | 5 | 1 | 1 | 5 | 5 | 1 | 1 | 1 | 41. (12.7) |
| G | 1 | 5 | 5 | 5 | 1/5 | 1/5 | 1 | 5 | 1/5 | 1/5 | 1/5 | 23. (7.1) |
| H | 1 | 5 | 5 | 5 | 1/5 | 1/5 | 5 | 1 | 1/10 | 1/5 | 1/5 | 22.9 (7.1) |
| I | 1 | 10 | 10 | 10 | 5 | 1 | 5 | 10 | 1 | 1 | 5 | 59. (18.3) |
| J | 1/5 | 5 | 10 | 5 | 1 | 1 | 5 | 5 | 1 | 1 | 5 | 39.2 (12.2) |
| K | 1 | 5 | 10 | 10 | 5 | 1 | 5 | 5 | 1/5 | 1/5 | 1 | 43.4 (13.5) |
| Column total | 7.7 | 56.2 | 86. | 51.3 | 14.9 | 6.8 | 32.6 | 37.6 | 5. | 10.1 | 14. | 322.2 Grand total |

**Table 3.10**   *Prioritization of Layout Alternatives Based on WIP Levels*

| WIP Levels | P | Q | R | S | T | Row totals (%) |
|---|---|---|---|---|---|---|
| | | | Layout | | | |
| P | 1 | 5 | 1 | 1/10 | 1/5 | 7.3 (9.9) |
| Q | 1/5 | 1 | 1/5 | 1/10 | 1/10 | 1.6 (2.2) |
| R | 1 | 5 | 1 | 10 | 5 | 22. (30.0) |
| S | 10 | 10 | 1/10 | 1 | 1/5 | 21.3 (29.0) |
| T | 5 | 10 | 1/5 | 5 | 1 | 21.2 (28.9) |
| Column total | 17.2 | 31 | 2.5 | 16.2 | 6.5 | 73.4 Grand total |

The weights typically used to compare the importance of each pair of criteria are

    1 = equally important
    5 = significantly more important      1/5 = significantly less important
  10 = extremely more important     1/10 = extremely less important

Note that the values in cells (i, j) and (j, i) are reciprocals. The resulting relative importance is presented in the last column in parenthesis. For this application, the most important criterion for facilities design selection is the impact on WIP levels (weight = 18.3), followed by the estimated cost of the solution (weight = 13.5).

This same methodology can be employed to compare all facilities design alternatives in each weighted criterion. For example, suppose five layout alternatives are generated, namely, P, Q, R, S, and T. Table 3.10 presents the ranking of the layout alternatives based on the impact of WIP levels criterion.

If we construct a similar table for the remaining ten criteria, we will be able to evaluate each layout alternative using the eleven criteria to identify the best layout. The format of this final table is presented in Table 3.11. The last column is computed as in Tables 3.9 and 3.10. The row totals (represented by $\Sigma$) are added to obtain the grand total, after which the percentages (%P, . . . , %T) are determined. These per-

**Table 3.11**   *Ranking of Layouts by All Criteria*

| | A | B | C | D | E | F | G | H | I | J | K | Row totals (%) |
|---|---|---|---|---|---|---|---|---|---|---|---|---|
| | | | | | | | Criteria | | | | | |
| P | | | | | | | | | $.099 \times .183 = .018$ | | | $\Sigma$ (%P) |
| Q | | | | | | | | | $.022 \times .183 = .004$ | | | $\Sigma$ (%Q) |
| R | | | | | | | | | $.300 \times .183 = .055$ | | | $\Sigma$ (%R) |
| S | | | | | | | | | $.290 \times .183 = .053$ | | | $\Sigma$ (%S) |
| T | | | | | | | | | $.289 \times .183 = .053$ | | | $\Sigma$ (%T) |
| Column | | | | | | | | | .183 | | | Grand Total |

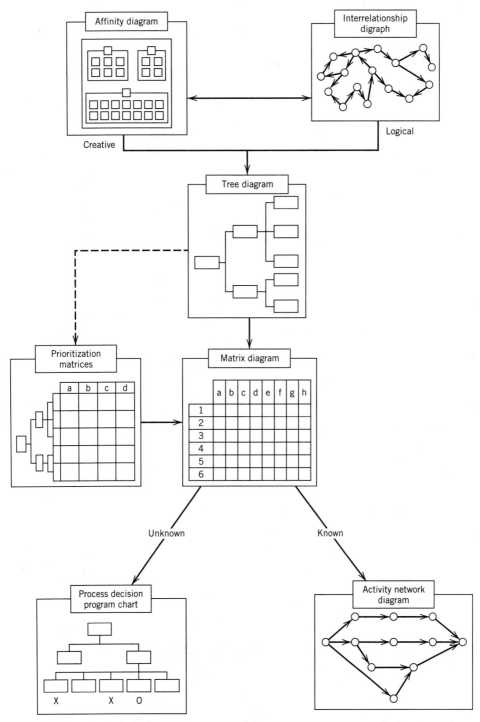

**Figure 3.21** Logical application sequence of the seven management and planning tools.

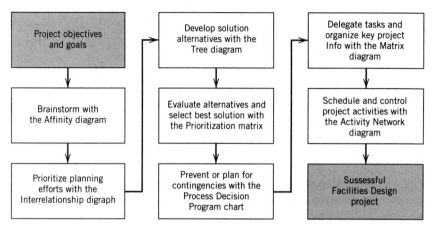

**Figure 3.22** How the seven management and planning tools facilitate the planning of a facilities design project.

centages tell us the relative goodness of each layout alternative. These results should be presented to plant management to facilitate final decisions regarding the layout.

The examples presented in this section are a subset of the possible uses of the seven management and planning tools. Interestingly, all the tools can be used to support an integrated facilities design process or they can also be used independently to support facilities design teams in the planning or resolution of specific issues. Figure 3.21 presents a graphical representation on the logical sequence when the tools are used in an integrated fashion. Figure 3.22 provides a flowchart of the application of the tools to a facilities design planning project. To summarize, Figure 3.23 presents a possible scenario for a facilities design team using the tools.

## *3.6* SUMMARY

Product, process, and schedule design decisions can have a significant impact on both the investment cost for a facility and the cost-effective performance of the activities assigned to the facility. The decisions made concerning product design, process planning, production schedules, and facilities planning must be jointly determined in order to obtain an integrated production system that achieves the firm's business objectives.

Even though facilities planners are not typically involved in product, process, and schedule design, it is important for the facilities planner to be familiar with these activities. Rather than simply reacting to requirements established by the product, process, and schedule design process, the facilities planner should be proactive and influence requirements definitions. Rather than being a passive observer, the facilities planner should be an active participant in decision making who influences the degrees of freedom for planning the facility.

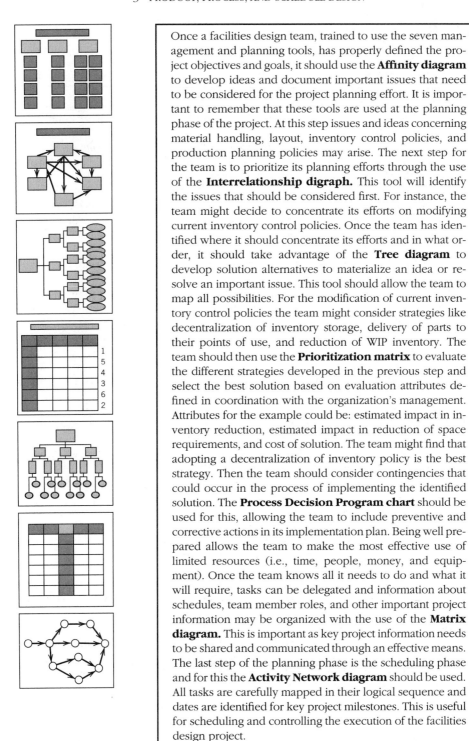

Once a facilities design team, trained to use the seven management and planning tools, has properly defined the project objectives and goals, it should use the **Affinity diagram** to develop ideas and document important issues that need to be considered for the project planning effort. It is important to remember that these tools are used at the planning phase of the project. At this step issues and ideas concerning material handling, layout, inventory control policies, and production planning policies may arise. The next step for the team is to prioritize its planning efforts through the use of the **Interrelationship digraph.** This tool will identify the issues that should be considered first. For instance, the team might decide to concentrate its efforts on modifying current inventory control policies. Once the team has identified where it should concentrate its efforts and in what order, it should take advantage of the **Tree diagram** to develop solution alternatives to materialize an idea or resolve an important issue. This tool should allow the team to map all possibilities. For the modification of current inventory control policies the team might consider strategies like decentralization of inventory storage, delivery of parts to their points of use, and reduction of WIP inventory. The team should then use the **Prioritization matrix** to evaluate the different strategies developed in the previous step and select the best solution based on evaluation attributes defined in coordination with the organization's management. Attributes for the example could be: estimated impact in inventory reduction, estimated impact in reduction of space requirements, and cost of solution. The team might find that adopting a decentralization of inventory policy is the best strategy. Then the team should consider contingencies that could occur in the process of implementing the identified solution. The **Process Decision Program chart** should be used for this, allowing the team to include preventive and corrective actions in its implementation plan. Being well prepared allows the team to make the most effective use of limited resources (i.e., time, people, money, and equipment). Once the team knows all it needs to do and what it will require, tasks can be delegated and information about schedules, team member roles, and other important project information may be organized with the use of the **Matrix diagram.** This is important as key project information needs to be shared and communicated through an effective means. The last step of the planning phase is the scheduling phase and for this the **Activity Network diagram** should be used. All tasks are carefully mapped in their logical sequence and dates are identified for key project milestones. This is useful for scheduling and controlling the execution of the facilities design project.

**Figure 3.23** A facilities design team using the seven management and planning tools.

# BIBLIOGRAPHY

1. Akao, Y. (ed.), *QFD: Integrating Customer Requirements into Product Design,* Productivity Press, Cambridge, MA, 1990.
2. Amirouche, F. M. L., *Computer-Aided Design and Manufacturing,* Prentice Hall, NJ, 1993.
3. Barnes, R. M., *Motion and Time Study Design and Measurement of Work,* John Wiley, New York, 1980.
4. Bourjault, A., *Contribution a une Approache Methodologique de l'Assemblage Automatise: Elaboration Automatique des Sequences Operatories,* Thesis for Grade de Docteur en Sciences Physiques at l'Universite de Franche Comte, November 12, 1984.
5. Brassard, M., *The Memory Jogger Plus,* GOAL/QPC, Methuen, MA, 1989.
6. Buffa, E. S., *Modern Production Management,* John Wiley, New York, 1977.
7. Burbidge, J. L., *The Introduction of Group Technology,* John Wiley, New York, 1975.
8. Camp, R. C., *Benchmarking: The Search for Industry Best Practices that Lead to Superior Performance,* Quality Press, White Plains, NY, 1989.
9. Chang, T., Wysk, R. A., Wang, H., *Computer Aided Manufacturing,* Prentice Hall, Englewood Cliffs, NJ, 1991.
10. El-Hadidi, T., *Simultaneous Evaluation Tool for Printed Circuit Boards,* Master Thesis in Industrial Engineering, Department of Industrial Engineering, North Carolina State University, Raleigh, NC, August 1991.
11. Francis, R. L., and White, J. A., *Facility Layout and Location; An Analytical Approach,* Prentice Hall, Englewood Cliffs, NJ, 1974.
12. Hales, H. L., *Computer-Aided Facilities Planning,* Marcel Dekker, New York, 1984.
13. Ham, I., *Introduction to Group Technology,* Pennsylvania State University, University Park, PA., 1975.
14. Hartley, J. R., *Concurrent Engineering,* Productivity Press, Cambridge, 1992.
15. Klein, C., *Generation and Evaluation of Assembly Sequence Alternatives,* Master of Science Thesis, MIT Mech. Eng. Dept., MIT, Boston, 1986.
16. Morris, W. T., *Engineering Economic Analysis,* Reston Publishing, Reston, VA., 1976.
17. Ranson, G. M., *Group Technology,* Pergamon, New York, 1970.
18. Shore, R. H., and Tompkins, J. A., *Designing Flexible Facilities,* North Carolina State Univ., Raleigh, NC, 1977.
19. Shore, R. H., and Tompkins, J. A., "Flexible Facilities Design," *AIIE Transactions,* Vol. 12, no. 2. pp. 200–205, June 1980.
20. Silva, D., *Concurrent Product and Process Design for Assembly,* Master Project in Integrated Manufacturing Systems Engineering, Integrated Manufacturing Systems Engineering Institute, North Carolina State University, Raleigh, NC, August 1990.
21. Subramaniam, S., *Design Evaluation and Cost Estimation Expert System,* Master Thesis in Industrial Engineering, Department of Industrial Engineering, North Carolina State Univ., Raleigh, NC, 1991.
22. Whitney, D. E., DeFazio, T. L., Gustavson, R. E., Graves, S. C., Abell, T., Cooprider, K., and Pappu, S., *Tools for Strategic Product Design, First Report, CSDL-R-2115,* MIT and The Charles Stark Draper Laboratory, Cambridge, MA, November 1988.
23. Zorowski, C. F., *PDM—A Product Assemblability Merit Analysis Tool,* 1986 National Design Engineering Technical Conference, ASME, Columbus, OH, 1986.

# PROBLEMS

3.1 You have been asked to plan a facility for a bank. Explain in detail your first step.

3.2 Why is it important to integrate product, process, quality, scheduling, and facilities design decisions? Who should be involved in this integration? Which techniques are available to support this integrated approach?

3.3 Identify from papers and books at least five companies that have designed or redesigned their facilities considering new manufacturing approaches (i.e., multiple receiving docks, decentralized storage areas, cellular manufacturing, kanbans). For each company, describe in detail the methodology employed, approaches used, and results obtained.

3.4 Identify from papers and books at least five companies that have used benchmarking and/or quality function deployment to identify competition and customer needs. For each company, describe in detail the methodology employed and results obtained.

3.5 Identify from papers and books at least two companies that are implementing concurrent engineering techniques for integrated product and process design using the team approach. For each company, explain the team composition, techniques used, methodology employed, and results obtained. Did they consider facilities design in the product and process integration?

3.6 Identify at least five CAD systems available to support the facilities design process. Prepare a comparison of these systems and recommend the best one.

3.7 Develop a bill of materials, an assembly chart, and an operation process chart for a cheese hamburger and a taco with everything on it. Identify the components that are purchased and the ones that are prepared internally.

3.8 Mention three differences between the assembly chart and the operation process chart.

3.9 Part X requires machining on a milling machine (operations A and B are required). Find the number of machines required to produce 2500 parts per week. Assume the company will be operating 5 days per week, 18 hours per day. The following information is known:

| Operation | Standard Time | Efficiency | Reliability | Scrap |
|---|---|---|---|---|
| A | 2 min. | 95% | 95% | 2% |
| B | 4 min. | 95% | 90% | 5% |

Note. The milling machine requires tool changes and preventive maintenance after every lot of 400 parts. These changes require 30 min.

3.10 Form a team and apply the seven management and planning tools for planning the redesign of any of the following:
   a. a bank,
   b. a kindergarden,
   c. a toolroom,
   d. a cafeteria,
   e. a laundry facility,
   f. a passenger area from a local airport, or
   g. a bookstore.

3.11 Given the following, what are the machine fractions for machines A and B to produce parts X and Y?

|                        | Machine A   | Machine B   |
|------------------------|-------------|-------------|
| Part X standard time   | 0.15 hr     | 0.2 hr      |
| Part Y standard time   | 0.10 hr     | 0.10 hr     |
| Part X scrap estimate  | 5%          | 5%          |
| Part Y scrap estimate  | 5%          | 5%          |
| Historical efficiency  | 90%         | 90%         |
| Reliability factor     | 90%         | 90%         |
| Equipment availability | 1600 hr/yr  | 1600 hr/yr  |

Part X routing is machine A, then B, and then A again; 110,000 parts are to be produced per year. Part Y routing is machine B, then A, and then B again; 250,000 parts are to be produced per year. Set-up times for parts X and Y are 20 min and 40 min, respectively.

**3.12** During one 8-hr shift, 500 nondefective parts are desired from a fabrication operation. The standard time for the operation is 20 min. Because the machine operators are unskilled, the actual time it takes to perform the operation is 25 min and, on the average, one-fifth of the parts that start to be fabricated are scrapped. Assuming that each of the machines used for this operation will not be available for 1 hr each shift, determine the number of machines required.

**3.13** Part A is produced in machines 1 and 2. Part B is produced in machines 3 and 4. Part A and B are assembled in workstation 5 to create C. Assembly C is painted in process 6. Given the scrap percentages P1,. . ., P6 for each operation and the desired output O6 for C, determine an equation that can be used to calculate the required input I1 to machine 1.

**3.14** Choose a simple recipe consisting of no more than ten ingredients. Examine carefully how the recipe is made. Develop a parts list, a bill of materials, a route sheet, an assembly chart, an operations process chart, and a precedence diagram for this recipe so that someone could follow the recipe without additional instructions.

**3.15** Take the parts list, bill of materials, route sheet, assembly chart, operations process chart, and precedence diagram from Question 3.14 and give it to another individual unfamiliar with the recipe. Have this individual decompose these charts into a written form of the recipe. Examine how close this derived recipe is to the original.

**3.16** In Figure 3.17, the second item under scheduling is "do not authorize products for which the needed parts are not available." Construct an affinity diagram of actions to provide visibility to this issue.

**3.17** Construct an interrelationship digraph for the results of Problem 3.16.

**3.18** Develop a tree diagram to identify the specifics on developing an operator certification program (first item under equipment issues in Figure 3.17).

**3.19** In a facilities design lab assignment, your instructor might be expecting specific deliverables (select one of your lab assignments). For such deliverables, identify the activities that must be performed. Construct a matrix diagram with deliverables (in rows) and activities (in columns). Using some symbols (i.e., star = extremely important, circle = important, triangle = less important, empty = no impact), specify the impact of each activity on the deliverables. An activity might impact more than one deliverable.

**3.20** Develop a contingency diagram for the field trips to nearby industries that the local IIE student chapter might be organizing for the year.

**3.21** Develop an implementation plan, using the activity network diagram, for your facilities design final project.

**3.22** Prepare a summary of the computer-based tools that can be employed in the design of products and processes.

**3.23** Identify from papers and books at least two other methodologies and/or models to support make-or-buy decisions. Prepare a summary of these methodologies and/or models. Based on this information, propose your own methodology and/or model for make-or-buy decisions.

**3.24** In a just-in-time and total quality management environment, scrap is reduced or eliminated. Describe at least five strategies to eliminate scrap and explain the impact of scrap elimination on facilities design.

**3.25** Expand Equation 3.4 to include:
   a.  number of shifts,
   b.  set-up times,
   c.  scrap, and
   d.  preventive maintenance.

# 4

# *FLOW, SPACE, AND ACTIVITY RELATIONSHIPS*

## *4.1* INTRODUCTION

In determining the requirements of a facility, three important considerations are flow, space, and activity relationships. **Flow** depends on lot sizes, unit load sizes, material handling equipment and strategies, layout arrangement, and building configuration. **Space** is a function of lot sizes, storage system, production equipment type and size, layout arrangement, building configuration, housekeeping and organization policies, material handling equipment, and office, cafeteria, and restroom design. **Activity relationships** are defined by material or personnel flow, environmental considerations, organizational structure, continuous improvement methodology (teamwork activities), control issues, and process requirements.

As indicated in previous chapters, facilities planning is an iterative process. The facilities planning team or the facilities planner needs to interact not only with product, process, and schedule designers but also with top management to identify alternative issues and strategies to consider in the analysis (i.e., lot sizes, storage-handling strategies, office design, organizational structure, environmental policy).

The facilities planner also needs to continually investigate the impact of modern manufacturing approaches on flow, space, and activity relationships. For example, concepts like decentralized storage, multiple-receiving docks, deliveries to points of use, decentralized management and support functions, quality at the source, cellular manufacturing, lean organizational structures, and small lot purchasing and production could challenge traditional activity relationships and reduce flow and space requirements.

Some of the traditional activity relationships to be challenged are centralized offices, a centralized storage area, a single-receiving area, and a centralized rework area. Flow requirements are reduced with external and internal deliveries to points of use, storage of inventories in decentralized storage areas close to points of use, movement of materials controlled by pull strategies with kanbans, and manufacturing cells. Less space is required for inventories; production, storage, and handling equipment; offices; parking lots; and cafeterias.

Different procedures, forms, and tables to determine flow, space, and activity relationships are covered in this chapter. Additionally, layout types are explained in Section 4.2 (including a consideration of cellular manufacturing), the logistics system is described in Section 4.3, and visual management is covered in Section 4.7. Activity relationships are considered in Section 4.3. Sections 4.4 through 4.6 address flow relationships. Space relationships are the subject of Section 4.7.

## *4.2* DEPARTMENTAL PLANNING

To facilitate the consideration of flow, space, and activity relationships, it is helpful to introduce the subject of departmental planning. At this point in the facilities planning process, we are not so concerned with organizational entities. Rather, we are interested in forming **planning departments.** Planning departments can involve production, support, administrative, and service areas (called production, support, administrative, and service planning departments).

Production planning departments are collections of workstations to be grouped together during the facilities layout process. The formulation of organizational units should parallel the formation of planning departments. If for some reason the placement of workstations violates certain organizational objectives, then modifications should be made to the layout.

As a general rule, planning departments may be determined by combining workstations that perform "like" functions. The difficulty with this general rule is the definition of the term "like." "Like" could refer to workstations performing operations on similar products or components or to workstations performing similar processes.

Depending on product volume-variety, production planning departments can also be classified as product, fixed materials location, product family (or group technology), and process planning departments (see Figure 4.1). As examples of production planning departments that consist of a combination of workstations performing operations on similar products or components, consider engine block production line departments, aircraft fuselage assembly departments, and uniform flat sheet metal departments. These are referred to as product planning departments because they are formed by combining workstations that produce similar products or components. **Product planning departments** may be further subdivided by the characteristics of the products being produced.

Suppose a large, stable demand for a standardized product, like an engine block, is to be met by production. In such a situation, the workstations should be combined into a planning department so that all workstations required to produce the product are combined. The resulting product planning department may be referred to as **production line department.**

Next, suppose a low, sporadic demand exists for a product that is very large and awkward to move, for example, and aircraft fuselage. The workstations should be combined into a planning department that includes all workstations required to

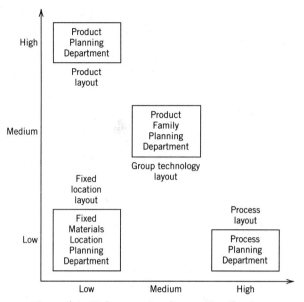

**Figure 4.1** Volume-variety layout classification.

produce the product and the staging area. This type of product planning department may be referred to as a **fixed materials location department.**

A third type of product planning department may be identified when there exists a medium demand for a medium number of similar components. Similar components form, a family of components that, in group technology terminology, may be produced via a "group" of workstations. The combination of the group of workstations results in a product planning department that may be referred to as a **product family department.**

Examples of planning departments based on the combination of workstations containing "similar" processes are metal cutting departments, gear cutting departments, and hobbing departments. Such planning departments are referred to as **process departments** because they are formed by combining workstations that perform "similar" processes.

The difficulty in defining process departments is in the interpretation of the word "similar." For example, in a facility specializing in the production of gears, it is quite conceivable that gear hobbing, gear shaping, and shaft turning would not be considered similar and each would be grouped into their own planning departments. However, in a facility producing mechanical switching mechanisms, it is conceivable that these same processes would be grouped into two, not three, planning departments: a gear cutting department, containing similar gear hobbing and shaping processes, and a turning department. Even more extreme, in a furniture facility, all metalworking may take place in a metalworking planning department. Therefore, the same three processes may be seen to be similar and grouped into a single process planning department. The determination of which workstations are to be considered similar is dependent on not only the workstations but also the relationships among workstations and between workstations and the overall facility.

Most facilities consist of a mixture of product and process planning departments. For example, in a facility consisting of mainly process planning departments producing a large variety of rather unrelated products, the detailed placement of individual workstations within a process department might be based on a product planning department philosophy. (As an illustration, all painting activities might be grouped together in a painting process department. However, the layout of the painting department can consist of a painting line designed on the basis of a product planning department philosophy). Conversely, in a facility consisting mainly of product planning departments producing a few, high-volume, standard products, it would not be surprising to find several "specialized" components produced in process planning departments.

A systematic approach should be used in combining workstations into departments. Each product and component should be evaluated and the best approach determined for combining workstations into planning departments. Table 4.1 summarizes the bases for combining workstations into planning departments.

Support, administrative, and service planning departments include offices and areas for storage, quality control, maintenance, administrative processes, cafeterias, restrooms, lockers, and so on. Traditionally, support, administrative, and service planning departments have been treated as "process" departments (similar activities performed within a certain area).

Organizations using modern manufacturing approaches are combining production, support, administrative, and service planning departments to create integrated production-support-administrative-service planning departments. For example, a manufacturing cell dedicated to the production of a family of parts and with dedicated support and administrative personnel and services (e.g., maintenance,

**Table 4.1** *Procedural Guide for Combining Workstations in Planning Departments*

| If the Product Is | The Type of Planning Department Should Be | And the Method of Combining Workstations into Planning Departments Should Be |
|---|---|---|
| Standardized and has a large stable demand. | Production line, product department. | Combine all workstations required to produce the product. |
| Physically large, awkward to move, and has a low sporadic demand | Fixed materials location, product department | Combine all workstations required to produce the product with the area required for staging the product |
| Capable of being grouped into families of similar parts that may be produced by a group of workstations | Product family, product department | Combine all workstations required to produce the family of products |
| None of the above | Process department | Combine identical work stations into initial planning departments and attempt to combine similar initial planning departments without obscuring important interrelationships within departments |

quality, materials, engineering, tooling, purchasing, management, vending machines, restrooms, lockers) could be an integrated planning department.

Many companies train their operators in most of the support, administrative, and management functions so that they can become autonomous. In these cases, the operators (called technicians or associates) plus a facilitator–coordinator can manage the operation of the manufacturing cell with minimum external support.

Activity relationships and flow and space requirements in a facility with self-managing teams will be totally different than in a facility with traditional production, support, and administrative planning departments (less material, personnel, tooling, and paperwork flow requirements and less space needs).

Many companies using modern manufacturing approaches are converting their facilities to combinations of product and product family (group technology) planning departments. Group technology layouts are combined with *Just-In-Time* (JIT) concepts in cellular manufacturing arrangements. A more detailed explanation of group technology and cellular manufacturing is presented as follows.

## Manufacturing Cells

Product family or group technology departments aggregate medium volume-variety parts into families based on similar manufacturing operations or design attributes. The machines that are required to manufacture the part family are grouped together to form a "cell."

*Manufacturing cells* group machines, employees, materials, tooling, and material handling and storage equipment to produce families of parts.

The most important benefits of cellular manufacturing are achieved when manufacturing cells are designed, controlled, and operated using Just-In-Time (JIT), Total Quality Management (TQM), and Total Employee Involvement (TEI) concepts and techniques (see Chapter 10). For example, *JIT manufacturing cells* use small lots, kanbans, standardized containers, simple material handling systems, short set-ups, decentralized storage areas, deliveries to the points of use, horizontal organizational structures, bakayokes, andons, and total productive maintenance.

Additionally, *manufacturing cells that use TQM principles* are designed to satisfy customer needs, and use process inspection, prevention measures, and feedback and quick reaction mechanisms (in process verification, quality at the source, self-inspection, individual responsibility, SPC, parameter design, design for quality, permanent solution of problems, and self-correction of errors).

Finally, manufacturing cells are ideal for cross-trained and flexible operators with multifunctional capabilities that are involved and empowered in teamwork.

When manufacturing cells are designed, controlled, and operated using JIT, TQM, and TEI principles, the following benefits can be achieved: *reduction of* inventories, space, machine breakdowns, rework, paperwork, warranty claims, storage and handling equipment, employee turnover and absenteeism, production leadtimes, cost, and stockouts; *simplification of* communication, handling, and production scheduling; and *improvement of* productivity, flexibility, inventory turnover, quality, and customer and employee satisfaction.

Successful implementation of manufacturing cells requires addressing selection, design, operation, and control issues. Selection refers to the identification of machine and part types for a particular cell. Cell design refers to layout and production and material handling requirements. Operation of a cell involves determining lot sizes, scheduling, number of operators, type of operators, and type of production control (push vs. pull). Finally, control of a cell refers to the methods used to measure the performance of the cell.

Several approaches have been proposed to address selection issues of manufacturing cells. The most popular approaches are classification and coding, production flow analysis, clustering techniques, heuristic procedures, and mathematical models [1, 2, 5, 7, 10, 12, 13, 15, 16, 17, 18, 19, 24, 28].

Classification is the grouping of parts into classes or part families based on design attributes and coding is the representation of these attributes by assigning numbers or symbols to them.

Production Flow Analysis [5] is a procedure for forming part families by analyzing the operation sequences and the production routing of a part or component through the plant.

Clustering methodologies are used to group parts together so that they can be processed as a family [7, 10, 12, 16, 17, 19, 28]. This methodology lists parts and machines in rows and columns, and interchanges them based on some criterion like similarity coefficients. For example, the Direct Clustering Algorithm (DCA) [7] forms clustered groups based on sequentially moving rows and columns to the top and left.

There are several heuristic procedures that have been developed for the formation of cells [2, 13, 15]. For example, a heuristic developed by Ballakur and Steudel [2] assigns machines to cells based on work load factors and assigns parts to cells based on a percentage of operations of a part processed within a cell.

Several other mathematical models have also been developed [1, 18, 25]. For example, a method proposed by Al-Qattan [1] to form machine cells and families of parts is based on the branch and bound method.

## Example 4.1

To develop an understanding of the cell formation task, the DCA [7] methodology will be applied to the part-machine matrix shown in Figure 4.2. This matrix is a representation of the machines visited by a set of parts during production. Again, our objective is to group parts and machines together to form cells.

The DCA methodology accomplishes this by first ranking each row and column by its number of occurrences, that is, "1"s. This ranking is represented by the number to the far right of each row and the bottom of each column. Once the ranking is performed, the rows and columns are to be sorted in descending order as shown in Figure 4.3. The next step in the methodology is to, starting with the first column, transfer all rows with occurrences to the top of the matrix. Thus, parts 1, 2, 3, 4, 5, 6, and 7 are moved above part 10. The result of this action is illustrated in Figure 4.4. Once the rows have been transferred, the columns are adjusted in a similar manner. As such, machine 6 is transferred in front of machine 24 as shown in Figure 4.5.

Since the modified part-machine matrix shows distinct clusters of parts and machines, no further modifications to the matrix are required. Therefore, cell 1 consists of machines 1, 3, 8, 9, 11, 17, 18, 22, 5, and 6, which will produce the part family of parts 1, 2, 3, 4, 5, 6, and 7. Cell 2 is composed of machines 1, 3, 24, 25, 26, 13, 14, 15, 16, and 21 which make parts 10, 11, 12, and 13. Finally, the last part family of parts 8 and 9 are produced in cell 3 by machines 2, 4, 7, 10, 12, 19, 20, and 23.

Note that both cell 1 and cell 2 require the use of machine 1 and machine 3. Ideally,

M/C #

| Part | 1 | 2 | 3 | 4 | 5 | 6 | 7 | 8 | 9 | 10 | 11 | 12 | 13 | 14 | 15 | 16 | 17 | 18 | 19 | 20 | 21 | 22 | 23 | 24 | 25 | 26 | Total |
|---|---|---|---|---|---|---|---|---|---|---|---|---|---|---|---|---|---|---|---|---|---|---|---|---|---|---|---|
| 1 | 1 |  | 1 |  | 1 |  |  | 1 | 1 |  | 1 |  |  |  |  |  | 1 | 1 |  |  |  | 1 |  |  |  |  | 9 |
| 2 | 1 |  | 1 |  | 1 |  |  | 1 | 1 |  | 1 |  |  |  |  |  | 1 | 1 |  |  |  | 1 |  |  |  |  | 9 |
| 3 | 1 |  | 1 |  | 1 |  |  | 1 | 1 |  | 1 |  |  |  |  |  | 1 | 1 |  |  |  | 1 |  |  |  |  | 9 |
| 4 | 1 |  | 1 |  | 1 |  |  | 1 | 1 |  | 1 |  |  |  |  |  | 1 | 1 |  |  |  | 1 |  |  |  |  | 9 |
| 5 | 1 |  | 1 |  |  | 1 |  | 1 | 1 |  | 1 |  |  |  |  |  | 1 | 1 |  |  |  | 1 |  |  |  |  | 9 |
| 6 | 1 |  | 1 |  |  | 1 |  | 1 | 1 |  | 1 |  |  |  |  |  | 1 | 1 |  |  |  | 1 |  |  |  |  | 9 |
| 7 | 1 |  | 1 |  |  | 1 |  | 1 | 1 |  | 1 |  |  |  |  |  | 1 | 1 |  |  |  | 1 |  |  |  |  | 9 |
| 8 |  | 1 |  | 1 |  |  | 1 |  |  | 1 |  | 1 |  |  |  |  |  |  | 1 | 1 |  |  | 1 |  |  |  | 8 |
| 9 |  | 1 |  | 1 |  |  | 1 |  |  | 1 |  | 1 |  |  |  |  |  |  | 1 | 1 |  |  | 1 |  |  |  | 8 |
| 10 | 1 |  | 1 |  |  |  |  |  |  |  |  |  | 1 | 1 | 1 | 1 |  |  |  |  | 1 |  |  | 1 | 1 | 1 | 10 |
| 11 | 1 |  | 1 |  |  |  |  |  |  |  |  |  | 1 | 1 | 1 | 1 |  |  |  |  | 1 |  |  | 1 | 1 | 1 | 10 |
| 12 | 1 |  | 1 |  |  |  |  |  |  |  |  |  | 1 | 1 | 1 | 1 |  |  |  |  | 1 |  |  | 1 | 1 | 1 | 10 |
| 13 | 1 |  | 1 |  |  |  |  |  |  |  |  |  | 1 | 1 | 1 | 1 |  |  |  |  | 1 |  |  | 1 | 1 | 1 | 10 |
|  | 11 | 2 | 11 | 2 | 4 | 3 | 2 | 7 | 7 | 2 | 7 | 2 | 4 | 4 | 4 | 4 | 7 | 7 | 2 | 2 | 4 | 7 | 2 | 4 | 4 | 4 |  |

**Figure 4.2**   A part-machine matrix.

M/C #

| Part | 1 | 3 | 8 | 9 | 11 | 17 | 18 | 22 | 5 | 24 | 25 | 26 | 13 | 14 | 15 | 16 | 21 | 6 | 2 | 4 | 7 | 10 | 12 | 19 | 20 | 23 | Total |
|---|---|---|---|---|---|---|---|---|---|---|---|---|---|---|---|---|---|---|---|---|---|---|---|---|---|---|---|
| 10 | 1 | 1 |  |  |  |  |  |  |  | 1 | 1 | 1 | 1 | 1 | 1 | 1 | 1 |  |  |  |  |  |  |  |  |  | 10 |
| 11 | 1 | 1 |  |  |  |  |  |  |  | 1 | 1 | 1 | 1 | 1 | 1 | 1 | 1 |  |  |  |  |  |  |  |  |  | 10 |
| 12 | 1 | 1 |  |  |  |  |  |  |  | 1 | 1 | 1 | 1 | 1 | 1 | 1 | 1 |  |  |  |  |  |  |  |  |  | 10 |
| 13 | 1 | 1 |  |  |  |  |  |  |  | 1 | 1 | 1 | 1 | 1 | 1 | 1 | 1 |  |  |  |  |  |  |  |  |  | 10 |
| 1 | 1 | 1 | 1 | 1 | 1 | 1 | 1 | 1 | 1 |  |  |  |  |  |  |  |  |  |  |  |  |  |  |  |  |  | 9 |
| 2 | 1 | 1 | 1 | 1 | 1 | 1 | 1 | 1 | 1 |  |  |  |  |  |  |  |  |  |  |  |  |  |  |  |  |  | 9 |
| 3 | 1 | 1 | 1 | 1 | 1 | 1 | 1 | 1 | 1 |  |  |  |  |  |  |  |  |  |  |  |  |  |  |  |  |  | 9 |
| 4 | 1 | 1 | 1 | 1 | 1 | 1 | 1 | 1 | 1 |  |  |  |  |  |  |  |  |  |  |  |  |  |  |  |  |  | 9 |
| 5 | 1 | 1 | 1 | 1 | 1 | 1 | 1 | 1 |  |  |  |  |  |  |  |  |  | 1 |  |  |  |  |  |  |  |  | 9 |
| 6 | 1 | 1 | 1 | 1 | 1 | 1 | 1 | 1 |  |  |  |  |  |  |  |  |  | 1 |  |  |  |  |  |  |  |  | 9 |
| 7 | 1 | 1 | 1 | 1 | 1 | 1 | 1 | 1 |  |  |  |  |  |  |  |  |  | 1 |  |  |  |  |  |  |  |  | 9 |
| 8 |  |  |  |  |  |  |  |  |  |  |  |  |  |  |  |  |  |  | 1 | 1 | 1 | 1 | 1 | 1 | 1 | 1 | 8 |
| 9 |  |  |  |  |  |  |  |  |  |  |  |  |  |  |  |  |  |  | 1 | 1 | 1 | 1 | 1 | 1 | 1 | 1 | 8 |
|  | 11 | 11 | 7 | 7 | 7 | 7 | 7 | 7 | 4 | 4 | 4 | 4 | 4 | 4 | 4 | 4 | 4 | 3 | 2 | 2 | 2 | 2 | 2 | 2 | 2 | 2 |  |

**Figure 4.3**   A sorted part-machine matrix.

M/C #

| Part | 1 | 3 | 8 | 9 | 11 | 17 | 18 | 22 | 5 | 24 | 25 | 26 | 13 | 14 | 15 | 16 | 21 | 6 | 2 | 4 | 7 | 10 | 12 | 19 | 20 | 23 |
|---|---|---|---|---|---|---|---|---|---|---|---|---|---|---|---|---|---|---|---|---|---|---|---|---|---|---|
| 1 | 1 | 1 | 1 | 1 | 1 | 1 | 1 | 1 | 1 |  |  |  |  |  |  |  |  |  |  |  |  |  |  |  |  |  |
| 2 | 1 | 1 | 1 | 1 | 1 | 1 | 1 | 1 | 1 |  |  |  |  |  |  |  |  |  |  |  |  |  |  |  |  |  |
| 3 | 1 | 1 | 1 | 1 | 1 | 1 | 1 | 1 | 1 |  |  |  |  |  |  |  |  |  |  |  |  |  |  |  |  |  |
| 4 | 1 | 1 | 1 | 1 | 1 | 1 | 1 | 1 | 1 |  |  |  |  |  |  |  |  |  |  |  |  |  |  |  |  |  |
| 5 | 1 | 1 | 1 | 1 | 1 | 1 | 1 | 1 |  |  |  |  |  |  |  |  |  | 1 |  |  |  |  |  |  |  |  |
| 6 | 1 | 1 | 1 | 1 | 1 | 1 | 1 | 1 |  |  |  |  |  |  |  |  |  | 1 |  |  |  |  |  |  |  |  |
| 7 | 1 | 1 | 1 | 1 | 1 | 1 | 1 | 1 |  |  |  |  |  |  |  |  |  | 1 |  |  |  |  |  |  |  |  |
| 10 | 1 | 1 |  |  |  |  |  |  |  | 1 | 1 | 1 | 1 | 1 | 1 | 1 | 1 |  |  |  |  |  |  |  |  |  |
| 11 | 1 | 1 |  |  |  |  |  |  |  | 1 | 1 | 1 | 1 | 1 | 1 | 1 | 1 |  |  |  |  |  |  |  |  |  |
| 12 | 1 | 1 |  |  |  |  |  |  |  | 1 | 1 | 1 | 1 | 1 | 1 | 1 | 1 |  |  |  |  |  |  |  |  |  |
| 13 | 1 | 1 |  |  |  |  |  |  |  | 1 | 1 | 1 | 1 | 1 | 1 | 1 | 1 |  |  |  |  |  |  |  |  |  |
| 8 |  |  |  |  |  |  |  |  |  |  |  |  |  |  |  |  |  |  | 1 | 1 | 1 | 1 | 1 | 1 | 1 | 1 |
| 9 |  |  |  |  |  |  |  |  |  |  |  |  |  |  |  |  |  |  | 1 | 1 | 1 | 1 | 1 | 1 | 1 | 1 |

**Figure 4.4**   A row transfer of part-machine matrix.

M/C #

|   |   | 1 | 3 | 8 | 9 | 11 | 17 | 18 | 22 | 5 | 6 | 24 | 25 | 26 | 13 | 14 | 15 | 16 | 21 | 2 | 4 | 7 | 10 | 12 | 19 | 20 | 23 |
|---|---|---|---|---|---|----|----|----|----|---|---|----|----|----|----|----|----|----|----|---|---|---|----|----|----|----|----|
|   | 1 | 1 | 1 | 1 | 1 | 1 | 1 | 1 | 1 | 1 |   |   |   |   |   |   |   |   |   |   |   |   |   |   |   |   |   |
|   | 2 | 1 | 1 | 1 | 1 | 1 | 1 | 1 | 1 | 1 |   |   |   |   |   |   |   |   |   |   |   |   |   |   |   |   |   |
|   | 3 | 1 | 1 | 1 | 1 | 1 | 1 | 1 | 1 | 1 |   |   |   |   |   |   |   |   |   |   |   |   |   |   |   |   |   |
|   | 4 | 1 | 1 | 1 | 1 | 1 | 1 | 1 | 1 | 1 |   |   |   |   |   |   |   |   |   |   |   |   |   |   |   |   |   |
| P | 5 | 1 | 1 | 1 | 1 | 1 | 1 | 1 | 1 |   | 1 |   |   |   |   |   |   |   |   |   |   |   |   |   |   |   |   |
| a | 6 | 1 | 1 | 1 | 1 | 1 | 1 | 1 | 1 |   | 1 |   |   |   |   |   |   |   |   |   |   |   |   |   |   |   |   |
| r | 7 | 1 | 1 | 1 | 1 | 1 | 1 | 1 | 1 |   | 1 |   |   |   |   |   |   |   |   |   |   |   |   |   |   |   |   |
| t | 10 | 1 | 1 |   |   |   |   |   |   |   |   | 1 | 1 | 1 | 1 | 1 | 1 | 1 | 1 |   |   |   |   |   |   |   |   |
|   | 11 | 1 | 1 |   |   |   |   |   |   |   |   | 1 | 1 | 1 | 1 | 1 | 1 | 1 | 1 |   |   |   |   |   |   |   |   |
| # | 12 | 1 | 1 |   |   |   |   |   |   |   |   | 1 | 1 | 1 | 1 | 1 | 1 | 1 | 1 |   |   |   |   |   |   |   |   |
|   | 13 | 1 | 1 |   |   |   |   |   |   |   |   | 1 | 1 | 1 | 1 | 1 | 1 | 1 | 1 |   |   |   |   |   |   |   |   |
|   | 8 |   |   |   |   |   |   |   |   |   |   |   |   |   |   |   |   |   |   | 1 | 1 | 1 | 1 | 1 | 1 | 1 | 1 |
|   | 9 |   |   |   |   |   |   |   |   |   |   |   |   |   |   |   |   |   |   | 1 | 1 | 1 | 1 | 1 | 1 | 1 | 1 |

**Figure 4.5**  A column transfer of part-machine matrix.

the machines in a cell are to be dedicated to its associated part family. However, this is not the case since machine 1 and machine 3 operate on the part families of cell 1 and cell 2. These machines are commonly called "bottleneck" machines since they bind the two cells together.

A common practice is to either address the parts of a particular part family that go to these machines or the bottleneck machines themselves. The parts can be removed from the part family, redesigned to operate on other machines within its corresponding cell, or outsourced. Unfortunately, every part in the part family of cell 2 visits the bottleneck machines and, for the sake of discussion, the operations of machine 1 and machine 3 cannot be performed by any other machine. As such, it is more practical to address the bottleneck machines. Therefore, a decision must be made as to whether it is appropriate to purchase another machine 1 and machine 3.

The cellular manufacturing system can be designed once the cells have been formed. The system can be either decoupled or integrated. Typically, a decoupled cellular manufacturing system uses a storage area to store part families after a cell has finished operating on them. Whenever another cell or department is to operate on the parts, they are retrieved from the storage area. Thus, the storage area acts as a decoupler making the cells and departments independent of each other. Unfortunately, this leads to excessive material handling and poor responsiveness.

To eliminate such inefficiencies, many companies are adapting an integrated approach to the design and layout of cellular manufacturing systems. Here, cells and departments are linked through the use of kanbans or cards.

Shown in Figure 4.6, production cards (POK) are used to authorize production of more components or subassemblies and withdrawal cards (WLK) are used to authorize delivery of more components, subassemblies, parts, and raw materials.

To understand what kanbans are, the motivation behind their development needs to be discussed. Traditionally, when a workstation completes its set of operations, it pushes its finished parts onto the next workstation irrespective of its need for those parts. This is referred to as "push" production control. For a situation where the supplying workstation operates at a rate faster than its consuming workstation, parts will begin to build up. Eventually, the consuming workstation will be overwhelmed with work.

To prevent this from happening, it would make sense if the supplying workstation did not produce any parts until its consuming workstation requested parts. This "pull" production control is typically called *kanban*. Kanban means signal and commonly uses cards to signal the supplying workstation that its consuming workstation requests more parts.

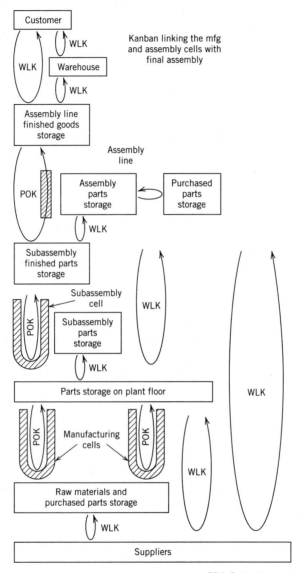

**Figure 4.6** Integrated cellular manufacturing system.

The next phase in the design of a cellular manufacturing system is the layout of each cell. Figure 4.7 illustrates an assembly cell layout at the Hewlett-Packard, Greely Division [4]. The U-shaped arrangement of workstations significantly enhances visibility since the workers are aware of everything that is occurring within the cell. Notice that the agenda easel ensures that all workers know what are the daily production requirements of the cell. Materials flow from workstation to workstation via kanbans. Also, red and yellow lights or andons are used to stop production whenever a workstation has a problem. Problems as they occur are tabulated on the "Problem" display. This helps the workers by indicating what potential problems they should be aware of.

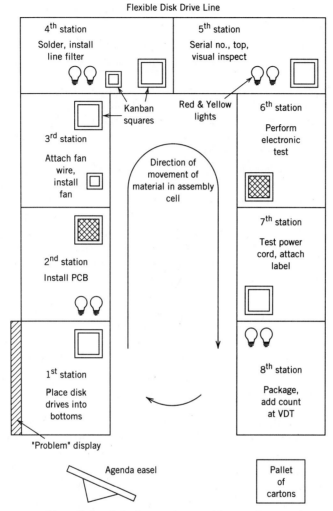

**Figure 4.7** An assembly cell for disk drives, designed by workers at the Hewlett-Packard, Greely Division.

# 4.3  ACTIVITY RELATIONSHIPS

Activity relationships provide the basis for many decisions in the facilities planning process. The primary relationships considered are

1.  Organizational relationships, influenced by span of control and reporting relationships.
2.  Flow relationships, including the flow of materials, people, equipment, information, and money.
3.  Control relationships, including centralized versus decentralized materials control, real time versus batch inventory control, shop floor control, and levels of automation and integration.

4. Environmental relationships, including safety considerations and temperature, noise, fumes, humidity, and dust.

5. Process relationships other than those considered above, such as floor loadings, requirements for water treatment, chemical processing, and special services.

Several relationships can be expressed quantitatively; others must be expressed qualitatively. Flow relationships, for example, are typically expressed in terms of the number of moves per hour, the quantity of goods to be moved per shift, the turnover rate for inventory, the number of documents processed per month, and the monthly expenditures for labor and materials.

Organizational relationships are usually represented formally by an organization chart. However, informal organizational relationships often exist and should be considered in determining the activity relationships for an organization. As an example, quality control may seem to have a limited organizational relationship with the receiving function in a warehouse; however, because of their requirement to interact closely, informal organizational relationships generally develop between the two functions. Organizational restructurings based on modern manufacturing approaches and motivated by increased international competition are decentralizing and relocating functions. This creates a new scenario that the facilities design planner needs to be aware of.

Flow relationships are quite important to the facilities planner, who views flow as the movement of goods, materials, information, and/or people. The movement of refrigerators from the manufacturer through various levels of distribution to the ultimate customer is an important flow process. The transmission of sales orders from the sales department to the production control department is an example of an information flow process. The movement of patients, staff, and visitors through a hospital are examples of flow processes involving people.

All four situations described are discrete flow processes where individual, discrete items are moved through the flow process. A continuous flow process differs from a discrete flow process in that movement is perpetual. Examples of continuous flow processes would include the flow of electricity, chemicals flowing through a processing facility, and oil flowing through a pipeline. Although many of the concepts described in this text are applicable to continuous flow processes, the primary emphasis herein is on discrete flow processes.

A flow process may be described in terms of the **subject** of flow, the **resources** that bring about flow, and the **communications** that coordinate the resources. The subject is the item to be processed. The resources that bring about flow are the processing and transporting facilities required to accomplish the required flow. The communications that coordinate the resources include the procedures that facilitate the management of the flow process. The perspective adopted for a flow process is dependent on the breadth of subjects, resources, and communications that exist in a particular situation.

If the flow process being considered is the **flow of materials into a manufacturing facility,** the flow process is typically referred to as a **materials management system.** The subjects of material management systems are the materials, parts, and supplies purchased by a firm and required for the production of its product. The resources of material management systems include:

1. The production control and purchasing functions.
2. The vendors.
3. The transportation and material handling equipment required to move the materials, parts, and supplies.
4. The receiving, storage, and accounting functions.

The communications within material management systems include production forecasts, inventory records, stock requisitions, purchase orders, bills of lading, move tickets, receiving reports, kanbans, electronic data interchange (EDI), and order payment. A schematic of the material management system is given in Figure 4.8.

If the flow of materials, parts, and supplies ***within* a manufacturing facility** is to be the subject of the flow process, the process is called the **material flow system.** The type of material flow system is determined by the makeup of the activities or planning departments among which materials flow. As noted previously, there are four types of production planning departments.

1. Production line departments.
2. Fixed material location departments.
3. Product family departments.
4. Process departments.

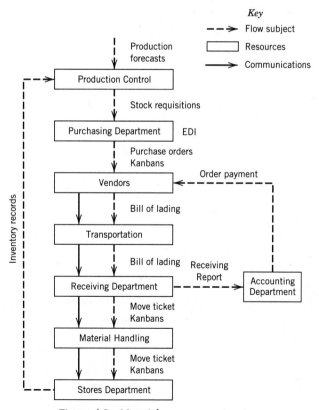

**Figure 4.8** Material management system.

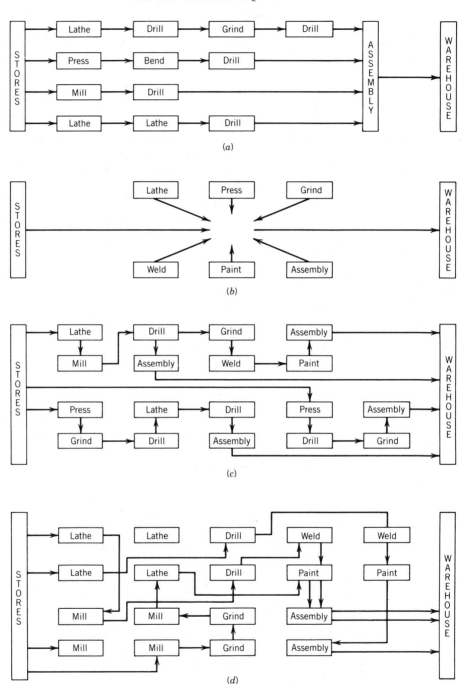

**Figure** 4.9   Material flow systems for various types of departments. (*a*) Product planning departments. (*b*) Fixed materials location departments. (*c*) Product family departments. (*d*) Process planning departments.

Typical material flow systems for each department type are shown in Figure 4.9. The subjects of material flow systems are the materials, parts, and supplies used by a firm in manufacturing its product. The resources of material flow systems include:

1.    The production control and quality control departments.
2.    The manufacturing, assembly, and storage departments.
3.    The material handling equipment required to move materials, parts, and supplies.
4.    The warehouse.

Communication within the material flow system includes: production schedules, work order releases, move tickets, kanbans, bar codes, route sheets, assembly charts, and warehouse records. A schematic of the material flow system is given in Figure 4.10.

     If the **flow of products from a manufacturing facility** is to be the subject of the flow, the flow process is referred to as the **physical distribution system.** The subject of physical distribution systems are the finished goods produced by a firm. The resources of physical distribution systems include:

1.    The customer.
2.    The sales and accounting departments and warehouses.
3.    The material handling and transportation equipment required to move the finished product.
4.    The distributors of the finished product.

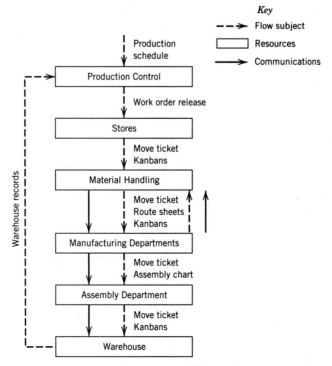

**Figure 4.10**  Material flow system.

The communications within the physical distribution system include: sales orders, packing lists, shipping reports, shipping releases, kanbans, EDI invoices, and bills of lading. A schematic of the physical distribution system is given in Figure 4.11.

The material management, material flow, and physical distribution system may be combined into one overall flow system. Such an overall flow process is referred to as the logistics system. A schematic of the **logistics system** is given in Figure 4.12.

Modern manufacturing approaches are impacting the logistics system in different ways. For example, some suppliers are locating facilities closer to the customer to deliver smaller lot sizes; customers are employing electronic data interchange systems and kanbans to request materials just-in-time; customers and suppliers are using continuous communication technologies with transportation system operators to prevent contingencies; products are being delivered to multiple receiving docks; products are being received in decentralized storage areas (supermarkets) at the points of use; in many cases, no receiving inspection is being performed (suppliers have been certified) and no paperwork is needed; production operators are retrieving the materials from the supermarkets when needed; products are being moved short distances in manufacturing cells and/or product planning arrangements; and simpler material handling and storage equipment alternatives are being employed to receive, store, and move materials (production operators perform retrieval and handling operations from supermarkets and among processes). These changes are creating efficient logistics systems with shorter lead times, lower cost, and better quality.

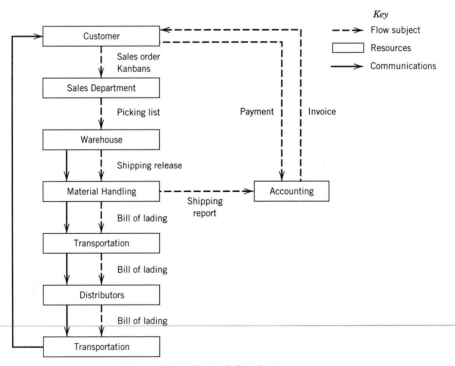

**Figure 4.11**   Physical distribution system.

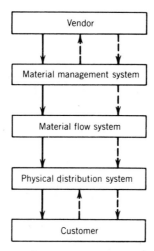

**Figure 4.12**  Logistics system.

# 4.4  FLOW PATTERNS

The macroflow considerations of material management, material flow, physical distribution, and logistics are of value to the facilities planner in that they define the overall flow environment within which movement takes place. Within the overall flow environment, a critical consideration is the pattern of flow. Patterns of flow may be viewed from the perspective of flow within workstations, within departments, and between departments.

## Flow Within Workstations

Motion studies and ergonomics considerations are important in establishing the flow within workstations. For example, flow within a workstation should be simultaneous, symmetrical, natural, rhythmical, and habitual. Simultaneous flow implies the coordinated use of hands, arms, and feet. Hands, arms, and feet should begin and end their motions together and should not be idle at the same instant except during rest periods. Symmetrical flow results from the coordination of movements about the center of the body. The left and right hands and arms should be working in coordination. Natural flow patterns are the basis for rhythmical and habitual flow patterns. Natural movements are continuous, curved, and make use of momentum. Rhythmical and habitual flow implies a methodical, automatic sequence of activity. Rhythmical and habitual flow patterns also allow for reduced mental, eye and muscle fatigue, and strain.

## Flow Within Departments

The flow pattern within departments is dependent on the type of department. In a product and/or product family department, the flow of work follows the product flow. Product flows typically follow one of the patterns shown in Figure 4.13. End-

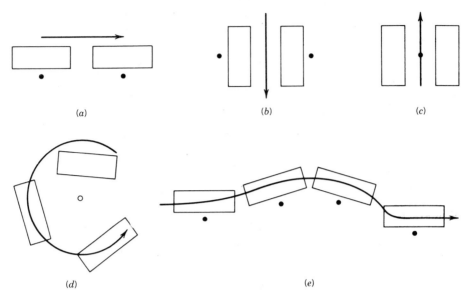

(a)  (b)  (c)

(d)  (e)

**Figure 4.13**  Flow within product departments. (*a*) End-to-end. (*b*) Back-to-back. (*c*) Front-to-front. (*d*) Circular. (*e*) Odd-angle.

to-end, back-to-back, and odd-angle flow patterns are indicative of product departments where one operator works at each workstation. Front-to-front flow patterns are used when one operator works on two workstations and circular flow patterns are used when one operator works on more than two workstations.

In a process department, little flow should occur between workstations within departments. Flow typically occurs between workstations and aisles. Flow patterns are dictated by the orientation of the workstations to the aisles. Figure 4.14 illustrates three workstation-aisle arrangements and the resulting flow patterns. The determination of the preferred workstation-aisle arrangement pattern is dependent on the interactions among workstation areas, available space, and size of the materials to be handled.

Diagonal flow patterns are typically used in conjunction with one-way aisles. Aisles that support diagonal flow patterns often require less space than aisles with either parallel or perpendicular workstation-aisle arrangements. However, one-way aisles also result in less flexibility. Therefore, diagonal flow patterns are not utilized often.

Flow within workstations and within departments should be enriched and en-

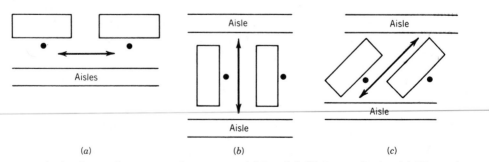

(a)  (b)  (c)

**Figure 4.14**  Flow within process departments. (*a*) Parallel. (*b*) Perpendicular. (*c*) Diagonal.

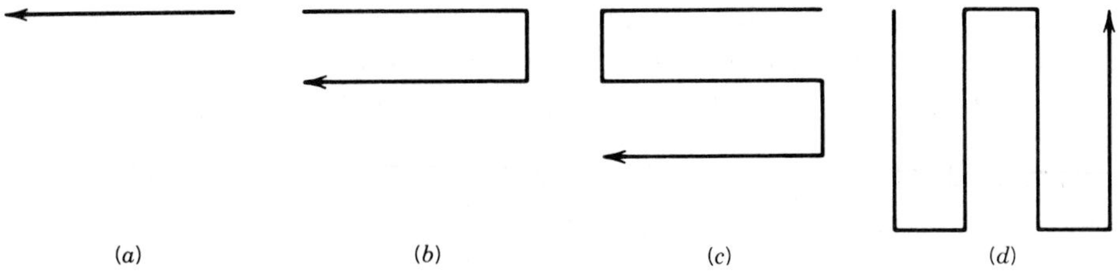

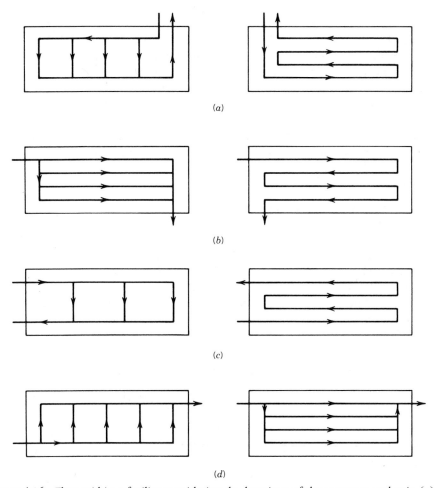

**Figure 4.15**  General flow patterns. (*a*) Straight line. (*b*) U-shaped. (*c*) S-shaped. (*d*) W-shaped.

(*a*)

(*b*)

(*c*)

(*d*)

**Figure 4.16**  Flow within a facility considering the locations of the entrance and exit. (*a*) At the same location. (*b*) On adjacent sides. (*c*) On the same side but at opposite ends. (*d*) On opposite sides.

larged to allow the operators not only to use their muscles but also their minds. Multifunctional operators can work on more than one machine if needed and can get involved in support and continuous improvement functions like quality, basic maintenance, material handling, record keeping, performance measurement tracking, and teamwork. This means that flow and location of materials, tools, paperwork, and quality verification devices should be considered in an integrated way.

## Flow Between Departments

Flow between departments is a criterion often used to evaluate overall flow within a facility. Flow typically consists of a combination of the four general flow patterns shown in Figure 4.15. An important consideration in combining the flow patterns shown in Figure 4.15 is the location of the entrance and exit. As a result of the plot plan or building construction, the location of the entrance (receiving department) and exit (shipping department) is often fixed at a given location and flow within the facility conforms to these restrictions. A few examples of how flow within a facility may be planned to conform to entrance and exit restrictions are given in Figure 4.16

An important design issue in Just-In-Time facilities is the determination of the appropriate number of receiving/shipping docks and decentralized storage areas (supermarkets) and their location. Each combination of number and location of receiving/shipping docks and supermarkets should be analyzed in detail considering integrated layout-handling alternatives to identify flow-time-cost-quality impact.

## 4.5   FLOW PLANNING

Planning effective flow involves combining the flow patterns given in Section 4.4 with adequate aisles to obtain a progressive movement from origination to destination. Effective flow within a facility includes the progressive movement of materials, information, or people *between departments*. Effective flow within a department involves the progressive movement of materials, information, or people *between workstations*. Effective flow within a workstation addresses the progressive movement of materials, information, or people *through the workstation*.

As noted, effective flow planning is a hierarchical planning process. The effec-

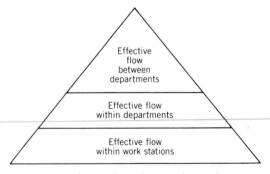

**Figure 4.17**   Flow planning hierarchy.

tive flow within a facility is contingent upon effective flow between departments. Such flow depends on effective flow within departments, which depends on effective flow within workstations. This hierarchy is shown in Figure 4.17. Planning for effective flow within the hierarchy requires the consideration of flow patterns and flow principles.

Morris [20] defines a principle as "simply a loose statement of something which has been noticed to be sometimes, but not always, true." The following principles have been observed to frequently result in effective flow: maximize directed flow paths, minimize flow, and minimize the costs of flow.

A directed flow path is an uninterrupted flow path progressing directly from origination to destination. An uninterrupted flow path is a flow path that *does not intersect with other paths*. Figure 4.18 illustrates the congestion and undesirable intersections that may occur when flow paths are interrupted. A directed flow path progressing from origination to destination is a flow path with no backtracking. As can be seen in Figure 4.19, backtracking increases the length of the flow path. The principle of minimizing flow represents the work simplification approach to material flow. The work simplification approach to material flow includes:

1. Eliminating flow by planning for the delivery of materials, information, or people directly to the point of ultimate use and eliminate intermediate steps.

2. Minimizing multiple flows by planning for the flow between two consecutive points of use to take place in as few movements as possible, preferably one.

3. Combining flows and operations wherever possible by planning for the movement of materials, information, or people to be combined with a processing step.

The principle of minimizing the cost of flow may be viewed from either of the following two perspectives.

1. Minimize manual handling by minimizing walking, manual travel distances, and motions.

2. Eliminate manual handling by mechanizing or automating flow to allow workers to spend full time on their assigned tasks.

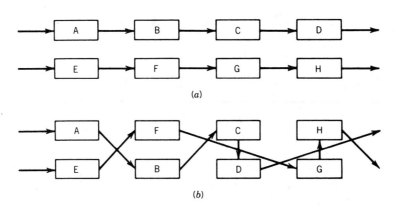

**Figure 4.18** The impact of interruptions on flow paths. (*a*) Uninterrupted flow paths. (*b*) Interrupted flow paths.

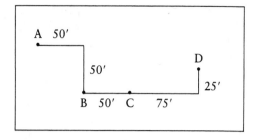

Flow Path A − B − C − D

$$(50' + 50') + 50' + (75' + 25') = 250 \text{ feet}$$

Flow Path A − B − A − C − D

$$(50' + 50') + (50' + 50') + (50' + 50' + 50') + (75' + 25') = 450 \text{ feet}$$

Backtrack Penalty

**Figure 4.19**   Illustration of how backtracking impacts the length of flow paths.

## 4.6   MEASURING FLOW

Flow among departments is one of the most important factors in the arrangement of departments within a facility. To evaluate alternative arrangements, a measure of flow must be established. Flows may be specified in a quantitative manner or a qualitative manner. Quantitative measures may include pieces per hour, moves per day, or pounds per week. Qualitative measures may range from an absolute necessity that two departments be close to each other to a preference that two departments not be close to each other. In facilities having large volumes of materials, information, and people moving between departments, a quantitative measure of flow will typically be the basis for the arrangement of departments. On the contrary, in facilities having very little actual movement of materials, information, and people flowing between departments, but having significant communication and organizational interrelations, a qualitative measure of flow will typically serve as the basis for the arrangement of departments. Most often, a facility will have a need for both quantitative and qualitative measures of flow and both measures should be used.

A chart that can be of use in flow measurement is the mileage chart shown in Figure 4.20. Notice the diagonal of the mileage chart is blank, since the question "How far is it from New York to New York?" makes little sense. Furthermore, the mileage chart is a symmetric matrix. (Distance charts do not have to be symmetric; if one-way aisles are used, distance between two points will seldom be symmetric.) In Figure 4.20, it is 963 miles from Boston to Chicago and also 963 miles from Chicago to Boston. When this occurs, the format of the mileage chart is often changed to a triangular matrix as shown in Figure 4.21.

| From \ To | Atlanta, GA | Boston, MA | Chicago, IL | Dallas, TX | New York, NY | Pittsburgh, PA | Raleigh, NC | San Francisco, CA |
|---|---|---|---|---|---|---|---|---|
| Atlanta, GA | | 1037 | 674 | 795 | 841 | 687 | 372 | 2496 |
| Boston, MA | 1037 | | 963 | 1748 | 206 | 561 | 685 | 3095 |
| Chicago, IL | 674 | 963 | | 917 | 802 | 452 | 784 | 2142 |
| Dallas, TX | 795 | 1748 | 917 | | 1552 | 1204 | 1166 | 1753 |
| New York, NY | 841 | 206 | 802 | 1552 | | 368 | 489 | 2934 |
| Pittsburgh, PA | 687 | 561 | 452 | 1204 | 368 | | 445 | 2578 |
| Raleigh, NC | 372 | 685 | 784 | 1166 | 489 | 445 | | 2843 |
| San Francisco, CA | 2496 | 3095 | 2142 | 1753 | 2934 | 2578 | 2843 | |

**Figure 4.20**  Mileage chart.

## Quantitative Flow Measurement

Flows may be measured quantitatively in terms of the amount moved between departments. The chart most often used to record these flows is a from-to chart. As can be seen in Figure 4.22, a from-to chart resembles the mileage chart given in Figure 4.20. The from-to chart is a square matrix, but is seldom symmetric. The lack of symmetry is because there is no definite reason for the flows from stores to milling to be the same as the flows from milling to stores.

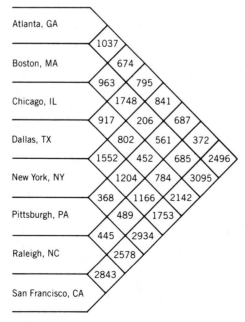

**Figure 4.21**  Triangular mileage chart.

| From \ To | Stores | Milling | Turning | Press | Plate | Assembly | Warehouse |
|---|---|---|---|---|---|---|---|
| Stores | | 12 | 6 | 9 | 1 | 4 | |
| Milling | | | | | 7 | 2 | |
| Turning | | 3 | | | 4 | | |
| Press | | | | | 3 | 1 | 1 |
| Plate | | 3 | 1 | | | 4 | 3 |
| Assembly | 1 | | | | | | 7 |
| Warehouse | | | | | | | |

**Figure 4.22**  From-to chart.

A from-to chart is constructed as follows:

1. List all departments down the row and across the column following the overall flow pattern. For example, Figure 4.23 shows various flow patterns that result in the departments being listed as in Figure 4.22.

2. Establish a measure of flow for the facility that accurately indicates equivalent flow volumes. If the items moved are equivalent with respect to ease of movement, the number of trips may be recorded in the from-to chart. If the items moved vary in size, weight, value, risk of damage, shape, and so on, then items may be established so that the quantities recorded in the from-to chart represent the proper relationships among the volumes of movement.

3. Based on the flow paths for the items to be moved and the established measure of flow, record the flow volumes in the from-to chart.

## Example 4.2

A firm produces three components. Components 1 and 2 have the same size and weight and are equivalent with respect to movement. Component 3 is almost twice as large and moving two units of either component 1 or 2 is equivalent to moving 1 unit of component 3. The departments included in the facility are A, B, C, D, and E. The overall flow path is A-B-C-D-E. The quantities to be produced and the component routings are as follows:

| Component | Production Quantities (per day) | Routing |
|---|---|---|
| 1 | 30 | A-C-B-D-E |
| 2 | 12 | A-B-D-E |
| 3 | 7 | A-C-D-B-E |

The first step in producing the from-to chart is to list the departments in order of the overall flow down the rows and across the columns. Then by considering each unit of component 3 moved to be equivalent to two moves of components 1 and 2, the flow volumes may be recorded as shown in Figure 4.24.

Notice that flow volumes below the diagonal represent backtracking and the closer the

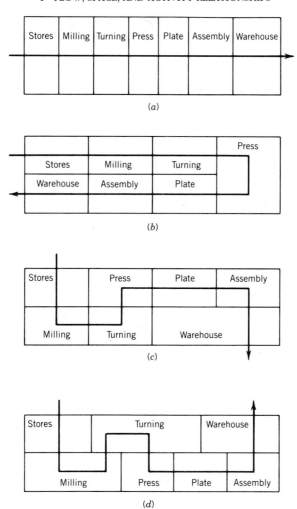

**Figure 4.23**  Flow patterns indicating the order of flow given in (*a*) Straight-line flow. (*b*) U-shaped flow. (*c*) S-shaped flow. (*d*) W-shaped flow.

flow volumes are to the main diagonal, the shorter will be the move in the facility. The moves below the diagonal in the from-to chart given in Figure 4.22 are turning to milling, plate to milling, plate to turning, and assembly to stores. If these moves are traced on the flow paths shown in Figure 4.23, it may be seen that they are all counter to the overall flow pattern. The diagonal 1 unit above the main diagonal in Figure 4.22 contains the moves: stores to milling, press to plate, plate to assembly, and assembly to warehouse. These moves may be seen in Figure 4.23 to be between adjacent departments, or departments one department away. The diagonal 2 units above the main diagonal in Figure 4.22 contains the moves: stores to turning, turning to plate, and press to assembly. These moves may be seen in Figure 4.23 to be of length two departments away along the overall flow path. In a similar manner, the fifth diagonal above the diagonal contains the move stores to assembly, and the fifth diagonal below the diagonal contains the move assembly to stores; these moves are five departments apart along the overall flow path, with the move from assembly to stores being counter to the direction of flow.

| From \ To | A | C | B | D | E |
|---|---|---|---|---|---|
| **A** | | ① 30 <br> ③2(7) = 14 | ②12 | | |
| | | 44 | 12 | 0 | 0 |
| **C** | | | ①30 | ③2(7) = 14 | |
| | 0 | | 30 | 14 | 0 |
| **B** | | | | ①30 <br> ②12 | ③2(7) = 14 |
| | 0 | 0 | | 42 | 14 |
| **D** | | | ③2(7) = 14 | | ①30 <br> ②12 |
| | 0 | 0 | 14 | | 42 |
| **E** | 0 | 0 | 0 | 0 | |

**Figure 4.24** From-to chart for Example 4. The circled numbers represent component numbers and the number following the circled numbers indicates the volume of equivalent flows for the component.

## Qualitative Flow Measurement

Flows may be measured qualitatively using the closeness relationships values developed by Muther [21] and given in Table 4.2. The values may be recorded in conjunction with the reasons for the closeness value using the relationship chart given in Figure 4.25.

A relationship chart may be constructed as follows:

1. List all departments on the relationship chart.
2. Conduct interviews or surveys with persons from each department listed on the relationship chart and with the management responsible for all departments.
3. Define the criteria for assigning closeness relationships and itemize and record the criteria as the reasons for relationship values on the relationship chart.
4. Establish the relationship value and the reason for the value for all pairs of departments.
5. Allow everyone having input to the development of the relationship chart to have an opportunity to evaluate and discuss changes in the chart.

**Table 4.2    *Closeness Relationship Values***

| Value | Closeness |
|---|---|
| A | Absolutely necessary |
| E | Especially important |
| I | Important |
| O | Ordinary closeness okay |
| U | Unimportant |
| X | Undesirable |

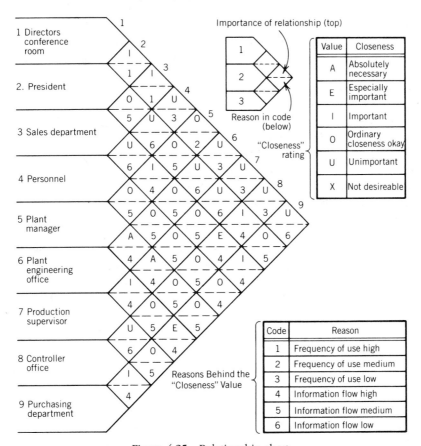

**Figure 4.25**   Relationship chart.

It is important that this procedure be followed in developing a relationship chart. If, instead of the facilities planner synthesizing the relationship among departments as described above, the department heads are allowed to assign the closeness relationships with other departments, inconsistencies may develop. The inconsistencies follow from the form of the chart. The relationship chart by definition requires that the relationship value between departments A and B be the same as the relationship value between departments B and A. If individual relationship values were assigned by department heads and the head of department A said the relationship with B was unimportant (U), and the head of department B said the relationship with A was of ordinary importance (O), an inconsistency would exist. It is best to avoid these inconsistencies by having the facilities planner assign relationship values based on input from the important parties and then have the same parties evaluate the final result.

It is important to emphasize the difference between relationship values U and X. Two departments can be placed adjacent to each other with a relationship value U but they cannot be placed adjacent to each other with a relationship value X (environmental, safety, and facilities constraints). Departments with relationship values U do not gain anything by being close to each other.

From a facilities planning perspective, activity relationships are often translated into **proximity requirements.** For example, if two activities have a strong, positive relationship, then they are typically located close together if not adjacent to one another. Likewise, if two activities have a strong, negative relationship they are typically separated and located far apart from each other.

It should be emphasized that activity relationships can frequently be satisfied in ways other than through physical separation. As an example, information relationships may be satisfied with communication links that include live television hookups, computer ties, pneumatic tube delivery systems, and so on. Likewise, noisy areas can be enclosed, fumes can be vented, and other environmental relationships can be dealt with using special facilities, rather than using distance separation.

Due to the multiplicity of relationships involved, it is advisable to construct separate relationship charts for each *major* relationship being measured. For example, different relationship charts might be constructed for material flow, personnel flow, information flow, organizational, control, environmental, and process relationships. A comparison of the resulting relationship charts will indicate the need for satisfying the relationships by using nondistance-related solutions.

## *4.7*  SPACE REQUIREMENTS

Perhaps the most difficult determination in facilities planning is the amount of space required in the facility. The design year for a facility is typically 5 to 10 years in the future. Considerable uncertainty generally exists concerning the impact of technology, changing product mix, changing demand levels, and organizational designs for the future. Because of the numerous uncertainties that exist, people in the organization tend to "hedge their bets" and provide inflated estimates of space requirements. The facilities planner then has the difficult task of projecting **true space requirements** for the uncertain future.

To further complicate matters, there exists Parkinson's law. Loosely translated, it states that things will expand to fill all available capacity sooner than you plan. Hence, even though the facility might be constructed with sufficient space for the future, when the future arrives there will be no space available for it!

Because of the nature of the problem involving the determination of space requirements, we recommend that it be approached systematically. Specifically, space requirements should be developed "from the ground up."

In determining space requirements for storage warehousing activities, inventory levels, storage units, storage methods and strategies, equipment requirements, building constraints, and personnel requirements must be considered. Space requirements for storage of materials and supplies will be addressed in more depth in subsequent chapters.

In manufacturing and office environments, space requirements should be determined first for individual workstations; next, departmental requirements should be determined, based on the collection of workstations in the department.

As explained before, modern manufacturing approaches are changing drastically space requirements in production, storage areas, and offices. Specifically, space requirements are being reduced because (1) products are delivered to the points of use in smaller lot and unit load sizes; (2) decentralized storage areas are located at the points of use; (3) less inventories are carried (products are "pulled" from preceding processes using kanbans and internal and external inefficiencies have been eliminated); (4) more efficient layout arrangements (i.e., manufacturing cells) are used; and (5) companies are downsizing (focused factories, leaner organizational structures, decentralization of functions, multifunctional employees, high-performance team environments).

The determination of production space requirements is presented in the following sections. A section is also included to address the issue of visual management and space requirements. Personnel and storage space requirements are presented in Chapters 5 and 9, respectively.

## Workstation Specification

Recall that a facility was defined in Chapter 1 as including the fixed assets required to accomplish a specific objective. Because a workstation consists of the fixed assets needed to perform specific operations, a workstation can be considered to be a facility. Although it has a rather narrow objective, the workstation is quite important. Productivity of a firm is definitely related to the productivity of each workstation.

A workstation, like all facilities, includes space for equipment, materials, and personnel. The equipment space for a workstation consists of space for:

1. The equipment.
2. Machine travel.
3. Machine maintenance.
4. Plant services.

Equipment space requirements should be readily available from machinery data sheets. For machines already in operation, machinery data sheets should be available from either the maintenance department's equipment history records or the accounting department's equipment inventory records. For new machines, machinery data sheets should be available from the equipment supplier. If machinery data sheets are not available, a physical inventory should be performed to determine at least the following:

1. Machine manufacturer and type.
2. Machine model and serial number.
3. Location of machine safety stops.
4. Floor loading requirement.
5. Static height at maximum point.
6. Maximum vertical travel.
7. Static width at maximum point.
8. Maximum travel to the left.
9. Maximum travel to the right.
10. Static depth at maximum point.

11.   Maximum travel toward the operator.
12.   Maximum travel away from the operator.
13.   Maintenance requirements and areas.
14.   Plant service requirements and areas.

Floor area requirements for each machine, including machine travel, can be determined by multiplying total width (static width plus maximum travel to the left and right) by total depth (static depth plus maximum travel toward and away from the operator). To the floor area requirement of the machine add the maintenance and plant service area requirements. The resulting sum represents the total machinery area for a machine. The sum of the machinery areas for all machines within a workstation gives the machinery area requirement for the workstation.

The materials areas for a workstation consist of space for:

1.   Receiving and storing materials.
2.   In-process materials.
3.   Storing and shipping materials.
4.   Storing and shipping waste and scrap.
5.   Tools, fixtures, jigs, dies, and maintenance materials.

To determine the area requirements for receiving and storing materials, in-process materials, and storing and shipping materials, the dimensions of the unit loads to be handled and the flow of material through the machine must be known. Sufficient space should be allowed for the number of inbound and outbound unit loads typically stored at the machine. If an inventory holding zone is included within a department for incoming and outgoing materials, one might provide space for only two unit loads ahead of the machine and two unit loads after the machine. Depending on the material handling system, the minimum requirement for space might include that required for one unit load to be worked next, one unit load being worked from, one unit load being worked to, and one unit load that has been completed. Additional space may be needed to allow for in-process materials to be placed into the machine, for material such as bar stock to project beyond the machine, and for the removal of material from the machine. Space for the removal of waste (chips, trimmings, etc.) and scrap (defective parts) from the machine and storage prior to removal from the workstation must be provided.

Organizations that use kanbans will require less space for materials. Typically, only one or two containers or pallet loads of materials are kept close to the workstation and the rest of the materials (regulated by the number of kanbans) will be located in a decentralized storage area (supermarket) located close by.

The only remaining material requirement to be added to those previously mentioned in determining the total material area requirement is the space required for tools, fixtures, jigs, dies, and maintenance materials. A decision with respect to the storage of tools, fixtures, jigs, dies, and maintenance materials at the workstation or in a central storage location will have a direct bearing on the area requirement. At the very least, space must be provided for the accumulation of tools, fixtures, jigs, dies, and maintenance materials required while altering the machine setup.

As the number of setups for a machine increases, so do the work station area requirements for tools, fixtures, jigs, dies, and maintenance materials. Also, from a security, damage, and space viewpoint, the desirability of a central storage location increases. Employee empowerment approaches based on trust, equality, and involvement favor decentralization of anything required to assure production schedule, quality at the source, autonomous maintenance, housekeeping and organization, continuous improvement, communication, and recognition (i.e., kanban boards, mistake proof mechanisms (bakayokes), quality evaluation tools, quick setup tools, basic maintenance materials and spare parts, housekeeping items, problem boards, information-communication-recognition centers, meeting tables, andons).

Consequently, organizations implementing these approaches should allocate space to the workstations according to the concepts to be employed.

The personnel area for a workstation consists of space for:

1. The operator.
2. Material handling.
3. Operator ingress and egress.

The specification of the space requirements for the operator and for material handling can be obtained directly from the method of performing the operation. The method should be determined using a motion study of the task and an ergonomics study of the operator. The following general guidelines are given to illustrate the types of factors to be considered.

1. Workstations should be designed so the operator can pick up and discharge materials without walking or making long or awkward reaches.
2. Workstations should be designed for efficient and effective utilization of the operator.
3. Workstations should be designed to minimize the time spent manually handling materials.
4. Workstations should be designed to maximize operator safety comfort and productivity.
5. Workstations should be designed to minimize hazards, fatigue, and eye strain.

In addition to the space required for the operator and for material handling, space must be allowed for operator ingress and egress. A minimum of a 30-in. aisle is needed for operator travel past stationary objects. If the operator walks between a stationary object and an operating machine, a minimum of a 36-in. aisle is required. If the operator walks between two operating machines, a minimum of a 42-in. aisle is needed.

Figure 4.26 illustrates the space requirements for a workstation. A sketch like Figure 4.26 should be drawn for each workstation in order to visualize the operators' activities. The facilities planner should simulate the operator reporting to the job, performing the task, changing the setup, maintaining the machine, reacting to emergency situations, going to lunch and breaks, cleaning the workstation, evaluating quality, working in teams, responding to feedback boards, and leaving at the end of the shift. Such a simulation will assure the adequacy of the space allocation and may aid in significantly improving the overall operation.

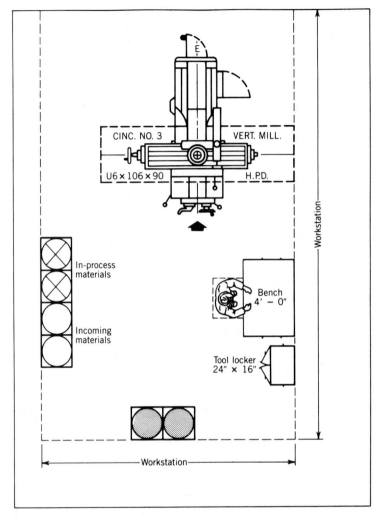

**Figure 4.26** Workstation sketch required to determine total area requirements.

## Department Specification

Once the space requirements for individual workstations have been determined, the space requirements for each department can be established. To do this, we need to establish the departmental service requirements. Departmental area requirements are not simply the sum of the areas of the individual workstations included within the department. It is quite possible tools, dies, equipment maintenance, plant services, housekeeping items, storage areas, operators, spare parts, kanban boards, information-communication-recognition boards, problem boards, and andons may be shared to save space and resources (see Figure 4.7). However, care must be taken to ensure that operational interferences are not created by attempting to combine areas needed by individual workstations. Additional space is required within each department for

Table 4.3  *Aisle Allowance Estimates*

| If the Largest Load Is | Aisle Allowance Percentage Is[a] |
|---|---|
| Less than 6 ft² | 5–10 |
| Between 6 and 12 ft² | 10–20 |
| Between 12 and 18 ft² | 20–30 |
| Greater than 18 ft² | 30–40 |

[a]Expressed as a percentage of the net area required for equipment, material, and personnel.

material handling within the department. Aisle space requirements still cannot be determined exactly, as the department configurations, workstation alignments, and material handling system have not been completely defined. However, at this point we can approximate the space requirement for aisles, since the relative sizes of the loads to be handled are known. Table 4.3 provides a guide for use in estimating aisle space requirements.

Departmental service requirements equal the sum of the service requirements for the individual workstations to be included in a department. These requirements, as well as departmental area requirements, should be recorded on a departmental service and area requirements sheet. Such a sheet is shown in Figure 4.27.

## Example 4.3

A planning department for the ABC Company consists of 13 machines that perform turning operations. Five turret lathes, six automatic screw machines, and two chuckers are included in the planning department. Bar stock, in 8-ft bundles, is delivered to the machines. The "footprints" for the machines are 4 × 12 ft for turret lathes, 4 × 14 ft for screw machines, and 5 × 6 ft for chuckers. Personnel space footprints of 4 × 5 ft are used. Materials storage requirements are estimated to be 20 ft² per turret lathe, 40 ft² per screw machine, and 50 ft² per chucker. An aisle space allowance of 13% is used. The space calculations are summarized in Figure 4.27. A total of 1447 ft² of floor space is required for the planning department. If space is to be provided in the planning department for a supervisor's desk, it must be added to the total for equipment, materials, personnel, and aisles.

Notice that in this example, Company ABC has organized the machines in a process planning department and that provisions for tooling, feedback boards, autonomous maintenance, quick changeovers, quality assurance, and team meetings have not been included in the design. It is also assumed that flexible material handling equipment alternatives are used to move materials, in-process inventory, and finished goods.

## Aisle Arrangement

Aisles should be located in a facility to promote effective flow. Aisles may be classified as departmental aisles and main aisles. Consideration of departmental aisles will be deferred until departmental layouts are established. Recall that space was allotted for departmental aisles on the department service and area requirement sheet.

## DEPARTMENTAL SERVICE AND AREA REQUIREMENT SHEET

Company _____ A.B.C., Inc.     Prepared by _____ J.A.

Department _____ Turning     Date _____    

| Work Station | Quantity | Service Requirements | | | Floor Loading | Ceiling Height | Area (square feet) | | | |
| --- | --- | --- | --- | --- | --- | --- | --- | --- | --- | --- |
| | | Power | Compressed Air | Other | | | Equipment | Material | Personnel | Total |
| Turret lathe | 5 | 440 V AC | 10 CFM @ 100 psi | | 150 PSF | 4' | 240 | 100 | 100 | 440 |
| Screw machine | 6 | 440 V AC | 10 CFM @ 100 psi | | 190 PSF | 4' | 280 | 240 | 120 | 640 |
| Chucker | 2 | 440 V AC | 10 CFM @ 100 psi | | 150 PSF | 5' | 60 | 100 | 40 | 200 |

Net area required    <u>1280</u>

13% aisle allowance    <u>167</u>

Total area required    <u>1447</u>

**Figure 4.27** Department service and area requirements sheet.

Table 4.4  *Recommended Aisle Widths for Various Types of Flow*

| Type of Flow | Aisle Width (feet) |
|---|---|
| Tractors | 12 |
| 3-ton Forklift | 11 |
| 2-ton Forklift | 10 |
| 1-ton Forklift | 9 |
| Narrow aisle truck | 6 |
| Manual platform truck | 5 |
| Personnel | 3 |
| Personnel with doors opening into the aisle from one side | 6 |
| Personnel with doors opening into the aisle from two sides | 8 |

Planning aisles that are too narrow may result in congested facilities having high levels of damage and safety problems. Conversely, planning aisles that are too wide may result in wasted space and poor housekeeping practices. Aisle widths should be determined by considering the type and volume of flow to be handled by the aisle. The type of flow may be specified by considering the people and equipment types using the aisle.

Table 4.4 specifies aisle widths for various types of flow. If the anticipated flow over an aisle indicates that only on rare occasions will flow be taking place at the same time in opposite directions, the aisle widths for main aisles may be obtained from Table 4.4. If, however, the anticipated flow in an aisle indicates that two-way flow will occur frequently, the aisle width should equal the sum of the aisle widths required for the types of flow in each direction.

Curves, jogs, or nonright angle intersections should be avoided in planning for aisles. Aisles should be straight and lead to doors. Aisles along the outside wall of a facility should be avoided unless the aisle is used for entering or leaving the facility. Column spacing should be considered when planning aisle spacing. When column spacing is not considered, the columns will often be located in the aisle. Columns are often used to border aisles, but rarely should be located in an aisle.

## Visual Management and Space Requirements

New manufacturing-management approaches are impacting dramatically the way facilities are designed. At the department level, people empowerment is creating a visual communication revolution often referred to as visual management (see Figure 4.28). A visual management system could include:

A.  Identification, housekeeping, and organization (one place for everything and everything in place). Teams need to relate to a place they can identify as their own. A clean environment where they work, meet, review indicators of the status of the work, post information, display team identity symbols and exam-

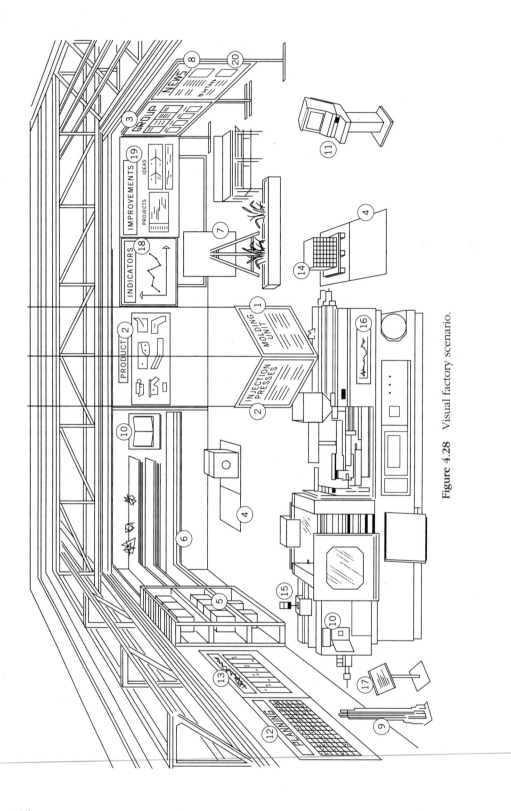

**Figure 4.28** Visual factory scenario.

ples of their products, display standard production methods, and identify properly locations for materials, tools, dies, fixtures, etc. Figure 4.28 illustrates

1. identification of the department;
2. identification of activities, resources, and products;
3. identification of the team;
4. markings on the floor (kanban squares, dedicated location for material handling equipment);
5. markings of tools, racks, fixtures;
6. technical area;
7. communication and rest area;
8. information and instructions; and
9. housekeeping tools.

B.  Visual documentation (tolerances, work instructions, operating instructions for machinery, self-inspection instructions, auditing procedures, plant layout, and floor charts). Figure 4.28 illustrates

10. manufacturing instructions and technical procedures area.

C.  Visual production, maintenance, inventory, and quality control (wall-size schedule charts, kanban boards, autonomous maintenance boards, alarm lamps for machine malfunctions-andons, visual pass/fail templates for quick reading of gauges, charts generated by statistical process control methods, record of every problem encountered-problem boards). Figure 4.28 illustrates

11. computer terminal,
12. production schedule,
13. maintenance schedule,
14. identification of inventories and work-in-process,
15. monitoring signals for machines,
16. statistical process control, and
17. record of problems.

D.  Performance measurement (objectives, goals, indicators) to show the actual game score. Supervisors–facilitators and workers together look at the situation and examine ways to improve it. Figure 4.28 illustrates

18. objectives, results, and differences.

E.  Progress status (visual mechanisms for tracking and celebrating progress and improvement). Figure 4.28 illustrates

19. improvement activities and
20. company project and mission statement.

It is obvious that a visual management system will make a department look better and will help production and support personnel achieve production and maintenance schedules; control inventories, spare parts, and quality; conform to standards; focus on objectives and goals; and provide follow up to continuous improvement process. It is also obvious that to use space efficiently, facilities planners need to use walls and aisles to display as much information as possible and need to allow for dedicated areas for materials, dies, housekeeping and maintenance tools, team meetings, and computer terminals.

## *4.8* SUMMARY

Activity relationships and space requirements are essential elements of a facilities plan. In this chapter we have emphasized the importance of their determination, as well as indicated that determination will not be a simple process. In a sense, the activity relationships and space requirements used in facilities planning provide the foundation for the facility plan. How well the facility achieves its objectives is dependent on the accuracy and completeness of activity relationships and space requirements.

Among the activity relationships considered, flow relationships are of considerable importance to the facilities planner. Included in the consideration of flow relationships are the movement of goods, materials, information, and people. Flow relationships were viewed on a microlevel as the flow within a workstation and on a macrolevel as a logistics system. Flow relationships may be specified by defining the subject, resources, and communications comprising a flow process. Flow relationships may be conceptualized by considering overall flow patterns and may be analyzed by considering general flow principles. Both quantitative and qualitative measures of flow were considered. The evaluation of flow relationships is a primary criterion for good facilities plans and serves as the basis for the development of most facilities layouts.

We have also emphasized in this chapter the impact that cellular manufacturing, visual management, and many other modern manufacturing approaches have on flow, space, and activity relationships determination. The use of these concepts could change dramatically, building size and shape; external-internal flow; and location of production, support, and administrative areas.

## BIBLIOGRAPHY

1. Al-Qattan, I., "Designing Flexible Manufacturing Cells Using a Branch and Bound Method," *International Journal of Production Research,* Vol. 28, No. 2, 1990, pp. 325–336.
2. Ballakur, A., and Steudel, H. J., "A Within-Cell Utilization Based Heuristic for Designing Cellular Manufacturing Systems," *International Journal of Production Research,* Vol. 25, No. 5, 1987, pp. 639–665.
3. Black, J T., "Cellular Manufacturing Systems Reduce Setup Time, Make Small Lot Production Economical," *Industrial Engineering,* November, 1983, pp. 36–48.
4. Black, J T., *The Design of the Factory with a Future,* McGraw-Hill, 1991, New York.
5. Burbidge, J. L., "Production Flow Analysis," The *Production Engineer,* April-May, 1971, pp. 139–152.
6. Burbidge, J. L., *The Introduction of Group Technology,* John Wiley, New York, 1975.
7. Chan, H. M., and Milner, D. A., "Direct Clustering Algorithm for Group Formation in Cellular Manufacturing," *Journal of Manufacturing Systems,* Vol. 1, No. 1, 1982, pp. 65–74.
8. Dumolien, W. J., and Santen, W. P., "Cellular Manufacturing Becomes Philosophy of

Management at Components Facility," *Industrial Engineering,* November 1983, pp. 72–76.

9.  Greif, Michael, *The Visual Factory,* Productivity Press, Cambridge, MA, 1991.

10. Gupta, T., and Seifoddini, H., "Production Data Based Similarity Coefficient for Machine-Component Cellular Manufacturing System, Grouping Decisions the Design of a Cellular Manufacturing System," *International Journal of Production Research,* Vol. 28, No. 7, 1990, pp. 1247–1269.

11. Ham, I., *Introduction to Group Technology,* Pennsylvania State University, University Park, PA, 1975.

12. Han, C., and Ham, I., "Multiobjective Cluster Analysis for Part Family Formations," *Journal of Manufacturing Systems,* Vol. 5, No. 4, 1986, pp. 223–229.

13. Harhalakis, G., Nagi, R., and Proth, J. M., "An Efficient Heuristic in Manufacturing Cell Formation for Group Technology Applications," *International Journal of Production Research,* Vol. 28, No. 1, 1990, pp. 185–198.

14. Hirano, H., *JIT Factory Revolution: a pictorial guide to factory design of the future,* Productivity Press, Cambridge, MA, 1989.

15. Khator, S. K., and Irani, S. A., "Cell Formation in Group Technology: A New Approach," *Computers in Industrial Engineering,* Vol. 12, No. 2, 1987, pp. 131–142.

16. King, J. R., "Machine-Component Grouping in Production Flow Analysis: An approach using a Rank Order Clustering Algorithm," *International Journal of Production Research,* Vol. 18, No. 2, 1980, pp. 213–232.

17. Kusiak, A., "The Generalized Group Technology Concept," *International Journal of Production Research,* Vol. 25, No. 4, 1987, pp. 561–569.

18. Logendran, R., and West, T. M., "A Machine-Part Based Grouping Algorithm in Cellular Manufacturing," *Computers in Industrial Engineering,* Vol. 19, No. 1–4, 1990, pp. 57–61.

19. McAuley, J., "Machine Grouping for Efficient Production," *The Production Engineer,* February 1972, pp. 53–57.

20. Morris, W. T., *Analysis for Material Handling Management,* Richard D. Irwin, Homewood, IL, 1962.

21. Muther, R., *Systematic Layout Planning,* Cahners Books, Boston, 1973.

22. Ranson, G. M., *Group Technology,* Pergamon, New York, 1970.

23. Schonberger, R. J., "Plant Layout Becomes Product-Oriented with Cellular, Just-In-Time Production Concepts," *Industrial Engineering,* November 1983, pp. 66–71.

24. Seifoddini, H., "A Probabilistic Model for Machine Cell Formation," *Journal of Manufacturing Systems,* Vol. 9, No. 1, 1990, pp. 69–75.

25. Srinivasan, G., Narendran, T. T., and Mahadevan, B., "An Assignment Model for the Part Families Problem in Group Technology," *International Journal of Production Research,* Vol. 28, No. 1, 1990, pp. 215–222.

26. Stoner, D. L., Tice, K. J., and Asthon, J. E., "Simple and Effective Cellular Approach to a Job Shop Machine Shop," *Manufacturing Review,* Vol. 2, June 1989, pp. 119–125.

27. Tsuchiya, S., *Quality Maintenance: Zero Defects Through Equipment Management,* Productivity Press, Cambridge, MA, 1992.

28. Wu, H. L., "Design of a Cellular Manufacturing System: A Syntactic Pattern Recognition Approach," *Journal of Manufacturing System,* Vol. 5, No. 2, 1986, pp. 81–87.

29. Wantuck, K. A., *Just-In-Time for America,* The Forum, Ltd., Milwaukee, WI, 1989.

30. Wemmerlov, U., and Hyer, N. L., "Cellular Manufacturing in the U.S. Industry: A Survey of Users," *International Journal of Production Research,* Vol. 27, No. 9, 1989, pp. 1511–1530.

31. *The Idea Book,* edited by Japan Human Relations Association, Productivity Press, Cambridge, MA, 1988.

# PROBLEMS

**4.1** Explain the meaning of a material management system, a material flow system, and a physical distribution system for a hospital.

**4.2** When would you recommend a group technology layout?

**4.3** The paper flow to be expected through one section of city hall consists of the following:

> 10  medical records from records to marriage licenses per day
>
> 7  certificates from printing to marriage licenses per day
>
> 4  blood samples from marriage licenses to lab per day
>
> 4  blood sample reports from lab to marriage licenses per day
>
> 1  box of medical records from marriage licenses to records per week

The following load-equivalence conversions can be used:

> One medical record is equivalent to one certificate.
>
> One medical record is equivalent to one-half of a blood sample.
>
> One medical record is equivalent to one blood sample report.
>
> One medical record is equivalent to one-tenth of a box of medical records.

Develop a from-to chart for this section of city hall and then develop a relationship chart.

**4.4** Which factors should be considered when determining the space requirements for a workstation, assuming that visual management is used (list at least ten).

**4.5** When would you recommend a fixed position layout?

**4.6** Which layout type is very popular in Just-In-Time facilities? Explain why.

**4.7** Mention three limitations of the process layout type.

**4.8** Visit a local fast-food restaurant. Describe in detail the arrangement of aisles for the customers (the service line, the booth area, etc.) and the arrangement of aisles for the workers (the area behind the counter) by sketching a layout. Draw the flow of the customers and employees on the layout. Explain how the flow of the workers and customers would change if the layout was different. Is there adequate space for the customers and the employees to operate efficiently? Can any improvement in the layout create a more efficient flow? Describe any visual management approaches being used.

**4.9** Choose ten main components of a kitchen, (i.e., oven, sink). Develop a relationship chart of these ten components. Sketch a kitchen containing these ten components and arrange them based upon the relationship chart findings.

**4.10** What is the impact of "backtracking" in a manufacturing process? Discuss several methods to prevent "backtracking" from occurring.

**4.11** What are the pros and cons of having multiple input/output points (receiving and shipping areas) in a given manufacturing facility (list at least three of each)? What types of considerations should a facilities designer take into account when determining to use multiple input/output points?

**4.12** Which information does the facilities designer need from top management, product designer, process designer, and schedule designer? Describe at least ten modern manufacturing approaches that impact drastically the facilities design process.

**4.13** When do you justify the use of integrated production-support-administrative-service planning departments?

**4.14** Describe activity relationships and flow and space requirements in a facility with self-managing teams.

4.15  Provide at least ten benefits of using manufacturing cells. Which modern manufacturing approaches are usually employed in conjunction with manufacturing cells (include at least 10)?

4.16  Identify from papers and books at least two cell formation approaches. Provide a summary of these approaches.

4.17  Develop a part-machine matrix like the one in Figure 4.2 for a situation with ten parts and 15 machines. Use the DCA methodology to identify clusters (cells) of parts and machines. Do you have any bottleneck machine? Recommend at least three different ways to resolve this situation.

4.18  Define a kanban. Which are the different types of kanbans? List at least five benefits of using kanbans. Why are kanbans important in cellular manufacturing?

4.19  Mention at least seven benefits of using U-shape arrangements in manufacturing cells.

4.20  Many organizations are using reengineering approaches to redesign production and administrative processes. Identify at least two papers dealing with this subject and prepare a summary of important issues described in the papers. Using this information, describe the impact of reengineering on activity relationships and flow and space requirements.

4.21  Many organizations are using restructuring or downsizing approaches to redesign their organizational structures. Identify at least two papers dealing with this subject and prepare a summary of important issues explained in the papers. Using this information, describe the impact of restructuring or downsizing on activity relationships and flow and space requirements.

4.22  Many organizations are using reengineering and restructuring or downsizing approaches to redesign their processes and organizational structures. Identify at least three papers dealing with applications of these approaches. Using the information provided in these cases, prepare a summary of the impact of these approaches on facilities design.

4.23  Describe the impact of modern manufacturing approaches on the logistics system.

4.24  Which new functions are being managed by multifunctional operators? Describe the impact of this new roles on activity relationships and flow and space requirements.

4.25  Which visual management approaches impact most activity relationships and floor and space requirements (provide at least five)? Explain why.

4.26  Is it important for a facilities planner to consider the logistics system? Explain.

4.27  Go to a local quick food hamburger restaurant and assess the impact of the order taker's workstation on the overall facility flow. Describe the process and make recommendations for improvement.

4.28  What are the trade-offs involved in parallel, perpendicular, and diagonal flow within parking lots?

4.29  How are the flow principles taken into consideration by you from the time you wake up in the morning to go to school until you go to bed that evening? (Assume you did not cut your classes and you did not stay up all night studying.)

4.30  Given the spatial schematic below, evaluate the flow path lengths for the following components:

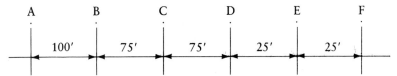

a.  Component 1 routing A-B-C-D-E-F.
b.  Component 2 routing A-C-B-D-E-F.
c.  Component 3 routing A-F-E-D-C-B-A-F.

**4.31**  Evaluate the aisle widths in the building in which this class is taught.

**4.32**  What is the impact on a city having streets that do not meet at right angles? Relate this to aisles.

**4.33**  A waiter at the famous French restaurant Joe's is interested in obtaining your help in improving the layout of the serving area. After establishing the space needs for the salad serving, beverage serving, dessert serving, soup serving, entree serving, and bill writing areas, you decide your next step is to develop a from-to chart. To do this you need to establish equivalency of loads. Establish these equivalencies and explain your reasoning.

**4.34**  Given the following from-to chart, can you recommend a change in the overall flow pattern that will reduce flow?

| From \ To | A | B | C | D |
|---|---|---|---|---|
| A |  |  | 4 |  |
| B |  |  |  | 4 |
| C |  | 4 |  |  |
| D |  |  |  |  |

**4.35**  Develop a relationship chart for the relationships between you, your professor, college dean, university president, and your steady date. Develop another relationship chart from your professor's viewpoint for the same people.

**4.36**  Design a survey that is to be used to determine the relationships among departments in a local hospital.

# 5

# *PERSONNEL*
# *REQUIREMENTS*

## *5.1*  INTRODUCTION

The planning of personnel requirements includes planning for employee parking, locker rooms, restrooms, food services, drinking fountains, and health services. This challenge has been compounded by the advent of the 1989 American with Disabilities Act (ADA). The facilities planner must integrate barrier-free designs in addressing the personnel requirements of the facility.

Personnel requirements can be among the most difficult to plan because of the number of philosophies relating to personnel. For example:

1. "Our firm is responsible for our employees from the moment they leave their home until they return. We must provide adequate methods of getting to and from work."
2. "Employees should earn their parking locations; all spaces should be assigned to specific individuals."
3. "Employees spend one third of their life within our facility; we must help them enjoy working here."
4. "A happy worker is a productive worker."
5. "A hot lunch makes a worker more productive since it supplies them energy."
6. "Workers who do not feel well are unsafe workers; we should provide medical care to maintain health."
7. "Our company has an obligation to our personnel; we will make all our facilities ADA compliant."
8. "Executives may drink coffee and smoke at their desks, but such activities are not allowed for clerks and secretaries."

9. "Our employees work hard; the least we can do is provide a place for them to unwind after a hard day's work. Employees who play together will work better together."

10. "Personnel considerations are of little importance in our facility. We pay people to work, not to have a good time."

All of these philosophies are debatable and none of them are universally accepted. Nevertheless, if the management of a facility firmly adopts one of these philosophies, little can be done by the facilities planner other than plan the facility to conform with these philosophies. This chapter presents how to plan personnel requirements into a facility, given the desires of management.

# 5.2   THE EMPLOYEE-FACILITY INTERFACE

An interface between an employee's work and nonwork activities must be provided. The interface functions as a storage area for personal property of the employee during work hours. Personal property typically includes the automobile and the employee's personal belongings, such as coats, clothes, purses, and lunches.

## Employee Parking

Planning employee parking areas is very similar to planning stores or warehouse areas. The procedure to be followed is

1. Determine the number of automobiles to be parked.
2. Determine the space required for each automobile.
3. Determine the available space for parking.
4. Determine alternative parking layouts for alternative parking patterns.
5. Select the layout that best utilizes space and maximizes employee convenience.

Care must be used when determining the number of automobiles to be parked. General rules of thumb may not be applicable. For remote sites not being serviced by public transportation, a parking space may be required for every 1.25 employees. At the other extreme, a centralized location served by public transportation may require a parking space for every three employees. The number of parking spaces to be provided must be specifically determined for each facility and must be in accordance with local zoning regulations. Attention should be paid to the requirement for handicapped parking. Although minimum requirements can be as low as two handicapped spaces per 100 parking spaces, five handicapped spaces per 100 parking spaces is not uncommon. The key is to check with your local or state building agency for the required standard. Surveys of similar facilities in the area of the new facility will provide valuable data with respect to the required number of parking spaces. If considerable variability exists between the employee-parking space requirement ratio for similar facilities the reasons for this variability should be deter-

Recommended range of stall widths (SW)

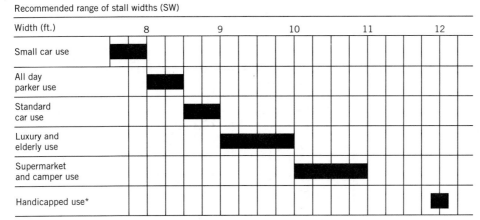

* Minimum requirements = 1 or 2 per 100 stalls or as specified by local, state, or federal law; convenient to destination

**Figure 5.1**   Recommended packing dimensions. *Source:* Ramsey et al. [8]

mined. The new facility should be planned in accordance with existing facilities with which it has the greatest similarity.

The size of a parking space for an automobile can vary from 7 × 15 ft to 9.5 × 19 ft, depending on the type of automobile and the amount of clearance to be provided. The total area required for a parked automobile depends on the size of the parking space, the parking angle, and the aisle width. Figure 5.1 shows the recommended range of stall widths, parking angle options, and typical parking dimensions in feet.

The factors to be considered in determining the specification for a specific parking lot are

1.  The percentage of automobiles to be parked that are compact automobiles. As a planning guideline, if more specific data are not available, 33% of all parking is often allocated to compact automobiles [5].

2.  Increasing the area provided for parking decreases the amount of time required to park and de-park.

3.  Angular configurations allow quicker turnover; perpendicular parking yields greater space utilization.

4.  As the angle of a parking space increases, so does the required space allocated to aisles.

Automobiles parked in employee parking lots will typically be parked for an entire shift. Therefore, parking and de-parking time is not as important as space utilization. A parking space width of 7.5 ft for compact automobiles and 8.5 ft for standard-sized automobiles is recommended. In Table 5.1, there are 3 car groups (G1—small cars, G2—standard cars, and G3—large cars). For a given group, there are corresponding stall width (SW) options. Group I has 1 SW option (8'-0"), Group II has 3 (8'-6", 9'-0", and 9'-6") and similarly, Group III has 3 (9'-0", 9'-6", and 10'-0"). For each stall width option, there are 4 configurations, W1, W2, W3, and W4. These configurations are as shown in Figure 5.2. Each configuration for a given SW has 10 corresponding angles of park, "θ," and an associated "W" dimension. The

Table 5.1   *Parking Dimensions for Each Car Group as a Function of Single and Double Loaded Module Options*

| | SW | W | θ ANGLE OF PARK | | | | | | | | | |
|---|---|---|---|---|---|---|---|---|---|---|---|---|
| | | | 45° | 50° | 55° | 60° | 65° | 70° | 75° | 80° | 85° | 90° |
| Group I: small cars | 8'0" | 1 | 25'9' | 26'6" | 27'2" | 29'4" | 31'9" | 34'0" | 36'2" | 38'2" | 40'0" | 41'9" |
| | | 2 | 40'10" | 42'0" | 43'1" | 45'8" | 48'2" | 50'6" | 52'7" | 54'4" | 55'11" | 57'2" |
| | | 3 | 38'9" | 40'2" | 41'5" | 44'2" | 47'0" | 49'6" | 51'10" | 53'10" | 55'8" | 57'2" |
| | | 4 | 36'8" | 38'3" | 39'9" | 42'9" | 45'9" | 48'6" | 51'1" | 53'4" | 55'5" | 57'2 |
| Group II: standard cars | 8'6" | 1 | 32'0" | 32'11" | 34'2" | 36'2" | 38'5" | 41'0" | 43'6" | 45'6" | 46'11" | 48'0" |
| | | 2 | 49'10" | 51'9" | 53'10" | 56'0" | 58'4" | 60'2" | 62'0" | 63'6" | 64'9" | 66'0" |
| | | 3 | 47'8" | 49'4" | 51'6" | 54'0" | 56'6" | 59'0" | 61'2" | 63'0" | 64'6" | 66'0" |
| | | 4 | 45'3" | 46'10' | 49'0" | 51'8" | 54'6" | 57'10" | 60'0" | 62'6" | 64'3" | 66'0" |
| | 9'0" | 1 | 32'0" | 32'9" | 34'0" | 35'4" | 37'6" | 39'8" | 42'0" | 44'4" | 46'2" | 48'0" |
| | | 2 | 49'4" | 51'0" | 53'2" | 55'6" | 57'10" | 60'0" | 61'10" | 63'4" | 64'9" | 66'0" |
| | | 3 | 46'4" | 48'10" | 51'4" | 53'10" | 56'0" | 58'8" | 61'0" | 63'0" | 64'6" | 66'0" |
| | | 4 | 44'8" | 46'6" | 49'0" | 51'6" | 54'0" | 57'0" | 59'8" | 62'0" | 64'2" | 66'0" |
| | 9'6" | 1 | 32'0" | 32'8" | 34'0" | 35'0" | 36'10" | 38'10" | 41'6" | 43'8" | 46'0" | 48'0" |
| | | 2 | 49'2" | 50'6" | 51'10" | 53'6" | 55'4" | 58'0" | 60'6" | 62'8" | 64'6" | 65'11" |
| | | 3 | 47'0" | 48'2" | 49'10" | 51'6" | 53'11" | 57'0" | 59'8" | 62'0" | 64'3" | 65'11" |
| | | 4 | 44'8" | 45'10" | 47'6" | 49'10" | 52'6" | 55'9" | 58'9" | 61'6" | 63'10" | 65'11" |

Group III: large cars

| | | 32'7" | 33'0" | 34'0" | 35'11" | 38'3" | 40'11" | 43'6" | 45'5" | 46'9" | 48'0" |
|---|---|---|---|---|---|---|---|---|---|---|---|
| 9'0" | 1 | 32'7" | 33'0" | 34'0" | 35'11" | 38'3" | 40'11" | 43'6" | 45'5" | 46'9" | 48'0" |
| | 2 | 50'2" | 51'2" | 53'3" | 55'4" | 58'0" | 60'4" | 62'9" | 64'3" | 65'5" | 66'0" |
| | 3 | 47'9" | 49'1" | 52'3" | 53'8" | 56'2" | 59'2" | 61'11" | 63'9" | 65'2" | 66'0" |
| | 4 | 45'5" | 46'11" | 49'0" | 51'8" | 54'9" | 58'0" | 61'0" | 63'2" | 64'10" | 66'0" |
| 9'6" | 1 | 32'4" | 32'8" | 33'10" | 34'11" | 37'2" | 39'11" | 42'5" | 45'0" | 46'6" | 48'0" |
| | 2 | 49'11" | 50'11" | 52'2" | 54'0" | 56'6" | 59'3" | 61'9" | 63'4" | 64'8" | 66'0" |
| | 3 | 47'7" | 48'9" | 50'2" | 52'4" | 55'1" | 58'4" | 60'11" | 62'10" | 64'6" | 66'0" |
| | 4 | 45'3" | 46'8" | 48'5" | 50'8" | 53'8" | 57'0" | 59'10" | 52'2" | 64'1" | 66'0" |
| 10'0" | 1 | 32'4" | 32'8" | 33'10" | 34'11" | 37'2" | 39'11" | 42'5" | 45'0" | 46'6" | 48'0" |
| | 2 | 49'11" | 50'11" | 52'2" | 54'0" | 56'6" | 59'3" | 61'9" | 63'4" | 64'8" | 66'0" |
| | 3 | 57'7" | 48'9" | 50'2" | 52'4" | 55'1" | 58'4" | 60'11" | 62'11" | 64'6" | 66'0" |
| | 4 | 45'3" | 46'8" | 48'5" | 50'8" | 53'8" | 57'0" | 59'10" | 62'2" | 64'1" | 66'0" |

*Source:* Ramsey et al. [8]

115

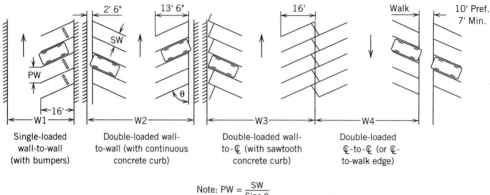

Note: PW = $\dfrac{SW}{Sine\ \theta}$

θ is the parking angle, PW is parking width and SW is the stall width. At angle of 90° (sine 90° = 1), PW = SW. As the parking angle decreases, PW increases accordingly.

**Figure 5.2**   Single- and double-loaded module options. *Source:* Ramsey et al. [8].

angle of park is as indicated in Figure 5.2. It is the angle defined by the parking lane and curb, as shown for example in Figure 5.2, W2. Once the car group and associated stall width is determined (let us say GII SW 9′-0″) a parking configuration (W1, W2, W3 or W4) can then be chosen as the initial iteration, depending on the lot size and lot design constraints. Figure 5.2 outlines typical parking configuration options available to the facilities planner.

Using the information from Figures 5.1, 5.2 and Table 5.1, the facilities planner can generate several parking layout alternatives that will optimize the space allocated for parking and maximize employee convenience.

An important issue related to parking lot planning is the location of facility entrances and exits or ingress and egress conditions. Employees should not be required to walk more than 500 ft from their parking place to the entrance of the facility. The entrances should be convenient not only to their parking location, but also to their place of work. If multiple parking lots and entrances are needed in order to accommodate employees, then employees should be assigned to specific lots and entrances. All plant entrances and exits must be carefully planned to meet appropriate insurance and safety codes. Sections 1910.36 and 1910.37 of the Occupational Safety and Health Act (OSHA) standards describe the requirements for entrances and exits.

## *Example 5.1*

A new facility is to have 200 employees. A survey of similar facilities indicates that one parking space must be provided for every two employees and that 40% of all automobiles driven to work are compact automobiles. Five percent of the spaces should be allocated for the handicapped. The available parking lot space is 180 ft and 200 ft deep. What is the best parking layout?

If the new facility were to have the same number of parking spaces as similar facilities, 100 spaces would be required. Of these 100 spaces, 40 could be for compact automobiles. However, not all drivers of compact cars will park in a compact space. Therefore, only 30 compact spaces will be provided. Begin the layout of the lot using 90° double-loaded, two-way traffic because of its efficient use of space to determine if the available lot is adequate. From Figure 5.2, W4 is the required module option. Using the W4 module and Table 5.1, we can obtain the following:

| | |
|---|---|
| Compact cars (8'-0") | Module width |
| 90°, W4 | 57'-2" |
| Standard cars (8'-6") | 66'-0" |
| 90°, W4 | |

Check to see if the depth of the lot (200 ft) can accommodate a parking layout consisting of 2 modules of standard cars and 1 compact module,

$$2 (66 \text{ ft}) + 1 (57' - 2") = 189' - 2"$$

$$189' - 2" < 200 \text{ ft, therefore depth requirement OK.}$$

Each compact module row will yield a car capacity based on the width of the lot (180 ft) divided by the width requirement per stall (8') times the rows per module (2).

$$\left(\frac{180}{8}\right) \times 2 = 44 \text{ potential number of compact cars}$$

Similarly, each standard module row will yield a car capacity based on the width of the lot (180 ft) divided by the width requirement per stall (8.5') times the number of rows per module (2) times the number of modules (2).

$$\left(\frac{180}{8.5}\right) \times 2 \times 2 = 84 \text{ potential number of standard cars}$$

Total possible = 44 + 84 = 128, which is greater than the required number. Therefore, module configuration (W4) is feasible. A possible alternative of (2 rows/modules × 2 standard rows) + (2 rows/module × 1 compact row) for a total of six rows is a starting point for the layout.

Modifying the layout to account for handicap requirements and circulation reveals the following:

Row 1 will handle all five handicap spaces = 5(12') = 60
The remaining space will be occupied by standard cars

(180 − 60)/8.5 = 14 spaces

Row adjusting for two circulation lanes of 15' each number 2 will handle

(180 − (15 × 2))/8.5 = 17 spaces

Rows 3 and 4 will yield the same number of spaces
Row 5 will have        (180 − 60)/8 = 18 spaces
Row 6 will handle      180/8 = 22 spaces

See Figure 5.3 for the recommended layout. The assignment of compact, standard and handicap spaces is as follows:

| Row | Compact | Standard | Handicap |
|---|---|---|---|
| #1 | | 14 | 5 |
| #2 | | 17 | |
| #3 | | 17 | |
| #4 | | 17 | |
| #5 | 18 | | |
| #6 | 22 | | |
| Total | 40 | 65 | 5 |

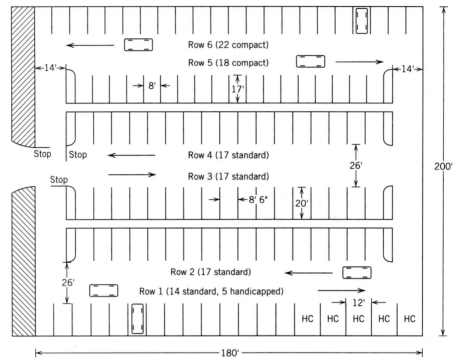

**Figure 5.3**  Parking lot for example 5.1.

## Storage of Employees Personal Belongings

A location for storage of employee personal belongings should be provided between the employee entrance and work area. Employees typically store lunches, briefcases, and purses at their place of work. Employees who are not required to change their clothes and who work in an environment where toxic substances do not exist need only to be provided with a coat rack. Employees who either change their clothes or work where toxic substances are present should be provided with lockers. The lockers may be located in a corridor adjacent to the employee entrance if clothes changing does not take place. More commonly, locker rooms are provided for each sex even if clothes changing is not required. Each employee should be assigned a locker. For planning purposes, 6 ft² should be allocated for each person using the locker room. If shower facilities are to be provided, they should be located in the locker room. Sinks and mirrors are also typically included there. If toilet facilities are to be included, they must be physically separated from the locker room area should lunches be stored in the lockers. Locker rooms are often located along an outside wall adjacent to the employee entrance. This provides excellent ventilation and employee convenience while not interfering with the flow of work within the facility. Figure 5.4 provides an example of a plant entrance and changing room layout.

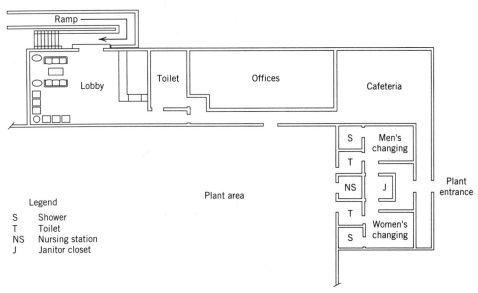

**Figure 5.4**  Plant entrance and changing room layout.

## 5.3  RESTROOMS

A restroom should be located within 200 ft of every permanent workstation. Decentralized restrooms often provide greater employee convenience than large, centralized restrooms. Mezzanine restrooms are common in production facilities as they may be located where employees are centered without occupying valuable floor space. However, access to restrooms must be available to handicapped employees. Hence, some restrooms must be at ground level. In any event, the location should comply with local zoning regulations.

Unless restrooms are designed for single occupancy, separate restrooms should be provided for each sex. The recommended minimum number of toilets for the number of employees working within a facility is given in Table 5.2. In restrooms for males a urinal may be substituted for a toilet, provided that the number of toilets is not reduced to less than two-thirds the minimum recommended in Table 5.2. For space planning purposes, 15 ft² should be allowed for each toilet and 6 ft² for each urinal. Toilets and urinals must be designed to accommodate wheelchairs for handicapped employees as well.

Lavatories and worksinks should be provided in accordance with the lavatory column in Table 5.2. In no restroom should less than one sink per three toilets be provided. When multiple users may use a sink at a time, 24 linear inches of sink or 20 inches or circular basin may be equated to one sink. For planning purposes, 6 ft² should be allowed for each sink.

Entrance doorways into restrooms should be designed such that the interior of the restroom is not visible from the outside when the door is open. A space allowance

**Table 5.2**   *Plumbing Fixture Requirements for Number of Employees*

### Business, Mercantile, Industrial Other than Foundry and Storage

| Water Closets | Employees | Lavatories | Employees |
|---|---|---|---|
| 1 | 1–15 | 1 | 1–20 |
| 2 | 16–35 | 2 | 21–40 |
| 3 | 36–55 | 3 | 41–60 |
| 4 | 56–80 | 4 | 61–80 |
| 5 | 81–110 | 5 | 81–100 |
| 6 | 111–150 | 6 | 101–125 |
| 7 | 151–190 | 7 | 126–150 |
|  |  | 8 | 151–175 |
| One additional water closet for each 40 in excess of 190 | | One additional lavatory for each 30 in excess of 175 | |

### Industrial, Foundries, and Storage

| Water Closets | Employees | Lavatories | Employees |
|---|---|---|---|
| 1 | 1–10 | 1 | 1–8 |
| 2 | 11–25 | 2 | 9–16 |
| 3 | 26–50 | 3 | 17–30 |
| 4 | 51–80 | 4 | 31–45 |
| 5 | 81–125 | 5 | 46–65 |
| One additional water closet for each 45 in excess of 125 | | One additional lavatory for each 25 in excess of 65 | |

### Assembly, Other Than Religious, and School

| Water Closets | Occupants | Urinal | Male Occupants | Lavatories | Occupants |
|---|---|---|---|---|---|
| 1 | 1–100 | 1 | 1–100 | 1 | 1–100 |
| 2 | 101–200 | 2 | 101–200 | 2 | 101–200 |
| 3 | 201–400 | 3 | 201–400 | 3 | 201–400 |
| 4 | 401–700 | 4 | 401–700 | 4 | 401–700 |
| 5 | 701–1100 | 5 | 701–1100 | 5 | 701–1100 |
| One additional water closet for each 600 in excess of 1100 | | One additional urinal for each 300 in excess of 1100 | | One additional lavatory for each 1500 in excess of 1100. Such lavatories need not be supplied with hot water. | |

**Assembly, Religious**
**One Water Closet and One Lavatory**
**Assembly Schools**
For pupils' use:

1. Water closets for pupils; in elementary schools, one for each 100 males and one for each 75 females; in secondary schools, one for each 100 males and one for each 45 females.
2. One lavatory for each 50 pupils.
3. One urinal for each 30 male pupils.
4. One drinking fountain for each 150 pupils, but at least one on each floor having classrooms.

Where more than five persons are employed, providing fixtures as required for group C1 occupancy.

Table 5.2   (continued)

## Institutional

(Persons whose movements are not limited.) Within each dwelling unit:

1.    One kitchen sink.
2.    One water closet.
3.    One bathtub or shower.
4.    One lavatory.

Where sleeping accommodations are arranged as individual rooms, provide the following for each six sleeping rooms:

1.    One water closet.
2.    One bathtub and shower.
3.    One lavatory.

Where sleeping accommodations are arranged as a dormitory, provide the following for each 15 persons (separate rooms for each sex except in dwelling units):

1.    One water closet.
2.    One bathtub or shower.
3.    One lavatory.

## Institutional, Other Than Hospitals

(Persons whose movements are limited.) On each story:

1.    Water closets: one for each 25 males and one for each 20 females.
2.    One urinal for each 50 male occupants.
3.    One lavatory for each ten occupants.
4.    One shower for each ten occupants.
5.    One drinking fountain for each 50 occupants.

Fixtures for employees the same as required for group C1 occupancy, in separate rooms for each sex.

## Institutional, Hospitals

For patients' use:

1.    One water closet and one lavatory for each ten patients.
2.    One shower or bathtub for each 20 patients.
3.    One drinking fountain or equivalent fixture for each 100 patients.

Fixtures for employees the same as required for group C1 occupancy, and separate from those for patients.

## Institutional, Mental Hospitals

For patients' use:

1.    One water closet, one lavatory, and one shower or bathtub, for eight patients.
2.    One drinking fountain or equivalent fixture for each 50 patients.

Fixtures for employees the same as required for group C1 occupancy, and separate from those for patients.

## Institutional, Penal Institutions

For inmate use:

1.    One water closet and one lavatory in each cell.
2.    One shower on each floor on which cells are located.
3.    One water closet and one lavatory for inmate use available in each exercise area.

Table 5.2   (continued)

Lavatories for inmate use need not be supplied with hot water. Fixtures for employees the same as required for group C1 occupancy, and placed in separate rooms from those used by inmates.

**Business (C-1 Occupancy)**

Building used primarily for the transaction of business with the handling of merchandise being incident to the primary use.

**Merchantile (C-2 Occupancy)**

Building used primarily for the display of merchandise and its sale to the public.

**Additional Requirements**

1.   One drinking fountain or equivalent fixture for each 75 employees.
2.   Urinals may be substituted for not more than one third of the required number of water closets when more than 35 males are employed.

**Industrial (C-3 Occupancy)**

Building used primarily for the manufacture or processing of products.

**Storage (C-4 Occupancy)**

Buildings used primarily for the storage of or shelter for merchandise, vehicles, or animals.

**Additional Requirements**

1.   One drinking fountain or equivalent fixture for each 75 employees.
2.   Urinals provided where more than 10 males are employed: one for 11–29; two for 30–79; one additional urinal for every 80 in excess of 79.

**Assembly (C-5 Occupancy)**

Building used primarily for the assembly for athlete educational, religious, social, or similar purposes.

**Additional Requirement**

One drinking fountain for each 1000 occupants but at least one on each floor.

**General Notes**

1.   Plumbing fixture requirements shown are based on New York State Uniform Fire Prevention and Building Code and can serve only as a guide. Consult codes in force in area of construction and state and federal agencies (Labor Department, General Services Administration, etc.) and comply with their requirements.
2.   Plumbing fixture requirements are to be based on the maximum legal occupancy and not on the actual or anticipated occupancy.
3.   Proportioning of toilet facilities between men and women is based on a 50–50 distribution. However, in certain cases conditions of occupancy may warrant additional facilities for men or women above the basic 50–50 distribution.

*Source:* Ramsey et al. [8]

of 15 ft$^2$ should be used for the entrance. A sample layout of a commercial restroom is shown in Figure 5.5.

Beds or cots should be provided in restrooms for women. The area should be segregated from the restroom by a partition or curtain. If between 100 and 250 women are employed, two beds should be provided. One additional bed should be provided for each additional 250 female employees. A space allowance of 60 ft$^2$ should be used for each bed.

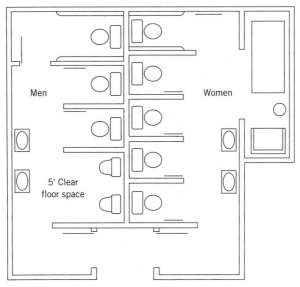

**Figure 5.5**   Typical commercial restroom layout.

# *5.4*   FOOD SERVICES

The shifting of new facilities away from central business districts, the shortening of meal breaks, OSHA forbidding consumption of foods where toxic substances exist, and the increased importance of employee fringe benefits have all contributed to making food services more important to the facilities planner. Food service activities may be viewed by a firm as a necessity, a convenience, or a luxury. The viewpoint adopted, as well as a firm's policy on off-premises dining, subsidizing the costs of meals, and the amount of time allowed for meals, has a significant impact on the planning of food service facilities.

Food service facilities should be planned by considering the number of employees who eat in the facilities during peak activity time. Kitchen facilities, on the other hand, should be planned by considering the total number of meals to be served. If employees eat in shifts, the first third of each shift will typically be used by the employee preparing to eat and obtaining the meal. The remainder of the time will be spent at a table eating. Therefore, if a 30-min meal break is planned, dining shifts, as shown in Table 5.3, may begin every 20 min. In a like manner, if a 45-min. break is planned, shifts may begin every 30 min.

Food services requirements may be satisfied by any of the following alternatives:

1. Dining away from the facility.
2. Vending machines and cafeteria.
3. Serving line and cafeteria.
4. Full kitchen and cafeteria.

**Table 5.3**   *Shifting Timing for 30-min Lunch Breaks*

| Beginning of Lunch Break | Time Sat Down in Chair | End of Lunch Break |
|---|---|---|
| 11:30 A.M. | 11:40 A.M. | 12:00 Noon |
| 11:50 A.M. | 12:00 Noon | 12:20 P.M. |
| 12:10 P.M. | 12:20 P.M. | 12:40 P.M. |
| 12:30 P.M. | 12:40 P.M. | 1:00 P.M. |

The first alternative certainly simplifies the task of the facilities planner. However, requiring employees to leave the facility for meals results in the following disadvantages:

1.   Meal breaks must be longer.
2.   Employee supervision is lost, which may result in
    a.   Employees not returning to work.
    b.   Horseplay.
    c.   Returning to work intoxicated.
    d.   Returning to work late.
3.   A loss of worker interaction
4.   A loss of worker concentration on the tasks to be performed.

In addition, for many facilities off-site restaurants are not sufficient to handle employee meal breaks conveniently. For these reasons it is typically recommended that employees not leave the facility for meals and that space be planned within the facility for dining.

For each of the three remaining food service alternatives, a cafeteria is required. Cafeterias should be designed so that employees can relax and dine conveniently. Functional planning, ease of cleaning, and aesthetic considerations should all be considered, as the cafeteria will be utilized for purposes other than standard employee meals. (Cafeterias are often used as auditoriums.) Also, movable partitions may be used to create conference rooms and more private luncheon meeting rooms.

An integral part of the cafeteria is the food preparation or serving facilities. The option of a serving line or full-service kitchen will be contingent on the number of employees to be served. If a facility employs over 200 people, a serving line is a feasible alternative. Figure 5.6 shows an efficient walk-through serving line which can be used for sandwiches in combination with catered food service.

Space requirements for cafeterias should be based on the maximum number of employees to eat in the cafeteria at any one time. Table 5.4 gives general area requirement guidelines. The space allocated within the ranges given in Table 5.4 should be determined by the type of tables to be utilized. Popular table sizes are 36-, 42-, and 48-in. square tables and rectangular tables 30 in. wide and 6, 8, and 10 ft long. Square tables require more aisle space than rectangular tables, but result in more attractive cafeterias. Table sizes depend on whether or not employees retain their trays during the meal. A 36-in. square table is adequate for four employees if they do not retain their trays. Standard trays are 14 × 18 in. therefore, a 48-in. square table is most suitable if employees retain their trays. Square tables (42 in. × 42 in.) are quite spa-

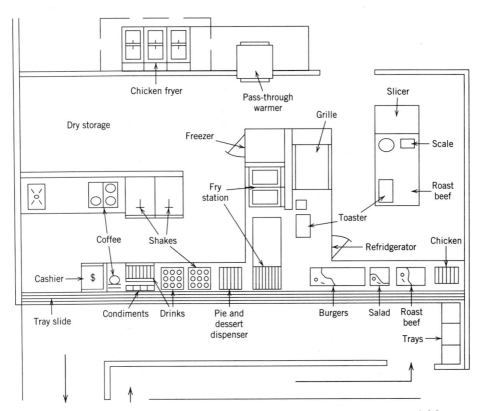

**Figure 5.6**  Serving line with caterer's preparation area. *Source:* Ramsey et al. [8].

cious when trays are not retained or when trays are smaller than standard used. If 36-in. square tables are used, an average space figure from Table 5.4 should be utilized. If 42-in. square tables are used, space requirements between the average and highest values stated in Table 5.4 should be utilized. If 48-in. square tables are used, the highest values stated in Table 5.4 should be utilized for space planning.

Rectangular tables 6, 8, and 10 ft long adequately seat three, four, and five employees, respectively, on each side of the table with no end seats. If individual 6-ft rectangular tables are to be utilized, an average space figure from Table 5.4 should be utilized. Figures between the average and lowest figures given in Table 5.4 should be used for tables placed end-to-end that seat between 6 and 12 people. If more than 12 people may be seated in a row of tables, the lowest figures given in Table 5.4 may be utilized.

The use of vending machines is the least troublesome way of providing food services for employees. It is also the most flexible of on-site food service alternatives.

**Table 5.4** *Space Requirements for Cafeterias*

| Classification | Square Footage Allowance per Person |
| --- | --- |
| Commercial | 16–18 |
| Industrial | 12–15 |
| Banquet | 10–11 |

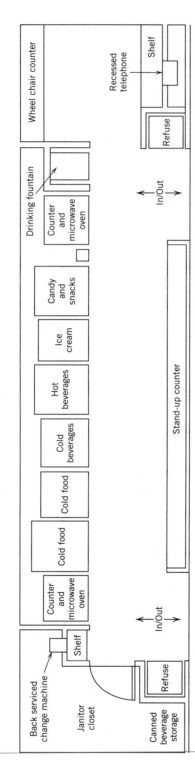

**Figure 5.7** Typical institutional vending area.

*Note.* Refrigerated vending machines and microwave ovens require a rear wall clearance of up to 8 in. to permit cooling. No hot water is required. Some beverage dispensing units require a cold water line with a shut-off valve. Overflow waste disposes into internal bucket or tray. A separate, 115-V electrical circuit for each machine is suggested. Delivery amperages range from approximately 2 to 20 amperes (A) per machine.

*Source:* Ramsey et al. [8].

Employees not wishing to purchase their lunches typically feel more at ease if vending machines are utilized than if serving lines or full kitchens are provided. For space planning purposes 1 ft$^2$ per person should be allowed for the vending machine area, based on the maximum number of persons eating at one time. Figure 5.7 shows a typical institutional vending area layout.

If a facility employs over 200 people, a serving line is a feasible alternative. When a serving line is utilized, a caterer is frequently contracted to prepare all food off site and to serve the food to employees. The advantage of serving lines is that they offer employees the benefits of a full kitchen but require little effort by the management of the facility. The cost of a meal is often quite competitive with the cost of running a full kitchen for facilities employing less than 400 people. A typical industrial serving line requires 300 ft$^2$ and can service seven employees per minute [4]. A sufficient number of service lines should be provided so that during peak demand for service employees receive their meals in one third of the lunch period.

When a full kitchen is used, a serving line (see Figure 5.6) and a kitchen must be included in the facility. A full kitchen can usually be justified economically only if there are over 400 employees within a facility. A kitchen allows the management of a facility to have full control over food service and be able to respond to employee desires. Such control is often a costly and troublesome undertaking; it is one that many firms do not wish to undertake. Space planning for kitchens to include space for food storage, food preparation, and dishwashing should be based on the total number of meals to be prepared. Space estimates for full kitchen are given in Table 5.5.

Food services should be located within a facility using the following guidelines.

1. Food services should be located within 1000 ft of all permanent employee workstations. If employees are required to travel more than 1000 ft, decentralized food services should be considered.

2. Food services should be centrally located and positioned so that the distance from the furthest workstation is minimized.

3. Food services should be located to allow delivery of food and trash pickup.

4. Food services should be located to allow employees an outside view while eating.

5. Food services should be located so that adequate ventilation and odors and exhaust do not interfere with other activities within the facility.

Table 5.5  *Space Required for Full Kitchens*

| Number of Meals Served | Area Requirements (ft$^2$) |
| --- | --- |
| 100–200 | 500–1000 |
| 200–400 | 800–1600 |
| 400–800 | 1400–2800 |
| 800–1300 | 2400–3900 |
| 1300–2000 | 3250–5000 |
| 2000–3000 | 4000–6000 |
| 3000–5000 | 5500–9250 |

*Source:* Kotschevar et al. [5].

## *Example 5.2*

If a facility employs 600 people and they are to eat in three equal 30-min shifts, how much space should be planned for a cafeteria with vending machines, serving lines, or a full kitchen?

If 48-in. square tables are to be utilized, Table 5.5 indicated 12 ft$^2$ are required for each of the 200 employees to eat per shift. Therefore, a 2400-ft$^2$ cafeteria should be planned.

If a vending machine area is to be used in conjunction with the cafeteria, an area of 200 ft$^2$ should be allocated for vending machines. Thus, a vending machine food service facility would require 2600 ft$^2$.

A serving line may serve 70 employees in the first third of each meal shift. Therefore, three serving lines of 300 ft$^2$ each should be planned. A total of 3300 ft$^2$ would be required for a food service facility using serving lines.

A full kitchen will require 3300 ft$^2$ for serving lines plus (from Table 5.6) 2400 ft$^2$ for the kitchen. Therefore, a total of 5700 ft$^2$ would be required for a full service food service facility.

Two issues related to food services are drinking fountains and break areas. Drinking fountains should be located within 200 ft of any location where employees are regularly engaged in work. Local building codes should be consulted for determining their exact location. Drinking fountains should be conveniently located but must not be located where employees using drinking fountains may be exposed to a hazard. Drinking fountains are often located near restrooms and locker rooms for the convenience of employees and because plumbing is readily available.

If the cafeteria is within 400 ft of most employees, rest breaks should be confined to the cafeteria. If a cafeteria serving line exists and it may be used to dispense drinks and snacks, it should be utilized for breaks. If a serving line may not be used or if one does not exist, drinks and snacks should be available in the cafeteria from vending machines during break periods.

Locating vending machines throughout a facility may cause supervision, food containment, and trash disposal problems. Nevertheless, if the cafeteria is more than 400 ft from most employees, some vending machines should be located within the facility to reduce employee travel time. When vending machines are used, break areas should be conveniently located for employees while at the same time not interfering with other activities.

## *5.5*  HEALTH SERVICES

It is very difficult to predict facility requirements for health services. Some facilities have little more than a well-supplied first aid kit, while other facilities have small hospitals. Therefore, local building codes should be checked in establishing a facility's requirements. The types of health services that may be provided within a facility include

1.  pre-employment examinations,
2.  first aid treatment,
3.  major medical treatment,
4.  dental care, and
5.  treatment of illnesses.

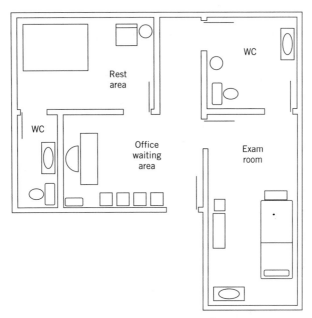

**Figure 5.8**   Nursing station layout.

The facilities planner should check the firm's operating procedure to determine what types of services are to be offered and what health services staff is to be housed within the facility. At the very least, a small first aid room should be included. The minimal requirements for a first aid room are an approved first aid kit, a bed, and two chairs. A minimum of 100 ft$^2$ is required. If a nurse is to be employed, the first aid room should have two beds and should be expanded to 250 ft$^2$. In addition, a 75 ft$^2$ waiting room should be included. For each additional nurse to be employed, 250 ft$^2$ should be added to the space requirements for the first aid room and 25 ft$^2$ should be added to the space requirements for the waiting room. Figure 5.8 shows a typical nurse/first aid area for a plant facility. If a physician is to be employed on a part-time basis to perform pre-employment physicals, a 150 ft$^2$ examination room should be provided. If physicians are to be employed on a full-time basis the space requirement should be planned in conjunction with a physician, based on the types of services to be offered.

Health services should be located such that examination rooms are adjacent to first aid rooms and close to the most hazardous tasks. Health services should include toilet facilities and should either be soundproofed or located in a quiet area of the plant. They are often located next to restrooms or locker rooms.

## 5.6   BARRIER-FREE COMPLIANCE

Facilities planners must incorporate the intent of the Americans with Disabilities Act (ADA). The intent is to ensure that disabled persons shall have the same right as the able-bodied to the full and free use of all facilities that serve the public. To this extent, all barriers that would impede the use of the facility by the disabled person must be removed, thereby making the facility barrier free. As facilities planners, the definition of barriers must be understood. What are considered barriers? A barrier is a physical object that impedes a disabled person's access to the use of a facility, for example, a door that is not wide enough to accommodate a wheel chair or stairs without ramp access to a facility. Note also that communication barriers that are an integral part of the physical structure of the facility are included, for example, permanent signage. The facilities planner must recognized that this applies to all public facility use groups:

- Assembly
- Business
- Educational
- Factories and Industrial
- Institutional

The ADA will fundamentally impact the way industrial engineers approach the design of a facility—from the parking lot to entering and exiting the facility, moving

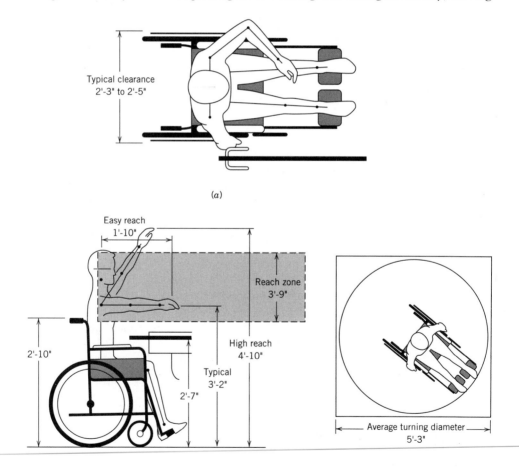

(a)

(b)                                          (c)

**Figure 5.9**  Wheel chair dimensions and turning radius.

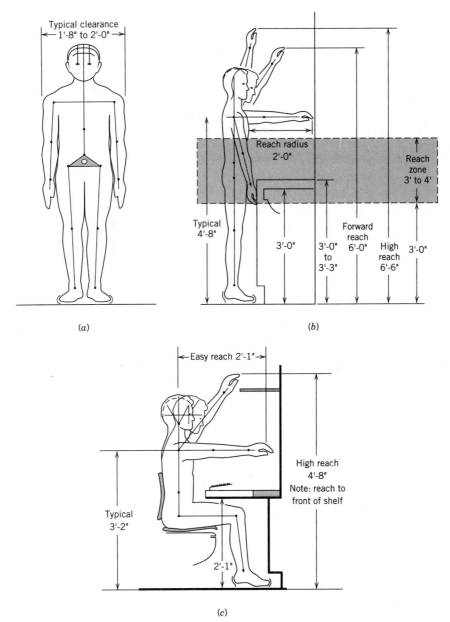

**Figure 5.10**   Able-bodied anthropomorphic clearance and reach requirements in standing and sitting positions.

within the interior of the facility, workstations, offices, and restrooms. To remain effective as facilities planners we must account for the handicapped person's space requirements versus that of an able-bodied person. Consider Figure 5.9 and the wheel chair dimension's reach and maneuverability requirements.

Compare these dimensions in Figure 5.9 with the dimension of an able-bodied person's typical clearance and reach requirements as given in Figure 5.10. Although there are significant physical differences between able-bodied and physically disadvantaged individuals, there exists a reach zone where both groups can comfortably access objects placed in this zone (see Figures 5.9*b* and 5.10*b*). This zone, as shown, is typically 3 ft to 4 ft above floor level. These upper and lower limits were defined by the reach limits of both groups. The upper limit is dictated by the height that a physically disadvantaged person can easily reach an object, whereas the lower limit is defined by the reach of the average able-bodied person without bending.

From analyzing the clearances for both groups, it is obvious that facility entrances, doors, hallways, and so on must be wide enough to accommodate the wheel chair, typically a 3 ft. minimum. Also, fixed facility elements in laboratories or other work or study areas using workbenches requires a minimum clear width of 3 ft. This width should be increased to a 3 ft-6 in. minimum if the aisle is to accommodate access simultaneously by an able-bodied person and a wheel-chaired person, for example in library stalls. Additionally, by mapping the reach requirement of the average person against that of the handicapped, there exists a zone 3 ft. to 4 ft. where both groups can comfortably access items. Thus, the placement of telephones, towel racks, trash, cup disposers, switches, thermostat, fire alarm, emergency call box, door latches, elevator control panel, elevator call button, and so on must be typically a maximum of 40 in. above floor level to effectively accommodate both groups.

# 5.7   SUMMARY

Guidelines to determine facility requirements for parking lots, locker rooms, restrooms, food services, health services, and ADA compliance have been provided. Answers to the following types of questions are *not* the responsibility of the facilities planner:

1.   Should employees be assigned parking locations or should parking be random?
2.   Should locker rooms be provided even though clothes changing does not take place?
3.   Should restrooms be centralized or decentralized?
4.   Should food be provided to employees from vending machines or serving lines?
5.   Should a physician be employed to provide health services or is a nurse adequate?

The answers to these questions are generally the responsibility of the human resources department, industrial relations department, or personnel department. Once these answers are provided, the facilities planner can utilize the guidelines set forth in this chapter to plan personnel services within the facility. However, because the individuals who are responsible for answering such questions might not understand the cost impact of their decisions, the facilities planner should play an active role in the overall facilities planning decision-making process.

# BIBLIOGRAPHY

1. Baerwald, J. E., *Transportation And Traffic Engineering Handbook,* Prentice-Hall, Englewood Cliffs, NJ, 1976.
2. Callendar, J. H., *Time Saver Standards For Architectural Design Data,* 6th ed., McGraw-Hill, New York, 1982.
3. Kearney, D. S., *The New ADA: Compliance And Costs,* R. S. Means Company, Inc., Kingston, MA., 1992.
4. Kotschevar, L. H., and Terrall, M. E., *Food Service Planning: Layout And Equipment,* 2nd ed., John Wiley, New York, 1977.
5. Mumford S., "The ADA: Striving To Comply," *Buildings,* Vol. 87, No. 4, bb. 46–48, April 1993
6. Stokes, J. W., *Food Service In Industry And Institutions,* Brown, Dubuque, IA., 1973.
7. Tompkins, J. A., Bailey, D. C., and Gupta, R. M., "Raleigh Capital Area Parking Study," Department of Industrial Engineering, North Carolina State University, Raleigh, NC, 1977.
8. Ramsey, C. G. and Sleeper, H. R., *Architectural Graphic Standards,* John Wiley, New York, 1988.
9. *Boca National Building Code,* 11th ed., Building Officials & Code Administrators International, Inc., Country Club Hills, IL, 1990.
10. De Chiara J., and Callender, J., *Time Saver Standards For Building Types,* 3rd ed., McGraw-Hill, New York, 1990.

# PROBLEMS

**5.1** Is it true that a well-fed employee is a happy employee? Is a happy employee a productive employee? Explain.

**5.2** Compare the bulkiness of the nonwork belongings that are often brought to a facility by an employee.

**5.3** What is the impact on the space required in a parking lot if spaces are assigned or if filled randomly?

**5.4** A parking lot is to be 400 ft wide and 370 ft deep. How may standard-sized cars fit in this lot?

**5.5** What is the impact of allocating 25% of the spaces in a company parking lot to compact cars?

**5.6** A facility is to house 50 female and 50 male employees. How much space should be planned for the restrooms?

**5.7** The Ajax Manufacturing Company has decided to allow its employees 1 hour for lunch. The lunch breaks are to start at 11:00 A.M. and to end at 1:00 P.M. How many lunch starting times can they accommodate if lunch breaks must begin and end on the hour, on the quarter hour, or on the half hour?

**5.8** What is the impact on space requirements for a vending machine and cafeteria food service for the following luncheon patterns?

> 800 employees—one shift
> 400 employees—two shifts
> 200 employees—four shifts

**5.9**  What is the impact on space requirements for a serving line and cafeteria food service for the luncheon patterns given in Problem 5.8?

**5.10**  What is the impact on space requirements for a full kitchen and cafeteria food service for the luncheon patterns given in Problem 5.8?

**5.11**  If two nurses and a part-time physician are to be employed in a health services area, how much space should be allowed in the facilities plan?

**5.12**  Describe the largest personnel service problem on campus. How could this problem be resolved.

**5.13**  Are personnel services as important in office facilities as they are in production facilities? Describe in detail.

**5.14**  Is it a good idea to consult with employees who are to utilize personnel services before planning these services? Why or why not?

**5.15**  Discuss the impact on personnel services of multilevel versus single-level facilities.

**5.16**  What are the ADA implications on
   a.  The building in which this class is taught.
   b.  Your school's Football Stadium
   c.  Your school's Student Center

# Part Two

# DEVELOPING ALTERNATIVES: CONCEPTS AND TECHNIQUES

# 6

# *MATERIAL HANDLING*

## *6.1*  INTRODUCTION

As stated in Chapter 1, the design of the material handling system is an important component of the facilities design problem. There exists a strong relationship between the layout design function and the material handling design function. It is seldom the case that one can be considered without also considering the other. In facilities planning in particular, consideration *must* be given to the material handling requirements of an organization.

Whether the material being handled is mail in a postal system, money in a banking system, units of product in a production system, refuse in a refuse disposal system, or people in a transportation system, many of the same principles apply. Material handling problems arise in a wide variety of contexts and a host of alternative solutions usually exist. For this reason, it is important for a systematic approach to be taken in designing the system.

In this chapter we define material handling and then discuss a number of material handling principles. The analysis of material handling problems and the design of material handling systems are also treated. The importance of the selection of the handling unit is also emphasized. Finally, we consider a number of material handling equipment alternatives.

## *6.2*  MATERIAL HANDLING DEFINED

In a typical factory, material handling accounts for 25% of all employees, 55% of all factory space, and 87% of production time [9]. Material handling is estimated to represent between 15 and 70% of the total cost of a manufactured product. Certainly, material handling is one of the first places to look for cost reductions.

Material handling is also one of the first places to look for quality improvements. It has been estimated that between 3 and 5% of all material handled becomes damaged. The scrapes and scratches on the walls and floors of our manufacturing facilities are evidence enough of the quality problems that arise from careless material handling.

At this point, the conclusion might be reached that the answer to our competitive problem might be to minimize or even eliminate material handling. The cost savings in labor and equipment would be tremendous. In addition, the fewer times a product is handled, the less likelihood there is that the product will become damaged.

A policy of minimizing material handling was followed by General Electric in the redesign of their dishwasher facility in Louisville, Kentucky. Their results, provided in Table 6.1, were impressive.

In addition, General Electric received a large increase in market share, and in the summer of 1983, *Consumer Reports* rated the GE dishwasher as the best value among dishwashers. By what appears to be simply handling less, General Electric has set an admirable example for all manufacturers in industrial competitiveness.

Simply handling less, however, is not the answer. Material handling is a very valuable activity. Material handling is a means by which *total manufacturing costs are reduced* through reduced inventories, improved safety, reduced pilferage, and improved material control. Material handling is a means by which *manufacturing quality is improved* by reducing inventory and damage through improved handling practices. Finally, material handling is the means by which any production strategy is executed.

Clearly then, simply handling less is not the answer. Handling smarter and with an increased awareness of the vital role material handling plays in manufacturing and distribution is the key to improving the firm's ability to compete in a global economy.

## Understanding Material Handling

Any technology playing such a vital role in the economy as material handling deserves our attention and comprehension. The balance of the chapter is written with the intent of enhancing your understanding of material handling. Several definitions, a description of the scope of material handling, and a discussion of the guidelines and principles that govern material handling systems design and operation are included.

Table 6.1 *General Electric Dishwasher Facility Redesign {9}*

| Performance Measure | Before | After |
|---|---|---|
| Unit Cost (Index) | 100 | 70 |
| Inventory Turns | 13 | 25 |
| Reject Rates | 10% | 3% |
| Output per Employee (Index) | 100 | 133 |
| No. of Times Unit Handled | 27 | 3 |

Our consideration of material handling equipment is purposefully placed at the end of the chapter. It is essential that those who are striving to become material handling system designers grasp the fundamentals of material handling before considering material handling equipment alternatives. It is essential that the requirements for material handling be understood before material handling solutions are contemplated!

***Definitions.***  There is no unique, all encompassing definition of material handling. In fact, the *Materials Handling Handbook* [23] lists nine definitions on its opening page. Nonetheless, the following two definitions should convey the key points.

1.  Material handling is the art and science of moving, storing, protecting, and controlling material.

    - *Art:* Material handling can be described as an art because material handling problems and material handling systems cannot be explicitly solved/designed solely with scientific formulas or math models. Material handling requires an "appreciation" for right and wrong, which accompanies significant practical experience in the field.

    - *Science:* Material handling can be described as a science because the engineering method—define the problem, collect and analyze data, generate alternative solutions, evaluate alternatives; select and implement preferred alternative—is an integral part of the material handling problem solving and system design process. In addition, mathematical models and quantitative techniques of analysis are a very valuable part of the process.

    - *Moving:* Moving material is required to create time and place utility (i.e., the value of having the material at the right time and at the right place). Any movement of material requires that the size, shape, weight, and condition of the material, as well as the path and frequency of the move be analyzed.

    - *Storing:* Storing material provides a buffer between operations, facilitates the efficient use of people and machines, and provides for efficient organization of material. Material storage considerations include the size, weight, condition, and stackability of the material; the required throughput; and building constraints such as floor loading, floor condition, column spacing, and clear height.

    - *Protecting:* The protection of material includes both the packaging, packing, and unitizing done to protect against damage and theft of material, as well as the use of safeguards on the information system to include protection against the material being mishandled, misplaced, misappropriated, and processed in the wrong sequence. Continuous improvement programs strive to eliminate "inspecting quality into a product, by designing quality into the product." In a similar fashion, the material handling system should be designed to minimize the need for inspections and costly methods of protecting the material.

    - *Controlling:* Truly controlling material requires physical and status material control. Physical control is control of the orientation of, sequence of, and space between material. Status control is the real-time awareness of the location, amount, destination, origin, ownership, and schedule of material. Care must be taken to ensure that too much control is not exerted over the material handling system. Maintaining the correct degree of control is a challenge, because the right amount of control depends upon the culture of the organization and the people who manage and perform the material handling functions.

- *Material:* We define "material" very broadly, including bulk material and unit loads, in any form—solid, liquid, or gas. Furthermore, occasionally, we will include paperwork and information as material, depending upon the application setting. From potato chips to semiconductor chips, from coin, currency, and negotiable securities to sides of beef, from sailing ships to printed circuit boards, the principles and approaches we present are applicable in the movement, storage, control, and protection of material.

2. *Material handling means providing the right amount of the right material, in the right condition, at the right place, at the right time, in the right position, in the right sequence, and for the right cost, by using the right method(s).* Naturally, if the right methods are being used, then the material handling system will be safe and damage free.

***Right Amount.***   What does it mean to provide the *right amount* of material? Just-in-Time (JIT) inventory management focuses attention on the *right amount* of inventory to have in both manufacturing and distribution. In too many instances, inventories are used as expensive insurance policies in an attempt to insure against an unlikely event, but at a cost substantially greater than the cost the inventories are attempting to avoid! Such practices are called Just-in-Case (JIC) inventory management.

A positive aspect of JIT is its questioning attitude regarding inventory levels. Instead of automatically thinking of having inventories, with JIT you automatically think in terms of not having inventories. Because of JIT, production lot sizes have been reduced substantially, in some cases lots of size one have become a reality. However, despite the attention given to lot sizes, less attention has been given to the issue of load sizes.

Even if the material handling system designer is unable to influence production lot sizes, load sizes can have a significant influence on the overall inventory level in the system. Regardless of the lot size, the load size can range from single units to pallet quantities. Whereas it was once advocated that the size of the unit load should be increased to reduce the cost of handling, today smaller load sizes are being used to reduce the cost of inventory.

To have the *right amount* of material it is also necessary to coordinate issue quantities with stocking quantities. Unfortunately, in cases where issue quantities greater than one are used, the issue quantity is seldom related to the stocking quantity. An example is the issuance of quantities in multiples of 5 or 10 when cartons contain quantities in multiples of dozens. We recall an instance in which parts were stocked in cartons of 50, but they were typically issued in quantities of 30. In this case, because one partial carton already existed, the storeroom attendant had to count the parts one-by-one, including the balance taken from a second carton. How much quicker and more accurately it would have been to issue a full carton of 50; alternately, if cartons had contained 25 units cost might have been further reduced.

In determining the right amount of material to stock in a warehouse, it is often required to determine the right amount to stock in the active picking area and the right amount to stock in reserve storage. Likewise, one must determine the right amount to provide as safety stock. In each case, the answer must be a function of the replenishment, overage, and underage costs, as well as the overall impact on the system. The point to be gained from this is that the *right amount* is not always zero, despite what some may claim.

***Right Material.***  Why does the definition of material handling include an emphasis on providing the right material? The two most common errors made in order picking are picking the wrong amount and picking the wrong material. Picking the right material is not an easy task, especially in warehouses that are designed to complicate order picking, not simplify it.

Walk through any warehouse and look at the part numbering system used, the labels used, how item status information is maintained, where material is located, and the picking documents used by the order pickers. If you are still unconvinced that the opportunity is great for picking the *wrong material,* look at the number of different people involved in generating the picking document, picking the order, packing it, and shipping it to the customer. The fact that so many orders are picked correctly is miraculous.

For the material handling system to move, store, protect, and control the right material, it must include an accurate identification system. And, automatic identification is the *key* to accurate identification. While situations might exist where automatic identification cannot be justified, the speed, accuracy, and economics of automatic identification available today generally cannot be matched by manual approaches.

The proliferation of bar coding technology, the emergence and maturation of radio frequency identification, and the continued development of magnetic and other identification technologies make it possible to provide the *right material* consistently.

In addition to using automatic identification, simplify the part numbering system. Reduce the number of part numbers by standardizing on parts and removing obsolete parts from the database. Next, simplify the part number; part numbers often convey vast amounts of information by using multiple data fields. Unfortunately, many part numbers have not been redesigned to accommodate today's computer environment, which allows the use of a hierarchical part numbering system, with all but the first level resident in computer memory; all the alphanumeric part number needs to do is unlock the right "information doors" in the computer. Further, experience shows that part numbers should be designed for people to use even when it is planned to do all part identification automatically.

As noted by an order picker at a major catalog distribution center, locating "look alikes" together results in a man's medium red knit shirt being picked instead of a man's large red knit shirt. If more order picking systems were designed by people who had performed the function, perhaps we would find that providing the right material would occur quite naturally.

Finally, it is important to recognize that moving, storing, protecting, and controlling the right material requires a decision as to which material to move, store, protect, and control. In particular, not all material has to be controlled in the same way. Infrequently used material of low value does not require the same degree of control as frequently used material of high value. Similarly, not all material has to be stored, some can be shipped direct to the customer from the manufacturer and bypass the distribution center. In summary, each item of material should be examined carefully to ascertain if it is the right material for movement, storage, protection, and control by the material handling system.

***Right Condition.*** The third element in the definition of material handling emphasizes the need to provide material in the right condition. What do we mean by the right condition? The first thing that comes to mind is top quality, the absence of damage. Since the material handling system is frequently a major source of product damage, it is essential that *total quality* be embedded in the design and performance of the handling system.

While the right condition of material means quality, that is not all that it means. For example, we are also concerned about the status of the material, i.e., its location, the processing steps that have been performed, its physical characteristics, its availability for shipment, and the need for tests or inspections.

Suppose a particular part goes through a sequence of processing stages. Further, suppose the part is used in a number of different products, with the only difference being the color it is painted. If in-process storage occurs, what is the preferred condition of the stored parts? Unless there are unique requirements that do not readily come to mind, storage of unpainted parts is preferred. Otherwise, more inventory would be required.

Another example of the right condition of material being stored occurs with service parts distribution centers, which receive bulk shipments of replacement parts from multiple manufacturing plants. Parts are shipped to retail customers on demand in prespecified quantities. When parts are sold in packaged quantities, the first step following receipt of the bulk material is often a packaging operation; packaged parts are stored until ordered by retail customers. However, packaging adds cost to the item (and increases inventory carrying cost), it expands the number of stock keeping units tracked in inventory (due to the multiple pack quantities of the same part), and it explodes the space requirements for the parts. The right condition for the part in storage may well be in an unpackaged state, with packaging performed upon demand. (If all inventory cannot be maintained in an unpackaged state, then a mixture of packaged and unpackaged parts might be preferred to all parts being packaged.)

If quality is what the customer says it is, then we must identify the customers of the material handling system. Essentially, each person who depends on the material handling system to perform his or her job is a customer. Hence, it is important to ascertain what each customer requires in terms of the condition of the material served by the handling system.

***Right Sequence.*** Work simplification teaches that productivity can be increased by eliminating unnecessary steps in an operation and improving those that remain. Also, productivity improvement can occur by combining steps and changing the sequence of steps performed. The impact of the sequence of activities performed on the efficiency of an operation is very evident in material handling. Therefore, we include in the definition of material handling the need to move, store, protect, and control materials in the *right sequence.*

The impact of sequencing on movement, storage, protection, and control requirements was illustrated during a visit to a Toyota assembly plant in Japan. We noted the absence of material inventories along the assembly line. Whereas at that time sizeable inventories were provided for line feeding at the U.S. auto assembly plants we visited, small inventories were present at Toyota. Two differences were obvious, Toyota's suppliers replenished the line with small shipments going direct to the assembly line; also, the variety of material needed was much less at Toyota than at the U.S. assembly plants. Basically, Toyota attempted to schedule similar automobiles together to minimize the variety of materials required for assembly. The U.S. approach then was to balance the assembly line on the basis of labor content; Toyota factored inventory reduction into the assembly sequence.

Yet another reminder of the importance of sequence was provided by a computer manufacturer in one of its printed circuit board assembly plants. When they scheduled assembly to minimize the time required to assemble a batch of circuit boards, they found that circuit boards had to make a large number of moves between insertion machines; further, the storage buffers at each machine were sizeable. Basically, material handling was the bottleneck activity in circuit board assembly. When they changed their assembly criterion from minimizing time to minimizing material handling, they achieved a significant reduction in moves and inventory without increasing assembly time significantly!

A third reminder of the impact of sequence occurred when a retail food chain attempted to automatically palletize mixed loads of product. Soft drink beverage producers face a similar need. Unfortunately, the complexity of automatic palletizing mixed loads increases exponentially with the variety of carton sizes to be accommodated. On the other hand, if cartons can be presequenced then the potential for automatic palletizing increases dramatically.

A fourth reminder occurred when we visited the automation laboratory of a copier manufacturer and talked with the engineers who programmed the robots to perform automatic assembly. Because of the time required to change end-of-arm tooling on the robot, the conventional assembly sequence had to be changed. The result was a decrease in end-of-arm tooling required across several robots, and a reduction in overall assembly time.

An overlooked opportunity for systems improvements through sequence changes is often the design of the control system. At what point or points controls are used in the process can have a dramatic impact on the complexity of the control system. For example, moving upstream or downstream the point at which bar codes are read can have a significant impact on the cost and success of the control system. Unfortunately, the control system is usually designed to mimic the manual system in place previously; little thought is given to taking advantage of the technological advantages provided through automatic identification either in terms of the location of control points or the frequency of control checks.

***Right Orientation.***  Of all the "right" components in the material handling definition, the *right orientation* usually receives the least amount of attention from material handling systems designers. Notice how much of what people do in manufacturing and distribution is physically orienting material. Regaining the orientation of material represents a significant portion of people's activities. Precision machined parts are dumped into tote boxes; subsequently someone sorts out the parts and reorients them for the next operation. Had the parts been oriented or "positionally controlled" then automatic transfers might have been justified.

In general, the physical orientation of material either can be *retained* through the material handling system or it can be *regained*. Either way, costs are involved. Which is cheaper, retaining or regaining physical orientation? Of course the answer depends on the particular application. However, many overlook the option of retaining the physical orientation. Instead, they routinely regain orientation by using either manual methods, vibratory bowl feeders, guides on conveyors, or some other mechanical or automatic method.

It is difficult to analyze an existing manufacturing, assembly, or distribution operation without finding improvement opportunities involving the orientation of material. What might be accomplished if material is properly oriented? Examples include automatic identification, sortation, assembly, packaging, and unitizing of parts and cartons.

Physical orientation is often accommodated by changing the design of a part. By adding locator holes or pins, automatic orientation of parts might become feasible. Likewise, by changing the design of individual parts, automatic assembly might become viable. A number of firms have redesigned parts to allow automatic assembly; with the objective of achieving assembly along a single axis, the orientation of individual parts is critically important.

Not only do parts and cases need to be oriented properly, but so do palletized loads. Depending on the column spacing and location of doors and docks in an existing warehouse, the use of a 40″ × 48″ pallet versus a 48″ × 40″ pallet can make a big difference in the overall utilization of storage space available. Further, it is important for the material handling systems designer to maintain the proper orientation of unit loads as they make right-angle transfers on a conveyor system; we know of cases where special equipment was installed because the orientation of the load to the storage/retrieval machine was overlooked.

***Right Place.***   Since the 1950s, material handling has been defined in terms of providing material at the right place, at the right time. However, even though considerable attention has been given to the significance of place, an examination of a number of material handling systems suggests that it would be beneficial to revisit some of the fundamental issues concerning the role of place in material handling.

When material arrives at the receiving dock, what happens to it? Quite often it is placed in one or more "temporary" locations or staging points before it is eventually placed in storage. Regardless of whether material has assigned (dedicated) or randomized (floating) storage locations, it should be placed in the right location—it should be in the right place. Further, it is important for material locations to be entered quickly and accurately in the locator system.

When material arrives on the manufacturing floor, what happens to it? Not unlike the warehouse, it tends to be stacked on the floor or placed in "buffers" awaiting further processing. In fact, the warehouse is not nearly as culpable as manufacturing and assembly plants in failing to locate material in the right place.

In determining the right place for material to be stored, it is also important for decisions to be made as to whether central storage or distributed storage is best for a particular application. Likewise, decisions must be made in manufacturing as to where in the process storage should occur (if at all); in continuous flow manufacturing, raw material storage is preferred to storage of in-process material, which is preferred to finished goods storage, i.e., upstream, not downstream, is more likely the right place if storage occurs.

Yet another example of material needing to be at the right place involves the location of material in a human operator's "golden zone." From an ergonomics point of view, material should be placed within easy reach; hence, stooping, bending, and stretching should be avoided. This applies in the design of a work station for an assembly operator, as well as the assignment of material to storage locations for an order picker.

Finally, while there might be more than one right place for material, the number of wrong places far exceeds the number of right places. And, an aisle is never a right place for material to be stored, staged, or queued.

**Right Time.**  The need for the material handling system to move, store, protect, and control material at the right time is increasingly important due to time-based competition. Next-day delivery is a "standard" in many mail-order and service parts operations; software products, for example, are routinely shipped using an overnight service. Quick response systems reduce the time required to manufacture and deliver carpet products to customers.

In order for the material handling system to be able to satisfy the requirements for timely responses, excess capacity in the system is generally required. Because of the variation inherent in the requirements placed on the material handling system, it is essential that waiting times be avoided by having sufficient capacity available to respond to unexpected demands.

Cycle-time reduction is a primary target in continuous improvement programs Total quality management (TQM), for example, places considerable emphasis on reducing non-value-adding activities in the process. The notion of supply chain management hinges on reducing the length of the supply pipeline (from suppliers through customers) by reducing the time required to move, store, protect, and control material throughout the pipeline.

Although we advocated shorter overall system time, we did not advocate moving, storing, protecting, and controlling material using faster equipment. In fact, to reduce overall system time often means using slower speeds, not faster speeds. While this may seem a contradiction in terms, it has been shown repeatedly that slower average speeds can be preferable to faster average speeds if accompanied by reductions in the variation in speeds; variance reduction is the key! Being too soon can be worse than being too late in moving, storing, protecting, and controlling material. The emphasis is on the *right time,* not the *fastest time.*

**Right Cost.**  What is the *right cost* in material handling? Remember, the objective of the firm is to maximize the value provided to the shareholders; it is not to achieve a minimum cost of material handling. We believe the material handling system can be a revenue enhancer, rather than a cost contributor; our belief is predicated on the competitive advantage of a material handling system that meets our definition. Today, firms compete on the basis of product functionality, product quality, service quality, time, and cost. To do so, the material handling system must be both effective (does the right things) and efficient (does things right).

Few firms know how much they spend on material handling. If an accountant is asked to provide an estimate, be careful! You might be quoted depreciation charges on the lift truck fleet. Unfortunately, traditional accounting systems do not accurately capture the systems costs associated with material handling. Although a significant percentage of direct labor time is devoted to moving, storing, protecting, and controlling material, generally, the majority of the material handling cost is buried in a firm's overhead costs. Consequently, it is not a simple undertaking to measure the cost of material handling.

In measuring material handling costs, both costs incurred and costs foregone should be measured. The former include investment and operating costs for material handling technology and personnel. The latter include the costs of inventories, space, inspectors, expediters, and other personnel not needed because of the installation of the material handling system. Reduction in losses due to damage and pilfering need to be included. However, perhaps the most significant monetary impact of the system will be on the revenue side of the equation, because of the increased market share that will result from doing the right things right with the material handling system.

In sum, minimizing material handling costs is the wrong objective. The right objective is maximizing overall shareholder value. For that reason, increased investments in handling technology might be needed. The bottom line for material handling: the *right cost* is not necessarily the *lowest cost.*

***Right Methods.***   Finally, to do all the right things right, we need to employ the *right methods.* There are three aspects of the right methods that merit our attention. First, if there are right methods, then there must also be *wrong methods.* Second, it is important to recognize what makes methods *right* and what makes them *wrong.* Third, note that it is *methods,* not *method;* using more than one method is generally the *right* thing to do. We will consider each of these, albeit briefly.

In comparison with material handling, we know of no other technologies that are chosen on the basis of so little emphasis on requirements. Since the early 1970s, we have advocated requirements-driven materials handling systems over solution-driven systems. Solution-driven systems are those in which technologies are chosen without consideration for how the technologies match requirements; instead of defining requirements and matching technology options to requirements, the solution-driven approach force-fits a technology on the system.

What makes a method *right?* Before providing the answer, consider what it isn't! It isn't necessarily the most sophisticated method under consideration; it isn't necessarily the newest method; and it is not necessarily the least expensive method. Simply stated, a method is right if it satisfies the requirements of providing the right amount of the right materials, in the right condition, in the right sequence, in the right orientation, at the right place, at the right time, and at the right cost.

Clearly, material handling is much more than simply handling material. As described, material handling is an art and a science. It involves the movement, storage, control, and protection of material, with the objective of providing time and place utility.

## *The Scope of Material Handling*

Apple [1] gives three possible interpretations of the scope of material handling activities. The first, and narrowest, interpretation is the "conventional" interpretation; here, the primary emphasis in on the movement of materials from one location to another, usually within the same plant. Very little attention is given to the interrelationships of the individual handling tasks. Unfortunately, this is a very common interpretation of the scope of material handling. A second interpretation is the "contemporary" point of view, in which attention centers on the overall flow of materials in a plant or warehouse; interrelationships between handling tasks are analyzed, and an effort is made to develop an integrated material handling plan. A third interpretation, the broadest, is the "progressive" interpretation; this is the systems perspective, defining material handling as all activities involved in handling material from all suppliers, handling material within the manufacturing or distribution facility, and distributing finished goods to customers. It is the broad definition of the scope of material handling that this book embraces.

## 6.3  MATERIAL HANDLING PRINCIPLES

In 1966 the College-Industry Council on Material Handling Education (CICMHE) adopted 20 principles of material handling. The principles were revised in 1981 to reflect changes in industrial operations in the years since they were first adopted. The resulting material handling principles are given in Table 6.2 on page 158.

The material handling principles provide concise statements of the fundamentals of material handling practice. Condensed from decades of expert material handling experience, not unlike the material handling equation, they provide guidance and perspective to material handling system designers. As with all design tools, the applicability of a material handling principle depends on the conditions that exist; due to the many different conditions that might exist, it is unlikely that any principle will always be applicable. Rather than being design axioms, the principles serve as rough guides or rules of thumb for material handling systems design; hence, the principles should not be considered substitutes for sound judgment.

Based on the principles, a number of checklists have been developed to facilitate the identification of improvement opportunities for existing systems. An example of a checklist for material handling is provided in Figure 6.1. Checklists can serve a useful purpose in designing new and improved systems. Namely, they can be used to ensure that "nothing falls through the cracks" and that "all the t's have been crossed and all the i's have been dotted in designing the system. There are so many detailed considerations involved in designing a material handling system that it is easy to overlook some aspects. Hence, checklists can make a valuable contribution to the design process.

## 6.4  MATERIAL HANDLING SYSTEM DESIGN

In designing new or improved material handling systems, the six-phased engineering design process should be used. (The engineering design process was described in

**Sheet 1**

Dept. _____ Building _____ Plant _____

Date _____ Surveyed by _____

MATERIAL HANDLING AUDIT CHECK SHEET

| Conditions indicating possible productivity improvement opportunities | Condition exists here (√) | To correct this, we need: | | | | Other (for comments) |
|---|---|---|---|---|---|---|
| | | Supervisor attention (√) | Management attention (√) | Analytical study (√) | Capital investment (√) | |
| 1. Delays in material moving | | | | | | |
| 2. Excessive material on hand | | | | | | |
| 3. Production equipment idle for material shortage | | | | | | |
| 4. Long hauls | | | | | | |
| 5. Cross traffic | | | | | | |
| 6. Manual handling | | | | | | |
| 7. Outmoded handling equipment | | | | | | |
| 8. Inadequate handling equipment | | | | | | |
| 9. Insufficient handling equipment | | | | | | |
| 10. Unbalanced sequence of operations | | | | | | |
| 11. Idle handling equipment | | | | | | |
| 12. Obstacles to material flow | | | | | | |

|  |  | | | | | |
|-----|------|--|--|--|--|--|
| 13. | Material piled directly on floor | | | | | |
| 14. | Poor workplace layout for material | | | | | |
| 15. | Disorderly storage | | | | | |
| 16. | Cluttered aisles | | | | | |
| 17. | Cluttered workspace | | | | | |
| 18. | Crowded dockspace | | | | | |
| 19. | Motor truck and railroad car tieup | | | | | |
| 20. | Manual loading techniques | | | | | |
| 21. | Excessive wasted "cube" in storage | | | | | |
| 22. | Excessive aisles | | | | | |
| 23. | Operations unduly scattered | | | | | |
| 24. | Poor locations of service areas | | | | | |
| 25. | Lack of in-plant container standardization | | | | | |
| 26. | Lack of unit load technique | | | | | |
| 27. | Excessive MH equipment maintenance cost | | | | | |
| 28. | Rehandling | | | | | |
| 29. | Handling done by direct labor | | | | | |

**Figure 6.1** Material handling productivity audit checklist. *(From [41] with permission.)*

MATERIAL HANDLING AUDIT CHECK SHEET (cont'd)

**Sheet 2**

Dept. _____ Building _____ Plant _____

Date _____ Surveyed by _____

| Conditions indicating possible productivity improvement opportunities | Condition exists here ($\checkmark$) | To correct this, we need: | | | | Other (for comments) |
|---|---|---|---|---|---|---|
| | | Supervisor attention ($\checkmark$) | Management attention ($\checkmark$) | Analytical study ($\checkmark$) | Capital investment ($\checkmark$) | |
| 30. Operators traveling for supplies, material | | | | | | |
| 31. Supplies moved by poor techniques | | | | | | |
| 32. High indirect payroll | | | | | | |
| 33. Material waiting for papers | | | | | | |
| 34. Excessive demurrage | | | | | | |
| 35. Unexplained delays | | | | | | |
| 36. Idle labor | | | | | | |
| 37. Inspection not properly located | | | | | | |
| 38. Excessive scrap | | | | | | |
| 39. Hazardous lifting by hand | | | | | | |
| 40. Misdirected material | | | | | | |
| 41. Clumsy, dangerous "homemade" handling rigs | | | | | | |

| | | | | | | |
|---|---|---|---|---|---|---|
| 42. | Lack of standardization on handling equipment | | | | | |
| 43. | Long travel distances for material, equipment, and personnel | | | | | |
| 44. | Backtracking of material | | | | | |
| 45. | Nonstandard process routing | | | | | |
| 46. | Opportunity for group technology layout | | | | | |
| 47. | Opportunity for product layout | | | | | |
| 48. | Opportunity for process layout | | | | | |
| 49. | No real-time dispatching of equipment | | | | | |
| 50. | No modular MH system | | | | | |
| 51. | No modular work stations | | | | | |
| 52. | Automatic identification system not used | | | | | |
| 53. | No one-way aisles | | | | | |
| 54. | MH equipment running empty | | | | | |
| 55. | Different things treated same | | | | | |
| 56. | Excessive trash removal | | | | | |
| 57. | Centralized storage | | | | | |
| 58. | Decentralized storage | | | | | |

**Figure 6.1** (continued)

## Sheet 3

Dept. _____ Building _____ Plant _____
Date _____ Surveyed by _____

| Conditions indicating possible productivity improvement opportunities | Condition exists here (√) | To correct this, we need: | | | | |
|---|---|---|---|---|---|---|
| | | Supervisor attention (√) | Management attention (√) | Analytical study (√) | Capital investment (√) | Other (for comments) |
| 59. No incentive system for MH labor | | | | | | |
| 60. Low usage of automated MH equipment | | | | | | |
| 61. Variable path equipment used for fixed-path handling | | | | | | |
| 62. System not capable of expansion and/or change | | | | | | |
| 63. Low usage of industrial robots | | | | | | |
| 64. No parts preparation performed prior to manufacturing | | | | | | |
| 65. No prekitting of work | | | | | | |
| 66. Lack of automated loading/unloading of trailers | | | | | | |
| 67. Poor MH at the work station | | | | | | |
| 68. Lack of industrialized truck attachments and below hook lifters | | | | | | |

| | | | | | |
|---|---|---|---|---|---|
| 69. | Equipment capacity not matched to load requirement | | | | |
| 70. | Manual palletizing/depalletizing | | | | |
| 71. | Lack of equipment for unitizing and stabilizing loads | | | | |
| 72. | Lack of a long-range MH plan | | | | |
| 73. | No short-interval scheduling of MH equipment | | | | |
| 74. | Lack of narrow-aisle and very-narrow-aisle storage equipment | | | | |
| 75. | Low bay storage areas | | | | |
| 76. | Poor utilization of overhead space | | | | |
| 77. | Single-sized pallet rack openings | | | | |
| 78. | No palletless handling of unit loads | | | | |
| 79. | Storage in part number sequence | | | | |
| 80. | Randomized storage | | | | |
| 81. | Dedicated storage | | | | |
| 82. | Crushed loads in block stacking | | | | |
| 83. | No ABC storage classification | | | | |

**Figure 6.1**  (continued)

153

**Sheet 4**

Dept. _____ Building _____ Plant _____
Date _____ Surveyed by _____

| Conditions indicating possible productivity improvement opportunities | Condition exists here (✓) | To correct this, we need: | | | |
|---|---|---|---|---|---|
| | | Supervisor attention (✓) | Management attention (✓) | Analytical study (✓) | Capital investment (✓) | Other (for comments) |
| 84. Obsolete and inactive material | | | | | |
| 85. Floor-stacked material in receiving, QC, and shipping | | | | | |
| 86. Aisles and storage locations not clearly marked. | | | | | |
| 87. Manual stock locator system | | | | | |
| 88. Lack of standardization in part numbers | | | | | |
| 89. Cycle counting for physical inventory | | | | | |
| 90. No formal audit program in use | | | | | |
| 91. No guards to protect racks and columns | | | | | |
| 92. Guided aisles without guide rail entry | | | | | |
| 93. Loads overhanging pallet | | | | | |

154

|  |  | | | | | | | |
|---|---|---|---|---|---|---|---|---|
| 94. | Excessive floor, rack, and structured loading | | | | | | | |
| 95. | Equipment operating at excessive speed | | | | | | | |
| 96. | Front-to-back rack members not provided | | | | | | | |
| 97. | MH equipment does not fit through doors | | | | | | | |
| 98. | No sprinklers and smoke detectors | | | | | | | |
| 99. | Hazardous and flammable material not segregated and identified | | | | | | | |
| 100. | Lack of ventilation in battery charging area | | | | | | | |
| 101. | Entrances and exits not secured | | | | | | | |
| 102. | Waste and trash containers located near docks | | | | | | | |
| 103. | Inadequate number of fire extinguishers | | | | | | | |
| 104. | No contingency plan for fire loss | | | | | | | |
| 105. | Sagging load beams and bent trusses on racks | | | | | | | |
| 106. | No formal training for MH equipment operators | | | | | | | |

**Figure 6.1** (continued)

*155*

**Sheet 5**

Dept. _____

Date _____

Building _____

Surveyed by. _____

Plant _____

MATERIALS HANDLING AUDIT CHECK SHEET (cont'd)

| Conditions indicating possible productivity improvement opportunities | Condition exists here (√) | To correct this, we need: | | | | Other (for comments) |
|---|---|---|---|---|---|---|
| | | Supervisor attention (√) | Management attention (√) | Analytical study (√) | Capital investment (√) | |
| 107. No preventive maintenance program | | | | | | |
| 108. No equipment replacement program | | | | | | |
| 109. No dock levelers | | | | | | |
| 110. Unscheduled arrival of outbound and inbound carriers | | | | | | |
| 111. Decentralized receiving and shipping | | | | | | |
| 112. Inbound material not unitized | | | | | | |
| 113. Inadequate number of dock doors | | | | | | |
| 114. Receiving numbers not preassigned | | | | | | |
| 115. Picking lists not printed in picking sequence | | | | | | |
| 116. Orders picked one at a time | | | | | | |

156

| | | | | | |
|---|---|---|---|---|---|
| 117. Aisle lengths unplanned | | | | | |
| 118. Excessive honeycombing in storage | | | | | |
| 119. Poor quality pallets, not standardized | | | | | |
| 120. Manual sorting in order accumulation | | | | | |
| 121. Poor work-in-process control | | | | | |
| 122. Energy-inefficient lighting | | | | | |
| 123. Lights, heaters, and fans poorly located | | | | | |
| 124. No dock enclosures | | | | | |
| 125. Excessive heating, ventilation, and air conditioning for material stored | | | | | |
| 126. Poorly insulated walls and roof | | | | | |
| 127. Poorly designed enclosures for environmentally controlled areas | | | | | |
| 128. Lack of scheduled energy use to reduce peak loads | | | | | |
| 129. Unclean floors | | | | | |
| 130. Battery charging too frequently | | | | | |
| 131. Other | | | | | |

**Figure 6.1** (continued)

**Table 6.2**  *Material Handling Principles*

1. *Orientation Principle:* Study the problem thoroughly prior to preliminary planning in order to identify existing methods and problems, physical and economic constraints, and to establish future requirements and goals.

2. *Planning Principle:* Establish a plan to include basic requirements, desirable options, and the consideration of contingencies for all material handling and storage activities.

3. *Systems Principle:* Integrate those handling and storage activities that are economically viable into a coordinated system of operations, including receiving, inspection, storage, production, assembly, packaging, warehousing, shipping, and transportation.

4. *Unit Load Principle:* Handle product in as large a unit load as practical.

5. *Space Utilization Principle:* Make effective utilization of all cubic space.

6. *Standardization Principle:* Standardize handling methods and equipment wherever possible.

7. *Ergonomic Principle:* Recognize human capabilities and limitations by designing material handling equipment and procedures for effective interaction with the people using the system.

8. *Energy Principle:* Include energy consumption of the material handling systems and material handling procedures when making comparisons or preparing economic justifications.

9. *Ecology Principle:* Use material handling equipment and procedures that minimize adverse effects on the environment.

10. *Mechanization Principle:* Mechanize the handling process where feasible to increase efficiency and economy in the handling of materials.

11. *Flexibility Principle:* Use methods and equipment that can perform a variety of tasks under a variety of operating conditions.

12. *Simplification Principle:* Simplify handling by eliminating, reducing, or combining unnecessary movements and/or equipment.

13. *Gravity Principle:* Utilize gravity to move material wherever possible, while respecting limitations concerning safety, product damage, and loss.

14. *Safety Principle:* Provide safe material handling equipment and methods that follow existing safety codes and regulations in addition to accrued experience.

15. *Computerization Principle:* Consider computerization in material handling and storage systems, when circumstances warrant, for improved material and information control.

16. *System Flow Principle:* Integrate data flow with physical material flow in handling and storage.

17. *Layout Principle:* Prepare an operation sequence and equipment layout for all viable system solutions, then select the alternative system which best integrates efficiency and effectiveness.

18. *Cost Principle:* Compare the economic justification of alternate solutions in equipment and methods on the basis of economic effectiveness as measured by expense per unit handled.

19. *Maintenance Principle:* Prepare a plan for preventive maintenance and scheduled repairs on all material handling equipment.

20. *Obsolescence Principle:* Prepare a long-range and economically sound policy for replacement of obsolete equipment and methods with special consideration to after-tax life cycle costs.

Chapter 1.) As it relates to designing material handling systems, the engineering design process consists of the following steps or phases:

1. Define the objectives and scope for the material handling system.
2. Analyze the requirements for moving, storing, protecting, and controlling material.
3. Generate alternative designs for meeting material handling system requirements.
4. Evaluate alternative material handling system designs.
5. Select the preferred design for moving, storing, protecting, and controlling material.
6. Implement the preferred design, including the selection of supplies, training of personnel, installation, debug and startup of equipment, and periodic audits of system performance.

To stimulate the development of alternative system designs, the "ideal systems approach" should be considered. As proposed by Nadler [27] and depicted in Figure 6.2, the ideal systems approach consists of the following phases:

1. AIM for the theoretical ideal system.
2. CONCEPTUALIZE the ultimate ideal system.
3. DESIGN the technologically workable ideal system.
4. INSTALL the recommended system.

The **theoretical ideal system** is a perfect system having zero cost, perfect quality, no safety hazards, no wasted space, and no management inefficiencies. The **ultimate ideal system** is a system that probably would be achievable at some point in the future, but is not achievable at the present time because of a lack of available technology. The **technologically workable ideal system** is a system for which the required technology is available; however, costs or other conditions may prevent some components from being installed now. The **recommended system** is a cost-effective system that will work now without obstacles to its successful implementation.

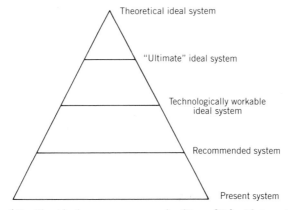

Theoretical ideal system

"Ultimate" ideal system

Technologically workable
ideal system

Recommended system

Present system

**Figure 6.2** The ideal systems approach. *(From [27] with permission.)*

Throughout the design process, a questioning attitude should prevail. The basic questions of why, what, where, when, how, who, and which should be asked constantly. In particular, the following questions should be addressed as a minimum:

1. Why
   a. Is handling required?
   b. Are the operations to be performed as they are?
   c. Are the operations to be performed in the given sequence?
   d. Is material received as it is?
   e. Is material shipped as it is?
   f. Is material packaged as it is?

2. What
   a. Is to be moved?
   b. Data are available and required?
   c. Alternatives are available?
   d. Are the benefits and disbenefits (costs) for each alternative?
   e. Is the planning horizon for the system?
   f. Should be mechanized/automated?
   g. Should be done manually?
   h. Shouldn't be done at all?
   i. Other firms have related problems?
   j. Criteria will be used to evaluate alternative designs?
   k. Exceptions can be anticipated?

3. Where
   a. Is material handling required?
   b. Do material handling problems exist?
   c. Should material handling equipment be used?
   d. Should material handling responsibility exist in the organization?
   e. Will future changes occur?
   f. Can operations be eliminated, combined, simplified?
   g. Can assistance be obtained?
   h. Should material be stored?

4. When
   a. Should material be moved?
   b. Should I automate?
   c. Should I consolidate?
   d. Should I eliminate?
   e. Should I expand (contract)?
   f. Should I consult vendors?
   g. Should a postaudit of the system be performed?

5.   How
   a.  Should material be moved?
   b.  Do I analyze the material handling problem?
   c.  Do I sell everyone involved?
   d.  Do I learn more about material handling?
   e.  Do I choose from among the alternatives available?
   f.  Do I measure material handling performance?
   g.  Should exceptions be accommodated?

6.   Who
   a.  Should be handling material?
   b.  Should be involved in designing the system?
   c.  Should be involved in evaluating the system?
   d.  Should be involved in installing the system?
   e.  Should be involved in auditing the system?
   f.  Should be invited to submit equipment quotes?
   g.  Has faced a similar problem in the past?

7.   Which
   a.  Operations are necessary?
   b.  Problems should be studied first?
   c.  Type equipment (if any) should be considered?
   d.  Materials should have real-time control?
   e.  Alternative is preferred?

As shown in Figure 6.3, liberal use of the question "Why?" is essential to separate what must be from what has been; asking what and why defines the correct materials to be handled; asking where, when, and why identifies the necessary moves to be performed; asking how, who, and why establishes the correct methods to be used; and asking which and why yields the preferred design. Factors to be considered in analyzing material handling problems include the types of materials, their physical

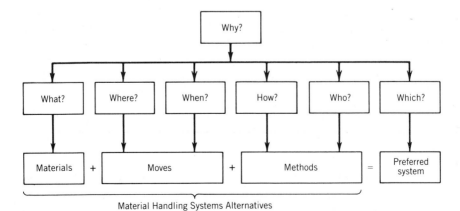

**Figure 6.3**   Material handling system equation.

characteristics, the quantities to be moved, the sources and destinations for each move, the frequencies or rates at which moves must be made, equipment alternatives, and the units to be handled, among others.

The combination of material characteristics and move or flow requirements is referred to as **material flow.** Hence, to develop material flow system requirements, one should focus on the material to be handled, stored, and controlled and the flow or throughput requirements for the system. Material flow is transformed to material handling by the *method* of handling, storing, and controlling the material.

Some frequently encountered reasons for considering changes in the way material is handled by an organization include:

**1.** Reduce costs.
**2.** Reduce damage.
**3.** Increase space and equipment utilization.
**4.** Increase throughput.
**5.** Increase productivity.
**6.** Improve working conditions.

While it is true that material handling improvements can result in the benefits listed above, it is also the case that a number of so-called "improvements" in the material handling system can result in the following disbenefits:

**1.** Increased capital requirements.
**2.** Decreased flexibility.
**3.** Decreased reliability, maintainability, and operability.

Hence, one must consider both the short-range and long-range effects of the alternatives considered.

An alternative method of representing the importance of the questioning attitude in designing material handling systems is given by the following expression:

$$\sum_{moves} [\text{Why} (\text{Where} + \text{What} + \text{When})]$$

The expression within brackets defines the method of performing each move within a facility. The moves are considered individually and without reference to other moves within a facility. The initial consideration for a particular move is "Why should this move be performed?" The multiplication by "Why" represents the initial consideration and work simplification approach to material handling. Each move should be evaluated by asking the following questions:

**1.** Can the move be eliminated?
**2.** Can the move be combined with another or within an in-transit operation?
**3.** Can the move be simplified?
**4.** Can the sequence of moves be changed to advantage?

Once it has been determined that a move will take place, the move must be studied to determine the best method of performing the move. Such study is represented by the "where + what + when" expression in parentheses. A chart that can be extremely useful when investigating each move is the **material handling planning chart** (*MHPC*), depicted in Figure 6.4.

## MATERIAL HANDLING PLANNING CHART

Company: A.R.C., Inc.  
Product: Air Speed Control Valve  
Prepared by: J.A.  
Date: _____  
Layout Alternative: 1  
Sheet: 1 of 8

| Step No. | O | T | S | I | Description | Oper. No. | Dept. | Cont. Type | Size | Wt. | Qty. Per Cont. | Freq. | Dist. | Method of Handling |
|---|---|---|---|---|---|---|---|---|---|---|---|---|---|---|
| 1 |  |  | X |  | Bar stock in storage (2200) |  | Stores |  |  |  |  |  |  |  |
| 2 |  | X |  |  | From Stores to Saw Dept. |  |  | LDDSE (FK.TRK) | 2.5" × 3.5 × 16" | 75 lb | 10 bars | 3 times daily | 16 ft | Fork lift |
| 3 |  |  | X |  | Store in Saw Department |  | Saw |  |  |  |  |  |  |  |
| 4 | X |  |  |  | Cut to length | 0101 | Saw |  |  |  |  |  |  |  |
| 5 |  | X |  |  | From Saw to Grinding |  |  | TOTE pan | 15" × 12" × 7" | 30 lb | 30 | Twice daily | 10 ft | Platform hand truck |
| 6 |  |  | X |  | Store in Grinding |  | Grinding |  |  |  |  |  |  |  |
| 7 | X |  |  |  | Grind to length | 0201 | Grinding |  |  |  |  |  |  |  |
| 8 |  | X |  |  |  |  |  | TOTE pan | 15" × 12" × 7" | 30 lb | 30 | Twice daily | 13 ft | Platform hand truck |
| 9 |  |  | X |  | Store in Deburring |  | Deburring |  |  |  |  |  |  |  |
| 10 | X |  |  |  | Deburr | 0301 | Deburring |  |  |  |  |  |  |  |
| 11 |  | X |  |  | From Deburring to Dr. Prs |  |  | TOTE pan | 15" × 12" × 7" | 30 lb | 30 | Twice daily | 16 ft | Platform hand truck |
| 12 |  |  | X |  | Store in Drill Press |  | Drill Press |  |  |  |  |  |  |  |
| 13 | X |  |  |  | Dr. (4) holes tap, ream, c'sk | 0401 | Drill Press |  |  |  |  |  |  |  |
| 14 |  | X |  |  | From Dr. Press to Tur. Lathe |  |  | TOTE pan | 15" × 12" × 7" | 30 lb | 30 | Twice daily | 33 ft | Platform hand truck |

**Figure 6.4** Material handling planning chart for an air flow regulator. Key: Operation—O, transportation—T, storage—S, inspection—I.

The first eight columns of the MHPC should be completed in the same manner as a flow process chart. The flow process chart is an expansion of the operation process chart in that it includes operations, inspections, transportations, storages, and delays. The consideration of a delay is omitted in the MHPC because a delay and a storage usually have the same facility requirements when planning a facility.

Each operation, transport, storage, and inspection should be listed on the MHPC. A transport should be recorded before and after most operations, storages, and inspections. All movements, even if performed by the machine operator, and storages, even if only for a very short time, should be recorded on the MHPC. Transports should be recorded as "From _____ To _____." These first eight columns answer most of the "where questions" concerning moves.

The ninth through twelfth columns of the MHPC answer the question "What is to be moved?" A final decision as to what is to be moved usually cannot be made at this point. Unit loads of tote pans, tote boxes, or pallet boxes, for example, may all be under consideration. While considering alternative unit loads, the process information will remain fixed but the data concerning what is to be moved will vary. The space provided in the heading of the MHPC labeled "Method of Handling" should be utilized to represent the alternative unit loads being considered.

Utilizing the market estimate, the route sheet, the capacity of the unit loads, and the frequency of a move, the "when" in the symbolic expression can be determined. Frequencies should be recorded in the thirteenth column of the MHPC.

The fourteenth column of the MHPC, the length of the move, cannot be determined prior to the layout's completion. Such information is required to specify fully the "where" of a move. While developing the final facilities plan, several alternative layouts will be developed and the information concerning length of the moves should be determined and recorded on the MHPC for each handling alternative. Once this information is available, all of the information required to determine the best method of performing a move has been determined and the appropriate material handling equipment may be selected for the move.

## 6.5  UNIT LOADS

Although each of the 20 principles of material handling can be applied in a number of situations to achieve economies in handling, one that deserves special attention is the **unit load principle.** The concept of a unit load is derived from the unit size principle; a unit load can be defined simply as the unit to be moved or handled at one time. In some cases the unit load is one item of production; in other situations the unit load is several cartons, each containing numerous items of production.

The unit load includes the container, carrier, or support that will be used to move materials. Unit loads consist of material in, on, or grouped together by something. The primary advantage of using unit loads is the capability of handling more items at a time and reducing the number of trips, handling costs, loading and unloading times, and product damage.

The widespread use of containerization in physical distribution emphasizes the importance of the unit load concept. The size of the unit load can range from a single carton to an intermodal container (e.g., piggyback trailer). Additionally, the integrity of the unit load can be maintained in a number of ways. As examples, tote boxes, cartons, pallets, and pallet boxes can be used to "contain" the unit load. Likewise, strapping, shrinkwrapping, and stretchwrapping can be used to enclose the unit load.

The determination of the size of the unit load, as well as the method of containing it, is influenced by a number of factors. As examples, the material to be unitized, the number of times the unit load is handled before it must be deunitized, the quantity of material to be handled, the environmental conditions to which the unit load is exposed, the susceptibility of the material to damage, the security aspects of the material being handled, the method of receiving, storing, shipping, and handling the unit load, and the other unit loads to be utilized within the system, among others, will all affect the manner in which the unit load is sized and contained.

In designing the unit load, Apple [1] suggests that the following steps be taken:

1. Determine the applicability of the unit load concept.
2. Select the type of unit load to be used.
3. Identify the most remote source of a potential unit load.
4. Establish the farthest practicable destination for the unit load.
5. Determine the size of the unit load.
6. Establish the configuration of the unit load.
7. Determine the method of building the unit load.

Unit loads can be lifted in a variety of ways. For example, a forklift truck is frequently used to move unit loads placed on pallets; hence, unit loads can be *lifted* by a support beneath the load. Additionally, below-hook lifters are used in conjunction with cranes and hoists to lift or *suspend* unit loads. A third method of picking up a unit load is by *squeezing* the load, for example, a clamp truck. A fourth approach is to insert the lifting element into the unit load; an example of such an approach is a ram used to lift steel coils. A number of variations of the four approaches exist; as an example, the support of a unit load on a film of air allows it to be pushed along manually.

The most common methods of unitizing a unit load are containers, platforms, sheets, and racks; additionally, a number of loads are inherently unitized. Tote boxes, cartons, bins, crates, baskets, and bags can be used as containers for the unit load. Skids and pallets are the most popular platforms. Cardboard, plywood, and polyethylene slipsheets are examples of sheets used to support a unit load; slipsheets are typically used in conjunction with a push/pull attachment on an industrial truck. Racks and specially designed buggies and carts are often used to define the unit load.

A popular method of building a unit load is to place one or more items on a pallet. As depicted in Figure 6.5, pallets can be designed in a variety of shapes and sizes. As a result, the pallet selection problem is rather complex. Among the factors to be considered in determining the style and size of the pallet are the following:

1. Size of the carrier in shipping and receiving.
2. Size and weight of items to be placed on the pallet.
3. Dimensions of space for storing loaded/unloaded pallets.
4. Equipment used to move loaded/unloaded pallets.
5. Slave pallet versus nonslave pallet considerations.

**6.** Cost, supply, and maintenance considerations.

**7.** Aisle widths, door sizes, and clear stacking heights.

In the case of full trailer or rail car shipments, freight rates may be based on weight and/or volume. Hence, depending on who is paying the freight cost, material may or may not be shipped on pallets. If pallets are used and "shipping cube" utilization is to be maximized, then the pallet design becomes quite important. The size of the items to be placed on the pallet will influence the size of the pallet. As shown in Figure 6.6, for a given pallet size there exists a variety of different stacking patterns that can be used in placing cartons on the pallet. Cube utilization per pallet varies,

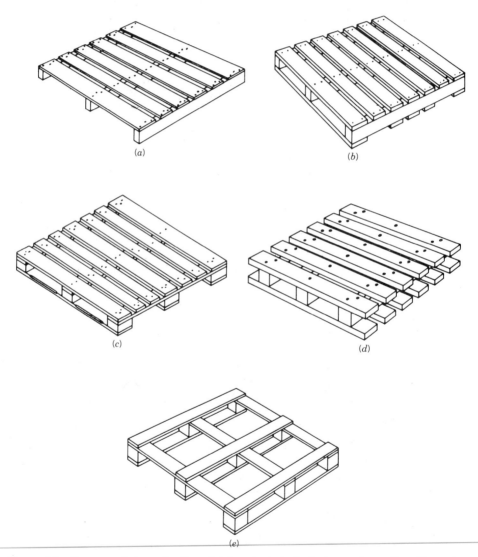

**Figure 6.5** Shapes and sizes of pallets. (*a*) Standard single-deck wooden pallet. (*b*) Double-faced nonreversible pallet for pallet truck handling. (*c*) Four-way block-leg pallet. (*d*) Double-wing-type (stevedore) pallet. (*e*) Three-board single-deck expendable shipping pallet. (*From [6] with permission.*)

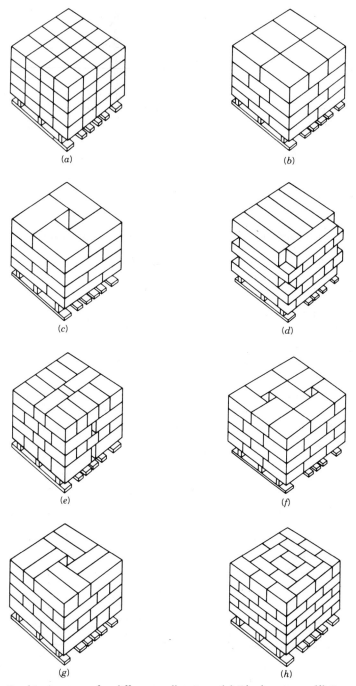

**Figure 6.6**  Stacking patterns for different pallet sizes. (*a*) Block pattern. (*b*) Row pattern. (*c*) Pinwheel pattern. (*d*) Honeycomb pattern. (*e*) Split-row pattern. (*f*) Split-pinwheel pattern. (*g*) Split-pinwheel pattern for narrow boxes. (*h*) Brick pattern. (*From [6] with permission.*)

depending on the pattern employed. Likewise, overall cube utilization is affected by pallet size determination.

Theoretically, for each carton size one could determine from among all possible combinations the pallet size and stacking pattern that maximizes cube utilization. However, both the number of such combinations and the impracticality of having numerous pallet sizes in the system precludes such an approach. Consequently, one generally restricts the number of alternative pallet sizes to two or three popular sizes and selects the size that yields a satisfactory utilization of cubic space for the most popular items to be placed on the pallet.

Among the most popular pallet sizes are the following:[1]

| | |
|---|---|
| 32 × 40 in. | 42 × 42 in. |
| 36 × 48 in. | 48 × 40 in. |
| 40 × 48 in. | 48 × 48 in. |

Others considered standard sizes by the American National Standards Institute are:

| | |
|---|---|
| 24 × 32 in. | 36 × 42 in. |
| 32 × 48 in. | 48 × 60 in. |
| 36 × 36 in. | 48 × 72 in. |

When the items being stored are quite heavy, weight considerations may be more important than volume considerations. Not only will weight considerations influence the structural design of the pallet, but they will also influence the capacities of the handling equipment and storage racks.

Pallet dimensions will also influence and be influenced by the dimensions of the space to be used for storing loaded and unloaded pallets. In the case of rack storage of a loaded pallet, the rack opening must be compatible with the cubic dimensions of the unit load (including the pallet and necessary clearances). When unit loads are to be stacked and stored without the use of storage racks, column spacings and layout of aisles will either influence or be influenced by pallet dimensions.

Unit load dimensions can influence (and be influenced by) the handling equipment to be used. For example, if tote boxes are to be transported by a roller conveyor, the dimensions of the conveyor and the tote box should be compatible.

An interesting design question is the following: "Should the material handling system be designed around the unit load or should the unit load be designed to fit the material handling system?" The obvious answer is "Neither! The unit load is an integral part of the material handling system and a *simultaneous* determination should be made." While the desire for a simultaneous determination is laudable, it remains the case that *sequential* decisions are generally employed in designing large-scale systems. As a result, in reality it is found that either one designs the unit load first or one designs the handling and storage system first. Of these two, the former seems preferable. Yet many find it difficult to accept the fact that the size of the tote box or the pallet should be among the first determinations made, rather than among the last.

In many cases, an existing system is to be improved rather than a new system designed. As a result, unit load specifications may be influenced by the existing build-

---

[1]It is conventional to let the first dimension correspond to the stringer length or depth of the pallet; the width of the pallet is given by the second number.

ing. Door widths, column spacings, aisle widths, turning radii of vehicles, and clear stacking heights are among the factors that will influence such a determination. There have been numerous instances of equipment purchased that could not be installed on schedule because it was too large to pass through the doors. Likewise, there have been instances of unit loads that were not sized for stacking in common carriers, passage in two-way aisles, and handling by the conveyor system.

# 6.6  MATERIAL HANDLING EQUIPMENT

For many, *material handling* is synonymous with *material handling equipment*. That is unfortunate because there is much more to material handling than the *method* employed. Furthermore, in some applications the best method of handling material may not require equipment. The scope of material handling was addressed in a previous section and will not be repeated here. Suffice it to say that the focus should be first on the *material,* second on the *move,* and third on the *method.* It is a natural reaction to encounter a problem situation and to immediately begin thinking of equipment to solve it. Such temptations should be resisted.

Having warned against a preoccupation with equipment in the early stages of facilities planning, it is now appropriate to consider the variety of equipment commonly used in satisfying the material handling requirements of a facility. The growth of technology in recent years has resulted in new generations of equipment being rapidly developed. Hence, anyone faced with a material handling equipment selection problem must constantly keep abreast of the alternatives available.

In this section we consider a number of types of material handling equipment. However, it must be noted that this treatment is not exhaustive. Furthermore, as with any rapidly developing area of technology, material handling equipment is undergoing continuous change. For this reason, a conscientious effort is required to keep up to date on new material handling equipment.

The four categories defined to facilitate this description of material handling equipment are (1) containers and unitizing equipment, (2) material transport equipment, (3) storage and retrieval equipment, and (4) automatic identification and communication equipment. This functional classification of material handling equipment is established to allow a description of alternative solutions to each of the four principal material handling challenges—containing (or protecting), moving, storing, and controlling material. A detailed outline of the equipment types in each category follows.

Containers and unitizing equipment initiate our description since they provide the common denominator in all material handling system designs. Material transport equipment—conveyors, industrial vehicles, and monorails, hoists, and cranes—follow as the foundation for all material handling system designs. Storage and retrieval equipment is used to store and retrieve large and small loads. Automatic identification and communication equipment including bar code readers, radio frequency data terminals, and light- and voice-directed communication devices are used with all of the other equipment types to coordinate and automate information handling requirements.

This section is not intended to provide an exhaustive description of material handling equipment. (More detailed descriptions are provided in the references at the end of the chapter.) It is intended to be a description of the functions, applications, benefits, and costs of the major classifications of material handling equipment.

I.   Containers and Unitizing Equipment
    A. Containers
        1. Pallets
        2. Skids and Skid Boxes
        3. Tote Pans
    B. Unitizers
        1. Stretchwrap
        2. Palletizers
II.  Material Transport Equipment
    A. Conveyors
        1. Chute Conveyor
        2. Belt Conveyor
            a. Flat Belt Conveyor
            b. Telescoping Belt Conveyor
            c. Troughed Belt Conveyor
            d. Magnetic Belt Conveyor
        3. Roller Conveyor
        4. Wheel Conveyor
        5. Slat Conveyor
        6. Chain Conveyor
        7. Tow Line Conveyor
        8. Trolley Conveyor
        9. Power and Free Conveyor
        10. Cart-on-Track Conveyor
        11. Sorting Conveyor
            a. Deflector
            b. Push Diverter
            c. Rake Puller
            d. Moving Slat Conveyor
            e. Pop-Up Skewed Wheels
            f. Pop-Up Belts and Chains
            g. Pop-Up Rollers
            h. Tilting Slat Conveyor
            i. Tilt Tray Sorter
            j. Cross Belt Sorter
            k. Bombardier Sorter

B. Industrial Vehicles
   1. Walking
      a. Hand Truck and Hand Cart
      b. Pallet Jack
      c. Walkie Stacker
   2. Riding
      a. Pallet Truck
      b. Platform Truck
      c. Tractor Trailer
      d. Counterbalanced Lift Truck
      e. Straddle Carrier
      f. Mobile Yard Crane
   3. Automated
      a. Automated Guided Vehicles
         i. Unit Load Carrier
         ii. Small Load Carrier
         iii. Towing Vehicle
         iv. Assembly Vehicle
         v. Storage/Retrieval Vehicle
      b. Automated Electrified Monorail
      c. Sorting Transfer Vehicle
C. Monorails, Hoists, and Cranes
   1. Monorail
   2. Hoist
   3. Cranes
      a. Jib Crane
      b. Bridge Crane
      c. Gantry Crane
      d. Tower Crane
      e. Stacker Crane

III.   Storage and Retrieval Equipment
   A. Unit Load Storage and Retrieval
      1. Unit Load Storage Equipment
         a. Block Stacking
         b. Pallet Stacking Frame
         c. Single-Deep Selective Rack
         d. Double-Deep Rack
         e. Drive-In Rack
         f. Drive-Thru Rack
         g. Pallet Flow Rack

        h. Push-Back Rack

        i. Mobile Rack

        j. Cantilever Rack

    2. Unit Load Retrieval Equipment

        a. Walkie Stacker

        b. Counterbalance Lift Truck

        c. Narrow Aisle Vehicles

           i. Straddle Truck

           ii. Straddle Reach Truck

           iii. Sideloader Truck

           iv. Turret Truck

           v. Hybrid Truck

        d. Automated Storage/Retrieval Machines

  B. Small Load Storage and Retrieval Equipment

    1. Operator-to-Stock—Storage Equipment

        a. Bin Shelving

        b. Modular Storage Drawers in Cabinets

        c. Carton Flow Rack

        d. Mezzanine

        e. Mobile Storage

    2. Operator-to-Stock—Retrieval Equipment

        a. Picking Cart

        b. Order Picker Truck

        c. Person-aboard Automated Storage/Retrieval Machine

        d. Robotic Retrieval

    3. Stock-to-Operator Equipment

        a. Carousels

           i. Horizontal Carousel

           ii. Vertical Carousel

           iii. Independent Rotating Rack

        b. Miniload Automated Storage and Retrieval Machine

        c. Vertical Lift Module

        d. Automatic Dispenser

IV. Automatic Identification and Communication Equipment

  A. Automatic Identification and Recognition

    1. Bar Coding

        a. Bar Codes

        b. Bar Code Readers

    2. Optical Character Recognition

    3. Radio Frequency Tag

4. Magnetic Stripe

5. Machine Vision

B. Automatic, Paperless Communication

1. Radio Frequency Data Terminal

2. Voice Headset

3. Light and Computer Aids

4. Smart Card

# Containers and Unitizing Equipment

Containers—pallets, skids and skid boxes, and tote pans—and unitizers—stretch-wrappers and palletizers—create a convenient unit load to facilitate and economize material handling and storage operations. These devices also protect and secure material.

## Containers

Containers are frequently used to facilitate the movement and storage of loose items. In grocery shopping we use boxes and bags to facilitate the handling tasks. In industrial applications, loose items are often placed in tote pans or in skid boxes; additionally, depending on the size and configuration of items to be moved or stored, they might be placed on a pallet or a skid to facilitate their movement and storage using lift trucks or other material handling equipment.

Consider, for example, the need to convey oats unitized in large burlap bags. While the bags might convey well using a belt conveyor, they would not do so using a roller conveyor. However, if the bags were placed on a pallet, then a roller conveyor could be used. Depending on how many degrees of freedom one has in designing the handling system, as well as the distribution and frequency of material handling movements to be performed, one might change the means of conveyance or the means of unitizing the load to be conveyed.

In general, decisions as to how material is to be moved, stored, or controlled are influenced by how the material is contained. The container decision, in essence, becomes the critical decision; it is the cornerstone for the material handling system. For this reason, if you are able to choose the container(s) to be used, then you should consider the impact of the decision(s) on subsequent choices of movement, storage, and control technologies.

**Pallets.** Pallets are by far the most common form of unitizing device and were described in detail in Section 6.5

**Skids and Skid Boxes.** Skids and skid boxes, illustrated in Figure 6.7, are used frequently in manufacturing plants. Often made of metal, they are quite rigid and are well-suited for unitizing a wide variety of items. Generally, skid boxes are too heavy to be lifted manually.

**Figure 6.7**    Skid boxes block stacked and accessed by a counterbalance lift truck.

*Tote Pans.*    Tote pans (see Figure 6.8) are used to unitize and protect loose items. Returnable tote pans have become a popular alternative to shipping in cardboard containers. When empty, tote pans should nest or collapse to insure high space utilization. Also, tote dimensions should coordinate with case and pallet dimensions to insure high utilization of material handling vehicles.

**Figure 6.8**    Tote pans used for parts storage in a microload ASRS. *(Courtesy of Litton UHS)*

***Unitizers.*** In addition to the use of containers (skid boxes and tote pans) and platforms (pallets and skids) to unitize a load, special equipment has been designed to facilitate the formation of a unit load. To motivate our consideration of unitizers, suppose you are faced with the need to move pallet loads of cartons. How do you ensure that the cartons placed on the pallet do not shift and become separated from the load? How could you automatically create the pallet load of cartons? How might you automatically remove cartons from the unit load?

To answer the first question, we consider the use of stretchwrap, shrinkwrap, strapping, and banding equipment. To answer the second and third questions, we consider the use of automatic palletizers and depalletizers.

***Stretchwrap.*** Shrinkwrap and stretchwrap equipment (Figure 6.9), as well as strapping and banding equipment, is used to unitize a load. Strapping can be performed using steel, fiber, and plastic materials; shrinkwrapping and stretchwrapping are performed using plastic film. The strapping process can be performed manually or mechanically and is best suited for compact loads. Shrinkwrapping is performed by placing a plastic bag over the load and applying heat and suction to enclose the load; shrinkwrapping is similar to the "blister packaging" process used for small articles. Stretchwrapping is performed by wrapping plastic film tightly around a load; multiple layers can be applied to obtain the same degree of weather protection provided by shrinkwrapping.

**Figure 6.9** Stretchwrap equipment. *(Courtesy of Infra Pak)*

**Figure 6.10(*a*)**  Palletizer induction station. *(Courtesy of Litton UHS)*

***Palletizers.***  Palletizers (Figure 6.10) and depalletizers are used for case goods handling as well as can and bottle handling. Palletizers receive products and place them on a pallet-according to prespecified patterns; depalletizers receive pallet loads and remove the product from the pallet automatically. A variety of sizes and styles are available for both palletizers and depalletizers.

## Material Transport Equipment

Material transport equipment—conveyors, industrial vehicles, monorails, hoists, and cranes—are distinguished from the other categories of material handling equipment by their primary function: material transport. Material transport equipment types are distinguished from one another by their

- degree of automation (walking, riding, and automated),
- flow pattern (continuous vs. intermittent, synchronous vs. nonsynchronous),
- flow path (fixed vs. variable),
- location (underground, in-floor, floor level, overhead), and
- throughput capacity.

Well-designed material transport systems achieve an effective match between material move requirements and these material move characteristics.

**Figure 6.10(*b*)**   Pattetizer. *(Courtesy of Litton UHS)*

## Conveyors

Conveyors are used when material is to be **moved frequently between specific points;** they are used to move material over a **fixed path.** Hence, there must exist a sufficient volume of movement to justify dedicating the equipment to the handling task. Depending on the materials to be handled and the move to be performed, a variety of conveyors can be used.

Conveyors can be categorized in several ways. For example, the type of product being handled (**bulk** or **unit**)[2] and the location of the conveyor (**overhead** or **floor**) have served as bases for classifying conveyors. Interestingly, such classification systems are not mutually exclusive. Specifically, a belt conveyor can be used for bulk and unit materials; likewise, a belt conveyor can be located overhead or on the floor.

Bulk materials, such as soybeans, grain, dry chemicals, and saw dust, might be conveyed using chute, belt, pneumatic, screw, bucket, or vibrating conveyors; unit materials such as castings, machined parts, and materials placed in tote boxes, on pallets, and in cartons might be conveyed using chute, belt, roller, wheel, slat, vibrating, pneumatic, trolley, or tow conveyors. Material can be transported on belt, roller, wheel, slat, vibrating, screw, pneumatic, and tow conveyors that are mounted either overhead or at floor level.

***Chute Conveyor.***   The chute conveyor (Figure 6.11) is one of the most inexpensive methods of conveying material. The chute conveyor is often used to link two powered

**Figure 6.11** Chute conveyors. (*a*) To provide accumulation in shipping areas. (*Courtesy of Lear Siegler/Rapistan Division*) (*b*) To convey items between floors. (*Courtesy of Standard Conveyor Company*)

conveyor lines; it is also used to provide accumulation in shipping areas; a spiral chute can be used to convey items between floors with a minimum amount of space required. While chute conveyors are economical, it is also difficult to control the items being conveyed by chutes; packages may tend to shift and turn so that jams and blockages occur.

*Belt Conveyors.* There are a wide variety of belt conveyors employed in modern material handling systems. The most popular types are flat belt conveyor, telescoping belt conveyor, trough belt conveyor, and magnetic belt conveyor.

*Flat Belt Conveyor.* A flat belt conveyor (Figure 6.12) is normally used for transporting light- and medium-weight loads between operations, departments, levels, and buildings. It is especially useful when an incline or decline is included in the conveyor path. Because of the friction between the belt and the load, the belt conveyor provides considerable control over the orientation and placement of the load; however, friction also prevents smooth accumulation, merging, and sorting on the belt. The belt is generally either **roller** or **slider bed** supported. If small and irregularly shaped items are being handled, then the slider bed would be used; otherwise, the roller support is usually more economical.

*Telescoping Belt Conveyor* A telescoping belt conveyor (see Figure 6.13) is a flat belt conveyor that operates on telescoping slider beds. They are popular at receiving and shipping docks where the conveyor is extended into inbound/outbound trailers for unloading/loading.

*Troughed Belt Conveyor* The troughed belt conveyor (Figure 6.14) is often used to transport bulk materials. When loaded, the belt conforms to the shape of the troughed rollers and idlers.

Figure 6.12(a)　Flat belt conveyor. *(Courtesy of Litton UHS)*

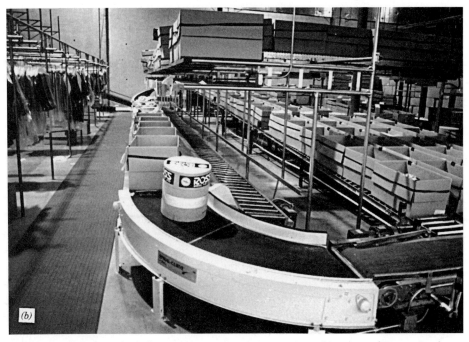

Figure 6.12(b)　Flat belt conveyor; power curve section. *(Courtesy of Portec, Inc.)*

**Figure 6.13**   Telescoping belt conveyor.

*Magnetic Belt Conveyor.*   A magnetic belt conveyor (Figure 6.15) consists of a steel belt and either a magnetic slider bed or a magnetic pulley. It is used to transport ferrous materials vertically, upside down, and around corners, as well as the separation of ferrous and nonferrous materials.

*Roller Conveyor.*   The roller conveyor (Figure 6.16) is a very popular type of material handling conveyor; it may be powered or nonpowered. The nonpowered roller conveyor is referred to as a "gravity" conveyor, as motion is achieved by inclining the roller section. Powered (or live) roller conveyors are generally either **belt** or **chain** driven. However, one manufacturer has employed a **revolving drive shaft** to power the rollers; rollers are connected individually to the drive shaft by an elastomeric belt. The roller conveyor is well suited for accumulating loads and merging/sorting operations. Because of the roller-surface, the materials being transported must have a rigid riding surface.

*Wheel Conveyor.*   The wheel or skate-wheel conveyor (Figure 6.17) is similar to the roller conveyor in design and function; a series of skate wheels are mounted on a shaft. Spacing of the wheels is dependent on the load being transported. Although wheel conveyors are generally more economical than the roller conveyor, they are limited to light-duty applications.

*Slat Conveyor.*   The slat conveyor (Figure 6.18) consists of discretely spaced slats connected to a chain. The slat conveyor functions much like a belt conveyor in that the unit being transported retains its position relative to the conveying surface. Because the conveying surface moves with the product, the orientation and placement of the load is controlled. Heavy loads with abrasive surfaces or loads that might damage a belt are typically conveyed using a slat conveyor. Additionally, bottling and canning plants use **flat chain** or slat conveyors because of wet conditions, temperature, and cleanliness requirements.

**Figure 6.14**  Troughed belt conveyors. *(Courtesy of Jervis B. Webb Co. and Lear Siegler/Rap-istan Division)*

**Figure 6.15**  Magnetic belt conveyor. *(Courtesy of Bunting Magnetics Co.)*

***Chain Conveyor.***  The chain conveyor (Figure 6.19) consists of one or more endless chains on which loads are carried directly. In transporting bulk materials, a chain is located in the bottom of a trough and pulls the material through the trough. Chain conveyors are often used to transport tote boxes and pallets, as only two or three chains typically are required to provide sufficient contact with the rigid support or container to effect movement.

***Tow Line Conveyor.***  The tow or towline conveyor (Figure 6.20) is used to provide power to wheeled carriers such as trucks, dollies, or carts that move along the floor.

**Figure 6.16(*a*)**  Roller conveyor.

**Figure 6.16(*b*)** Line shaft drive for a roller conveyor.

**Figure 6.17** Skate wheel conveyor.

**Figure 6.18** Slat conveyors. *(Courtesy of Fredenhagen, Inc. and Acco Babcock)*

Essentially, the tow conveyor provides power for fixed-path travel of carriers that have variable path capability. The towline can be located either overhead, flush with the floor, or in the floor. Towline systems often include selector-pin or pusher-dog arrangements to allow automatic switching between power lines or onto an unpowered spur line for accumulation. Tow conveyors are generally used when long distances and a high frequency of movement are involved.

***Trolley Conveyor.*** The trolley conveyor (Figure 6.21) consists of a series of trolleys supported from or within an overhead track. They are generally equally spaced in a closed loop path and are suspended from a chain. Specially designed carriers can be used to carry multiple units of product. They have been used extensively in processing, assembly, packaging, and storage operations.

***Power and Free Conveyor.*** The power-and-free conveyor (Figure 6.22) is similar to the trolley conveyor in that discretely spaced carriers are transported by an overhead

**Figure 6.19(*a*)**   Chain conveyor. *(Courtesy of Jervis B. Webb Co.)*

**Figure 6.19(*b*)**   Chain conveyor.

**Figure 6.20(*a*)**  In-floor tow line conveyor.

**Figure 6.20(*b*)**  Overhead tow line conveyor.

**Figure 6.21** Trolley conveyors. *(Courtesy of Lear Siegler/Rapistan Division and Acco Babcock)*

**Figure 6.22(a)**  Power and free conveyor. *(Courtesy of Jervis B. Webb Co.)*

**Figure 6.22(b)**  Inverted power and free conveyor. *(Courtesy of Jervis B. Webb Co.)*

**Figure 6.23** Cart-on-track conveyor. *(Courtesy of SI Handling Systems, Inc.)*

chain. However, the power-and-free conveyor utilizes two tracks: one powered and the other unpowered or free. Carriers are suspended from a set of trolleys that run on the free track. Linkage between the power chain and the trolleys on the free track is achieved using a "dog." The dogs on the power chain mate with similar extensions on the carrier trolleys and push the carriers forward on the free track. The advantage of the power-and-free design is that carriers can be disengaged from the power chain and accumulated or switched onto spurs.

*Cart-on-Track Conveyor.*   The cart-on-track conveyor (Figure 6.23) is used to transport a cart along a track. Employing the principle of a screw, a cart is transported by a rotating tube. Connected to the cart is a drive wheel that rests on the tube; the speed of the cart is controlled by varying the angle of contact between the drive wheel and the tube. The basic elements of the conveyor are the rotating tube, track, and cart. The carts are independently controlled to allow multiple carts to be located on the tube. Carts can accumulate on the tube because they will be stationary when the drive wheel is parallel to the tube.

*Sorting Conveyor*   A sorting conveyor (see Figure 6.24) is used to assemble material (i.e., cases, items, totes, garments) with a similar characteristic (i.e., destination, customer, store) by correctly identifying the similar merchandise and transporting it to the same location. The key performance attributes of a variety of popular sorting conveyors are summarized in Table 6.3. The major distinguishing features are the

**Figure 6.24**   Hand induction to a sorting conveyor. *(Courtesy of Litton UHS)*

**Table 6.3  Sorting Systems: Summary Comparison**

| | Manual | Push Divert | Pull Divert | Slat | Pop-Up Wheels | Pop-Up Rollers | Tilt-Tray |
|---|---|---|---|---|---|---|---|
| Max. Sorts per Minute | 15–25 | 30–35 | 30–40 | 50–150 | 65–150 | 15–20 | 65–300 |
| Load Range | 1–75 lb. | 1–75 lb. | 10–100 lb. | 1–200 lb. | 3–300 lb. | 10–200 lb. | 1–300 lb. |
| Min. Distance between Spurs | Touching | 5'–7' | 2'–3' | 4'–5' | 4'–5' | Almost Touching | 1' |
| Diverter Impact on Load | Gentle | Medium | Medium to Rough | Gentle | Gentle | Gentle | Medium to Rough |
| Initial Cost | Lowest | Medium | High | High | High | Low to Medium | High |
| Maintenance Cost | Lowest | Low | Medium | Medium to High | Medium | Low | Medium to High |

method used to divert items into accumulation lanes and the resulting sortation capacity.

*Deflector*  A deflector (see Figure 6.25) is a stationary or moveable arm which deflects product flow across a belt or roller conveyor to the desired location. A deflector is necessarily in position before the item to be sorted reaches the discharge point. Stationary arms remain in a fixed position and represent a barrier to items coming in contact with them. With the stationary arm deflector, all items are deflected in the same direction. Movable arm or pivoted paddle deflectors are impacted by the item to be sorted in the same manner as the stationary arm deflector. However, the element of motion has been added. With the movable arm deflector (i.e., the paddle), items are selectively diverted. Pivoted deflectors may be equipped with a belt conveyor flush with the surface of the deflector (a power face) to speed or control the divert. Paddle deflecting systems are sometimes referred to as steel belt sorters since at one time steel belts were used to reduce the friction encountered in diverting products across the conveyor. Deflectors can support medium (1200 to 2000 cartons per hour) throughput for up to 75-lb loads.

*Push Diverter*  A push diverter (see Figure 6.26) is similar to a deflector in that it does not contact the conveying surface but sweeps across to push the product off the opposite side. Push diverters are mounted beside (air or electric powered) or above the conveying surface (paddle pushers) and are able to move items faster and with greater control than a deflector. Overhead push diverters are capable of moving products to either side of the conveying surface whereas side-mounted diverters move conveyed items in one direction only to the side opposite that on which they are mounted. Push diverters have a capacity of 3600 cases per hour for loads up to 100 lb.

*Rake Puller*  A rake puller is best applied when the items to be sorted are heavy and durable. Rake puller tines fit into slots between powered or nonpowered roller conveyors. Upon command, a positioning stop device and the tines pop up

Figure 6.25   Deflector.

**Figure 6.26** Push diverter.

**Figure 6.27** Moving slat sorter. *(Courtesy of Litton UHS)*

from beneath the roller conveyor surface to stop the carton. The tines pull the carton across the conveyor, and then drop below the roller surface for a noninterference return to the starting position. During the return stroke, the next carton can be moving into position.

*Moving Slat Conveyor*   The moving slat conveyor (see Figure 6.27) is differentiated from the other sorting conveyors by the fact that the divert takes place in-line along the roller conveyor.

*Pop-Up Skewed Wheels*   Pop-up skewed wheels (see Figure 6.28) are capable of sorting flat-bottomed items. The skewed wheel device pops up between the rollers of a powered roller conveyor or between belt conveyor segments and directs sorted items onto a powered takeaway lane. Rates of between 5,000 and 6,000 cases per hour can be achieved.

*Pop-Up Belts and Chains*   Pop-up belt and chain sort devices (see Figure 6.29) are similar to pop-up skewed wheels in that they rise from between the rollers of a

**Figure 6.28**   Pop-up skewed skatewheels. *(Courtesy of Litton UHS)*

**Figure 6.29**   Pop-up chain. *(Courtesy of Litton UHS)*

powered roller conveyor to alter product flow. Belt and chain sortation devices are capable of handling heavier items than the wheeled devices.

*Pop-Up Rollers*   Pop-up rollers (see Figure 6.30) rise up between the chains or rollers of a chain or roller conveyor to alter the flow of product. Pop-up rollers provide relatively inexpensive means for sorting heavy loads at rates of 1000 to 1200 cases per hour.

*Tilting Slat Conveyor*   In a tilting slat conveyor, the product occupies the number of slats required to contain its length. The sort is executed by tilting the occupied slats. Hence, tilting slat sorters are best applied when a wide variety of product lengths will be handled. The tilting slat is capable of tilting in either direction. Slats may be arranged in a continuous over-and-under configuration.

*Tilt Tray Sorter*   Continuous chains of tilting trays are used to sort a wide variety of lightweight merchandise on a tilt tray sorter (see Figure 6.31). The trays may be fed manually, or by one of the many types of induction devices available. Tilt tray systems can sort to either side of the sorter. Tilt tray sorters do not discriminate for the shape of the product being sorted. Bags, boxes, envelopes, documents, software, etc. can all be accommodated. The tilt tray sorter is not appropriate for long items. The capacity of the tilt tray sorter, expressed in pieces or sorts per hour, is expressed as the ratio of the sorter speed to the pitch or length of an individual tray. Rates of 10,000 to 15,000 items per hour can be achieved.

*Cross-Belt Sorter*   A cross-belt sorter (see Figures 6.32 and 6.33) is so named because each item rests on a carrier equipped with a separate powered section of

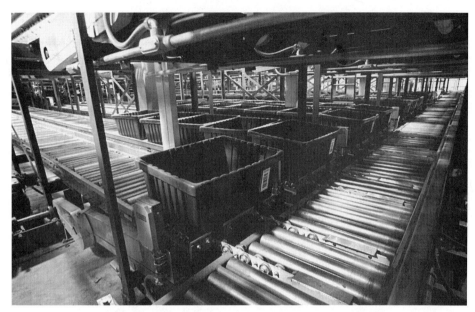

**Figure 6.30**  Pop-up roller. *(Courtesy of Litton UHS)*

**Figure 6.31(*a*)**  Tilt tray sorter. *(Courtesy of Litton UHS)*

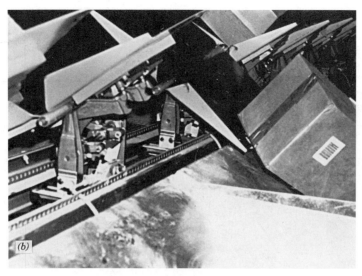

**Figure 6.31(b)**   Tilt tray sorter.

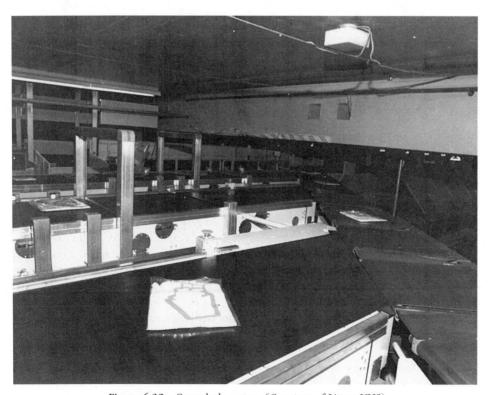

**Figure 6.32**   Cross-belt sorter. *(Courtesy of Litton UHS)*

**Figure 6.33** Accumulation lanes at the end of a cross-belt sorter. *(Courtesy of Litton UHS)*

belt conveyor which operates orthogonal to the direction of material transport. Hence, the sorting capacity is enhanced and the width of the accumulating chutes can be reduced.

***Bombardier Sorter*** A bombardier sorter is so-named because items are dropped down through swinging doors much like bombs are dropped through the belly of an airplane.

## Industrial Vehicles

Industrial vehicles represents a versatile method of performing material handling; it is referred to as *variable path equipment*. Industrial vehicles are generally used when movement is either intermittent or over long distances. Apple [1] notes that conveyors are used when the primary function is *conveying,* cranes and hoists are used when *transferring* is the primary function, and industrial trucks are used when the primary function is *maneuvering* or *transporting*.

Three categories of industrial vehicles are defined for this description—walking, riding, and automated.

***Walking Industrial Vehicles.*** There are three major classes of walking industrial vehicles: (1) hand trucks and hand carts, (2) pallet jacks, and (3) and walkie stackers. Their popularity stems from their simplicity and low price.

***Hand Truck and Hand Cart***   The hand truck or cart (Figure 6.34) is one of the simplest and most inexpensive types of material handling equipment. Used for small loads and short distances, the hand truck is a versatile method of moving material manually. A host of carts have been designed to facilitate manual movement of material, as well as the movement of material by conveyor and industrial vehicles.

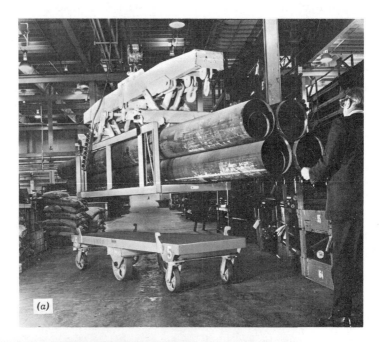

**Figure 6.34**   Hand truck/cart. *(Courtesy of Hamilton Caster & Mfg. Co.)*

***Pallet Jack***   The pallet jack (Figure 6.35) is used to lift, maneuver, and transport a pallet load of material short distances. The pallet jack can be either manual or battery powered for both lifting and transporting. The lifting capability is typically from 6 in. to 10 in.

***Walkie Stacker***   The walkie stacker (Figure 6.36) extends the lifting capability of the pallet jack to allow unit loads to be stacked or placed in storage racks. The walkie stacker can have either a straddle or reach design. The straddle type straddles the load with its outriggers; the reach type uses a pantograph or scissors device to allow the load to be retrieved from and placed in storage.

A walkie stacker allows a pallet to be lifted, stacked, and transported short distances. The operator steers from a walking position behind the vehicle. In a situation where there is low throughput, short travel distances, and low vertical storage height, and a low cost solution is desired, the walkie stacker may be appropriate.

***Riding Industrial Vehicles.***   Riding industrial vehicles allow the vehicle operator to ride to, from, and between locations. Hence, they are typically used for longer moves than walking vehicles. Riding vehicles also offer additional weight and storage height capacity.

***Pallet Truck.***   The pallet truck (Figure 6.37) extends the transporting capability of the pallet jack by allowing the operator either to ride or walk. The pallet truck is used when the distance to be traveled precludes walking.

***Platform Truck***   The platform truck (Figure 6.38) is a version of the industrial truck. Instead of having forks for lifting and supporting the pallets being transported, the platform truck provides a platform for supporting the load. It does not have lifting capability and is used for transporting; consequently, an alternative method of loading and unloading must be provided.

**Figure 6.35**   Pallet jack. *(Courtesy of Blue Giant Equipment Corporation)*

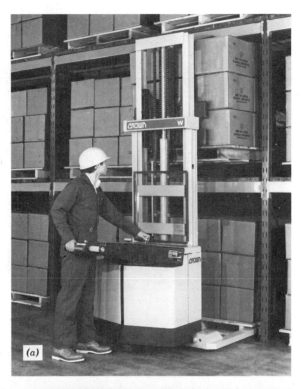

**Figure 6.36**   Walkie stacker. *(Courtesy of Crown Lift Trucks and Blue Giant Equipment Corporation)*

**Figure 6.36** (continued)

*Tractor Trailer* The tractor-trailer combination (Figure 6.39) extends the transporting capability of the hand truck by providing a powered, rider-type vehicle to pull a train of connected trailers. A wide variety of tractor-trailers are available, including walkie/rider and remote control alternatives.

*Counterbalanced Lift Truck* The "workhorse" of materials handling is the counterbalanced lift truck (Figure 6.40). Although it is usually referred to as a fork truck, not all designs use forks for lifting, for example, ram attachments and platform designs. As with most of the industrial trucks, counterbalanced lift truck can be either

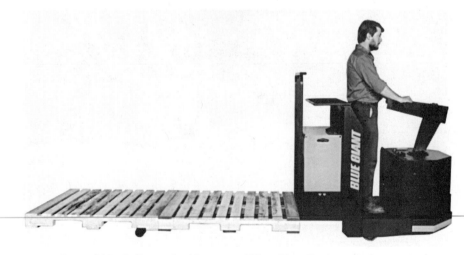

**Figure 6.37** Pallet truck. *(Courtesy of Blue Giant Equipment Corporation)*

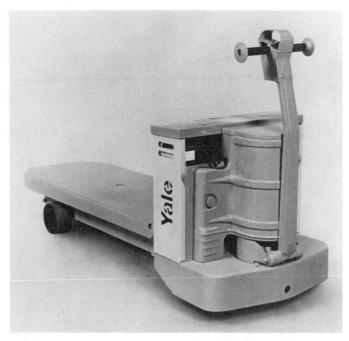

**Figure 6.38**   Platform truck. *(Courtesy of Yale Industrial Trucks)*

**Figure 6.39**   Tractor trailer. *(Courtesy of Litton UHS)*

**Figure 6.40** Counterbalanced lift trucks. *(Courtesy of Yale Industrial Trucks)*

**Figure 6.40** (continued)

battery powered (BP) or powered by an internal combustion engine (ICE); the latter can use either gasoline, propane, or diesel fuel. The tires used on the industrial truck can be either cushion (CT) for indoor operation or pneumatic (PT) for outdoor operation. The load-carrying capability of the counterbalanced lift truck can range from 1000 to over 100,000 lb.

As the name implies, counterbalance lift trucks employ a heavy counter balance over the rear wheels to achieve lift weight capacities of up to 100,000 lbs and lift height capacities between 25 and 30 ft. Counterbalance lift trucks may be gas or battery powered. A counterbalanced truck may not be used to store double deep.

As selective pallet rack is the benchmark pallet storage mode, the counterbalanced truck may be considered the benchmark storage/retrieval vehicle. When it is desirable to use the same vehicle for loading and unloading trucks and storing and retrieving loads, the counterbalanced truck is the logical choice. For use in block stacking, drive-in and drive-thru rack and pallet stacking frames, the operating aisles normally provided are suitable for counterbalanced trucks. Since counterbalance trucks must turn within a storage aisle to retrieve a pallet load, the aisle width required (10–13 ft) to operate is wider than required for some other lift truck alternatives. The relatively low cost and flexibility of counterbalance trucks are their main advantage.

Besides forks, other attachments may be used to lift unique load configurations on a vertical mast. A variety of lift truck attachments are described and illustrated as follows.

Industrial truck attachments are available for a wide variety of applications; indeed, the design of industrial truck attachments seems limited only by the ingenuity and creativity of the design engineer. Some of the more popular attachments are shown in Figures 6.41 through 6.44. The primary function of an industrial truck attachment is either to save time, conserve space, reduce product damage, save labor, or eliminate equipment. The sideshifter (Figure 6.41) permits the forks to be shifted from side to side so that the operator can pick up or spot a pallet without having to reposition the truck; hence, very slow and precise positioning of the truck is not necessary. The load push/pull attachment (Figure 6.42) allows a slipsheet to be substituted for the pallet. The cinder block attachment eliminates the need for a pallet, saves space, and reduces the weight of a load. Other attachments that eliminate the need for a pallet include the carton clamp (Figure 6.43), the paper roll clamp (Figure 6.44), and the ram attachment, among others.

*Straddle Carrier*   The straddle carrier (Figure 6.45), carries the load underneath the driver. The truck straddles a load, picks it up, and transports it to the desired location. Used primarily outdoors for long, bulky loads, the straddle carrier can handle up to 120,000 lb and can accommodate intermodal containers.

*Mobile Yard Crane*   The mobile crane (Figure 6.46) illustrates the relationship between crane operations and industrial vehicles. Numerous varieties of mobile cranes exist as well as crane-type attachments for industrial vehicles.

*Automated Industrial Vehicles.*   The category of automated industrial vehicles is distinguished from the other industrial vehicles by the elimination of human intervention from powering and guiding the movement of the vehicle. Instead, vehicles are automatically guided by electrified wires buried in the floor, magnetic tape lined

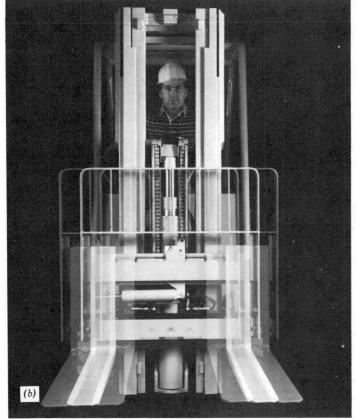

**Figure 6.41**   Sideshifters. *(Courtesy of Yale Industrial Trucks and Cascade Corporation)*

**Figure 6.42** Load push/pull attachment. *(Courtesy of Clark Equipment Company and Cascade Corporation)*

**Figure 6.43**  Carton clamps.  *(Courtesy of Cascade Corporaton)*

**Figure 6.44**   Paper roll clamps. *(Courtesy of Cascade Corporation)*

**Figure 6.45** Straddle carrier. *(Courtesy of Hyster Corporation)*

**Figure 6.46** Mobile Yard crane. *(Courtesy of Hyster Corporation)*

along the floor, rails mounted in the ceiling, cameras mounted on the vehicle, or inertial guidance systems. The types of vehicles included in this category are all types of automated guided vehicles (AGVs), automated electrified monorails (AEMs), and sorting transfer vehicles (STVs).

***Automated Guided Vehicles.***  An automated guided vehicle, or AGV, is essentially a driverless industrial truck. It is a steerable, wheeled vehicle, driven by electric motors using storage batteries, and it follows a predefined path along an aisle. AGVs may be designed to operate as a tractor, pulling one or more carts, or may be unit load carriers. Figure 6.47 illustrates a unit load AGV, which is the most common type in manufacturing and distribution.

The path followed by an AGV may be a simple loop or a complex network, and there may be many designated load/unload stations along the path. The vehicle incorporates a path-following system, typically electromagnetic, although some optical systems are in use.

With an electromagnetic path-following system, a guide wire that carries a radio frequency (RF) signal is buried in the floor. The vehicle employs two antennae, so that the guide wire can be bracketed. Changes in the strength of the received signal are used to determine the control signals for the steering motors so that the guide wire is followed accurately. When it is necessary for the vehicle to switch from one guide wire to another (e.g., at an intersection), two different frequencies can be used, with the vehicle being instructed to switch from one frequency to the other. Obviously, a significant level of both analog and digital electronic technology is incorporated into the vehicle itself. In addition, the vehicle routing and dispatching system will not only employ programmable controllers, but minicomputers or even mainframe computers.

The electromagnetic path-following system may also support communication between a host control computer and the individual vehicles. In systems with a number of vehicles, the host computer may be responsible for both the routing and dispatching of the vehicles and collision avoidance. A common method for collision avoidance is zone blocking, in which the path is partitioned into zones, and a vehicle is never allowed to enter a zone already occupied by another vehicle. This type of collision avoidance involves a high degree of active, host computer-directed control.

Optical path-following systems use an emitted light source and track the reflection from a special chemical stripe painted on the floor. In a similar fashion, codes can be painted on the floor to indicate to the vehicle that it should stop for a load/unload station. Simple implementations will require all actions of the vehicle to be preprogrammed, for example, stop and look for a load, stop and unload, or proceed to the next station. The vehicles typically will not be under the control of a host computer and must therefore employ some method for collision avoidance other than zone blocking. Proximity sensors on the vehicles permit several vehicles to share a loop without colliding.

The state of the art in AGVs is advancing rapidly. There are now "smart" vehicles that can navigate for short distances without an electromagnetic or optical path. Similarly, vehicles are being equipped with sufficient on-board computing capability to manage some of the routing control and dispatching functions. Current developments in the field are leading to path-free, or "autonomous" vehicles, which do not require a fixed path and are capable of "intelligent" behavior. Much of the work currently

**Figure 6.47(a)** Unit load AGV. *(Courtesy of Jervis B. Webb Co.)*

**Figure 6.47(b)** Unit load AGV. *(Courtesy of Harnischfeger Engineers, Inc.)*

**Figure 6.47(c)**   Unit load AGV. *(Courtesy of Mentor)*

**Figure 6.47(d)**   Unit load AGV. *(Courtesy of Elwell-Parker)*

under way in the AGV field is driven by the availability of small, powerful, but relatively inexpensive microcomputers.

The classifications of automated guided vehicles include

- unit load carriers (Figure 6.47) designed to carry a single unit load (e.g., pallet, roll),
- small load carriers (Figure 6.48) designed to carry a single small load (e.g., tote pan, case),
- towing vehicles (Figure 6.49) designed to tow load carrying carts,
- assembly vehicles (Figure 6.50) designed to transport products through an assembly process, and
- storage/retrieval vehicles (Figure 6.51) designed to lift and lower loads up to 25 ft high.

The key performance attributes of each system type are summarized in Table 6.4.

***Automated Electrified Monorails.***   Self-powered monorails (SPMs), (Figure 6.52) also referred to as automated electrified monorails (AEMs) resemble overhead power-and-free conveyors in much the same way that AGVs resemble towline conveyors. An SPM moves along an overhead monorail, but rather than being driven by a moving chain, it is driven by its own electric motor. The SPM provides many of the benefits of both overhead power-and-free conveyors and AGVs. It is overhead, so it does not create obstructions on the factory floor, and self-powered and program-

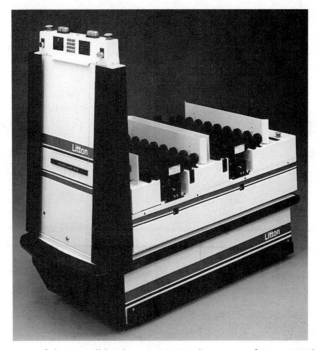

**Figure 6.48**   Small load carrier AGV. *(Courtesy of Litton UHS)*

**Figure 6.49** Towing AGV. *(Courtesy of Jervis B. Webb Co.)*

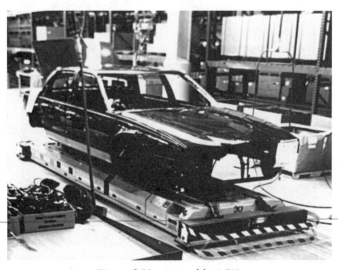

**Figure 6.50** Assembly AGV.

**Figure 6.51**   Storage/retrieval AGV.

mable, so it has a great deal of route flexibility. In contrast to the AGV, it may use a bus bar to provide the electric current required by the drive motors, rather than carrying a storage battery.

Another distinctive feature of SPMs is that they require no sophisticated path-following system, because the path must conform to the monorail structure. On the other hand, path selection, e.g., at an intersection or branch point, requires coordination between a vehicle-tracking function and a physical track-switching function. With an AGV using wire guidance, the path selection may be accomplished by the vehicle itself, simply by switching frequencies.

SPMs have been designed to interface with a wide variety of production equipment. The carriers can be switched from one monorail line to another and can even be raised or lowered between two different levels. As with AGVs, the SPM employs a high degree of computer control.

*Sorting Transfer Vehicles.*   Sorting transfer vehicles (Figure 6.53) automatically load and unload large and small unit loads at pickup and deposit points located around a conveyor loop. The vehicle's rapid and precise acceleration and decelera-

**Table 6.4**   *AGV System Comparisons [38]*

| System attribute | Unit Load | Towing | S/R | Light Load | Assembly |
|---|---|---|---|---|---|
| Weight capacity | 4000 lb | 50,000 lb | 6000 lb | 1000 lb | 2000 lb |
| Travel speed | 2 mph | 3 mph | 2 mph | 1 mph | 1 mph |
| Vehicle cost | $40,000 | $30,000 | $70,000 | $20,000 | $20,000 |
| System cost per vehicle | $80,000 | $70,000 | $100,000 | $50,000 | $40,000 |

*Notes.* Installation is an additional 10%. P/D stands cost between $1,000 and $6,000. Guide path costs between $20 and $50/ft.

**Figure 6.52(*a*)**   Self-propelled monorail carrier. *(Courtesy of Litton UHS)*

tion, up to and down from, 270 fpm (80 m/m) top speed, creates a high throughput unit load handling system.

## *Monorails, Hoists, and Cranes.*

Monorails and cranes are generally used to *transfer material from one point to another in the same general area;* hoists are used to *facilitate the positioning, lifting,* and *transferring of material within a small area.* Monorails, hoists, and cranes generally provide more flexibility in the movement path than do conveyors; however, they do not have the degree of flexibility provided by variable path equipment, such as industrial trucks.

Typically, the loads handled by monorails, hoists, and cranes are much more varied than those handled by a conveyor. Also, the movement of materials is generally much more intermittent when using monorails, hoists, and cranes than when using conveyors.

***Monorail.***   A monorail (Figure 6.54) consists of an overhead track on which a carrying device rides. The carrier can be either top running or underhung. The carrying

Figure 6.52(*b*)　Self-propelled monorail carrier. *(Courtesy of Litton UHS)*

Figure 6.53　Sorting transfer vehicle. *(Courtesy of Daifuku)*

**Figure 6.54**   Monorails. *(Courtesy of Acco Babcock and Mannesmann Demag)*

device can be either powered or unpowered. If powered, then the carrying device itself is generally powered electrically or pneumatically. Additionally, intelligent carriers have been developed through the application of microprocessors. The monorail functions like a trolley conveyor, except the carrying devices operate independently and the track need not be a closed loop.

*Hoist.*   A hoist (Figure 6.55) is a lifting device that is frequently attached to a monorail or crane. The hoist may be manually, electrically, or pneumatically powered.

### Cranes

*Jib Crane.*   A jib crane (Figure 6.56) has the appearance of an arm that extends over a work area. A hoist is attached to the arm to provide lifting capability. The arm may be mounted on a wall or attached to a floor-mounted support. The arm can rotate and the hoist can move along the arm to achieve a wide range of coverage.

*Bridge Crane*   As the name implies, a bridge crane (Figure 6.57) resembles a bridge that spans a work area. The bridge is mounted on tracks so that a wide area can be covered. The bridge crane and hoist combination can provide three-dimensional coverage of a department. The bridge can be either top riding or underhung. The top-riding crane can accommodate heavier loads. However, the underhung crane is considered to be more versatile than the top-riding crane because of its ability to transfer loads and interface with monorail systems.

*Gantry Crane*   The gantry crane (Figure 6.58) spans a work area in a manner similar to the bridge crane; however, it is generally floor supported rather than over-

**Figure 6.55**   Hoist. *(Courtesy of Acco Babcock)*

**Figure 6.56**   Jib crane. *(Courtesy of Acco Babcock)*

head supported on one or both ends of the spanning section. The support can either be fixed in position or travel on runways.

*Tower Crane*   A tower crane (Figure 6.59) is most often seen on construction sites, but is also usable for ongoing material handling operations. The tower crane consists of a single upright which may be fixed or on a track having a cantilever boom. A hoist operates on the boom which may be rotated 360° about the upright.

*Stacker Crane*   The stacker crane (Figure 6.60) is similar to a bridge crane. Instead of using a hoist, a mast is supported by the bridge; the mast is equipped with forks or a platform, which are used to lift unit loads. The stacker crane is often used for storing and retrieving unit loads in storage racks; however, it can also be used without storage racks when materials are stackable. The stacker crane can be controlled either remotely or with an operator on board in a cab attached to the mast. The stacker crane can operate in multiple aisles. They are often used in high-rise applications, with storage racks more than 50 ft high.

**Figure 6.57** Bridge cranes. *(Courtesy of Mannesmann Demag)*

## Storage and Retrieval Equipment

Storage and retrieval equipment is distinguished from material transport equipment by its primary function—to house material for staging or building inventory and to retrieve material for use. In some cases, the retrieval equipment is one of the material transport systems described above. In other cases a new equipment type will be introduced.

**Figure 6.58**   Gantry cranes. *(Courtesy of Harnischfeger Corporation)*

**Figure 6.58**   (continued)

**Figure 6.59**   Tower cranes. *(Courtesy of Potain/America, Inc.)*

**Figure 6.60**   Stacker crane. *(Courtesy of Harnischfeger Engineers, Inc.)*

For discussion purposes, two major classifications of storage and retrieval equipment are defined—unit load and small load systems. Unit load systems typically house large loads such as full pallets, large boxes, and/or large rolls of material. Their applications include housing inventory prior to full unit load shipping, housing reserve storage for replenishment of forward picking areas, and/or housing inventory for partial unit load picking.

Small load storage and retrieval systems typically house small inventory quantities of one or more items in a storage location. The maximum storage capacity for each storage location is typically less than 500 lbs.

## Unit Load Storage and Retrieval

For discussion purposes, unit load storage and retrieval systems are further subdivided into unit load storage systems which house unit loads and unit load retrieval systems which allow access to unit loads for retrieval.

Table 6.5  *Summary Comparison of Unit Load Storage Systems*

| | Floor Storage | Stacking Frames | Single-Deep | Double-Deep | Drive-In Rack | Drive-Thru | Flow Rack | Push-Back | Mobile Rack | Cantilever |
|---|---|---|---|---|---|---|---|---|---|---|
| Cost per position | n/a | $50 | $40 | $50 | $65 | $65 | $200 | $150 | $250 | |
| Potential storage density | A | B | D | C | B | B | B | B | A | B |
| Load access | F | F | A | C | B | B | B | A | F | A |
| Throughput capacity | B | D | B | C | C | C | A | C | F | C |
| Inventory and location control | F | F | A | C | D | D | C | C | D | B |
| FIFO maintenance | F | F | A | C | D | D | A | C | C | A |
| Ability to house variable load sizes | A | D | C | C | D | D | F | C | C | B |
| Ease of installation | A | A | C | C | C | C | F | C | F | B |

***Unit Load Storage Equipment.*** Unit load storage equipment types are distinguished from one another by their rack configuration, lane depth capacity, stacking capacity, unit load access, and capital expense. Table 6.5 details the distinguishing features of each unit load storage equipment alternative.

***Block Stacking*** Block stacking (see Figure 6.61) refers to unit loads that are stacked on top of each other and stored on the floor in storage lanes (blocks) 2 to 10 loads deep. Depending on the weight and stability of the loads, the stacks may range from 2 loads high to a height determined by acceptable safe limits or by the building clear height. Block stacking is particularly effective when there are multiple pallets per SKU (stock keeping unit) and when inventory is turned in large increments, i.e., several loads of the same SKU are received or withdrawn at one time.

A phenomenon referred to as honeycombing occurs with block stacking as loads are removed from a storage lane. Since only one SKU can be effectively stored in a lane, empty pallet spaces are created, which cannot be utilized effectively until an entire lane is emptied. Therefore, in order to maintain high utilization of the available storage positions, the lane depth (number of loads stored from the aisle) must be carefully determined. Because no investment in racks is required, block stacking is easy to implement and allows flexibility with floor space.

Obviously block stacking is not a type of equipment. However, since it is the

**Figure 6.61** Block stacking.

benchmark for evaluating all other types of unit load storage equipment, it is described here.

*Stacking Frame*   A stacking frame (see Figure 6.62) is portable and enables the user to stack material, usually in pallet sized loads, on top of one another. They are typically used with unit loads that do not conveniently stack upon themselves. The design issues are the same as those faced in block stacking systems.

Portable racks can either be frames that are attached to standard wooden pallets or self contained steel units made up of decks and posts. When not in use the racks can be disassembled and stored in a minimum of space.

*Single-Deep Selective Rack*   A single-deep selective rack (see Figure 6.63) is a simple construction of metal uprights and cross-members and provides immediate access to each load stored. Unlike block stacking, when a pallet space is created by the removal of a load, it is immediately available. Loads do not need to be stackable and may be of varying heights and widths. In instances where the load depth is highly variable, it may be necessary to provide load supports or decking.

Selective pallet rack might be considered as the "benchmark" storage mode, against which other systems may be compared for advantages and disadvantages. Most storage systems benefit from the use of at least some selective pallet rack for items whose storage requirement is less than six units.

*Double-Deep Rack*   A double-deep rack (see Figure 6.64) is merely single-deep selective rack that is two unit load positions deep. The advantage of the double-deep feature is that fewer aisles are needed which results in a more efficient use of floor space. In most cases a 50% aisle space savings is achieved versus selective rack. Double deep racks are used where the storage requirement for an SKU is six units or greater and when product is received and picked frequently in multiples of two unit loads. Since units loads are stored two deep, a double-reach fork lift is required for storage/retrieval.

**Figure 6.62**   Stacking frames.

**Figure 6.63**   Single-deep selective rack.

***Drive-In Rack***   A drive-in rack (see Figure 6.65) extends the reduction of aisle space begun with double-deep rack. Drive-in racks typically provide for storage lanes from 5 to 10 loads deep. They allow a lift truck to drive in to the rack several positions and store or retrieve a unit load. This is possible because the rack consists of upright columns that have horizontal rails to support unit loads at a height above that of the lift truck. This construction permits a second or even a third level of storage, with

**Figure 6.64** Double-deep rack.

**Figure 6.65** Drive-in/thru rack.

each level being supported independently of the other. A drawback of drive-in rack is the reduction of lift truck travel speed needed for safe navigation within the confines of the rack construction.

***Drive-Thru Rack*** A drive-thru rack is merely drive-in rack that is accessible from both sides of the rack. It is for staging loads in a flow-thru fashion were a unit load is loaded at one end and retrieved at the other end. The same design considerations for drive-in rack apply to drive-thru rack.

***Pallet Flow Rack*** Functionally, pallet flow rack (see Figure 6.66) is used like drive-thru rack, but loads are conveyed on wheels, rollers, or air cushions from one end of a storage lane to the other. As a load is removed from the front of a storage lane the next load advances to the pick face. The main purpose of the flow rack is to simultaneously provide high throughput and storage density. Hence, it is used for those items with high inventory turnover.

***Push-Back Rack*** With a rail-guided carrier provided for each pallet load, push back rack (see Figure 6.67) provides last-in-first-out deep lane storage. As a load is placed into storage, its weight and the force of the putaway vehicle pushes the other loads in the lane back into the lane to create room for the additional load. As a load is removed from the front of a storage lane, the weight of the remaining load automatically advances remaining loads to the rack face.

***Mobile Rack*** A mobile rack (see Figure 6.68) is essentially a single-deep selective rack on wheels or tracks. This design permits an entire row of racks to move away from adjacent rows. The underlying principal is that aisles are only justified

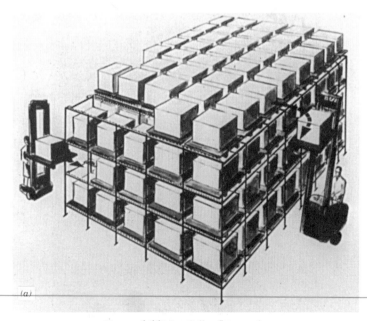

(a)

**Figure 6.66(*a*)** Pallet flow rack.

**Figure 6.66(*b*)** Pallet flow rack. *(Courtesy of Interroll Corporation)*

**Figure 6.66(*c*)** Pallet flow rack.

**Figure 6.67**   Push-back rack.

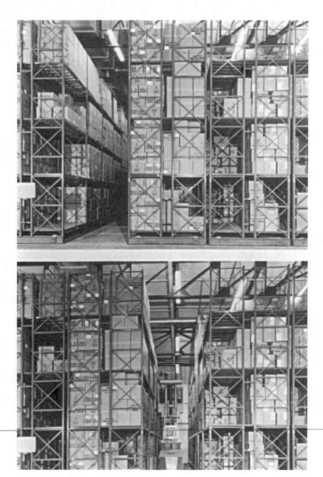

**Figure 6.68**   Mobile rack.

when they are being used. The rest of the time they are occupying valuable space. Access to a particular storage row is achieved by moving the adjacent row and creating an aisle in front of the desired row. Mobile racks are useful when space is scarce and inventory turnover is low.

*Cantilever Rack*   The load bearing arms of a cantilever rack (see Figure 6.69) are supported at one end, as the name implies. The racks consist of a row of single upright columns, spaced several feet apart, with arms extending from one or both sides of the uprights to form supports for storage. The advantage of cantilever racks is that they provide long unobstructed storage shelves with no uprights to restrict the use of horizontal space. The arms can be covered with decking of wood or metal or

**Figure 6.69**   Cantilever rack. *(Courtesy of Jervis B. Webb Co.)*

0

0

can be used without decking. They are applicable for long items such as sofas, rugs, rod, bar, pipe, and sheets of metal or wood.

*Unit Load Retrieval Equipment.* Unit load retrieval equipment types are distinguished from one another by their degree of automation, capital expense, lift height capacity, and aisle width requirements. Equipment offering greater lift height capacity, operating in narrower aisles, and offering greater degrees of automation come with higher prices. Those higher prices may be justified for the associated space and labor savings. The performance attributes of the most popular unit load retrieval equipment types are summarized in Table 6.6.

Two popular devices for retrieving unit loads are walkie stackers and counterbalance lift trucks. Their characteristics were described previously in the section headed "Industrial Vehicles." The other major categories of unit load retrieval equipment are narrow aisle vehicles and automated storage/retrieval (AS/R) machines.

*Walkie Stacker* (See section headed, "Industrial Vehicles.")

*Counterbalance Lift Truck* (See section headed, "Industrial Vehicles" above.)

*Narrow-Aisle Vehicles* Narrow-aisle vehicles are distinguished from walkie stackers and counterbalance lift trucks by their design to operate in space-efficient storage aisles between 5 and 9 ft wide and 25–60 ft tall. The equipment types typically included in this category include straddle trucks, straddle reach trucks, sideloader trucks, turret trucks, and hybrid trucks.

*Straddle Truck* A straddle truck (see Figure 6.70) is most often used in warehouses where aisle space is scarce and/or excessively expensive. The principle is to

**Figure 6.70** Narrow-aisle straddle trucks. *(Courtesy of Clark Equipment Co. and Raymond Corp.)*

Table 6.6 *A Comparison of Unit Load Retrieval Technologies*

| System Attribute | Counterbalance | Straddle | Straddle Reach | Sideloader | Turret | Hybrid | AS/RS |
|---|---|---|---|---|---|---|---|
| Vehicle cost | $30,000 | $35,000 | $40,000 | $75,000 | $95,000 | $125,000 | $200,000 |
| Lift height capacity | 22' | 21' | 30' | 30' | 40' | 50' | 75' |
| Aisle width | 10–13' | 7–9' | 6–8' | 5–7' | 5–7' | 5–7' | 4–5' |
| Weight capacity | 2–10k | 2–6k | 2–5k | 2–10k | 3–4k | 2–4k | 2–5k |
| Lift speed | 80 fpm | 60 fpm | 50 fpm | 50 fpm | 75 fpm | 60 fpm | 100 fpm |
| Travel speed | 550 fpm | 470 fpm | 490 fpm | 440 fpm | 490 fpm | 490 fpm | 500 fpm |

**Figure 6.71** Narrow-aisle reach trucks. *(Courtesy of Clark Equipment Co., Lansing Bagnall, and Raymond Corp.)*

provide load and vehicle stability using outriggers instead of counterbalanced weight thereby reducing the aisle width requirement to 7 to 9 feet. To access loads in storage, the outriggers are driven into the rack allowing the forks to come flush with the pallet face. Hence, it is necessary to support the floor level load on rack beams.

*Straddle Reach Truck*    The straddle reach truck (see Figure 6.71) was developed from the conventional straddle truck by shortening the outriggers on the straddle truck and providing a "reach" capability. In so doing, the outriggers do not have to be driven under the floor level load to allow access to the storage positions. Hence, no rack beam is required at the floor level, conserving rack cost and vertical storage requirements.

Two basic straddle reach truck designs are available: mast- and fork-reach trucks. The mast-reach design consists of a set of tracks along the outriggers that support the mast. The fork-reach design consists of a pantograph or scissors mounted on the mast.

The double-deep-reach truck, a variation of the fork-reach design, allows the forks to be extended to a depth that permits loads to be stored two deep.

*Sideloader Truck*    The sideloader truck (see Figure 6.72) loads and unloads from one side, thus eliminating the need to turn in the aisle to access storage positions. There are two basic sideloader designs. Either the entire mast moves on a set of tracks transversely across the vehicle, or the forks project from a fixed mast on a pantograph.

Aisle width requirements are less than that for straddle trucks and reach trucks. A typical aisle would be 5–7 feet wide, rail or wire guided. Sideloaders can generally access loads up to 30 ft high.

**Figure 6.72**    Narrow-aisle sideloader trucks. *(Courtesy of Raymond Corp.)*

**Figure 6.73**   Narrow-aisle turret trucks. (*a*) Swing fork. (*Courtesy of Raymond Corp.*) (*b*) Swing fork. (*Courtesy of Lansing Bagnall*) (*c*) Swing mast. (*Courtesy of Drexel Industries, Inc.*)

The sideloader truck must enter the correct end of the aisle to access a particular location, which adds an additional burden to routing the truck. A variety of load types can be handled using a sideloader. The vehicle's configuration particularly lends itself to storing long loads in cantilever rack.

*Turret Truck*   The turret truck (see Figure 6.73), swingmast truck, and shuttle truck are members of the modern family of designs which do not require the vehicle to make a turn within the aisle to store or retrieve a pallet. Rather, the load is lifted either by forks which swing on the mast, a mast which swings from the vehicle, or a shuttle fork mechanism.

Generally, these types of trucks provide access to load positions at heights up to 40 ft, which provides the opportunity to increase storage density where floor space is limited. They can also run in aisles 5–7 ft wide, further increasing storage density.

Narrow-aisle side-reach trucks generally have good maneuverability outside the aisle, and some of the designs with telescoping masts may be driven into a shipping trailer.

Since narrow-aisle side reach trucks do not turn in the aisle, the vehicle may be wire guided or the aisles may be rail guided, allowing for greater speed and safety in the aisle and reducing the chances of damage to the vehicle and/or rack.

*Hybrid Truck*   A hybrid truck (see Figure 6.74) (so-called because of its resemblance to an automated storage/retrieval (S/R) vehicles) is similar to a turret truck, except the operator's cab is lifted with the load. The turret vehicle evolved from the S/R machine used in automated storage/retrieval systems. Unlike the S/R machine, the hybrid truck is not captive to an aisle, but may leave one aisle and enter another.

**Figure 6.74**   Narrow-aisle storage/retrieval trucks *(Courtesy of Eaton-Kenway and Hartman Material Handling Systems)*

Present models available are somewhat clumsy outside the aisle, but operate within the aisle at a high throughput rate.

Hybrid trucks operate in aisle widths ranging from 5 to 7 ft, allow rack storage up to 60 ft high in a rack supported building, and may include an enclosed operator's cab, which may be heated and/or air conditioned. Sophisticated hybrid trucks are able to travel horizontally and vertically simultaneously to a load position.

Excellent floorspace utilization and the ability to transfer vehicles between storage aisles are the major benefits of hybrid trucks. The lack of reconfiguration flexibility, high capital expense, and high dimensional tolerance in the rack are the disadvantages of hybrid trucks.

*Automated Storage and Retrieval Machines* An automated storage and retrieval system (AS/RS) is defined by the AS/RS product section of the Material Handling Institute as a storage system that uses fixed-path storage and retrieval (S/R) machines running on one or more rails between fixed arrays of storage racks.

A unit load AS/RS (see Figure 6.75) usually handles loads in excess of 1000 pounds and is used for raw material, work-in-process, and finished goods. The number of systems installed in the United States is in the hundreds, and installations are commonplace in all major industries.

**Figure 6.75(*a*)** Unit load ASRS. *(Courtesy of Jervis B. Webb Co.)*

**Figure 6.75(b)**   Unit load ASRS. *(Courtesy of Jervis B. Webb Co.)*

A typical AS/RS operation involves the S/R machine picking up a load at the front of the system, transporting the load to an empty location, depositing the load in the empty location, and returning empty to the *input/output* (I/O) point. Such an operation is called a *single command* (SC) operation. Single commands accomplish either a storage or a retrieval between successive visits to the I/O point. A more efficient operation is a *dual command* (DC) operation. A DC involves the S/R machine picking up a load at the I/O point, traveling loaded to an empty location (typically the closest empty location to the I/O point), depositing the load, traveling empty to the location of the desired retrieval, picking up the load, traveling loaded to the I/O point, and depositing the load. The key idea is that in a DC, two operations, a storage and a retrieval, are accomplished between successive visits to the I/O point.

A unique feature of the S/R machine travel is that vertical and horizontal travel occur simultaneously. Consequently, the time to travel to any destination in the rack is the maximum of the horizontal and vertical travel times required to reach the destination from the origin. Horizontal travel speeds are up to 600 ft/min; vertical, 150 ft/min.

The typical unit load AS/RS configuration, if there is such a thing, would include unit loads stored one deep (i.e., single deep), in long narrow aisles, (4 to 5 ft wide) each of which contains a S/R machine. The one I/O point would be located at the lowest level of storage and at one end of this system.

More often than not, however, one of the parameters defining the system is atypical. The possible variations include the depth of storage, the number of S/R machines assigned to an aisle, and the number and location of I/O points. These variations are described in more detail as follows.

When the variety of loads stored in the system is relatively low, throughput requirements are moderate to high, and the number of loads to be stored is high, it is often beneficial to store loads more than one deep in the rack. Alternative configurations include

- Double deep storage with single-load width aisles. Loads of the same stock-keeping unit (SKU) are typically stored in the same location. A modified S/R machine is capable of reaching into the rack for the second load, (see Figure 6.76).

- Double deep storage with double-load-width aisles. The S/R machine carries two loads at a time and inserts them simultaneously into the double deep cubicle.

- Deep lane storage with single-load-width aisles. An S/R machine dedicated to storing will store material into the lanes on either side of the aisle. The lanes may hold up to 10 loads each. On the output side, a dedicated retrieval machine will remove material from the racks. The racks may be dynamic, having gravity or powered conveyor lanes (see Figure 6.77).

- Rack entry module (REM) systems in which a REM moves into the rack system and places/receives loads onto/from special rails in the rack (see Figure 6.78).

- Twin-shuttle systems in which the S/R machine is equipped with two shuttle tables and is capable of handling two unit loads per trip.

- Multiple S/R machines operating within the same aisle. Though rare, these systems are used to simultaneously achieve high throughput and storage density.

Another variation of the typical configuration is the use of transfer cars to transport S/R machines between aisles. Transfer cars are used when the storage requirement is high relative to the throughput requirement. In such a case, the throughput requirement does not justify the purchase of an S/R machine for each aisle, yet the number of aisles of storage must be sufficient to accommodate the storage requirement.

A third system variation is the number and location of I/O points. Throughput requirements or facility design constraints may mandate multiple I/O points at locations other than the lower-left-hand corner of the rack. Multiple I/O points might be used to separate inbound and outbound loads and/or to provide additional throughput capacity. Alternative I/O locations include the type of the system at the end of the rack (some AS/RS are built underground) and the middle of the rack.

## Small Load Storage and Retrieval Equipment

Small load storage and retrieval equipment is classified as operator-to-stock (OTS, sometimes referred to as man-to-part or in-the-aisle systems) if the operator travels

Figure 6.76  Double-deep ASRS. *(Courtesy of Eaton-Kenway, Inc.)*

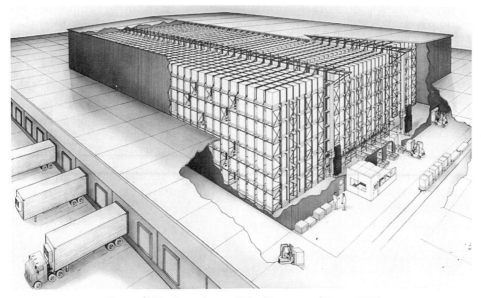

Figure 6.77  Deep-lane ASRS. *(Courtesy of Litton UHS)*

**Figure 6.78** Rack-entry module ASRS. *(Courtesy of Litton UHS)*

to the storage location to retrieve material and as stock-to-operator (sometimes referred to as part-to-man or end-of-aisle systems) if the material is mechanically transported to an operator for retrieval. Stock-to-operator equipment types often offer higher productivity, easier supervision, and better item security and protection than operator-to-stock alternatives. At the same time, stock-to-operator options often are more expensive, more difficult to reconfigure, and require more maintenance than operator-to-stock options. The performance attributes of the major equipment types are summarized in Table 6.7.

In OTS options, the design and selection of the storage mode may be separated from the design and selection of the retrieval mechanism. Hence, OTS storage equipment and OTS retrieval equipment are described separately.

*Operator-to-Stock—Storage Equipment* The three principal operator-to-stock equipment types for housing small loads are bin shelving, modular storage drawers in cabinets, and gravity flow rack. To improve space utilization, each of the storage systems can be incorporated into mezzanine or mobile storage configurations.

*Bin Shelving* Bin shelving (see Figure 6.79) is the oldest and still the most popular (in terms of sales volume and number of systems in use) equipment alternative in use for small parts order picking. The low initial cost, easy reconfigurability, easy installation, and low maintenance requirements are at the heart of this popularity.

It is important to recall that the lowest initial cost alternative may not be the most cost effective alternative, or the alternative which meets the prioritized needs of an operation. With bin shelving, savings in initial cost and maintenance may be offset by inflated space and labor requirements.

**Table 6.7  A Comparison of Several Small Part Storage Alternatives**

| System Attribute | Unit of Measure | Bin Shelving | Gravity Flow Racks | Storage Drawers | Horizontal Carousel | Vertical Carousel | Miniload ASRS | Automatic Dispensing |
|---|---|---|---|---|---|---|---|---|
| Gross system cost | Initial Cost/Purchased Ft3 | $5–$15 | $3–$5 | $25–$30 | $20–$35 | $40–$70 | $30–$40 | $300–$600 per dispenser |
| Net system cost | Initial Cost/Available Ft3. | $10–$30 | $9–$15 | $31–$38 | $40–$70 | $65–$100 | $38–$50 | |
| Floorspace requirements | Ft3 of inventory housed per Ft2 of floorspace. | 1–1.2 | 0.7–0.85 | 1.8–2.5 | 0.8–1.25 | 5.0–6.0 | 4.0–5.0 | |
| Human factors | Ease of retrieval. | Average | Average | Good | Average | Excellent | Excellent | Good |
| Maintenance requirements | | Low | Low | Low | Medium | Medium | High | High |
| Item security | | Average | Average | Excellent | Good | Excellent | Excellent | Average |
| Flexibility | Ease to reconfigure | High | High | High | Medium | Low | Low | Low |
| Pick rate | Order lines per person-hour | C: 25–125 T: 100–350 M: 25–250 W: 300–500 | C: 25–125 T: 100–350 M: 25–250 W: 300–500 | C: 25–125 T: 100–350 M: 25–250 W: 300–500 | 50–250 | 50–300 | 25–125 | 500–1000 |
| Key | T = Tote Picking | C = Cart Picking | M = Man Aboard ASRS | W = Wave Picking | | | | |

**Figure 6.79** Bin shelving. *(Courtesy of Equipto)*

Space is frequently underutilized in bin shelving, since the full inside dimensions of each unit are rarely usable. Also, since people are extracting the items, the height of shelving units is limited by the reaching height of a human being. As a result, the available building cube may also be underutilized.

The consequences of low space utilization are twofold. First, low space utilization means that a large amount of square footage is required to store the products. The more expensive it is to own and operate the space, the more expensive low space utilization becomes. Second, the greater the square footage, the greater the area that must be traveled by the order pickers, and thus, the greater the labor requirement and costs.

Two additional disadvantages of bin shelving are supervisory problems and item security/protection problems. Supervisory problems arise because it is difficult to supervise people through a maze of bin shelving units. Security and item protection problems arise because bin shelving is one of a class of open systems (i.e., the items are exposed to and accessible from the picking aisles).

As with all of the equipment types, these disadvantages must be evaluated and compared with the advantages of low initial cost and low maintenance requirements in order to make an appropriate equipment selection.

***Modular Storage Drawers in Cabinets*** Modular storage drawers/cabinets (see Figure 6.80) are called modular because each storage cabinet houses modular storage drawers which are subdivided into modular storage compartments. Drawer heights

**Figure 6.80**   Modular storage drawers in cabinets. *(Courtesy of Equipto)*

range from 3 in. to 24 in., and each drawer may hold up to 400 lb worth of material. The storage cabinets can be thought of as shelving units which house storage drawers.

The primary advantage of storage drawers/cabinets over bin shelving is the large number of SKUs which can be stored and presented to the order picker in a small area. The cabinet and drawer suppliers inform us that one drawer can hold from 1 to 100 SKUs (depending on the size, shape, and inventory levels of the items), and that a typical storage cabinet can store the equivalent of 2 to 4 shelving units worth of material. This dense storage stems primarily from the ability to create item housing configurations within a drawer which very closely match the cubic storage requirements of each SKU. Also, since the drawers are pulled out into the aisle for picking, space does not have to be provided above each SKU to provide room for the order picker's hand and forearm. This reach space must be provided in bin shelving storage, otherwise items deep in the unit could not be accessed.

Several benefits accrue from the high density storage characteristic of storage drawer systems. First, and obviously, the more material that can be packed into a smaller area, the smaller the space requirement. Hence, space costs are reduced. When the value of space is at a true premium, such as on a battle ship, on an airplane, or on the manufacturing floor, the reduction in space requirements alone can be enough to justify the use of storage drawers and cabinets. A second benefit resulting from a reduction in square footage requirements is a subsequent reduction in the travel time and hence labor requirements for order picking.

Additional benefits achieved by the use of storage drawers include improved picking accuracy and protection for the items from the environment. Picking accuracy is improved over that in shelving units because the order picker's sight lines to the items are improved, and the quantity of light falling on the items to be extracted is increased. With bin shelving, the physical extraction of items may occur anywhere from floor level to 7 ft off the ground, with the order picker having to reach into the shelving unit itself to achieve the pick. With storage drawers, the drawer is pulled out into the picking aisle for item extraction. The order picker looks down onto the contents of the drawer which are illuminated by the light source for the picking aisle. (The fact that the order picker must look down on the drawer necessitates that storage cabinets be less than 5 ft in height.) Item security and protection are achieved since the drawers can be closed and locked when items are not being extracted from them.

As one would expect, these benefits are not for free. Storage cabinets equipped with drawers are relatively expensive. Price is primarily a function of the number of drawers and the amount of sheet metal in the cabinet.

*Carton Flow Rack*   Carton flow rack (see Figure 6.81), or just flow rack, is another popular OTS equipment alternative. Flow rack is typically used for active items which are stored in fairly uniform sized and shaped cartons. The cartons are placed in the back of the rack from the replenishment aisle, and advance/roll towards the pick face as cartons are depleted from the front. This back to front movement insures first-in-first-out (FIFO) turnover of the material.

Essentially, a section of flow rack is a bin shelving unit turned perpendicular to the picking aisle with rollers placed on the shelves. The deeper the sections, the greater the portion of warehouse space that will be devoted to storage as opposed to aisle space. Further gains in space efficiency can be achieved by making use of the cubic space over the flow rack for full pallet storage.

Flow rack costs depend on the length and weight capacity of the racks. As is the case with bin shelving, flow rack has very low maintenance requirements and is available in a wide variety of standard section and lane sizes from a number of suppliers.

The fact that just one carton of each line item is located on the pick face means that a large number of SKUs are presented to the picker over a small area. Hence, walking and therefore labor requirements can be reduced with an efficient layout.

*Mezzanine*   Bin shelving, modular storage cabinets, flow rack, and even carousels can be placed on a mezzanine (see Figure 6.82). The advantage of using a mezzanine is that nearly twice as much material can be stored in the original square

**Figure 6.81(*a*)**   Gravity flow rack.

footage, inexpensively. The major design issues for a mezzanine are the selection of the proper grade of mezzanine for the loading that will be experienced, the design of the material handling system to service the upper levels of the mezzanine, and the utilization of the available clear height. At least 14 ft of clear height should be available for a mezzanine to be considered.

*Mobile Storage*   Bin shelving, modular storage cabinets, and flow rack can all be "mobilized" (see Figure 6.83). The most popular method of mobilization is the "train-track" method. Parallel tracks are cut into the floor, and wheels are placed on the bottom of the storage equipment to create "mobilized" equipment. The space savings accrue from the fact that only one aisle is needed between all the rows of storage equipment. The aisle is created by separating two adjacent rows of equipment. As a result, the aisle "floats" in the configuration between adjacent rows of equipment. The storage equipment is moved by simply sliding the equipment along the tracks, by turning a crank located at the end of each storage row, or by invoking electric motors which may provide the motive power. The disadvantage to this approach is the increased time required to access the items. Every time an item must be accessed, the corresponding storage aisle must be created.

(b)

**Figure 6.81(b)** Gravity flow rack. *(Courtesy of Kingston-Warren Corporation)*

**Figure 6.82** Mezzanine. *(Cortesy of Penco)*

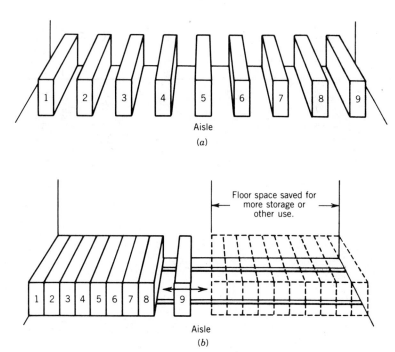

**Figure 6.83** An illustration of the space savings available via sliding racks. (*a*) Conventional arrangement. (*b*) Floor space savings via sliding racks.

*Operator-to-Stock Systems—Retrieval Equipment.* In operator-to-stock alternatives, the operator either walks or rides a vehicle to the pick location. The four retrieval options distinguished here include picking carts, order picker trucks, person-aboard AS/R machines, and robotic retrieval.

*Picking Cart* A variety of picking carts (see Figure 6.84) are available to facilitate the accumulation, sortation, and/or packing of orders as an order picker makes a picking tour. The carts are designed to allow an order picker to pick multiple orders on a picking tour, thus dramatically improving productivity as opposed to strict single order picking for small orders. The most conventional vehicles provide dividers for order sortation, a place to hold paperwork and marking instruments, and a step ladder for picking at levels slightly above reaching height. Additional levels of sophistication and cost bring powered carts, light-aided sortation, on-board computer terminals, and on-board weighing.

*Order Picker Truck* The order picker truck (see Figure 6.85), sometimes referred to as a cherry picker or stock picker, allows an operator to travel to retrieval locations well above floor level. In so doing the operator's productivity is reduced. However, productivity can be enhanced by minimizing vertical travel through popularity-based storage and/or intelligent pick tour construction.

*Person-aboard Automated Storage/Retrieval Machine* The person-aboard AS/R machine (Figure 6.86), as the name implies, is an automated storage and retrieval machine in which the picker rides aboard a storage/retrieval machine to, from and between retrieval locations. The storage modes may be stacked bin shelving

**Figure 6.84(*a*)**   Order picking cart.

units, stacked storage cabinets, and/or pallet rack. The storage/retrieval machine may be aisle captive or free roaming.

Typically, the picker will leave from the front of the system at floor level and visit enough storage locations to fill one or multiple orders, depending on order size. Sortation can take place on board if enough containers are provided on the storage/ retrieval (S/R) machine.

The person-aboard AS/R machine offers significant square footage and order picking time reductions over the previously described picker-to-part systems. Square footage reductions are available because storage heights are no longer limited by the reach height of the order picker. Shelves or storage cabinets can be stacked as high as floor loading, weight capacity, throughput requirements, and/or ceiling heights will permit. Retrieval times are reduced because the motive power for traveling is provided automatically, hence freeing the operator to do productive work while traveling and because search time is reduced since the operator is automatically delivered to the correct location.

**Figure 6.84(*b*)**  Order picking cart.

**Figure 6.85**  Narrow-aisle order picker trucks. *(Courtesy of Yale Industrial Trucks and Raymond Corporation)*

**Figure 6.86(*a*)**   Person-aboard ASRS. *(Courtesy of Demag)*

*Robotic Retrieval*   In rare instances, robotic retrieval equipment (see Figure 6.87) can be justified. Each robotic retrieval vehicle is equipped with a small carousel to permit order sortation, accumulation, and containment. The carousel travels up an down a mast on the robot as the robot traverses the picking aisle(s). As a storage drawer is pulled from a storage location onto the vehicle, the carousel rotates to the correct position.

*Stock-to-Operator Equipment.*   The major difference between stock-to-operator and operator-to-stock equipment is the answer to the question, Does the operator travel to the stock location, or does the stock travel to the operator? If the stock travels to the operator, the equipment is classified as stock-to-operator equipment.

In stock-to-operator options, the travel time component of total order picking time is shifted from the operator to a device for bringing locations to the picker. Also, the search time component of total order picking time is significantly reduced since the correct location is automatically presented to the operator. In well-designed systems, the result is a large increase in productivity. In poorly designed systems, the improvements may be negligible if the operator is required to wait on the device to present him or her with stock.

The two most popular classes of stock-to-operator equipment are carousels and

**Figure 6.86(b)**  Person-aboard ASRS. *(Courtesy of Demag)*

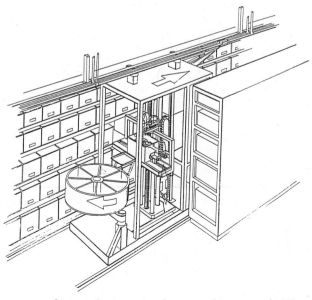

**Figure 6.87**  Robotic retrieval system. *(Courtesy of DTI)*

miniload automated storage and retrieval machines. A more expensive, yet highly productive class is automated dispensors. Each equipment type is described as follows.

*Carousels* Carousels, as the name implies, are mechanical devices which house and rotate items for storage and retrieval. Four classes of carousels are currently available for small load storage and retrieval applications—horizontal carousels, vertical carousels, and independently rotating racks.

*Horizontal Carousel* A horizontal carousel (see Figure 6.88) is a linked series of rotating bins of adjustable shelves driven on the top or on the bottom by a drive motor unit. Rotation takes place about an axis perpendicular to the floor at between 80 and 100 ft/min.

Items are extracted from the carousel by order pickers who occupy fixed positions in front of the carousel(s). Operators may be responsible for controlling the rotation of the carousel. Manual control is achieved via a keypad, which tells the carousel which bin location to rotate forward, and a foot pedal, which releases the carousel to rotate. Carousels may also be computer controlled, in which case the sequence of pick locations is stored in the computer and brought forward automatically.

A popular design enhancement is the installation of a light tree in front of a pair of carousels. The light tree has light displays for each level of the carousel. The correct retrieval location and quantity are displayed on the light. These displays significantly improve the productivity and accuracy of the storage and retrieval transactions.

A management option with carousels is the flexible scheduling of operators to carousels. If an operator is assigned to one carousel unit, he or she must wait for the carousel to rotate to the correct location between picks. If an operator is assigned to two or more carousels, he or she may pick from one carousel while the other is

**Figure 6.88** Horizontal carousel with lift/lower platform. *(Courtesy of White Storage & Retrieval, Inc.)*

rotating to the next pick location. Remember, the objective of stock-to-operator alternatives is to keep the operators busy extracting stock. Humans are excellent extractors of items; the flexibility of our limbs and muscles provides us with this capability. Humans are not efficient searchers, walkers, or waiters.

Horizontal carousels vary in length from 15 ft to 100 ft, and in height from 6 ft to 25 ft. The length and height of the units are dictated by the pick rate requirements and building restrictions. The longer the carousel, the more time required, on average, to rotate the carousel to the desired location. Also, the taller the carousel, the more time required to access the items. Heights over 6 ft necessitate the use of ladders or robot arms on vertical masts to access the items.

In addition to providing a high pick rate capacity, horizontal carousels make good use of the available storage space. Very little space is required between adjacent carousels, and the only lost space is that between parallel sections of bins on the same carousel unit.

One important disadvantage of horizontal carousels is that the shelves and bins are open. Consequently, item security and protection can be a problem.

A "twin-bin" horizontal carousel was recently introduced to the material handling market. In the twin-bin carousel, the traditional carousel carrier is split vertically in half and rotated 90°. This allows for more shallow carriers, thus improving the storage density for small parts.

*Vertical Carousel*  A vertical carousel (see Figure 6.89) is a horizontal carousel turned on its end and enclosed in sheet metal. As with horizontal carousels, an order picker operates one or multiple carousels. The carousels are indexed either automatically via computer control, or manually by the order picker working a keypad on the carousel's work surface.

**Figure 6.89(***a***)**   Vertical carousel. *(Courtesy of Remstar International)*

**Figure 6.89(*b*)**   Vertical carousel.

**Figure 6.89(*c*)**   Vertical carousel. *(Courtesy of Remstar International)*

Vertical carousels range in height from 8 feet to 35 ft. Heights (as lengths were for horizontal carousels) are dictated by throughput requirements and building restrictions. The taller the system, the longer it will take, on average, to rotate the desired bin location to the pick station.

Retrieval times for vertical carousels are theoretically less than those for horizontal carousels. The decrease results from the items always being presented at the order picker's waist level. This eliminates the stooping and reaching that goes on with horizontal carousels, further reduces search time, and promotes more accurate picking. (Some of the gains in item extract time are negated by the slower rotation speed of the vertical carousel. Recall that the direction of rotation is against gravity.)

Additional benefits provided by the vertical carousel include excellent item protection and security. In the vertical carousel, only one shelf of items is exposed at one time, and the entire contents of the carousel can be locked up.

On a per cubic foot of storage basis, vertical carousels are typically 40%–70% more expensive than horizontal carousels. The additional cost of vertical carousels over horizontal carousels is attributed to the sheet metal enclosure, and the extra power required to rotate against the force of gravity.

*Independent Rotating Rack*  An independent rotating rack carousel is like multiple one level horizontal carousels stacked on top of one another. As the name implies, each level rotates independently. As a result, several storage locations are ready to be accessed by an operator at all times. Consequently, the operator is continuously picking.

Remote order picking, assembly, and/or order or kit staging are more common applications of independent rotating racks. In those applications, a robot is positioned at the end of the IRR to store and retrieve inbound and outbound unit loads. In that way, the IRR is used to simultaneously achieve high throughput and storage density.

Clearly, for each level to operate independently, each level must have its own power and communication link. These requirements force the price of independent rotating rack well beyond that of vertical or horizontal carousels.

***Miniload Automated Storage and Retrieval Machine***  In the miniload automated storage and retrieval system (see Figure 6.90), a miniload storage/retrieval S/R machine travels horizontally and vertically simultaneously in a storage aisle, transporting storage containers to and from an order picking station located at one end of the system. The order picking station typically has two pick positions. As the order picker is picking from the container in the left pick position, the S/R machine is taking the container from the right pick position back to its location in the rack and returning with the next container. The result is the order picker rotating between the left and right pick positions. (A system enhancement used to improve the throughput of miniload operations is to assign multiple S/R devices per aisle.)

The sequence of containers to be processed is determined manually by the order picker keying in the desired line item numbers or rack locations on a keypad, or the sequence is generated and processed automatically by computer control.

Miniloads vary in height from 8 to 50 ft, and in length from 40 to 200 ft. As in the case with carousels, the height and length of the system are dictated by the throughput requirements and building restrictions. The longer and taller the system, the longer the time required to access the containers. However, the longer and taller the system, the fewer the aisles and S/R machines that will have to be purchased.

The transaction rate capacity of the miniload is governed by the ability of the S/R machine (which travels approximately 500 ft/min horizontally and 120 ft/min vertically) to continuously present the order picker with unprocessed storage containers. This ability, coupled with the human factors benefits of presenting the con-

**Figure 6.90(a)**   Miniload ASRS. *(Courtesy of Litton UHS)*

tainers to the picker at waist height in a well lit area, can produce impressive pick rates.

Square footage requirements are reduced for the miniload due to the ability to store material up to 50 ft high, the ability to size and shape the storage containers and the subdivisions of those containers to very closely match the storage volume requirements of each SKU, and an aisle width that is determined solely by the width of the storage containers.

The disadvantages of the miniload system are probably already apparent. As the most sophisticated of the system alternatives described thus far, it should come as no surprise that the miniload carries the highest price tag of any of the order picking system alternatives. Another result of its sophistication is the significant engineering and design effort that accompanies each system. Finally, greater sophistication leads to greater maintenance requirements. It is only through a disciplined maintenance program that miniload suppliers are able to advertise up time percentages between 97% and 99.9%.

Recently, pre-engineered, modular miniload system designs have been introduced to the U.S. material handling systems market. Pre-engineered systems offer the

**Figure 6.90(*b*)**  Miniload ASRS. *(Courtesy of Litton UHS)*

same range and degree of benefits as conventional miniload system designs, are less expensive, and are delivered and installed sooner.

*Vertical Lift Module*  The vertical lift module (see Figure 6.91) is a relatively new addition to family of small load storage and retrieval equipment. Simply put, a vertical lift module is a small miniload ASRS turned on its end. Systems range in height from 15 to 35 feet and can be interfaced with automated material transport systems. VLMs offer many of the benefits of miniload systems at a reduced price.

*Automatic Dispenser*  An automatic dispenser (see Figure 6.92) works much like vending machines for small items of uniform size and shape. Each item is allocated a vertical dispenser ranging from 2 to 6 inches wide and from 3 to 5 ft tall. (The width of each dispenser is easily adjusted to accommodate variable product sizes.) The dispensing mechanism acts to kick the unit of product at the bottom of the dispenser out onto a conveyor running between two rows of dispensers configured as an A-Frame over a belt conveyor. A tiny vacuum conveyor or small finger on a chain conveyor is used to dispense the items.

**Figure 6.90(c)**   Miniload ASRS. *(Courtesy of Litton UHS)*

Virtual order zones begin at one end of the conveyor and pass by each dispenser. If an item is required in the order zone it is dispensed onto the conveyor. Merchandise is accumulated at the end of the belt conveyor into a tote pan or carton. A single dispenser can dispense at a rate of up to 6 units per second. Automatic item pickers are popular in industries with high throughput for small items of uniform size and shape. Cosmetics, wholesale drugs, compact discs, videos, publications, and polybagged garments are some examples.

Replenishment is performed manually from the back of the system. This manual replenishment operation significantly cuts into the savings in picking labor requirements associated with pick rates on the order of 1500 picks per hour per pick head.

One new design for automated dispensing machines is an inverted A-Frame which streamlines the replenishment of automated dispensers and increases the storage density along the picking line. Another new design allows automated dispensing for polybagged garments. For a comparison several small part storage technologies, see Table 6.7.

(a)

**Figure 6.91(***a***)** Vertical lift module. *(Courtesy of Remstar International)*

## Automatic Identification and Communication Equipment

Automated status control of material requires that real-time awareness of the location, amount, origin, destination, and schedule of material be achieved automatically. This objective is in fact the function of automatic identification and communication technologies, technologies that permit real-time, nearly flawless data collection and communication. Examples of automatic identification and communication technologies at work include

- A vision system reading and interpreting labels to identify the proper destination for a carton traveling on a sortation conveyor
- A laser scanner to relay the inventory levels of a small parts warehouse to a computer via radio frequency communication
- A voice recognition system to identify parts received at the receiving dock

**Figure 6.91(b)**   Vertical lift module. *(Courtesy of Remstar International)*

- A radio frequency (RF) or surface acoustical wave (SAW) tag used to permanently identify a tote pan or pallet
- A card with a magnetic stripe that travels with a unit load to identify the load through the distribution channels

For discussion purposes we distinguish automatic identification systems from automatic communication systems. Automatic identification systems allow machines to identify material and capture key data concerning the status of the material. Automatic communication systems allow paperless communication of the captured data.

## Automatic Identification and Recognition

The list of automatic identification and recognition technologies is expanding and includes bar coding, optical character recognition, radio frequency tags, magnetic stripes, and machine vision.

***Bar Coding.***   A bar code system includes the bar code itself, bar code reader(s), and bar code printer(s). Each system component is described below. More comprehensive descriptions are provided by [17] and [58].

***Bar Codes***   A bar code (see Figure 6.93) consists of a number of printed bars and intervening spaces. The structure of unique bar/space patterns represents various alphanumeric characters. Examples appear on virtually every consumer item allowing check-out clerks in retail and grocery stores to scan the code on the item to automatically record its identification and price.

**Figure 6.92**   Automated dispensing system. *(Courtesy of ElectroCom Automation)*

The same bar/space pattern may represent different alphanumeric characters in different codes. Primary codes or symbologies for which standards have been developed include

- Code 39: An alphanumeric code adopted by a wide number of industry and government organizations for both individual product identification and shipping package/container identification.

- Interleaved 2 of 5 Code: A compact, numeric-only code still used in a number of applications where alphanumeric encoding is not required.

- Codabar: One of the earlier symbols developed, this symbol permits encoding of the numeric-character set, six unique control characters, and four unique stop/start characters that can be used to distinguish different item classifications. It is primarily used in nongrocery retail point of sale applications, blood banks and libraries.

- Code 93: Accommodating all 128 ASCII characters plus 43 alphanumeric characters and four control characters, Code 93 offers the highest alpha-numeric data density of the six standard symbologies. In addition to allowing for positive

**Figure 6.93** Sample automatic identification product codes. *(From [58] with permission.)*

switching between ASCII and alphanumeric, the code uses two check characters to ensure data integrity.

- Code 128: Provides the architecture for high density encoding of the full 128 character ASCII set, variable length fields, and elaborate character-by-character and full symbol integrity checking. Provides the highest numeric-only data density. Adopted in 1989 by the Uniform Code Council (U.S.) and the International Article Number Association (EAN) for shipping container identification.

- UPC/EAN: The numeric-only symbols developed for grocery supermarket point-of-sale applications and now widely used in a variety of other retailing environments. Fixed length code suitable for unique manufacturer and item identification only.

- Stacked Symbologies: Although a consensus standard has not yet emerged, the health and electronics industries have initiated programs to evaluate the feasibility of using Code 16K or Code 49, two microsymbologies that offer significant potential for small item encoding. Packing data in from two to 16 stacked rows, Code 16K accommodates the full 128-character ASCII set and permits the encoding of up to 77 characters in an area of less than 0.5 in$^2$. Comparable in terms of data density, Code 49 also handles the full ASCII character set. It encodes data in from two to eight rows and has a capacity of up to 49 alphanumeric characters per symbol.

A detailed comparison of these codes is provided in Table 6.8. Two-dimensional bar codes are a recent development in bar code symbology. Two-dimensional codes yield

Table 6.8  *A Comparison of Bar Code Symbologies [17]*

| | UPC | 39 | 2 of 5 | 128 | Codabar | 11 | 49 |
|---|---|---|---|---|---|---|---|
| Year developed | 1973 | 1974 | 1972 | 1981 | 1972 | 1977 | 1987 |
| Character set | Numeric | Alphanumeric | Numeric | Alphanumeric | Numeric | Numeric | Alphanumeric |
| Form | Continuous | Discrete | Continuous | Continuous | Discrete | Discrete | Continuous |
| Number of characters | 10 | 43/128 | 10 | 103/128 | 16 | 11 | 128 |
| Density: units per character | 7 | 13–16 | 7–9 | 11 | 12 | 8–10 | — |
| Density: characters per inch | 13.7 | 9.4 | 17.8 | 9.1 | 10 | 15 | 294/sq.in. |
| Application areas | Retail, auto supply | DoD | Industrial | Varied | Blood banks, libraries | Telecommunications | Varied |
| Data security | Moderate | High | High | High | High | High | — |

269

dramatic improvements in data density, the amount of information encoded per square inch.

***Bar Code Readers***   Bar codes are read by both contact and noncontact scanners. Contact scanners can be portable or stationary and use a wand or light pen. As the wand/pen is manually passed across the bar code, the scanner emits either white or infrared light from the wand/pen tip and reads the light pattern that is reflected from the bar code. This information is stored in solid-state memory for subsequent transmission to a computer. More sophisticated (and expensive) handheld units include fixed- and moving-beam scanners employing light emitting diodes (LEDs), low power helium-neon lasers, or the latest development, laser diodes. These devices can scan bar code labels from distances up to 6 ft.

The key performance attributes of a variety of bar code readers are summarized in Table 6.9.

*Contact Reader*   A contact reader is an excellent substitute for keyboard or manual data entry. Alphanumeric information is processed at a rate of up to 50 in./ min, and the error rate for a basic scanner connected to its decode is 1 in 1,000,000 reads. Light pens (see Figure 6.94) and wand scanners are examples of contact readers.

*Noncontact Reader*   A noncontact reader (see Figure 6.95) is usually stationary and includes fixed-beam, moving-beam scanners, and charged couple device (CCD) scanners. These scanners employ fixed-beam, moving beam, video camera or raster scanning technology to take from one to several hundred looks at the code as it passes. Most bar code scanners read codes bidirectionally by virtue of sophisticated decoding electronics, which distinguish the unique start/stop codes peculiar to each symbology and decipher them accordingly. Further, the majority of scanner suppliers now provide equipment with an autodiscrimination feature that permits recognition, reading and verification of multiple symbol formats with no internal or external adjustments. Finally, at least two suppliers have introduced omnidirectional scanners for industrial applications that are capable of reading a code at high speed throughout a large field of view regardless of its orientation.

Fixed-beam readers use a stationary light source to scan a bar code. They depend on the motion of the object to be scanned to move past the beam. Fixed-beam readers rely on consistent, accurate code placement on the moving object.

Moving-beam scanners employ a moving light source to search for codes on moving objects. Because the beam of light is moving, the placement of the code on the object is not critical. In fact, some scanners can read codes accurately on items moving at a speed of 1,000 ft/min.

CCD scanners have only recently been used to read bar codes. A CCD scanner is more like a camera than a bar code scanner. By changing camera lenses and focal lengths, the scanners read various bar codes at various distances. Like any camera system, the quality of the picture (the accuracy of the read) depends on the available light. When installed correctly, the CCD scanner is 99% accurate.

Omnidirectional scanners (see Figure 6.96) are a recent development in bar code reading technology. Omnidirectional scanners can successfully read and interpret a bar code regardless of its orientation.

***Bar Code Printers***   Bar code printers are the bar code system component that typically receive the least attention. However, the design and selection of the printer

Table 6.9   *A Comparison of Bar Code Readers [17]*

|  | Light Pen | Handheld Laser Scanner | Fixed Beam Stationary Scanner | Omnidirection Scanner |
|---|---|---|---|---|
| Range | Contact | 60″ | 60″ | 100″ |
| Depth of field | Contact | 24″ | 40″ | 30″–40″ |
| Scan rate | Manual | 32–100/sec. | 200–3450/sec. | 1440–2400/sec. |
| Resolution | 0.004–.0075 in. | 0.004–.0075 in. | 0.004–.040 in. | 0.004–.040 in. |
| Price | $100–$1,000 | $1,300–$2,500 | $1,000–$10,000 | $15,000–$40,000 |

**Figure 6.94**   Contact reader. *(Courtesy LXE)*

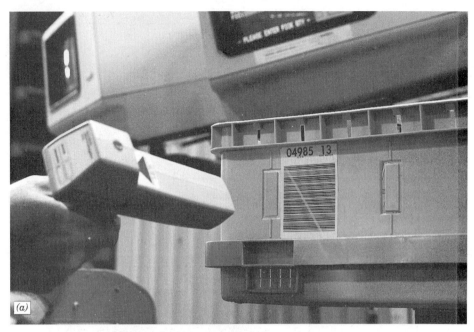

**Figure 6.95(a)** Hand-held bar code scanner.

**Figure 6.95(b)** Stationary bar code scanner.

**Figure 6.96** Omni-directional bar code scanner.

to a great extent determine the ultimate success of the bar code system since the quality of the printed code is critical to the acceptance and capacity of the system. There are five main classes of bar code printers—laser, thermal transfer, serial, impact, and ink jet. The capacity, print quality, typical price range, and relative cost of ownership are compared for the most popular printer classes in Table 6.10.

*Optical Character Recognition.* Optical characters are human and machine readable. An example is the account number printed along the bottom of most bank checks. Optical character recognition (OCR) systems read and interpret alphanumeric data so that people as well as computers can interpret the information contained in the data.

OCR labels are read with handheld scanners, much like bar codes. Optical character recognition systems operate at slower read rates than bar code systems and are priced about the same. OCR systems are attractive when both human- and machine-readable capabilities are required.

**Table 6.10** *A Comparison of Bar Code Printers*

| Printer Class | Print Quality | Throughput (in.$^2$/sec) | Typical Price Range | Relative Ownership Cost |
|---|---|---|---|---|
| Laser | Good to Excellent | 11–22 | $3,000–$10,000 | Medium |
| Thermal transfer | Excellent | 24–50 | $2,000–$6,000 | High |
| Serial | Poor to Good | 3–5 | $1,000–$3,000 | Low |
| Impact | Poor to Good | 5–20 | $5,000–$15,000 | Low |

*Source: How to Do It,* Quad II, Charlotte, NC, as seen in *Modern Materials Handling,* May 1993.

Until recently, the commercial applications of optical character recognition have been confined to document reading and limited use for merchandise tag reading at the retail point of sale. Without tight control of character printing and the reading environment, OCR's performance has not met the criteria established by other automatic identification techniques. A single printing anomaly, such as an ink spot or void can easily obscure or transpose an OCR character, rendering the label unreadable or liable to misreading. On the other hand, where encoding space is at a premium and the environment is relatively contaminant-free, OCR may be a viable alternative.

***Radio Frequency Tag.***   Both the radio frequency (RF) tag and the surface accoustical wave (SAW) tag encode data on a chip that is encased in a tag. When a tag is within range of a special antenna, the chip is decoded by a tag reader. Radio frequency tags can be programmable or permanently coded and can be read form up to 30 ft away. SAW tags are permanently coded and can be read only within a 6-ft range. A common example is the tag now placed in the windshield of automobiles that permits high-speed auto recognition and account debiting at high-speed toll booths.

Radio frequency and SAW tags are typically used for permanent identification of a container, where advantage can be taken of the tag's durability. RF and SAW technologies are also attractive in harsh environments where printed codes may deteriorate and become illegible.

***Magnetic Stripe.***   A magnetic stripe is used to store a large quantity of information in a small space. The stripe appears in common use on the back of credit cards and bank cards. The stripe is readable through dirt or grease, and the data contained in the stripe can be changed. The stripe must be read by contact, thus eliminating high-speed sortation applications. Magnetic stripe systems are generally more expensive than bar code systems.

***Machine Vision.***   Machine vision relies on cameras to take pictures of objects and codes and send the pictures to a computer for interpretation. Machine vision systems "read" at moderate speeds, with excellent accuracy at least for limited environments. Obviously, these systems do not require contact with the object or code. However, the accuracy of a read is strictly dependent on the quality of light. Machine vision is becoming less costly but is still relatively expensive.

## Automatic, Paperless Communication

Devices for automatically communicating with material handling operators include radio frequency data terminals, voice headsets, light and computer aids, and smart cards. Table 6.11 summarizes the key operating attributes of each system.

***Radio Frequency Data Terminal***   A handheld, arm-mounted, or lift-truck-mounted radio data terminal (RDT) (see Figure 6.97) is a reliable device for inventory and vehicle/driver management. The RDT incorporates a multicharacter display, full keyboard, and special function keys. It communicates and receives messages on a prescribed frequency via strategically located antennae and a host computer interface unit. Beyond the basic thrust toward tighter control of inventory, improved resource

**Table 6.11**   *A Comparison of Automatic, Paperless Communication Technologies*

|  | Real-Time? | Handsfree? | Eyesfree? | Cost |
|---|---|---|---|---|
| Bar code scanning | No | No | No | Labels, scanners |
| Batch terminals—scanning | No | No | No | Labels, scanners, terminals |
| Batch terminals—voice | No | Yes | Yes | Terminals |
| Pick-to-light | No | No | No | $100 per SKU |
| Smart cards | No | No | No | Readers, cards, displays |
| RF links—terminals | Yes | No | No | RF, terminals |
| RF links—headsets | Yes | Yes | Yes | RF, terminals, headsets |

**Figure 6.97(*a*)**   Hand-held RF terminal. *(Courtesy of Telxon Corporation)*

**Figure 6.97(***b***)**   Wrist-mounted RF terminal. *(Courtesy of Telxon Corporation)*

**Figure 6.97(***c***)**   Truck-mounted RF terminal. *(Courtesy of LXE)*

utilization is most often cited in justification of these devices. Further, the increasing availability of software packages that permit RDT linkage to existing plant or warehouse control systems greatly simplify their implementation. The majority of RDTs installed in the plant environment uses handheld wands or scanners for data entry, product identification and location verification. This marriage of technologies provides higher levels of speed, accuracy, and productivity than could be achieved by either technique alone. A recent advance places small radio frequency terminals and noncontact bar code readers on the wrists of warehouse operators. This allows hands-free operation of the technology. More detailed descriptions of the technology and its applications are provided in [17] and [58].

*Voice Headset*   Synthesized voice communication and human voice recognition (VR) systems are computer-based systems that translate computer data into synthesized speech and translate human speech into computer data. A headset (see Figure 6.98) with ear phones and an attached microphone is used to interact with the computer system. Radio frequencies are used to transmit communications to mobile operators. These systems are attractive when an operator's hands and eyes must be freed up for productive operations. Though VR systems are in their infancy, some systems recognize up to 1000 words and are 99.5% accurate. Systems are in use today for receiving inspection, lift truck storage and retrieval operations, and order picking.

**Figure 6.98**   Voice I/O system. *(Courtesy of Vocollect, Inc.)*

***Light and Computer Aids***   The objectives of light or computer aids (see Figure 6.99) are to reduce the search time, extract time, and documentation time portions of total order picking time and to improve order picking accuracy. Search time is reduced by a computer automatically illuminating a lighted display at the correct storage or retrieval location. Extract time is reduced since the lighted display also indicates the correct quantity to pick. Documentation time is reduced by allowing the order picker to push a button at the pick location to inform the host computer that the pick has been completed. The result is accurate order picking at a rate of up to 600 picks per man hour. Light displays are available for bin shelving, flow rack, and carousel systems.

Computer aids provide a computer display over each pick station. On the display is a picture of the configuration of the storage container in that pick station. A light shines, on the compartment within the container from which the pick is to be performed; the quantity to be picked is displayed on the computer screen. Systems of this type are also available for use with vertical carousels, miniload AS/RS, and vertical lift modules.

The basic solution approach of light and computer-aided systems is to take the thinking out of order picking. Consequently, an unusual human factors challenge is

**Figure 6.99**   Light-aided order picking. *(Courtesy of Kingston-Warren Corporation)*

**Figure 6.100**   Smart card. *(Courtesy of Kingston-Warren Corporation)*

introduced. The job may become too easy, and hence boring and even mentally degrading to some operators.

*Smart Card*   A smart card (essentially a credit card with a magnetic stripe) (see Figure 6.100) is now used to capture information ranging from employee identification, to the contents of a trailer load of material, to the composition of an order picking tour. For example, at a large cosmetics distribution center order picking tours are downloaded onto smart cards. The smart cards are in turn inserted into a smart card reader on each order picking cart. In so doing, the picking tour is illuminated on an electronic map of the warehouse appearing on the front of the cart.

## 6.7   Summary

In this chapter the subject of material handling was introduced. Following a definition of material handling, material handling principles were presented. Next, a systematic approach was suggested for designing material handling systems. The unit load concept was then explored.

A variety of types of material handling equipment was then presented to provide an introduction to the breadth of alternatives available and to emphasize the importance of staying abreast of equipment developments. Just as is the case in methods design, there is no "one best way" of performing material handling. In fact, there probably exist conditions for which each of the types of equipment presented is a preferred method. Unfortunately, too many individuals who "design" material handling systems are guilty of using the same solution for many different handling situations. The addage of "slotting the heads of nails in order to use a favorite screwdriver" applies to many material handling designs.

The equipment selection problem, in theory, is simple; namely, reduce the set of alternative approaches to those that are feasible based on the *material* and *move*. Unfortunately, in practice there generally exist numerous feasible alternatives as well as an abundant number of manufacturers for each equipment type. Consequently, the system designer must rely on judgment and experience in trimming the set of feasible alternatives to a manageable number; undoubtedly, the "optimum" design alternative may be eliminated in the process. However, it is generally the case that no single alternative is superior to all other alternatives by a wide margin. Rather, several good solutions will generally exist and one of them will be selected. The material handling equipment selection process is truly a "satisficing" rather than an "optimizing" process.

Our treatment of material handling equipment was not exhaustive. For example, people-mover systems, front-end loaders, over-the-road trucks, trash and waste disposal equipment, battery chargers, package sealers, labeling machines, elevators, guidance systems, bowl feeders, dock levelers, and conveyor transfers were not described. Additionally, not all types of conveyors, packaging equipment, industrial truck attachments, below-hook lifters, hoists, cranes, monorails, automated storage/retrieval systems, industrial trucks, and containers and supports were treated. Consequently, anyone involved in designing material handling systems must keep abreast of the field.

It is important for the facilities planner to cultivate and maintain a number of reliable information sources. The following are suggested.

1. Trade Associations

   a. The Material Handling Institute, Inc.
      8720 Red Oak Boulevard
      Suite 200
      Charlotte, North Carolina 28217

   b. Conveyor Equipment Manufacturers Association
      1000 Vermont Avenue, N.W.
      Washington, D.C. 20005

   c. Material Handling Equipment Distributors Association
      104 Wilmont Road
      Suite 208
      Deerfield, Illinois 60015

2. Technical and Professional Societies

   a. Institute of Industrial Engineers
      25 Technology Park—Atlanta
      Norcross, Georgia 30092

   b. Materials Handling and Management Society
      8720 Red Oak Boulevard
      Suite 224
      Charlotte, North Carolina 28217

   c. Council of Logistics Management
      2803 Butterfield Road
      Suite 380
      Oak Brook, Illinois 60521-1156

    d. Society of Manufacturing Engineers
      20501 Ford Road
      Dearborn, Michigan 48218

    e. Warehouse Education and Research Council
      1100 Jorie Boulevard
      Suite 170
      Oak Brook, Illinois 60521

3. Trade Journals

    a. *Modern Materials Handling*
      Cahners Publishing
      275 Washington Street
      Newton, Massachusetts 02158

    b. *Material Handling Engineering*
      Penton Publishing
      614 Superior Avenue
      Cleveland, Ohio 44113

    c. *Warehousing Management*
      Chilton Publishing
      1 Chilton Way
      Radnor, Pennsylvania 19089

# BIBLIOGRAPHY

1. Apple, J. M., *Material Handling Systems Design,* Ronald Press, New York, 1972.
2. Apple, J. M., *Lesson Guide Outline on Material Handling Education,* Material Handling Institute, Pittsburgh 1975.
3. Apple, J. M., *Plant Layout and Material Handling,* Ronald Press, 3rd ed., New York, 1977.
4. Apple, J. M. Jr., and Rickles, H. V., "Material Handling and Storage," *Production Handbook,* 4th ed., White, J. A. (ed.), John Wiley, New York, 1987, pp. 5.15–5.37.
5. Barger, B., "Materials Handling Equipment," *Production Handbook,* 4th ed., J. A. White (ed.), John Wiley, New York, 1987, pp. 4. 100–4. 127.
6. Bolz, H. A., and Hageman, G. E. (eds.), *Material Handling Handbook,* Ronald Press, New York, 1958.
7. Bonthuis, J. E., "Basic Pointers on Selecting Conveyor Transfers," *Plant Engineering,* vol. 29, no. 15, pp. 73–75, July 24, 1975.
8. Bowlen, R. W., "ABC's of Optical Scanning," *Industrial Engineering,* vol. 5, no. 10, pp. 14–22, October 1973.
9. Frazelle, E. H., "Material Handling: A Technology for Industrial Competitiveness," *Material Handling Research Center Technical Report,* Georgia Institute of Technology, Atlanta, April 1986.
10. Frazelle, E. H., *Material Handling Systems and Terminology,* Lionhart Publishing, Atlanta, 1992.
11. Frazelle, E. H., "Small Parts Order Picking: Equipment and Strategy," *Material Handling Research Center Technical Report Number 01-88-01,* Georgia Institute of Technology, Atlanta, 30332-0205.

12. Frazelle, E. H., and Apple, J. M. Jr., "Material Handling Technologies," *The Logistics Handbook,* W. A. Copacino (ed.), Macmillan, New York, 1993.

13. Frazelle, E. H., and Apple, J. M. Jr., "Warehouse Operations," *The Distribution Management Handbook,* J. A. Tompkins (ed.), McGraw-Hill, New York, 1993.

14. Frazelle, E. H., and McGinnis, L. F., "Automated Material Handling," *The Encyclopedia of Microcomputers,* A. Kent and J. G. Williams, eds., Marcel Dekker, New York and Basel, 1988.

15. Frazelle, E. H., and Ward, R. E., "Material Handling Technologies in Japan," *National Technical Information Service Report #PB93-128197,* Washington, D. C., 1992.

16. Groover, M. P., "Industrial Robots: A Primer on the Present Technology," *Industrial Engineering,* vol. 12, no. 11, pp. 54–61, 1980.

17. Hill, J. M., "Automatic Identification Systems," *Production Handbook,* 4th ed., J. A. White (ed.), John Wiley, New York, 1987, pp. 4.128–4. 142.

18. Hill, J. M., "Automatic Identification Perspective 1992, *Proceedings of the Material Handling Short Course,* Georgia Institute of Technology, Atlanta, 1992.

19. Hill, J. M., "Automatic Identification from A to Z," *Proceedings of the Material Handling Short Course,* Georgia Institute of Technology, Atlanta, 1993.

20. Horrey, R. J., "Sortation Systems: From Push to High-Speed Fully Automated Applications," *International Conference on Automation in Warehousing Proceedings,* Institute of Industrial Engineers, Atlanta, 1983, pp. 77–83.

21. Kulwiec, R. A., "A Selection and Application Overview: Package and Unit Handling Conveyors," *Plant Engineering,* vol. 28, no. 19, pp. 122–128, Sept. 19, 1974.

22. Kulwiec, R. A., "Material Handling Equipment Guide," *Plant Engineering,* August 21, 1980, pp. 88–99.

23. Kulwiec, R. A. (editor), *Materials Handling Handbook,* John Wiley, New York, 1985.

24. Miller, J. S., "Some Practical Pointers on Pallets and Containers for Automated Conveyors," *Plant Engineering,* vol. 30, no. 24, pp. 88–90, November 25, 1976.

25. Morgan, F. C., "Types and Applications for Switches for Monorail Conveyor Systems," *Plant Engineering,* vol. 30, no. 6, pp. 175–176, March 18, 1976.

26. Muther, R., *Systematic Layout Planning,* Boston, MA, CBI Publishing Co., Inc., 1973.

27. Nadler, G., "What Systems Really Are," *Modern Materials Handling,* vol. 20, no. 7, pp. 41–47, July 1965.

28. Schultz, G. A., "Selection and Application Guidelines for Belt Conveyors for Unit and Package Loads," *Plant Engineering,* vol. 29, no. 16, pp. 71–74, August 7, 1975.

29. Schultz, G. A., "Selection and Application Guidelines for Belt Conveyors for Semibulk Loads," *Plant Engineering,* vol. 29, no. 17, pp. 85–88, August 21, 1975.

30. Schultz, G. A., "Selection and Application Guidelines for Belt Conveyors for Bulk Loads," *Plant Engineering,* vol. 29, no. 18, pp. 95–98, September 4, 1975.

31. Schultz, G. A., "Basic Specification and Application Guidelines for Chain Conveyors and Elevators," *Plant Engineering,* vol. 29, no. 18, pp. 95–98, September 5, 1975.

32. Schultz, G. A., "Chain Conveyors for Unit and Package Handling: A Survey of Equipment Types and Applications," *Plant Engineering,* vol. 30, no. 3, pp. 54–57, February 5, 1976.

33. Schultz, G. A., "Basic Selection and Application Factors for Overhead Trolley Conveyors," *Plant Engineering,* vol. 30, no. 4, pp. 117–120, February 19, 1976.

34. Schultz, G. A., Guidelines for Selecting Chain Conveyors and Elevators for Bulk Materials," *Plant Engineering,* vol. 30, no. 8, pp. 73–76, April 15, 1976.

35. Schultz, G. A., "Factors to Keep in Mind When Selecting Vibrating Conveyors," *Plant Engineering,* vol. 30, no. 17, pp. 105–108, August 19, 1976.

36. Schultz, G. A., "A Practical Procedure for Planning and Budgeting a Package Conveyor System," *Plant Engineering,* vol. 30, no. 18, pp. 93–95, September 2, 1976.

37. Sims, E. R. Jr., "Narrow-Aisle, High-Rise Lift Trucks," *Plant Engineering,* vol. 31, no. 21, pp. 186–191, October 13, 1977.

38.  Suzuki, J., "Guide to the Installation of Automated Sorters," *Proceedings of the 10th International Conference on Automation in Warehousing,* Institute of Industrial Engineers, Atlanta. Oct, 1989.

39.  Tompkins, J. A., *Facilities Design,* North Carolina State University, Raleigh, NC, 1975.

40.  Tompkins, J. A., "Material Handling Systems," *Encyclopedia of Computers and Technology,* (J. Belzer, A. G. Holzman, and A. Kent, eds.), Vol. 10, Marcel Dekker, New York, 1978, pp. 254–273.

41.  White, J. A., *Yale Management Guide to Productivity,* Yale Industrial Truck Division, Eaton Corp., Philadelphia, 1978.

42.  White, J. A., "Determining How Much to Handle," *Modern Materials Handling,* Vol. 46, No. 2, February 1991, p. 31.

43.  White, J. A., "Providing the Right Materials," *Modern Materials Handling,* Vol. 46, No. 3, March 1991, p. 27.

44.  White, J. A., "Handle in the Right Condition," *Modern Materials Handling,* Vol. 46, No. 5, April 1991, p. 27.

45.  White, J. A., "The Right Stuff—in Sequence," *Modern Materials Handling,* Vol. 46, No. 6, May 1991, p. 27.

46.  White, J. A., "Pay Attention to Orientation," *Modern Materials Handling,* Vol. 46, No. 7, June 1991, p. 29.

47.  White, J. A., "Put Them in the Right Place!," *Modern Materials Handling,* Vol. 46, No. 8, July 1991, p. 27.

48.  White, J. A., "Handle at the Right Time!," *Modern Materials Handling,* Vol. 46, No. 9, August 1991, p. 25.

49.  White, J. A., "The Right Cost of Handling," *Modern Materials Handling,* Vol. 46, No. 10, September 1991, p. 27.

50.  White, J. A., "Selecting the Right Methods," *Modern Materials Handling,* Vol. 46, No. 12, October 1991, p. 27.

51.  White, J. A., and Apple, J. M. Jr., "Robots in Material Handling," *Handbook of Industrial Robotics,* S. Y. Nof (ed.), John Wiley, New York, 1985, pp. 955–970.

52.  White, J. A., and Apple, J. M. Jr., "Material Handling," *International Encyclopedia of Robotics Applications and Automation,* R. Dorf (ed.), John Wiley New York, 1988, pp. 873–879.

53.  Wilde, J., "Container Design Issues," *Proceedings of the 1992 Material Handling Management Course,* Institute of Industrial Engineers, Atlanta, June 1992.

54.  Zollinger, H. A., "Do It Yourself Guide to Costing Stacker Systems," *Automation,* vol. 21, pp. 90–93, September 1974.

55.  Zollinger, H. A., "Planning, Evaluating, and Estimating Storage Systems," Presented at Institute of Material Management Education First Annual Winter Seminar Series, Orlando, FL February 1982.

56.  "A Guide for Evaluating and Implementing a Warehouse Bar Code System," Warehousing Education and Research Council, Oak Brook, IL, 1992.

57.  *Basics of Material Handling,* Material Handling Institute, Charlotte, NC, 1973.

58.  *Automatic Identification Systems for Material Handling and Material Management,* Automatic Identification Manufacturers Product Section, Materials Handling Institute, Charlotte, NC, 1976.

59.  *Considerations for Planning and Implementing an Automated Storage/Retrieval System,* Material Handling Institute, Charlotte, NC, 1977.

60.  *Considerations for Planning and Installing an Automated Storage/Retrieval System,* Automated Storage/Retrieval Systems Product Section, Material Handling Institute, Charlotte, NC, 1977.

61.  *Considerations for Planning and Installing Automatic Guided Vehicle Systems,* Automatic Guided Vehicle Systems Section, Material Handling Institute, Charlotte, NC, 1980.

62. *Conveyor Terms and Definitions,* Conveyor Equipment Manufacturers Assn., Washington, D.C., 1966.

63. "Distribution Illustrated," *Handling and Shipping Management,* vol. 18, no. 10, pp. 60 70, 1977.

64. "How Companies Are Justifying High-Rise Storage and Retrieval Systems," *Material Handling Engineering,* vol. 31, no. 2, pp. 58–60, February 1976.

65. "Lift Truck Attachments," *Modern Materials Handling,* vol. 32, no. 8, pp. 55–64, August 1977.

66. "Material Handling: Specialized Handling Systems," *Plant Engineering,* Reprint Series 45-1.

67. "Material Handling: Unit Load Conveyors," *Plant Engineering,* Reprint Series 45-2.

68. "Material Handling: Warehousing," *Plant Engineering,* Reprint Series 45-3.

69. "Material Handling: Overhead Cranes and Hoists," *Plant Engineering,* Reprint Series 45-4.

70. "Material Handling: Storage Systems," *Plant Engineering,* Reprint Series 45-5.

71. "Material Handling: Lift Trucks," *Plant Engineering,* Reprint Series 45-6.

72. "Now You Can Cost-Estimate a High Rise Storage System," *Modern Materials Handling,* vol. 29, no. 2, pp. 52–53, February 1974.

73. "Production Engineering Report: Conveyors," *Production Engineering,* vol. 24, no. 18, pp. 50–59, October 1977.

74. "Radio Frequency Data Communications," Warehousing Education and Research Council, Oak Brook, IL, 1993.

75. *Warehouse Modernization and Layout Planning Guide,* U.S. Naval Supply Systems Command Publication 529, U.S. Naval Supply Systems Command, Richmond, VA, 1988.

# PROBLEMS

6.1 Discuss the relationship between quality and material handling.

6.2 Distinguish between the straddle truck and the straddle carrier as well as the reach truck and the side reach truck.

6.3 Compare the applications of a roller-supported belt conveyor with a belt-driven roller conveyor.

6.4 List the six steps of the engineering design process.

6.5 In designing a material handling system three basic elements must be integrated. What are they?

6.6 Define "material flow."

6.7 Briefly describe the following types of material handling equipment:
   a. Straddle truck.
   b. Slip sheets.
   c. Stacker crane.
   d. Storage carousel.
   e. Side shifter.

6.8 List three methods commonly used to power the rollers on a roller conveyor.

6.9 Distinguish between the following types of equipment and give specific reasons for choosing each.
   a. Reach truck.
   b. Straddle truck.

  **c.** Side-reach truck.

  **d.** Turret truck.

**6.10** Robots are being developed with the capability of sensing the orientation of randomly placed parts on a conveyor such that parts can be picked up and loaded into a machine tool automatically. What are three other approaches that can be used to overcome the problem of randomly oriented parts?

**6.11** What are three of the primary benefits obtained by using industrial truck attachments? For each, given an example of the attachment that provides the benefit.

**6.12** Briefly describe the circumstances that would lead you to consider each of the following conveyors in preference to the others:

  **a.** Belt conveyor, roller supported.

  **b.** Belt conveyor, slider bed supported.

  **c.** Powered roller conveyor.

  **d.** Towline conveyor.

**6.13** Name three types of operator-to-stock systems.

**6.14** Name three types of stock-to-operator picking systems.

**6.15** Name a device or communications technology that helps reduce search time and improves picking accuracy.

**6.16** In choosing a pallet storage system the fundamental trade-offs are product _____ and _____.

**6.17** Name three types of carousel storage/retrieval systems.

**6.18** With several levels of storage rotating independently, independently rotating racks can virtually eliminate picker _____ time.

**6.19** Name two picking systems in which picking always takes place at waist level.

**6.20** Flow racks are replenished from the _____ and offer _____ inventory rotation.

**6.21** A distinction between the storage/retrieval machine and the stacker crane is the _____ rides on a floor-mounted rail and the _____ is top supported.

**6.22** The stringer on a pallet has a length of forty-two inches and the board on top of the pallet supporting the load has a length of forty-eight inches. As a result, the pallet is _____ × _____.

# 7

# *LAYOUT*

## *7.1* INTRODUCTION

The generation of layout alternatives is a critical step in the facilities planning process, since the layout selected will serve to establish the physical relationships between activities. Recognizing that the layout ultimately selected will be either chosen from or based on one of the alternatives generated, it is important for the facilities planner to be both creative and comprehensive in generating layout alternatives.

In this chapter, our objective is to focus on the process of developing layout alternatives. We previously addressed the determination of requirements. Specifically, in Chapter 2 we treated the strategic relationships between facilities planning and manufacturing, distribution, and marketing. From that discussion we recognized the importance of taking a long-range viewpoint and coordinating the facilities plan with the plans of other organizational units. A facilities layout strategy should emerge from the overall strategic plan. Product, manufacturing, marketing distribution, management, and human resource plans will be impacted by and will impact on the facilities layout.

The facilities requirements resulting from product design, process design, and schedule design decisions were examined in Chapter 3. The impact of personnel requirements on space, proximities, and special facility features was treated in Chapter 5.

Chapter 4 provided a comprehensive treatment of activity relationships and space requirements as they affect facilities planning. From that discussion, as well as the emphasis in Chapter 1 on the establishment of activity relationships and determination of space requirements, it is clear that both are critically important in the design of a layout for a facility.

Before proceeding further, it seems appropriate to ask the following question, Which comes first, the material handling system or the facility layout? Many appear to believe the layout should be designed first and then the material handling system should be developed. Yet, material handling decisions can have a significant impact on the effectiveness of a layout. For example, the following decisions will affect the layout:

1.   Centralized versus decentralized storage of work-in-process (WIP), tooling, and, supplies.
2.   Fixed path versus variable path handling.
3.   The handling unit (unit load) planned for the systems.
4.   The degree of automation used in handling.
5.   The type of level of inventory control, physical control, and computer control of materials.

Each of these considerations affects the requirements for space, equipment, and personnel, as well as the degree of proximity required between functions.

Why do people tend to focus first on layout? Perhaps one reason is an over emphasis on the manufacturing process. For example, it seems perfectly logical to place department B next to department A, if process B occurs immediately after process A. In such a situation, the handling problem is reduced to, What is the best way to move materials from A to B? Conventional wisdom suggests that the handling problem can be addressed after the layout is finalized. However, if materials cannot flow directly from a machine in A to a machine in B, then WIP storage is required either in A, B, or elsewhere. Depending on the storage and control requirements, a centralized WIP storage area might be used, so that materials flow from A to S (storage), then from S to B. With such a centralized WIP storage system, materials do not flow from A to B and B no longer needs to be placed next to A. Furthermore, the centralized system provides added flexibility when the process sequence changes.

Another reason for the "layout first" syndrome could be a misapplication of the "handling less is best" adage. When the total system is considered including storage and control, then handling more might be better than handling less, if more and less relate to distance. Our experience indicates "handling less is best" in terms of the number of times materials are handled, but that handling distance is not always the major concern. So, which comes first, the material handling system or the facility layout? Our answer is, "Both!" The layout and the handling system should be designed simultaneously. However, the complexity of the design problem generally requires that a sequential process be used. For this reason, we recommend that a number of alternative handling systems be developed and the appropriate layout be designed for each. The preferred layout will be that which results from a consideration of the total system [25].

# 7.2   BASIC LAYOUT TYPES

In Chapter 4 we identified four types of planning departments:

1.   Fixed material location departments.
2.   Production line departments.

3. Product family departments.

4. Process departments.

Figure 4.1 illustrated the material flows for each type of department. Also provided in Figure 4.1 were the associated layouts common to each type planning department. This figure is repeated here as Figure 7.1.

The layout for a fixed material location department differs in concept from the other three. With the other layouts, material is brought to the workstation; in the case of the fixed material location departments the workstations are brought to the material. It is used in aircraft assembly, shipbuilding, and most construction projects. The layout of the fixed material location department involves the sequencing and placement of workstations around the material or product. Although fixed material location departments are generally associated with very large, bulky products, they certainly are not limited in their application. For example, in assembling computer systems it frequently occurs that the materials, subassemblies, housings, peripherals, and components are brought to a systems integration and test workstation and the finished product is "built" or assembled and tested at that single location. Such layouts will be refereed to as fixed product layouts.

The layout for a production line department is based on the processing sequence for the part(s) being produced on the line. Materials typically flow from one workstation directly to the next adjacent one. Nice, well-planned flow paths generally result in this high-volume environment. Such layouts will be referred to as product layouts.

The layout for a product family department is based on the grouping of parts to form product families. Nonidentical parts may be grouped into families based on common processing sequences, shapes, material composition, tooling requirements, handling/storage/control requirements, and so on. The product family is treated as a pseudoproduct and a pseudoproduct layout is developed. The processing equipment required for the pseudoproduct is grouped together and placed in a manufacturing cell. The resulting layout typically has a high degree of intradepartmental flow and little interdepartmental flow; it is variously referred to as a group layout and a product family layout.

The layout for a process department is obtained by grouping like processes together and placing individual process departments relative to one another based on flow between departments. Typically, there exists a high degree of interdepartmental flow and little intradepartmental flow. Such a layout is referred to as a process layout and is used when the volume of activity for individual parts or groups of parts is not sufficient to justify a product layout or group layout. Fixed product, product, group, and process layouts are compared in Table 7.1. Typically, one will find that a particular situation has some products that fit each of the layout types. Hence, a hybrid or combination layout will often result.

# 7.3 LAYOUT PROCEDURES

A number of different procedures have been developed to aid the facilities planner in designing layouts. These procedures can be classified into two main categories:

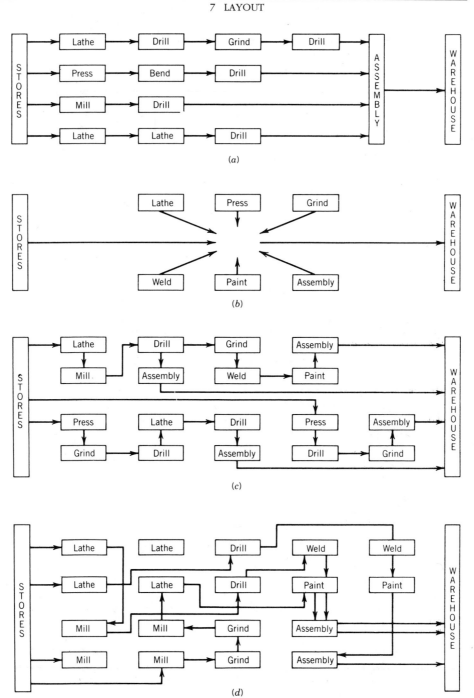

**Figure 7.1** Alternative types of layouts. (*a*) Production line product departments. (*b*) Fixed material location product departments. (*c*) Product family product departments. (*d*) Process departments.

**Table 7.1** *Advantages and Limitations of Fixed Product Layout, Product Layout, Group Layout, and Process Layout*

### Fixed Product Layout

| Advantages | Limitations |
|---|---|
| 1. Material movement is reduced. | 1. Personal and equipment movement is increased. |
| 2. When a team approach is used, continuity of operations and responsibility results. | 2. May result in duplicate equipment. |
| 3. Provides job enrichment opportunities. | 3. Requires greater skill for personnel. |
| 4. Promotes pride and quality because an individual can complete the "whole job." | 4. Requires general supervision. |
| 5. Highly flexible; can accommodate changes in product design, product mix, and production volume. | 5. May result in increased space and greater work-in-process. |
| | 6. Requires close control and coordination in scheduling production. |

### Product Layout

| Advantages | Limitations |
|---|---|
| 1. Smooth, simple, logical, and direct flow lines result. | 1. Machine stoppage stops the line. |
| 2. Small work-in-process inventories should result. | 2. Product design changes cause the layout to become obsolete. |
| 3. Total production time per unit is short. | 3. Slowest station paces the line. |
| 4. Material handling requirements are reduced. | 4. General supervision is required. |
| 5. Less skill is required for personnel. | 5. Higher equipment investment usually results. |
| 6. Simple production control is possible. | |
| 7. Special-purpose equipment can be used. | |

### Group Layout

| Advantages | Limitations |
|---|---|
| 1. By grouping products, higher machine utilization can result. | 1. General supervision required. |
| 2. Smoother flow lines and shorter travel distances are expected than for process layouts. | 2. Greater labor skills required for team members to be skilled on all operations. |
| 3. Team atmosphere and job enlargement benefits often result. | 3. Critically dependent on production control balancing the flows through the individual cells. |
| 4. Has some of the benefits of product layouts and process layouts; it is a compromise between the two. | 4. If flow is not balanced in each cell, buffers and work-in-process storage are required in the cell to eliminate the need for added material handling to and from the cell. |
| 5. Encourages consideration of general-purpose equipment. | 5. Has some of the disadvantages of product layouts and process layouts; it is a compromise between the two. |
| | 6. Decreases the opportunity to use special-purpose equipment. |

Table 7.1   (continued)

| Process Layout | |
|---|---|
| Advantages | Limitations |
| 1. Increased machine utilization. | 1. Increased material handling requirements. |
| 2. General-purpose equipment can be used. | 2. More complicated production control required. |
| 3. Highly flexible in allocating personnel and equipment. | 3. Increased work-in-process. |
| 4. Material handling requirements are reduced. | 4. Longer production lines. |
| 5. Diversity of tasks for personnel. | 5. Higher skills required to accommodate diversity of tasks required. |
| 6. Specialized supervision is possible. | |

construction type and improvement type. *Construction-type* layout methods basically involve developing a new layout "from scratch." *Improvement* procedures generate layout alternatives based on an existing layout.

Although the bulk of the current literature on facility layout concentrate on the development of construction-type procedures, most layout work still involves some form of improving the layout of existing facilities. As Immer [10, p. 32] observed as early as the 1950s,

> Much of (the) work will consist of making minor changes in existing layout, locating new machines, revising a section of the plant, and making occasional studies for material handling. The plans for a complete new production line or a new factory may make the headlines, but except for a war or a new expansion, the average layout (planner) will very seldom have to consider such a problem.

However, the passage on material handling can be a challenge since as we stated in our previous discussion that the layout and material handling system should be designed simultaneously and the process of revising a facility layout should be viewed as an opportunity to also review the effectiveness of existing material handling methods.

We begin our discussion of layout procedures by discussing some of the original approaches to the layout problem. The concepts in these approaches continue to serve as the foundation of many of the methodologies proposed today.

## Apple's Plant Layout Procedure

Apple [1] proposed the following detailed sequence of steps in producing a plant layout.

1.  Procure the basic data.
2.  Analyze the basic data.

3.  Design the productive process.
4.  Plan the material flow pattern.
5.  Consider the general material handling plan.
6.  Calculate equipment requirements.
7.  Plan individual workstations.
8.  Select specific material handling equipment.
9.  Coordinate groups of related operations.
10.  Design activity interrelationships.
11.  Determine storage requirements.
12.  Plan service and auxiliary activities.
13.  Determine space requirements.
14.  Allocate activities to total space.
15.  Consider building types.
16.  Construct master layout.
17.  Evaluate, adjust, and check the layout with the appropriate persons.
18.  Obtain approvals.
19.  Install the layout.
20.  Follow up on implementation of the layout.

Apple noted that these steps are not necessarily performed in the sequence given. As he put it,

> Since no two layout design projects are the same, neither are the procedures for designing them. And, there will always be a considerable amount of jumping around among the steps, before it is possible to complete an earlier one under consideration. Likewise, there will be some backtracking going back to a step already done—to re-check or possibly re-do a portion, because of a development not foreseen. [1, p. 14]

## Reed's Plant Layout Procedure

Reed [14] recommended the following "systematic plan of attack" as required steps in "planning for and preparing the layout."

1.  Analyze the product or products to be produced.
2.  Determine the process required to manufacture the product.
3.  Prepare layout planning charts.
4.  Determine workstations.
5.  Analyze storage area requirements.
6.  Establish minimum aisle widths.
7.  Establish office requirements.
8.  Consider personnel facilities and services.
9.  Survey plant services.
10.  Provide for future expansion.

Reed calls the layout planning chart "the most important single phase of the entire layout process" [14, p. 10]. It incorporates the following:

1. Flow process, including operations, transportation, storage, and inspections.
2. Standard times for each operation.
3. Machine selection and balance.
4. Manpower selection and balance.
5. Material handling requirements.

An example of a layout planning chart is given in Figure 7.2.

## Muther's Systematic Layout Planning (SLP) Procedure

Muther [12] developed a layout procedure he named systematic layout planning or SLP. The framework for SLP is given in Figure 7.3. It uses as its foundation the activity relationship chart described in Chapter 4 and illustrated in Figure 7.4.

Based on the input data and an understanding of the roles and relationships between activities, a material flow analysis (from-to-chart) and an activity relationship analysis (activity relationship chart) are performed. From the analyses performed, a relationship diagram is developed (Figure 7.5).

The relationship diagram positions activities spatially. Proximities are typically used to reflect the relationship between pairs of activities. Although the relationship diagram is normally two dimensional, there have been instances in which three dimensional diagrams have been developed when multistory buildings, mezzanines, and/or overhead space were being considered.

The next two steps involve the determination of the amount of space to be assigned to each activity. From Chapter 4, departmental service and area requirement sheets would be completed for each planning department. Once the space assignments have been made, space templates are developed for each planning department and the space is "hung on the relationship diagram" to obtain the space relationship diagram (Figure 7.6).

Based on modifying considerations and practical limitations, a number of layout alternatives are developed (Figure 7.7) and evaluated. The preferred alternative is then recommended.

While the process involved in performing SLP is relatively straightforward, it does not necessarily follow that difficulties do not arise in its application. We addressed in Chapter 4 such issues as the a priori assignment of activity relationships and the use of proximity as a criterion for measuring the degree of satisfaction of activity relationships. In addition to those concerns, it should be noted that alternative relationship diagrams can often be developed, with apparent equivalent satisfaction of activity relationships. Likewise, the shapes of the individual space templates used in constructing the space relationship diagram can influence the generation of alternatives. Finally, the conversion of a space relationship diagram into several feasible layout alternatives is not a mechanical process; intuition, judgment, and experience are important ingredients in the process.

The SLP procedure can be used sequentially to develop first a block layout and then a detailed layout for each planning department. In the latter application, rela-

LAYOUT PLANNING CHART

PART NO. 1 — PART NAME PLASTIC CONTACT PAD — PCS/ASSY ___ — PCS REQ/HR 70.4 — SHEET 1 OF 1
ASSY NO. — — ASSY NAME ___ — ASSY/PRODUCT 2 — PRODUCTION HRS/DAY 6.0 — PREPARED BY J.G.D. DATE 1-4-60
MATERIAL PLASTIC — SIZE 1¼ OD × ⅝ ID (FROM 4'×6'×¼ SHEETS) — PCS/DAY 422 — LOT SIZE 1 — APPROVED BY ___ DATE ___

| ST NO. | F M S I | DESCRIPTION | OPER NO. | DEPT NO. | TIME PER PIECE | MACHINE OR EQUIPMENT | MACH FRAC | COMB WITH | MACH REQD | OPER PER MACH | CREW FRAC | MAN FRAC | COMB WITH | MEN REQD | HOW MOVED | CONT TYPE | LOAD SIZE | DIST MOVED | REMARKS |
|---|---|---|---|---|---|---|---|---|---|---|---|---|---|---|---|---|---|---|
| 1 | | FROM MATERIALS STORAGE | | | | | | | | | | | | | FORK LIFT | PALLET | 4 SHEETS | 100' | |
| 2 | | ON PALLET BY SAW | | 2 | | | | | | | | | | | | | | | |
| 3 | | TO SAW TABLE | | | | — | | | | | | | | | | | | | |
| 4 | | SAW INTO STRIPS 2¼"×8' | 10 | 2 | .02 | TABLE SAW | .028 | 9-10 16-10 | 1 | 2 | .028 | .056 | 9-10 16-10 1-20 14-30 17-20 3-20 | 2 | | | | | |
| 5 | | TO RACK BY SAW | | 2 | | | | | | | | | | | | | | | |
| 6 | | IN RACK | | | | | | | | | | | 6-30 13-20 | | | | | | |
| 7 | | TO HEATER | | | | | | | | | | | 14-20 | | | | | | |
| 8 | | IN RACK BY HEATER | | 2 | | | | | | | | | | | | | | | |
| 9 | | FEED INTO HEATER | | | | | | | | | | | | | | | | | |
| 10 | | HEAT | 20 | 2 | .04 | HEATER | .055 | 9-20 16-20 | 1 | 1 | .055 | .055 | 6-20 3-10 2-10 9-20 16-20 13-10 14-10 17-10 | 1 | | | | | |
| 11 | | FEED TO PUNCH PRESS | | | | | | | | | | | | | | | | | |
| 12 | | PUNCH TO SHAPE | 30 | 2 | .04 | PUNCH PRESS | .055 | 4-10 4-20 9-30 16-30 | 1 | 1 | .055 | .055 | 4-10 4-20 9-30 16-30 | 1 | | | | | |
| 13 | | TO BIN BY PUNCH PRESS | | | | | | 19A-10 19B-10 | | | | | | | | | | | |
| 14 | | IN BIN | | 2 | | | | | | | | | 19A-20 5A-10 | | 4 WHEEL HAND TRUCK | TOTE BOX | 400 PER BOX | 150' | |
| 15 | | TO PARTS STORAGE | | 5 | | | | | | | | | | | | | | | |
| 16 | | IN PARTS STORAGE | | | | | | | | | | | | | | | | | |
| 17 | | TO ASSEMBLY | | | | | | | | | | | | | 4 WHEEL HAND TRUCK | TOTE BOX | 400 PER BOX | 60' | |
| 18 | | IN BIN IN ASSEMBLY | | 6 | | | | | | | | | | | | | | | |
| 19 | | TO TABLE | | | | | | | | | | | | | | | | | |

Figure 7.2  Layout planning chart.

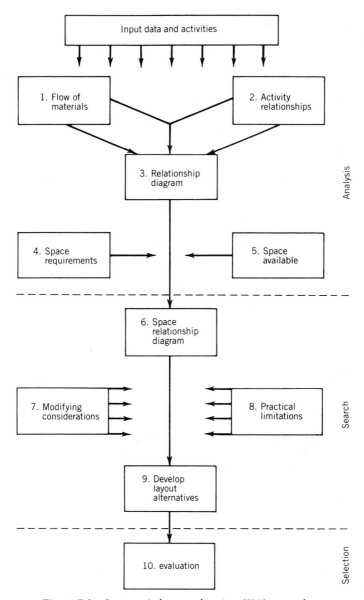

**Figure 7.3**   Systematic layout planning (SLP) procedure.

tionships between workstations, storage locations and entrances to and exits from the department are used to determine the relative location of activities.

## Algorithmic Approaches

The relative placement of departments on the basis of their "closeness ratings" or "material flow intensities" is one that can be reduced to an algorithmic process. In the discussion that follows, we will consider the following three methods:

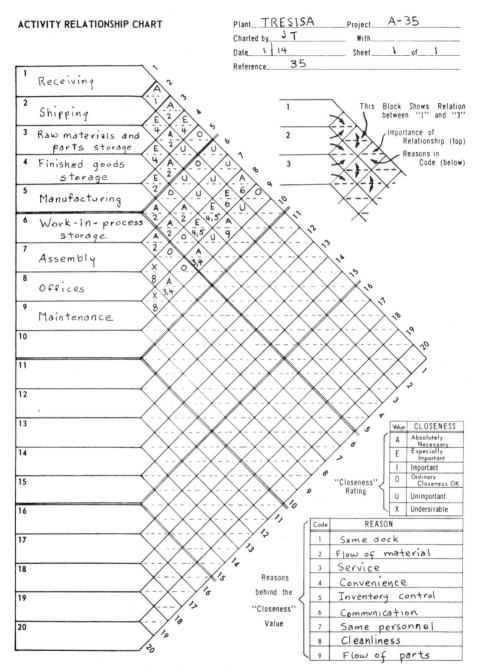

Figure 7.4   Activity relationship chart.

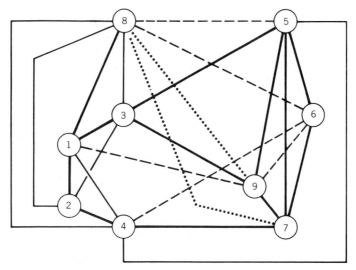

**Figure 7.5**  Relationship diagram.

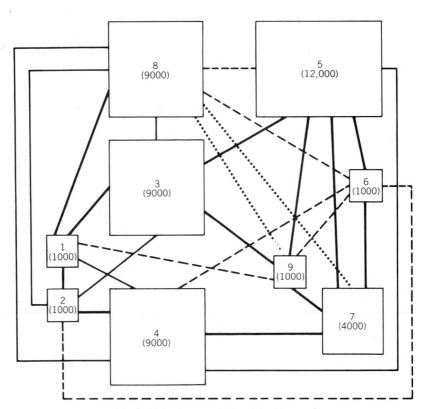

**Figure 7.6**  Space relationship diagram.

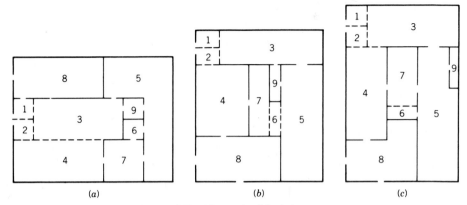

**Figure 7.7**   Alternative block layouts.

1.   Relationship Diagramming,

2.   Pairwise Exchange Method, and

3.   Graph-Based Construction Method.

Relationship Diagramming is a variation of SLP that utilizes some ideas suggested by Apple [1]. The Pairwise Exchange Method is based on the Travel Chart Method given by Reed [14] and the CRAFT procedure of Buffa, et al. [3]. The Graph-Based Construction Method has its roots in Graph Theory. We selected these methods for discussion based on their conceptual contributions to layout planning methods.

## Relationship Diagramming for New Layouts

To illustrate the relationship diagramming procedure, consider the information given in Table 7.2. The activity relationship chart for this illustration is shown in Figure 7.8. The information in this table is converted into a relationship diagramming worksheet as shown in Table 7.3 which will be used as the basis for constructing a relationship diagram and layout. In developing a relationship diagram, we will use a unit block as a template to represent each department.

**Table 7.2**   *Department Areas and Number of Unit Area Templates for the Example Problem*

| Code | Function | Area (Square feet) | Number of Unit Area Templates |
|------|----------|-------------------|-------------------------------|
| 1 | Receiving | 12,000 | 6 |
| 2 | Milling | 8,000 | 4 |
| 3 | Press | 6,000 | 3 |
| 4 | Screw machine | 12,000 | 6 |
| 5 | Assembly | 8,000 | 4 |
| 6 | Plating | 12,000 | 6 |
| 7 | Shipping | 12,000 | 6 |

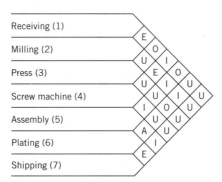

**Figure 7.8**  Activity relationship chart to illustrate a variation of the SLP.

The steps in constructing a relationship diagram are:

**Step 1.** *Select the first department to enter the layout.*

The department with the greatest number of "A" relationships is selected. If a tie exists, the tie-breaking rule is based on the following hierarchy of decisions: the greatest number of "E" relationships, the greatest number of "I" relationships, the fewest number of "X" relationships, and lastly, one of the remaining tied departments are selected randomly.

For the example, departments 5 and 6 are tied with the greatest number of "A" relationships. Department 6 is selected since it has more "E" relationships with another department. The selected department is placed in the center of the layout.

**Step 2.** *Select the second department to enter the layout.*

The second department selected should have an "A" relationship with the first department selected. Additionally, it should have the greatest number of "A" relationships with the other departments not yet selected. Ties are broken using the tie-breaking rules in Step 1.

**Table 7.3**  *Relationship Diagramming Worksheet*

| Rel. | Dept. 1 | Dept. 2 | Dept. 3 | Dept. 4 | Dept. 5 | Dept. 6 | Dept. 7 |
|------|---------|---------|---------|---------|---------|---------|---------|
| A    |         |         |         |         | 6       | 5       |         |
| E    | 2       | 1 4     |         | 2       |         | 7       | 6       |
| I    | 4       | 5 6     |         | 1 5     | 2 4 7   | 2       | 5       |
| O    | 3 5     |         | 1 6     |         | 1       | 3       |         |
| U    | 6 7     | 3 7     | 2 4 5 7 | 3 6 7   | 3       | 1 4     | 1 2 3 4 |
| X    |         |         |         |         |         |         |         |

For the example, department 5 is selected and it is placed in the layout adjacent to department 6.

**Step 3.** *Select the third department to enter the layout.*

The third department selected should have the highest combined relationship with the two departments already in the layout. The highest possible combined relationship would be an "A" relationship with the both of the departments already selected. The ranking hierarchy for the combined relationships is AA, AE, AI, A*, EE, EI, E*, II, and I*, where the notation "*" indicates that the relationship is an "O" or "U" relationships. If there is a tie, the tie-breaking rules in Step 1 apply.

From Table 7.3, we see that department 7 has an "E" relationship with department 6 and an "I" relationship with "E." At this stage, we now address the problem of where to locate department 7 relative to departments 5 and 6, which are already in the layout. The logic of the placement procedure is to locate the new department adjacent to the department with which it has the strongest relationship. Thus, department 7 is placed next to department 6. This is shown in Figure 7.9a.

**Step 4.** *Determine the fourth department to enter the layout.*

The fourth department selected is based on the same logic as in Step 3. The selection is based on the highest combined relationship with the three departments already in the layout. For this case, the ranking hierarchy is

> AAA, AAE, AAI, AA*, AEE, AEI, AE*, AII, AI*, A**, EEE, EEI, EE*, EII, EI*, E**, III, II*, and I**.

As one observes in Table 7.3, department 2 has a combined relationship II* with departments 5, 6, and 7, respectively. Since the strength of the relationship of 2 with departments 5 and 6 are equal, department 2 is placed so that it has a common edge with both 5 and 6. This is shown in Figure 7.9b.

**Step n.** *Department n is placed according to the rules described in Steps 3 and 4.*

The resulting relative locations of departments 1, 2, . . . , n are shown in Figure 7.9. In case of a tie, the tie-breaking rules given in Step 1 apply.

Given the relative location of block templates for this example, the corresponding final layout with the actual dimensions of the departments is shown in Figure 7.10. As a final note, we should emphasize that because of the subjective nature in which block and unit templates are located, several block template location layouts and final layouts should be developed.

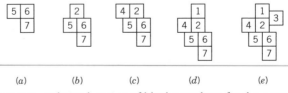

(a)        (b)        (c)        (d)        (e)

**Figure 7.9**  Relative location of block templates for the example.

|   |   | 1 | 1 | 1 |
|---|---|---|---|---|
| 4 | 4 | 1 | 1 |   |
| 4 | 4 | 3 | 1 |   |
| 4 | 2 | 3 | 3 |   |
| 4 | 2 | 2 | 2 |   |
| 5 | 6 | 6 | 6 |   |
| 5 | 6 | 6 | 6 |   |
| 5 | 7 | 7 | 7 |   |
| 5 | 7 | 7 | 7 |   |

**Figure 7.10** Final layout for the example using the relationship diagramming technique.

## Pairwise Exchange Method for Layout Improvement

As indicated in an earlier discussion, we suggested that majority of layout problems involves the redesign of an existing facility, which is typically triggered by the addition of new machines, changes in product mixes, decisions related to the contraction and expansion of storage areas, or a simple realization that the old layout is no longer adequate for its current needs. Thus, we are given an existing layout, and the problems is to come up with an improved layout. The measure of improvement is an important consideration. In this section, we will discuss one heuristic method for layout improvement based on minimizing the total cost of transporting materials among all departments in a facility. We will assume that the distance between departments is rectilinear and is measured from the department centroids. The pairwise exchange procedure is illustrated below through an example.

Consider four departments of equal sizes. The material flows between departments are given in Table 7.4. The existing layout is shown in Figure 7.11*a*.

A distance matrix can be obtained based on the existing layout as given in Table 7.5.

The total cost for the existing layout is computed as follows:

$$TC_{1234} = 10(1) + 15(2) + 20(3) + 10(1) + 5(2) + 5(1) = 125$$

The subscript notation indicates the order of the departments in the initial layout.

The pairwise exchange method simply states that for each iteration, all feasible exchanges in the locations of department pairs are evaluated and the pair that results in the largest reduction in total cost is selected. Since all departments areas are assumed to be of equal size, the feasible exchanges are 1-2, 1-3, 1-4, 2-3, 2-4, and 3-4.

Table 7.4 *Material Flow Matrix*

|   |   | To Department | | | |
|---|---|---|---|---|---|
|   |   | 1 | 2 | 3 | 4 |
| From | 1 | — | 10 | 15 | 20 |
| Department | 2 |   | — | 10 | 5 |
|   | 3 |   |   | — | 5 |
|   | 4 |   |   |   | — |

(a) Iteration 0    1 2 3 4

(b) Iteration 1    3 2 1 4

(c) Iteration 2    2 3 1 4

**Figure 7.11**   Layouts corresponding to each iteration.

The distance matrix is recomputed each time an exchange is performed. The total costs resulting from these exchanges are

$$TC_{2134}(1\text{-}2) = 10(1) + 15(1) + 20(2) + 10(2) + 5(3) + 5(1) = 105$$
$$TC_{3214}(1\text{-}3) = 10(1) + 15(2) + 20(1) + 10(1) + 5(2) + 5(3) = 95$$
$$TC_{4231}(1\text{-}4) = 10(2) + 15(1) + 20(3) + 10(1) + 5(1) + 5(2) = 120$$
$$TC_{1324}(2\text{-}3) = 10(2) + 15(1) + 20(3) + 10(1) + 5(1) + 5(2) = 120$$
$$TC_{1432}(2\text{-}4) = 10(3) + 15(2) + 20(1) + 10(1) + 5(2) + 5(1) = 105$$
$$TC_{1243}(3\text{-}4) = 10(1) + 15(3) + 20(2) + 10(2) + 5(1) + 5(1) = 125$$

Thus, we select the pair 1-3 and perform the exchange in the layout. This is shown in Figure 7.11*b*.

For the next iteration, we consider all feasible exchanges which consist of the same set as in iteration 1. The resulting total costs are

$$TC_{3124}(1\text{-}2) = 10(1) + 15(1) + 20(2) + 10(1) + 5(1) + 5(3) = 95$$
$$TC_{1234}(1\text{-}3) = 10(1) + 15(2) + 20(3) + 10(1) + 5(2) + 5(1) = 125$$
$$TC_{3241}(1\text{-}4) = 10(2) + 15(3) + 20(1) + 10(1) + 5(1) + 5(2) = 110$$
$$TC_{2314}(2\text{-}3) = 10(2) + 15(1) + 20(1) + 10(1) + 5(3) + 5(2) = 90$$
$$TC_{3412}(2\text{-}4) = 10(1) + 15(2) + 20(1) + 10(3) + 5(2) + 5(2) = 105$$
$$TC_{4213}(3\text{-}4) = 10(1) + 15(1) + 20(2) + 10(2) + 5(1) + 5(3) = 105$$

The pair 2-3 is selected with a total cost value of 90. Figure 7.11*c* shows the resulting layout after 2 iterations. Continuing on, the third iteration calculations are

$$TC_{3214}(1\text{-}2) = 10(1) + 15(2) + 20(1) + 10(1) + 5(2) + 5(3) = 95$$
$$TC_{1324}(1\text{-}3) = 10(2) + 15(1) + 20(3) + 10(1) + 5(1) + 5(2) = 120$$

**Table 7.5**   *Distance Matrix Based on Existing Layout*

|  |  | To Department | | | |
|---|---|---|---|---|---|
|  |  | 1 | 2 | 3 | 4 |
| From | 1 | — | 1 | 2 | 3 |
| Department | 2 |  | — | 1 | 2 |
|  | 3 |  |  | — | 1 |
|  | 4 |  |  |  | — |

$$TC_{3421}(1\text{-}4) = 10(1) + 15(3) + 20(2) + 10(2) + 5(1) + 5(1) = 125$$
$$TC_{2134}(2\text{-}3) = 10(1) + 15(1) + 20(2) + 10(2) + 5(3) + 5(1) = 105$$
$$TC_{3142}(2\text{-}4) = 10(2) + 15(1) + 20(1) + 10(3) + 5(1) + 5(2) = 100$$
$$TC_{4123}(3\text{-}4) = 10(1) + 15(2) + 20(1) + 10(1) + 5(2) + 5(3) = 95$$

Since the lowest total cost for this iteration, 95, is worse than the total cost value of 90 in the second iteration, then the procedure is terminated. Thus, the final layout arrangement is 2-3-1-4 as shown in Figure 7.11*c*.

***Remarks.*** The pairwise exchange procedure described above is not guaranteed to yield the optimal layout solution because the final outcome is dependent on the initial layout, that is, a different initial layout can result in another solution. Thus, we can only claim local optimality. Also, you may have observed that it is possible to cycle back to one of the alternative layout arrangements from a previous iteration. For instance, the layout arrangement 1-2-3-4 is what we started with and we see the same arrangement in the second iteration when departments 1 and 3 were exchanged based on the solution from iteration one, that is, layout arrangement 3-2-1-4. Additionally, symmetric layout arrangements may also occur, for example, 4-3-2-1 in iteration 3 is identical to 1-2-3-4.

Second, the pairwise exchange can be easily accomplished only if the department pair considered are of equal size (as we assumed in this example) or if they share a common border. Otherwise, we would have to expend some effort in rearranging all the departments between the two departments being exchanged. Given this common border condition, we can expand the concept of pairwise exchange to a three-department exchange, or two first then three and vice versa.

The observations made above are important considerations in designing efficient improvement-type layout algorithms.

## Graph-Based Construction Method for New Layouts

The recognition of the usefulness of graph theory as a mathematical tool in the solution of facilities planning problems dates back to the late 1960s, Krejcirik [11], and early 1970s, Seppanen and Moore [15]. The use of graph theory methods has strong similarities with the SLP method developed by Muther [12].

***Basic Concepts.*** Consider the block plan layout in Figure 7.12*a*. We first construct an adjacency graph where each node represents a department and a connecting arc between two nodes indicate that two departments share a common border. A similar graph is constructed for the alternative block plan layout shown in Figure 7.12*b*. We observe that the two graphs shown in Figure 7.12 are subgraphs of the relationship diagram shown in Figure 7.13*b*, which is derived from the relationship chart in Figure 7.13*a*. The relationship chart displays numerical "weights" rather than alphabetic closeness ratings.

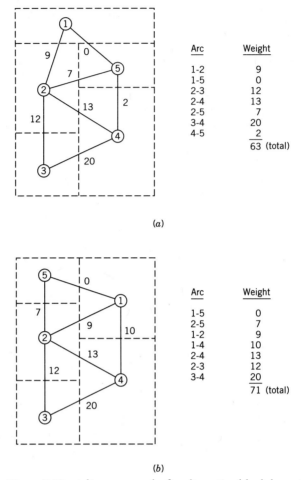

(a)

(b)

**Figure 7.12** Adjacency graphs for alternative block layouts.

Which block plan layout is better? We can score each block plan layout by summing the numerical weights assigned to each arc. On this basis, block plan layout (b) is better than block plan layout (a) with scores of 71 and 63, respectively. Thus, finding a maximally weighted block plan layout is equivalent to obtaining an adjacency graph with the maximum sum of arc weights.

Before we describe a method for determining adjacency graphs, we first make the following observations:

a. The score does not account for distance, nor does it account for relationships other than those between adjacent departments.

b. Dimensional specifications of departments are not considered; the length of common boundaries between adjacent departments are also not considered.

c. The arcs do not intersect; this property of graphs is called planarity. We note that the relationship diagram is usually a nonplanar graph.

d. The score is very sensitive to the assignment of numerical weights in the relationship chart.

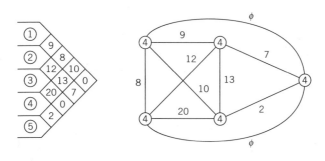

(*a*) Relationship Chart          (*b*) Relationship Diagram

**Figure 7.13**  Relationship chart and relationship diagram for graph-based example.

***Procedure.***  There are two strategies we can follow in developing a maximally weighted planar adjacency graph. One way is to start with the relationship diagram and selectively prune connecting arcs while making sure that the final graph is planar. A second approach is to iteratively construct an adjacency graph via a node insertion algorithm while retaining planarity at all times. A heuristic procedure based on the second approach is described below.

**Step 1.**  From the relationship chart in Figure 7.13*a*, select the department pair with the largest weight. Thus, departments 3 and 4 are selected to enter the graph.

**Step 2.**  Next, select the third department to enter. The third department is selected based on the sum of of the weights with respect to departments 3 and 4. From Figure 7.14*a* department 2 is chosen with a value of 25. The columns in this figure correspond to the departments already in the adjacency graph and the rows correspond to departments not yet selected. The last column gives the sum of the weights for each unassigned department.

**Step 3.**  We then pick the fourth department to enter by evaluating the value of adding one of the unassigned departments represented by a node on a face of the graph. A face of a graph is a bounded region of a graph. For instance, a triangular face is the region bounded by arcs 2-3, 3-4, and 4-2 in Figure 7.14*a*. We will denote this face as 2-3-4. The outside region is referred to as the external face. For our example, the value of adding departments 1 and 5 are 27 and 9, respectively. Department 1 is selected and placed inside the region 2-3-4, as shown in Figure 7.14*b*.

**Step 4.**  The remaining task is to determine on which face to insert department 5. For this step, department 5 can be inserted on faces 1-2-3, 1-2-4, 1-3-4, and 2-3-4. Inserting 5 on faces 1-2-4 and 2-3-4 yields identical values of 9. We arbitrarily select 1-2-4. The final adjacency graph is given in Figure 7.14*c*. This solution is optimal with a total sum of arc weights equal to 81.

**Step 5.**  Having determined an adjacency graph, a final step is to reconstruct a corresponding block layout. A block layout based on the final adjacency graph is shown in Figure 7.15. The manner by which we constructed the

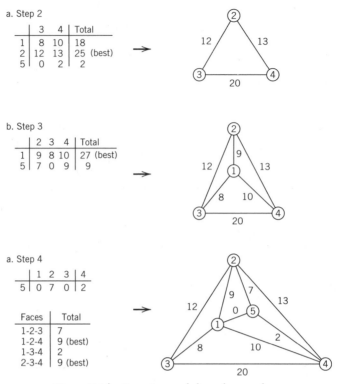

a. Step 2

|   | 3 | 4 | Total |
|---|---|---|---|
| 1 | 8 | 10 | 18 |
| 2 | 12 | 13 | 25 (best) |
| 5 | 0 | 2 | 2 |

b. Step 3

|   | 2 | 3 | 4 | Total |
|---|---|---|---|---|
| 1 | 9 | 8 | 10 | 27 (best) |
| 5 | 7 | 0 | 9 | 9 |

a. Step 4

|   | 1 | 2 | 3 | 4 |
|---|---|---|---|---|
| 5 | 0 | 7 | 0 | 2 |

| Faces | Total |
|---|---|
| 1-2-3 | 7 |
| 1-2-4 | 9 (best) |
| 1-3-4 | 2 |
| 2-3-4 | 9 (best) |

**Figure 7.14** Steps in graph-based procedure.

block layout is analogous to the SLP method. We should note that in constructing the block layout, the original department shapes had to be altered significantly in order to satisfy the requirements of the adjacency graph. In practice, we may not have as much latitude in making such alterations since department shapes are generally derived from the geometry of the individual machines within the department and the internal layout configuration.

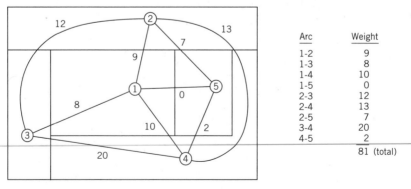

| Arc | Weight |
|---|---|
| 1-2 | 9 |
| 1-3 | 8 |
| 1-4 | 10 |
| 1-5 | 0 |
| 2-3 | 12 |
| 2-4 | 13 |
| 2-5 | 7 |
| 3-4 | 20 |
| 4-5 | 2 |
| | 81 (total) |

**Figure 7.15** Block layout from the final adjacency graph.

Finally, we should point out that there are algorithmic methods for performing this step as demonstrated by Giffin, et al. [7] and Hassan and Hogg [9].

# 7.4  THE IMPACT OF CHANGE

The need for a facility layout study can arise under a variety of circumstances. For example, some of the more common situations that arise in the context of plant layout include the following:

1.  Changes in the design of existing product, the elimination of products from the product line, and the introduction of new products.
2.  Changes in the processing sequences for existing products, replacements of existing processing equipment, and changes in the use of general-purpose and special-purpose equipment.
3.  Changes in production quantities and associated production schedules, resulting in the need for capacity changes,
4.  Changes in the organizational structure as well as changes in management philosophies concerning production strategies such as the adoption of Just-in-Time concepts, Total Quality Management, etc.

If requirements change frequently, then it is desirable to plan for change and to develop a flexible layout—one that can be easily modified, expanded, or contracted. Robert L. Propst [13, p. 16], in addressing the impact of change on facilities planning for electronics manufacturing, noted:

> Our ancestors could deal with changes as an evolutionary factor. Change could be digested in small steps or ignored for a lifetime. Today, it is the dominating reality; our new natural state of affairs.

> Curiously, it is the lack of flexibility in our physical facilities that is proving to be the bottleneck in electronics production. A great many irritations stem from services and facilities that respond too slowly, or not at all, to our buildings, furnishings, and service that have to be revitalized and revisualized.

Flexibility can be achieved by utilizing modular office equipment, workstations, and material handling equipment; installing general-purpose production equipment; utilizing a grid-based utilities and services system; and using modular construction. Additionally, the design of the facility can have a significant impact on the ease and cost of expansion. We will discuss some of these issues in more detail below.

## Adapting to Change and Planning for Facility Reorganization

Before getting too specific about how to plan for change, it would be worthwhile to step back and observe that in many manufacturing organizations there are cycles of expansion and decline due to the very nature of the business environment. In short, the manufacturing environment is very dynamic. The facility layout should also be treated as dynamic. In as much as businesses should have long-term business strategies, we must also have a master plan for facility layout. This master plan should attempt to anticipate future requirements and make provisions for adapting to changes in facility requirements. Oftentimes, prior decisions impose severe constraints which make it very difficult to institute innovative changes in the facility layout. A few examples include placements of shipping and receiving docks, locations of heavy machineries, clear building heights, floors with low load bearing capacities, load bearing walls, location of utilities, and so on.

The master layout plan must also provide the means for a facility to react quickly to change, adding capacity in a short period of time or be able to operate efficiently at scaled-down operating levels. The facility design must be flexible in order to provide this high level of responsiveness. We note, however, that variations in production requirements should not be interpreted as signaling a need to change the facility layout. Often, opportunities for adjustments can be made by seeking more efficient machine schedules, better maintenance of equipment, smoother material flows, and closer coordination with customers and suppliers in identifying the criticality of deliver dates. When all else fails, it may be time to do a new layout of the production facility. Or in some cases, it may be the easiest alternative to implement particularly if the cost of change is marginal. And when these occasions arise, then the layout must be flexible enough to quickly accommodate these changes.

How do we develop such a flexible layout? Harmon and Peterson [8] suggest the use of the following objectives:

1. Reorganize factory subplants to achieve superior manufacturing status.

2. Provide maximum perimeter access for receiving and shipping materials, components, and products as close to each subplant as practical.

3. Cluster all subplants dedicated to a product or product family around the final process subplant to minimize inventories, shortages, and improve communication.

4. Locate supplier subplants of common component subplants in a central location to minimize component travel distances.

5. Minimize the factory size to avoid wasted time and motion of workers.

6. Eliminate centralized storage of purchased materials, components, and assemblies and move storage to focused subplants.

7. Minimize the amount of factory reorganization that will be made necessary by future growth and change.

8. Avoid locating offices and support services on factory perimeters.

9. Minimize the ratio of aisle space to production process space.

The idea of breaking up a factory into smaller enterpreneurial units is not a new concept, see Skinner [17]. Each subunit, or a subplant, can be organized much more efficiently. Alternative layout configurations can be designed for each subplant to take advantage of their specific product and process requirements. An illustration of a factory, which is organized along the subplant concept, is given in Figure 7.16. The illustration shows a major flow structure revolving around a spine flow and material

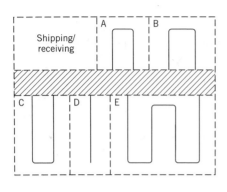 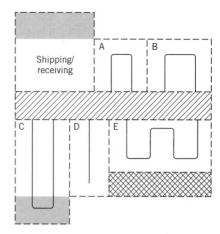

(a) Spine Layout with U, I and S Flows
Within Departments

(b) Layout After Expanding and Contracting
Some Departments

**Figure 7.16**   Illustration of a flexible layout.

flows within each subplant organized along U-shaped, I-shaped and S-shaped flow structures,[1] see Figure 7.16a. You will note that facility expansions and contractions are easily achieved with this modular configuration by simply extending outwards or contracting inwards within each sub-unit as shown in Figure 7.16b. A critical consideration, however, is the central flow structure, which could end up as the bottleneck when the total material handling requirements between subplants exceeds the capacity limits of the spine flow server. Sufficient capacity must be incorporated in the initial design in anticipation of future material flow requirements.

## Volvo Facilities

Volvo Skovdeverken's factory for the manufacture of four-cylinder engines is designed on the basis of the modular concept. According to Volvo, the factory

> . . . . is characterized by a new concept of layout, environment, technology and work organization. The company's own technicians have worked together with representatives of the employees and with outside experts during the planning and execution. This has resulted in a further development of the striving for a better working environment worker satisfaction and well-being which has been typical for Volvo's activities in various places for many years.
>       Since practical considerations require that the machining, assembly and test departments shall adjoin each other the factory layout is of basic importance. The objective was to create the atmosphere of a small workshop while retaining the advantages of rational production and a flexible flow of materials that a large factory makes possible. The result is a layout quite different from the traditional pattern and giving many practical and environmental advantages.

---

[1]For a comprehensive discussion on alternative material flow structures, see Tanchoco [18].

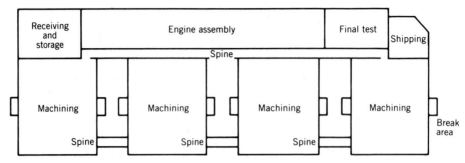

**Figure 7.17**   Block layout of the Volvo Skovdeverken factory.

The new factory consists of a body containing the assembly and test departments with four arms at right angles which contain the machining departments. These latter are separated from one another by generous garden areas. The method of assembly of the engine is one of the most interesting features, both from the technical and the work organizations points of view. The conventional assembly line principle has been replaced by an extremely flexible system of group assemblies. Electrically driven assembly trucks (AGVS) which are controlled by the assembly personnel, together with other technically advanced solutions, have given the work of building the engine a new interest. The different teams, in the machining departments as well as the assembly, have cooperated in the design of the workplaces and are now taking part in the development of new forms of work organization [21]

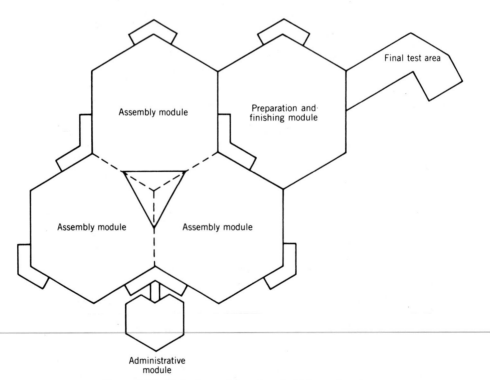

**Figure 7.18**   Volvo's modular plant in Kalmar, Sweden.

As shown in Figure 7.17, the four manufacturing modules are connected by a material flow/personnel flow spine at each end of the module. The plant is approximately 400,000 ft$^2$ in size. Windows in each manufacturing module and break area overlook landscaped gardens.

Volvo Kalmarverken's automobile assembly plant provides another example of a modular design. As shown in Figure 7.18, the assembly plant consists of four equal-sized hexagonal modules, with three two-story assembly modules and one one-story preparation and finishing module. Additionally, a smaller two-story hexagonal module housing administrative and engineering support offices is located in front of and connected to the assembly building.

The assembly operations are performed adjacent to the outer walls of the three assembly modules. The assembly paths are indicated in Figure 7.19. From Figure 7.20, notice the location of the material storage area in the center of the plant. Lift trucks are used to store and retrieve materials in the storage racks, as well as to transport materials between the storage area and pickup-and-deposit stations located on each floor. The lift truck operates on the first floor and lifts/lowers materials to/from the second floor.

The Volvo Kalmar plant is recognized internationally for its pioneering work on job enlargement and the team approach to automotive assembly. When the plan-

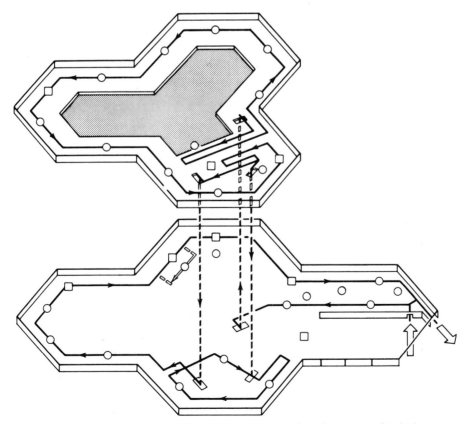

**Figure 7.19**  Assembly path at Volvo's Kalmar plant. (*Courtesy of Volvo.*)

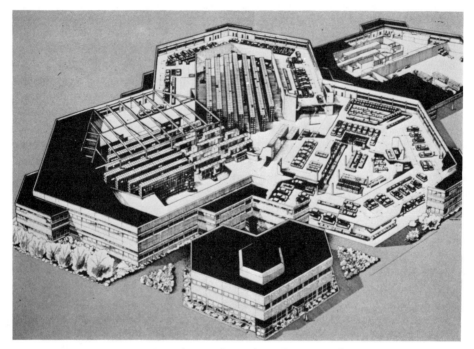

**Figure 7.20** Volvo's Kalmar plant. *(Courtesy of Volvo)*

ning for the Kalmar plant began in the early 1970s, Pehr G. Gyllenhammar, managing director of the Volvo Group, gave the following general directive:

> It has to be possible to create a working place which meets the need of the modern human being for motivation and satisfaction in his daily work. It must be possible to accomplish this objective without reducing efficiency. [21]

The team concept was one of the basic objectives established at the beginning of the facilities planning process. The organization was to be built around the team concept. The team members were to collaborate on a common set of tasks and work within an established production framework. They were to be allowed to switch jobs among themselves, vary their work pace, carry joint responsibility for quality, and have the possibility to influence their working environment. It was felt by Volvo's management that the larger number of work tasks, combined with team membership, would provide greater meaning and satisfaction for each employee.

As a result of the planning objectives, each work team was provided its own entrance, its own changing room and saunas, its own break area, and its own assembly area. Each assembly area is approximately 10,000 ft$^2$ in size and is viewed as the team's own small workshop.

Assembly is performed by 20 different teams. Each team completes one system in the car, for example, the electrical system, instruments, and safety equipment. Assembly is carried out on special wire-guided, battery-powered, computer-controlled carriers (automated guided vehicle system, AGVS).

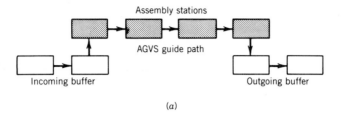

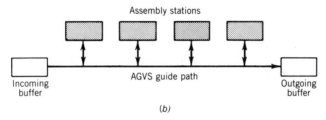

**Figure 7.21** Assembly approaches used at Volvo's Kalmar plant. (*a*) Straight-line assembly. (*b*) Dock assembly.

Two assembly approaches are used. As illustrated in Figure 7.21, straight-line assembly and dock assembly methods are used. With straight-line assembly the work to be performed by the team is divided among four or five workstations. The workers work in pairs and follow a car from station to station carrying out the entire work assignment belonging to their team. When a pair of workers complete work on one car, they walk back to the beginning station in their area and repeat the process. Typically, the two-person team members trade work assignments to provide additional variety to their work.

The dock assembly approach is used when the entire assembly task for an area is carried out at one of the four assembly stations by a team of two or three workers. With the dock assembly approach, the AGVS brings a car to an assembly dock where the entire work cycle is performed. The content and quality of the work performed is no different than that for the straight-line assembly approach.

## 7.5 DEVELOPING LAYOUT ALTERNATIVES

Because the final layout will result from the generation of alternative facility layouts, it is important that the number, quality, and variety of alternative designs be as large as feasible. If one agrees that facilities planning involves both art and science, then there would surely be agreement that "the art" comes in generating layout alternatives. It is at this stage that one's creativity is truly tested.

We have previously emphasized the need to divorce one's thinking from the present method. The ideal system approach was intended to facilitate the creative generation of alternatives. There exist a number of other aids to improve one's ability to generate more and better design alternatives. The following are some suggested approaches that have proven to be beneficial.

1. Exert the necessary effort.
2. Set a time limit.
3. Seek many alternatives.
4. Establish a goal.
5. Make liberal use of the questioning attitude.
6. Don't get bogged down in details too soon.
7. Don't "fail to see the forest for the trees."
8. Think big, then think little.
9. Don't be conservative.
10. Avoid premature rejection.
11. Avoid premature acceptance.
12. Refer to analogous problems of others.
13. Consult the literature.
14. Consult peers in other organizations.
15. Use the brainstorming technique.
16. Divorce your thinking from the existing solution.
17. Spread the effort out over time.
18. Involve operating people.
19. Involve management.
20. Involve experienced people.
21. Involve inexperienced people.
22. Involve those who oppose change.
23. Involve those who promote change.
24. Be aware of what the competition is doing.
25. Recognize your own limitations.
26. Look for trends.
27. Do your homework first.
28. Understand the requirements.
29. Don't overlook an improved present method.
30. Think long range.

The number of possible facility layouts for any reasonably sized problem can be quite large. Furthermore, the specification of constraints and the establishment of an accurate objective function are tasks that can be accomplished "fuzzily" at best. For this reason, the design process does not ask for the selection of the best design from among all possible feasible designs. Rather, it asks for a selection to be made from among several reasonable designs—it asks for the designer to "satisfice" not "optimize."

The problem of designing a facility layout can be distinguished from a number of other problem solving activities. As observed by Simon, "In ordinary language . . . we apply the term 'design' only to problem solving that aims at synthesizing new objects. If the problem is simply to choose among a given set of alternatives, e.g., to choose the location or site for a plant, we do not usually call it a design problem, even if the set of available alternatives is quite large, or possibly infinite [16, p. 295]."

Simon gives two reasons for calling the layout problem, but not, a math programming problem, a design problem. The negative reason he gives is that no simple finite algorithm exists for obtaining directly a solution to the layout problem. The positive reason given by Simon is that the process used for attacking the layout problem involves synthesizing the solution from intermediate, or component, decisions that are "*selective, cumulative,* and *tentative* [16]."

Because of the immense number of possible layout designs available to the facilities planner, the search process becomes a search over a space of design components and partially completed designs, rather than a search over a space of designs. In following the design process, very few *complete* designs are generated, compared, and evaluated. Instead, one generates design components, for example, receiving and shipping component, assembly component, and manufacturing component. Even at the component level, the comparisons and evaluations are typically performed at a macrodesign level. Quite often, only one complete design is produced. The sequential process of generating and evaluating component designs is sufficiently effective as a filter that the complete design is the final design.

## 7.6  SUMMARY

The objectives of this chapter were fourfold. First, we felt it important to cover the "tried-and-true" material concerning the basic types of layout and alternative layout planning procedures. Second, we wanted to discuss more contemporary methods and show the strong linkage between the classical and contemporary methods. Third, we wanted to emphasize the importance of planning for change. Change is the only thing that is certain to occur in the future and planning for change will help guarantee that an organization is able to adapt and remain competitive. Fourth, we wished to encourage you to "enter the world of the architect."

Aspects of form and style as they relate to facilities planning should not be dismissed out of hand. The architect typically approaches the layout planning problem from a different perspective than does the facilities planning engineer. It is important for the engineer to understand "where the architect is coming from." Furthermore, it has been our experience that the success of the engineer in performing facilities planning is frequently totally dependent on the ability to work together with the architect.

In this chapter, we have focused on a number of pragmatic issues related to layout planning. In the next chapter, we consider some computer aided design tools for developing facility layouts.

# BIBLIOGRAPHY

1.  Apple, J. M., *Plant Layout and Material Handling,* 3rd ed., John Wiley, New York, 1977.
2.  Askin, R. G. and Standridge, C. R. *Modeling and Analysis of Manufacturing Systems,* John Wiley, New York, 1993.
3.  Buffa, E. S., Armour, G. C., and Vollman, T. E., "Allocating Facilities with CRAFT," *Harvard Business Review,* vol. 42, no. 2, pp. 136–159, March/April 1964.
4.  Carrie, A. S., Moore, J. M., Roczniak, R., and Seppanen, J. J., "Graph Theory and Computer Aided Facilities Design," *OMEGA, The International Journal of Management Science,* vol. 6, no. 4, pp. 353–361, July/August 1978.
5.  Francis, R. L., McGinnis, L. F., and White, J. A., *Facility Layout and Location: An Analytical Approach,* 2nd ed., Prentice-Hall, Englewood Cliffs, NJ, 1992.
6.  Foulds, L. R., Gibbons, P. B., and Giffin, J. W., "Facilities Layout Adjacency Determination: An Experimental Comparison of Three Graph Theoretic Heuristics," *Operations Research,* vol. 33, no. 5, pp. 1091–1106, 1985.
7.  Giffin, J. W., Foulds, L. R., and Cameron, D. C., "Drawing a Block Plan From a Relationship Chart with Graph Theory and Microcomputer," *Computers and Industrial Engineering,* vol. 10, pp. 109–116, 1986.
8.  Harmon, R. L., and Peterson, L. D., *Reinventing the Factory: Productivity Breakthroughs in Manufacturing Today,* The Free Press, New York, 1990.
9.  Hassan, M. M. D., and Hogg, G. L., "On Constructing a Block Layout by Graph Theory," *International Journal of Production Research,* vol. 29, no. 6, pp. 1263–1278, 1991.
10. Immer, J. R., *Layout Planning Techniques,* McGraw-Hill, New York, 1950.
11. Krejcirik, M., *Computer Aided Plant Layout,* Computer Aided Design, vol. 2, pp. 7–19, 1969.
12. Muther, R., *Systematic Layout Planning,* 2nd ed., Cahners Books, Boston, 1973.
13. Propst, R. L., *The Action Factory System: An Integrated Facility for Electronics Manufacturing,* Herman Miller Corp., Zeeland, MI, 1981.
14. Reed, R. Jr., *Plant Layout: Factors, Principles and Techniques,* Richard D. Irwin, IL, 1961.
15. Seppannen, J., and Moore, J. M., "Facilities Planning with Graph Theory," *Management Science,* vol. 17, no. 4, pp. 242–253, December 1970.
16. Simon, H. A., "Style in Design," in C. M. Eastman (ed.), *Spatial Synthesis in Computer-Aided Building Design,* John Wiley, New York, 1975.
17. Skinner, W., "The Focused Factory," *Harvard Business Review,* pp. 113–121, May–June 1974.
18. Tanchoco, J. M. A., ed., *Material Flow Systems in Manufacturing,* Chapman and Hall, London, U.K., 1994.
19. Tompkins, J. A. "Modularity and Flexibility: Dealing with Future Shock in Facilities Design," *Industrial Engineering,* vol. 12, no. 9, pp. 78–81, September 1980.
20. Tompkins, J. A. and Moore, J. M. *Computer Aided Layout: A User's Guide,* American Institute of Industrial Engineers, Norcross, GA, 1977.
21. *Volvo: The Engine Factory That is Different,* a Volvo publication describing the Volvo Skovde plant, Gothenberg, Sweden.
22. *Montering vid AB Volvo, Lamarverken,* a Volvo publication (in Swedish) describing the Volvo Kalmar plant, Gothenberg, Sweden.
23. *Volvo Kalmerverken,* a Volvo publication describing the Volvo Kalmar plant, Gothenberg, Sweden.
24. White, J. A., "Modular Material Handling Systems," presentation to the AIIE/MHI Seminar, Long Beach, CA, March 1979.
25. White, J. A., "Layout: The Chicken or the Egg?," *Modern Materials Handling,* vol. 35, no. 9, p. 39, September 1980.

# PROBLEMS

**7.1**   What are some of the important factors that should be taken into consideration when a layout is being designed?

**7.2**   What kind of impacts do the material handling decisions have on the effectiveness of a facility layout in a manufacturing environment?

**7.3**   What evaluation measures can be used to evaluate material handling performance?

**7.4**   What kind of manufacturing environments are the following types of layout designs best suited for?

a.  Fixed Product Layout          b.  Product Layout
c.  Group Layout                  d.  Process Layout

**7.5**   Compare the primary layout design objectives for the following situations:

a.  Soda bottler                  b.  Printing shop
c.  Meat-processing plant         d.  Furniture manufacturing plant
e.  Computer chip maker           f.  Shipyard
g.  Refinery plant                h.  College campus

**7.6**   What are the basic differences between construction type and improvement type layout procedures?

**7.7**   Contrast and compare the plant layout procedures proposed by Apple, Reed, and Muther.

**7.8**   Four departments are to be located on a building of 600 ft × 1000 ft. The expected personnel traffic flows and area requirement for departments are shown in the tables below.
a.  Develop a block layout design using SLP, and
b.  Develop a block layout design using Relationship Diagramming.

| Dept. | A | B | C | D |
|-------|-----|-----|-----|-----|
| A | 0 | 250 | 25 | 240 |
| B | 125 | 0 | 400 | 335 |
| C | 100 | 0 | 0 | 225 |
| D | 125 | 285 | 175 | 0 |

| Department: | Department Dimension: |
|-------------|----------------------|
| A | 200 ft. × 200 ft. |
| B | 400 ft. × 400 ft. |
| C | 600 ft. × 600 ft. |
| D | 200 ft. × 200 ft. |

**7.9**   XYZ Inc. has a facility with six departments (A, B, C, D, E, and F). A summary of the processing sequence for ten products and the weekly production forecasts for the products are given in the tables below.
a.  Develop the from-to chart based on the expected weekly production.
b.  Develop a block layout design using SLP.
c.  Develop a block layout design using Relationship Diagramming.

| Product | Processing Sequence | Weekly Production |
|---------|---------------------|-------------------|
| 1 | A B C D E F | 960 |
| 2 | A B C B E D C F | 1,200 |
| 3 | A B C D E F | 720 |
| 4 | A B C E B C F | 2,400 |
| 5 | A C E F | 1,800 |
| 6 | A B C D E F | 480 |
| 7 | A B D E C B F | 2,400 |
| 8 | A B D E C B F | 3,000 |
| 9 | A B C D F | 960 |
| 10 | A B D E F | 1,200 |

| Dept. | Dimension (ft. × ft.) |
|-------|------------------------|
| A | 40 × 40 |
| B | 45 × 45 |
| C | 30 × 30 |
| D | 50 × 50 |
| E | 60 × 60 |
| F | 50 × 50 |

**7.10** A toy manufacturing company makes ten different types of products. There are fifteen equal sized departments involved. Given the following product routings and production forecasts.
  **a.** construct a from-to chart for the facility,
  **b.** develop a block layout design using SLP, and
  **c.** develop a block layout design using Relationship Diagramming.

| Product | Processing Sequence | Weekly Production |
|---------|---------------------|-------------------|
| 1 | A B C D B E F C D H | 500 |
| 2 | M G N O N O | 350 |
| 3 | H L H K | 150 |
| 4 | C F E D H | 200 |
| 5 | N O N | 100 |
| 6 | I J H K L | 150 |
| 7 | G N O | 200 |
| 8 | A C F B E D H D | 440 |
| 9 | G M N | 280 |
| 10 | I H J | 250 |

**7.11** Shown below is the activity relationship chart along with the space requirements for each of the six cells in a small auto parts manufacturing facility.
   **a.** Construct relationship and space relationship diagrams.
   **b.** Design the corresponding block layout using SLP.
   **c.** Construct a block layout using Relationship Diagramming.

| CELL A | 2,100 |
|--------|-------|
| CELL B | 2,100 |
| CELL C | 2,100 |
| CELL D | 2,800 |
| CELL E | 1,500 |
| CELL F | 1,500 |
| CELL G | 2,900 |

**7.12** An activity relationship chart is shown below for the American Mailbox Company. Construct a relationship diagram for the manufacturing facility. Given the space requirements (in sq. ft.),
   **a.** construct a block layout using SLP and
   **b.** develop a block layout using Relationship Diagramming.

| RECEIVING | 2,500 |
|-----------|-------|
| PUNCH PRESS | 5,500 |
| PRESS BENDING | 2,500 |
| PRESS FORMING | 2,500 |
| RIVETING | 1,500 |
| POWER SAWING | 2,500 |
| POWER DRAW | 2,000 |
| WELDING ROBOT | 1,000 |

**7.13** In an assembly plant, material handling between departments is performed using a closed-loop conveyor system. The figure below shows the facility layout plan.

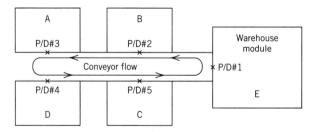

The building consists of three equal-sized assembly modules (A, B, and C), one administrative module (D), and one warehouse module (E). P/D points for each module are shown below. The administrative and warehouse activities are not to be moved; however, assembly areas A, B, C can be relocated. The distance between P/D points and the number of pallet loads move between departments are given below.

| Distance between P/Ds | | | | Pallet flow per day | | | | | |
|---|---|---|---|---|---|---|---|---|---|
| From | To | Distance | | From/To | A | B | C | D | E |
| P/D1 | P/D 2 | 60 ft. | | A | 0 | 0 | 5 | 0 | 30 |
| P/D2 | P/D 3 | 90 ft. | | B | 10 | 0 | 25 | 0 | 0 |
| P/D3 | P/D 4 | 30 ft. | | C | 25 | 5 | 0 | 0 | 0 |
| P/D4 | P/D 5 | 90 ft. | | D | 0 | 0 | 0 | 0 | 0 |
| P/D5 | P/D 1 | 60 ft. | | E | 5 | 20 | 5 | 0 | 0 |

Using the pairwise exchange method, determine new locations for assembly modules A, B, and C, which minimizes the sum of the products of pallet flows and conveyor travel distances.

7.14  A gantry crane serves as the material handling system in a five-cell manufacturing facility as shown below. The new monthly production forecast has been made. The corresponding from-to material flow chart is given in the table below. How should the five cells be rearranged to minimize the total travel distance of the gantry crane. Assume that the cost of relocating the cells are negligible.

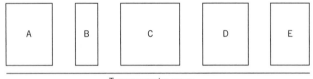

Two-way gantry crane

| Cell | A | B | C | D | E |
|---|---|---|---|---|---|
| A | — | 15 ft. | 33 ft. | 53 ft. | 71 ft. |
| B | | — | 18 ft. | 38 ft. | 56 ft. |
| C | | | — | 20 ft. | 38 ft. |
| D | | | | — | 18 ft. |
| E | | | | | — |

| Cell | A | B | C | D | E |
|---|---|---|---|---|---|
| A | — | 250 | 0 | 25 | 100 |
| B | 25 | — | 0 | 225 | 50 |
| C | 0 | 0 | — | 100 | 0 |
| D | 300 | 0 | 50 | — | 0 |
| E | 50 | 0 | 100 | 0 | — |

7.15  Four equal-sized machines are served by an automated guided vehicle (AGV) on a linear bidirectional track as shown in the figure below. Each machine block is 30 ft × 30 ft. The product routing information and required production rate are given in the table below. Determine a layout arrangement based on the pairwise exchange method. Assume that the pick-up/delivery stations are located at the midpoint of the machine edge along the AGV track.

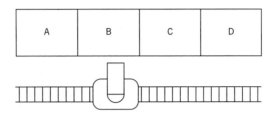

| Product | Processing Sequence | Weekly Production |
|---|---|---|
| 1 | B D C A C | 300 |
| 2 | B D A C | 700 |
| 3 | D B D C A C | 900 |
| 4 | A B C A | 200 |

**7.16** Using the data from Problem 15 but assuming that the locations of the P/D points for machines A and B are 5 ft from the lower-right-hand corner of each machine and the P/D points for machines C and D are 10 ft from the lower-left-hand corner of machines C and D, develop an improved layout using the pairwise exchange method.

**7.17** A mobile robot is serving two cells located at either sides of the AGV track as shown by the figure below. There are three machines placed in each cell. Given the from-to chart in the table below, find the best machine arrangements for both cells. Rearrangement is limited only to machines within each cell. Assume that the P/D point of each machine is located at the midpoint of the machine edge along the AGV track.

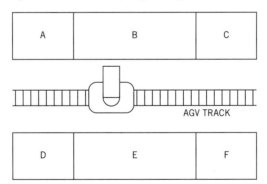

| M/C | A | B | C | D | E | F |
|---|---|---|---|---|---|---|
| A | — | 10 | 50 | 30 | 0 | 60 |
| B | 5 | — | 45 | 40 | 30 | 0 |
| C | 40 | 30 | — | 35 | 5 | 20 |
| D | 40 | 25 | 50 | — | 40 | 50 |
| E | 0 | 55 | 40 | 50 | — | 0 |
| F | 20 | 0 | 60 | 20 | 10 | — |

| M/C | Distance | M/C | Distance |
|-----|----------|-----|----------|
| A-B | 30 | D-E | 30 |
| A-C | 60 | D-F | 60 |
| B-C | 30 | E-F | 30 |

7.18   Five machines located in a manufacturing cell are arranged in an "U" configuration as shown in the layout below. The material handling system employed is a bidirectional conveyor system. Determine the best machine arrangement given the product routing information and production rates in the table.

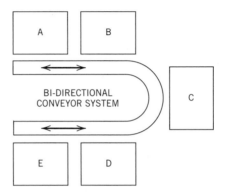

| Product | Machine Sequence: | Prod. Rate |
|---------|-------------------|------------|
| 1 | B-E-A-C | 100 |
| 2 | C-E-D | 200 |
| 3 | B-C-E-A-D | 500 |
| 4 | A-C-E-B | 150 |
| 5 | B-C-A | 200 |

| M/C | Distance (ft.) | M/C | Distance (ft.) |
|-----|----------------|-----|----------------|
| A-B | 20 | B-D | 100 |
| A-C | 70 | B-E | 120 |
| A-D | 120 | C-D | 50 |
| A-E | 140 | C-E | 70 |
| B-C | 50 | D-E | 20 |

7.19   The ABC Cooling and Heating Company manufactures several different types of air conditioners. Five departments are involved in the processing required for the products. A summary of the processing sequences required for the five major products and the weekly production volumes for the products are shown in the table below along with the department area. Based on the graph-based construction method, develop a block layout design.

| Product: | Process Sequence: | Weekly Production |
|---|---|---|
| 1 | A B C | 150 |
| 2 | A B E D | 200 |
| 3 | A C E | 50 |
| 4 | A C B E | 200 |
| 5 | A D E | 250 |

| Department | Area (sq. ft.) |
|---|---|
| A | 1,500 |
| B | 1,500 |
| C | 1,000 |
| D | 2,000 |
| E | 2,000 |

**7.20** The activity relationship chart for Walter's machine shop is shown in the figure below. The space requirements are in square feet. Construct the relationship diagram and design a block layout using the graph-based method.

| Cell | Area |
|---|---|
| CELL A | 1,000 |
| CELL B | 1,200 |
| CELL C | 1,000 |
| CELL D | 1,200 |
| CELL E | 2,000 |
| CELL F | 1,000 |
| CELL G | 1,600 |
| CELL H | 800 |

Relationship chart values (read in diagonal order):
9
7 4
8 0 1
5 6 5
0 0 1 8
2 7 0
5 6 0
0 0 8
0 1
0 0
7
0

**7.21** The from-to material flow matrix for an eight-department facility is given in the table below. Construct a relationship diagram based on the material flow matrix and construct a block layout using the graph-based method.

| Dept. | A | B | C | D | E | F | G | H |
|---|---|---|---|---|---|---|---|---|
| A | — | 302 | 0 | 0 | 0 | 66 | 0 | 68 |
| B | 0 | — | 504 | 20 | 136 | 154 | 56 | 40 |
| C | 0 | 0 | — | 76 | 352 | 0 | 122 | 94 |
| D | 0 | 0 | 0 | — | 0 | 0 | 180 | 8 |
| E | 0 | 0 | 0 | 0 | — | 122 | 0 | 282 |
| F | 0 | 0 | 0 | 0 | 0 | — | 188 | 24 |
| G | 0 | 0 | 0 | 0 | 0 | 0 | — | 296 |
| H | 0 | 0 | 0 | 0 | 0 | 0 | 0 | — |

| Dept. | Area Required (ft²) |
|---|---|
| A | 2,800 |
| B | 2,100 |
| C | 2,600 |
| D | 400 |
| E | 600 |
| F | 400 |
| G | 2,300 |
| H | 1,800 |

**7.22**  Consider the layout of five equal-sized departments. The material flow matrix is given in the figure below.

   **a.** Develop the final adjacency graph using the graph-based procedure.

   **b.** Develop a block layout based on the final adjacency graph obtained in part a.

|   | A | B | C | D | E |
|---|---|---|---|---|---|
| A | — | 0 | 5 | 25 | 15 |
| B | 0 | — | 20 | 30 | 25 |
| C | 0 | 25 | — | 40 | 30 |
| D | 30 | 5 | 20 | — | 0 |
| E | 20 | 30 | 5 | 10 | — |

**7.23**  The material flow matrix for ten departments is given below.

|   | A | B | C | D | E | F | G | H | I | J |
|---|---|---|---|---|---|---|---|---|---|---|
| A | — | 0 | 12 | 0 | 132 | 16 | 0 | 220 | 20 | 24 |
| B | 0 | — | 176 | 0 | 216 | 0 | 144 | 128 | 0 | 0 |
| C | 0 | 0 | — | 0 | 0 | 184 | 0 | 0 | 28 | 0 |
| D | 212 | 136 | 240 | — | 36 | 0 | 236 | 0 | 164 | 0 |
| E | 0 | 0 | 140 | 0 | — | 0 | 192 | 0 | 0 | 160 |
| F | 0 | 180 | 0 | 188 | 108 | — | 248 | 228 | 0 | 0 |
| G | 172 | 0 | 156 | 0 | 0 | 0 | — | 112 | 224 | 152 |
| H | 0 | 0 | 32 | 40 | 204 | 0 | 0 | — | 0 | 0 |
| I | 0 | 168 | 0 | 0 | 104 | 156 | 0 | 148 | — | 200 |
| J | 0 | 124 | 196 | 120 | 0 | 116 | 0 | 108 | 0 | — |

The area requirements are:

| Dept. | Area (sq. ft.) |
|-------|----------------|
| A | 400 |
| B | 1,000 |
| C | 2,600 |
| D | 400 |
| E | 2,400 |
| F | 1,000 |
| G | 3,600 |
| H | 1,200 |
| I | 400 |
| J | 2,400 |

a. Determine a final adjacency graph using the graph-based procedure.
b. Construct a block layout based on the adjacency graph in part a.

# 8

# *COMPUTER-AIDED LAYOUT*

## *8.1* INTRODUCTION

Chapter 7 provided approaches that can be used to produce facility layout alternatives manually. In this chapter, we are concerned with the use of a computer to facilitate the generation of facility layouts. As such, we will present various facility layout algorithms and models that we believe are suitable for computer implementation. Currently available computer-based layout algorithms cannot replace human judgment and experience, and they generally do not capture the qualitative characteristics of a layout. However, computerized layout algorithms can significantly enhance the productivity of the layout planner and the quality of the final solution by generating and numerically evaluating a large number of layout alternatives in a very short time. Computerized layout algorithms are also very effective in rapidly performing "what if" analyses based on varying the input data or the layout itself.

The algorithms presented in this chapter, for the most part, represent the outgrowth of university research. As such, commercial versions of the algorithms we present either do not exist or they must be obtained through the original authors. Also, due to limited space, we are only able to present what we hope is a diverse and representative subset of many computerized layout algorithms that have been developed to date. For further information on such algorithms, the reader may refer to Section 8.11 at the end of this chapter.

A number of commercial packages are available for facility layout. However, with some exceptions (see Section 8.10), such packages are either intended only for presentation purposes (that is, they are electronic drafting tools which facilitate the drawing or maintenance of a given layout) or they are designed strictly as a layout evaluation tool (that is, they can evaluate a layout provided that one has been supplied by the layout planner). While such tools can also significantly enhance the productivity of the layout planner, in this chapter we will focus on those algorithms that can actually improve a given layout or develop a new layout from scratch.

## *8.2* ALGORITHM CLASSIFICATION

Most layout algorithms can be classified according to the type of input data they require. Some algorithms accept only qualitative "flow" data (such as a relationship chart) while others work with a (quantitative) flow matrix expressed as a from-to chart. Some algorithms (such as BLOCPLAN) accept both a relationship chart and a from-to chart; however, the charts are used only one at a time when evaluating a layout. The current trend appears to be towards algorithms that use a from-to chart, which generally requires more time and effort to prepare but provides more information on parts flow (or material handling trips) when completed. Given that flow values can be converted to relationship ratings and vice versa, most algorithms can be used with either type of data. Of course, if a relationship chart is converted to a from-to chart (by assigning numerical values to the closeness ratings), the "flow" values picked by the layout planner represent only an ordinal scale.

Layout algorithms can also be classified according to their objective functions. There are two basic objectives: One aims at minimizing the sum of flows times distances while the other aims at maximizing an adjacency score. Generally speaking, the former, that is, the "distance-based" objective—which is similar to the classical Quadratic Assignment Problem (QAP) objective—is more suitable when the input data is expressed as a from-to chart, and the latter, that is, the "adjacency-based" objective, is more suitable for a relationship chart.

Consider first the distance-based objective. Let $m$ denote the number of departments, $f_{ij}$ denote the flow from department $i$ to department $j$ (expressed in number of unit loads moved per unit time), and $c_{ij}$ denote the cost of moving a unit load one distance unit from department $i$ to department $j$. The objective is to minimize the cost per unit time for movement among the departments. Expressed mathematically, the objective can be written as

$$\min z = \sum_{i=1}^{m} \sum_{j=1}^{m} f_{ij} c_{ij} d_{ij}, \tag{8.1}$$

where $d_{ij}$ is the distance from department $i$ to $j$. In many layout algorithms $d_{ij}$ is measured rectilinearly between department centroids; however, it can also be measured according to a particular aisle structure (if one is specified).

Note that the $c_{ij}$ values in Equation 8.1 are implicitly assumed to be independent of the utilization of the handling equipment, and they are linearly related to the length of the move. In those cases where the $c_{ij}$ values do not satisfy the above assumptions, one may set $c_{ij} = 1$ for all $i$ and $j$ and focus only on *total unit load travel* in the facility, i.e., the product of the $f_{ij}$ and the $d_{ij}$ values. In some cases, it may also be possible to use the $c_{ij}$ values as relative "weights" (based on unit load attributes such as size, weight, bulkiness, etc.) and minimize the weighted sum of unit load travel in the facility.

Consider next the adjacency-based objective where the adjacency score is computed as the sum of all the flow values (or relationship values) between those departments that are adjacent in the layout. Letting $x_{ij} = 1$ if departments $i$ and $j$ are adjacent (that is, they share a border) in the layout, and 0 otherwise, the objective is to maximize the adjacency score; that is,

$$\max z = \sum_{i=1}^{m} \sum_{j=1}^{m} f_{ij} x_{ij}. \tag{8.2}$$

Although the adjacency score obtained from Equation 8.2 is helpful in comparing two or more alternate layouts, it is often desirable to evaluate the *relative* efficiency of a particular layout with respect to a certain lower or upper bound. For this purpose, the layout planner may use the following "normalized" adjacency score:

$$z = \frac{\sum_{i=1}^{m} \sum_{j=1}^{m} f_{ij} x_{ij}}{\sum_{i=1}^{m} \sum_{j=1}^{m} f_{ij}}. \tag{8.3}$$

Note that the normalized adjacency score (which is also known as the *efficiency rating*) is obtained simply by dividing the adjacency score obtained from Equation 8.2 by the total flow in the facility. As a result, the normalized adjacency score is always between zero and 1. If the normalized adjacency score is equal to 1, it implies that all department pairs with positive flow between them are adjacent in the layout.

In some cases, the layout planner may represent an X relationship between departments $i$ and $j$ by assigning a negative value to $f_{ij}$. The exact negative value to be used should be determined with respect to the "real" (i.e., positive) flow values in the from-to chart. If such "negative flow" values are used, the normalized adjacency score must be modified as follows:

$$z = \frac{\sum_{(i,j) \in F} f_{ij} x_{ij} - \sum_{(i,j) \in \bar{F}} f_{ij} (1 - x_{ij})}{\sum_{(i,j) \in F} f_{ij} - \sum_{(i,j) \in \bar{F}} f_{ij}}, \tag{8.4}$$

where $F$ and $\bar{F}$ represent the set of department pairs with positive and negative flow values, respectively.

The adjacency-based objective has been used in a number of algorithms; see, for example, [12], [13], and [34], among others. Although such an objective is easy to use and intuitive (i.e., department pairs with high closeness ratings or large flows need to be adjacent in the layout), it is generally not a complete measure of layout efficiency since it disregards the distance or separation between non-adjacent departments. Therefore, as remarked in [4], it is possible to construct two layouts that have identical or similar adjacency scores but different travel distances in terms of parts flow.

Computerized layout algorithms can further be classified according to the format they use for layout representation. Most layout algorithms use the discrete representation (shown in Figure 8.1*a*) which allows the computer to store and manipulate the layout as a matrix. With such a representation, the area of each department is rounded off to the nearest integer number of grids. If the grid size is too large (too small), small (large) departments may have too few (too many) grids. Also, the grid size determines the overall "resolution" of the layout; a smaller grid size yields a finer resolution which allows more flexibility in department shapes. However, since the total area is fixed, a smaller grid size results in a larger number of

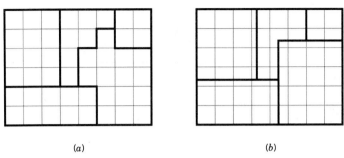

*(a)*            *(b)*

**Figure 8.1**   Discrete versus continuous layout representation. *(Adapted from [14].)*

grids (or a larger matrix) which can considerably increase the computational burden. Hence, in algorithms that use the discrete representation, selecting the appropriate grid size is an important decision that must be made early in the planning process.

The alternate representation (see Figure 8.1*b*) is the continuous representation where there is no underlying grid structure. (The grid in Figure 8.1*b* was retained only for comparison purposes.) Although such a representation is theoretically more flexible than its discrete counterpart, it is also more difficult to implement on a computer. In fact, except for one case (where the shapes of certain departments are "adjusted" to accommodate a nonrectangular building), present computerized layout algorithms that use the continuous representation are restricted to a rectangular building and rectangular department shapes.

While it is possible to model nonrectangular buildings by using fixed "dummy" departments (see Section 8.3), generally speaking, defining L-shaped, U-shaped, and other arbitrary, nonrectangular departments with the continuous representation is not straightforward. If a department is rectangular and we know the area it requires, then we need to know only the $x,y$-coordinate of its centroid and the length of its side in the north-south direction to specify its exact location and shape. (How would you specify the exact location and shape of an L-shaped department or a T-shaped department with a known area? Among alternative specifications, which one requires the minimum amount of data? Which one makes it easier to identify and avoid overlapping departments?)

Department shapes play an important role in computerized layout algorithms. Although we discuss specific shape measures in Section 8.8, first we need to present some basics. Since, by definition, a department represents the smallest indivisible entity in layout planning, a layout algorithm should not "split" a department into two or more pieces. If some departments are "too large," the layout planner can go back and reconsider how the departments were defined and change some of them as necessary. In the process, one large department may be redefined as two smaller departments. Once all the departments have been defined, however, a layout algorithm cannot change them.

Although the human eye is adept at judging shapes and readily identifying split departments, for the computer to "recognize" a split department we need to develop formal measures that can be incorporated into an algorithm. Consider, for example, the discrete representation, where a department is represented as a collection of grids. Suppose there is a "dot" that can move *only from one grid to an adjacent grid*. (Two grids are adjacent only if they share a border of positive length; two grids that "touch" each other at the corners are not considered adjacent.) We say that department *i* is not split if the above dot can start at any grid assigned to department *i* and travel to any other grid assigned to department *i* *without* visiting any grid that has not been assigned to department *i*. In other words, given the restrictions imposed on the movements of the dot, any grid assigned to department *i* must be "reachable" from any other such grid.

For example, the department shown in Figure 8.2*a* and 8.2*b* is split, while those shown in Figures 8.2*c* and 8.2*d* are not. However, according to the above definition, the department shown in Figure 8.2*e* is also not a split department. Departments such as the one shown in Figure 8.2*e* are said to contain an "enclosed void" [14] and as a rule-of-thumb are not considered practical or reasonable for facility layout purposes. (Of course, one possible exception to this rule is an "atrium." However, an atrium should generally be modeled by placing a fixed "dummy department" inside the building; refer to Section 8.3 for dummy departments.) The shape measures we present in Section 8.8 are devised to generally avoid department shapes such as those shown in Figures 8.2*d* and 8.2*e*. In the above discussion of department shapes we focused on the discrete representation. Although it has been used effectively only with rectangular departments, our comments apply to the continuous representation as well.

Lastly, computer-based layout algorithms can be classified according to their primary function; that is, layout *improvement* versus layout *construction*. Improvement-type algorithms generally start with an initial layout supplied by the analyst and seek to improve the objective function through "incremental" changes in the layout. Construction-type layout algorithms generally develop a layout "from scratch." They can be further divided into two categories: those that assume the building dimensions are given and those that assume they are not. The first type of construction algorithm

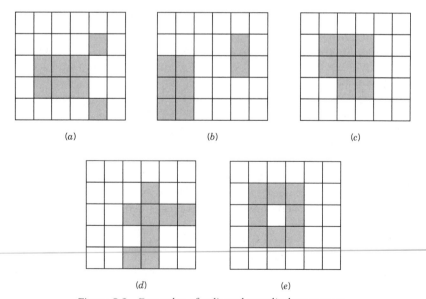

(a)                         (b)                         (c)

(d)                         (e)

**Figure 8.2** Examples of split and unsplit departments.

is suitable when an operation is being moved into an existing, vacant building. The second type of construction algorithm is suitable for "green field" applications.

Even with "green field" applications, however, there is usually a *site plan,* which shows the property, surrounding roads, etc. Given the constraints imposed by the site plan, it is often necessary to construct the new building within a certain "envelope." If a construction algorithm of the second type is used, it is often difficult to ensure that the resulting building will properly fit into such an envelope. With construction algorithms of the first type, on the other hand, one may model the above envelope as an "existing building" to obtain a proper fit between the actual building and the envelope. Primarily for the above reason, the construction algorithms we present in this chapter are of the first type; i.e., they assume the building dimensions are given.

In the following sections, we describe the overall modeling techniques and/or methods used in various computerized layout algorithms; namely, CRAFT, BLOC-PLAN, MIP, LOGIC, and MULTIPLE. (For ease of reference, in those cases where the original authors did not use an acronym, we created our own.) Detailed information such as input data format or output variables are not presented since such information is relevant and available only if the reader obtains a copy of the program. However, for each algorithm we present sufficient detail so that not only the fundamental concepts are covered but interested readers may develop their own basic implementation of the algorithm. All the algorithms we present are relatively recent ones except for CRAFT, which we consider because it is one of the first computerized layout algorithms developed and it provides a good platform for the others. Furthermore, all the algorithms we present require a basic understanding of heuristic search techniques (such as the "steepest descent" procedure) and the difference between "locally" and "globally" optimal solutions.

## *8.3* **CRAFT**

Introduced in 1963 by Armour, Buffa, and Vollman (see [1] and [7]), CRAFT *(Computerized Relative Allocation of Facilities Technique)* is one of the earliest layout algorithms presented in the literature. It uses a from-to chart as input data for the flow. Layout "cost" is measured by the distance-based objective function shown in Equation 8.1. Departments are not restricted to rectangular shapes and the layout is represented in a discrete fashion.

Since CRAFT is an improvement-type layout algorithm, it starts with an initial layout, which typically represents the actual layout of an existing facility but may also represent a prospective layout developed by another algorithm. CRAFT begins by determining the centroids of the departments in the initial layout. It then calculates the rectilinear distance between pairs of department centroids and stores the values in a distance matrix. The initial layout cost is determined by multiplying each entry in the from-to chart with the corresponding entries in the unit cost matrix (i.e., the $c_{ij}$ values) and the distance matrix.

CRAFT next considers all-possible two-way or three-way department exchanges and identifies the best exchange, that is, the one that yields the largest reduction in the layout cost. (No department can be split as a result of a two-way or three-way exchange.) Once the best exchange is identified, CRAFT updates the layout according to the best exchange, and computes the new department centroids as well as the new layout cost to complete the first iteration. The next iteration starts with CRAFT once again identifying the best exchange by considering all-possible two-way or three-way exchanges in the (updated) layout. The process continues until no further reduction in layout cost can be obtained. The final layout obtained in such a manner is also known as a two-opt (three-opt) layout since no two-way (three-way) exchanges can further reduce the layout cost.

Since computers were relatively "slow" in the 60s, the original implementation of CRAFT deviated slightly from the above description. When the program considered exchanging the locations of departments $i$ and $j$, instead of actually exchanging the department locations to compute their new centroids and the actual layout cost, it computed an *estimated* layout cost simply by temporarily treating the centroid of department $i$ in the current layout as the centroid of department $j$ and vice versa; that is, it simply swapped the *centroids* of departments $i$ and $j$.

The error incurred in estimating the layout cost as described above depends on the relative size of the two departments being exchanged. If the departments differ in size, the estimated centroids may deviate significantly from their correct locations. (Of course, if the departments are equal in area, no error will be incurred.) As a result, the actual reduction in the layout cost may be overestimated or underestimated. Although it does not fully address the above error, once the best exchange *based on the estimated layout cost* is identified, CRAFT exchanges the locations of the departments and computes their new centroids (and the actual layout cost) before continuing to the next iteration.

A more important refinement we need to make in the above description of CRAFT is concerned with the exchange procedure. Although at first it may seem "too detailed," this refinement is important from a conceptual point of view since it demonstrates an intricate aspect of using computers for layout purposes. When CRAFT considers exchanging two departments, instead of examining all-possible exchanges as we stated above, it actually considers exchanging only those departments that are either adjacent (i.e., share a border) or equal in area. Such a restriction is not arbitrary. Given that departments cannot be split, it would be impossible to exchange two departments without "shifting" the location of the other departments in the layout, *unless the two departments are either adjacent or equal in area*. (Why?) Since CRAFT is not capable of "automatically" shifting departments in such a manner, it considers exchanging only those that are adjacent or equal in area.

Obviously, two departments with equal areas, whether they are adjacent or not, can always be exchanged without shifting the other departments in the layout. However, if two departments are not equal in area, then adjacency is a *necessary but not sufficient* condition for being able to exchange them without shifting the other departments. That is, in certain cases, even if two (unequal-area) departments are adjacent, it may not be possible to exchange them without shifting the other departments. We will later present an example for such a case.

We also need to stress that, while searching for a better solution, CRAFT picks only the best (estimated) exchange at each iteration, which makes it a "steepest descent" procedure. It also does not "look back" or "look forward" during the above search. Therefore, CRAFT will terminate at the first two-opt or three-opt solution that it encounters during the search. Such a solution is very likely to be only locally optimal. Furthermore, with such a search procedure, the termination point (or the final layout) will be strongly influenced by the starting point (or the initial layout). Consequently, CRAFT is a highly "path-dependent" heuristic and to use it effectively we generally recommend trying different initial solutions (if possible) or trying different exchange options (two-way vs. three-way) at each iteration.

CRAFT is generally flexible with respect to department shapes. As long as the department is not split, it can accommodate virtually any department shape. Theoretically, due to the centroid-to-centroid distance measure, the optimum layout (which has an objective function value of zero) consists of concentric rectangles! Of course, the above problem stems from the fact that the centroid of some O-shaped, U-shaped, and L-shaped departments may lie outside the department itself. Unless the initial layout contains concentric departments, CRAFT will not construct such a layout. However, some department shapes may be irregular, and the objective function value may be underestimated due to the centroid-to-centroid measure.

CRAFT is normally restricted to rectangular buildings. However, through "dummy" departments, it can be used with nonrectangular buildings as well. Dummy departments have no flows or interaction with other departments but require a certain area specified by the layout planner. In general, dummy departments may be used to:

1. Fill building irregularities.

2. Represent obstacles or unusable areas in the facility (such as stairways, elevators, plant services, and so on).

3. Represent extra space in the facility.

4. Aid in evaluating aisle locations in the final layout.

Note that, when a dummy department is used to represent an obstacle, its location must be fixed. Fortunately, CRAFT allows the user to fix the location of any department (dummy or otherwise). Such a feature is especially helpful in modeling obstacles as well as departments such as receiving and shipping in an existing facility.

One of CRAFT's strengths is that it can capture the initial layout with reasonable accuracy. This strength stems primarily from CRAFT's ability to accommodate nonrectangular departments or obstacles located anywhere in a possibly nonrectangular building. However, in addition to being highly path dependent, one of CRAFT's weaknesses is that it will rarely generate department shapes that result in straight, uninterrupted aisles as is desired in the final layout. Fixing some departments to specific locations, and in some cases placing dummy departments in the layout to represent main aisles, may lead to more reasonable department shapes. Nevertheless, as is the case with virtually all computerized layout algorithms, the final layout generated by the computer should not be presented to the decision maker before the layout planner "molds" or "massages" it into a practical layout.

As we demonstrate in the following example, molding or massaging a layout involves relatively minor adjustments made to the department shapes and/or department areas in the final layout. The fact that such adjustments are almost always necessary does not imply that computerized layout algorithms are of limited use. To the contrary, by considering a large number of alternatives in a very short time, a computerized layout algorithm narrows down the solution space for the layout planner who can concentrate on further evaluating and massaging a few "promising" solutions identified by the computer.

# Example 8.1

Consider a manufacturing facility with seven departments. The department names, their areas, and the from-to chart are shown in Table 8.1. We assume that all the $c_{ij}$ values are set equal to one. The building and the current layout (which we supply as the initial layout to CRAFT) are shown in Figure 8.3, where each grid is assumed to measure $20' \times 20'$. Since the total available space (72,000 ft²) exceeds the total required space (70,000 ft²), we generate a single dummy department (H) with an area of 2000 ft². (In most cases, depending on the amount of excess space available, it is highly desirable to use two or more dummy departments with more or less evenly distributed space requirements. Note that the allocation of excess space in the facility plays a significant role in determining future expansion options for many departments.) For practical reasons, we assume that the locations of the receiving (A) and shipping (G) departments are fixed.

CRAFT first computes the centroid of each department which are shown in Figure 8.3. For each department pair, it then computes the rectilinear distance between their centroids and multiplies it by the corresponding entry in the from-to chart. For example, the rectilinear distance between the centroids of departments A and B is equal to six grids. CRAFT multiplies 6 by 45 and adds the result to the objective function. Repeating the above calculation for all department pairs with non-zero flow yields an initial layout cost of 2974 units. (We caution the reader that CRAFT computes the distances in grids, not in feet. Therefore, the actual layout cost is equal to $2974 \times 20 = 59,480$ units.)

Subsequently, CRAFT performs the first iteration and exchanges departments E and F to obtain the layout shown in Figure 8.4. Departments E and F are not equal in area; however, since they are adjacent, one can draw a single "box" around E and F such that it contains both departments E and F but no other departments. (Note that, if E and F were not adjacent, drawing such a box would not be possible.) Since the above box contains no other departments, CRAFT will exchange departments E and F without shifting any other department. (Naturally, CRAFT does not "draw boxes." We introduced the box analogy only to clarify our description of CRAFT.)

There are several ways in which departments E and F may be exchanged within the above box. Comparing the locations of departments E and F in Figures 8.3 and 8.4, it is evident that CRAFT started with the left-most column of department F (the larger of the two departments) and labeled the first 20 grids of department F as department E. To complete

Table 8.1 *Departmental Data and From-To Chart for Example 8.1*

| Department Name | Area (ft²) | No. of Grids | FLOW | | | | | | | |
| | | | A | B | C | D | E | F | G | H |
|---|---|---|---|---|---|---|---|---|---|---|
| 1. A: Receiving | 12,000 | 30 | 0 | 45 | 15 | 25 | 10 | 5 | 0 | 0 |
| 2. B: Milling | 8,000 | 20 | 0 | 0 | 0 | 30 | 25 | 15 | 0 | 0 |
| 3. C: Press | 6,000 | 15 | 0 | 0 | 0 | 0 | 5 | 10 | 0 | 0 |
| 4. D: Screw m/c. | 12,000 | 30 | 0 | 20 | 0 | 0 | 35 | 0 | 0 | 0 |
| 5. E: Assembly | 8,000 | 20 | 0 | 0 | 0 | 0 | 0 | 65 | 35 | 0 |
| 6. F: Plating | 12,000 | 30 | 0 | 5 | 0 | 0 | 25 | 0 | 65 | 0 |
| 7. G: Shipping | 12,000 | 30 | 0 | 0 | 0 | 0 | 0 | 0 | 0 | 0 |
| 8. H: Dummy | 2,000 | 5 | 0 | 0 | 0 | 0 | 0 | 0 | 0 | 0 |

```
     1  2  3  4  5  6  7  8  9 10 11 12 13 14 15 16 17 18
 1 | A  A  A  A  A  A  A  A  A  A  G  G  G  G  G  G  G  G |
 2 | A           •              A  G           •        G |
 3 | A  A  A  A  A  A  A  A  A  A  G  G  G              G |
 4 | B  B  B  B  B  C  C  C  C  C  E  E  G  G  G  G  G  G |
 5 | B     •     B  C     •     C  E  E  E  E  E  E  E  E |
 6 | B           B  C  C  C  C  C  E  E  E  E  E  E  E  E |
 7 | B  B  B  B  B  D  D  D  D  F  F  F  F  F  F  F  F  E  E |
 8 | D  D  D  D  D  D        D  F              •     F  F  F |
 9 | D           •        D  D  F  F  F  F  F              F |
10 | D  D  D  D  D  D  D  D  H  H  H  H  H  H  F  F  F  F  F |
```

**Figure 8.3**   Initial CRAFT layout and department centroids for Example 8.1 ($z = 2974 \times 20 = 59{,}480$ units).

the exchange, all the grids originally labeled with an E have been converted to an F. Of course, one may implement the above scheme starting from the rightmost column (or, say, the top row) of department F. Regardless of its exact implementation, however, such an exchange scheme generally leads to poor department shapes. In fact, the department shapes in CRAFT often have a tendency to deteriorate with the number of iterations even if all the departments in the initial layout are rectangular.

The *estimated* reduction in the layout cost obtained by exchanging (the centroids of) departments E and F is equal to 202 units. Upon exchanging departments E and F and computing the new department centroids, CRAFT computes the actual cost of the layout shown in Figure 8.4 as 2953 units. Hence, the actual reduction in the layout cost is 21 units as opposed to 202 units. The reader may verify that the above significant deviation is largely due to the fact that the new centroid of department F (after the exchange) deviates substantially from its estimated location.

In the next iteration, based on an estimated reduction of 95 units in the layout cost, CRAFT exchanges departments B and C to obtain the layout shown in Figure 8.5. The layout cost is equal to 2833.50 units, which represents an actual reduction of 119.50 units. This clearly illustrates that the error in estimation may be in either direction. Using estimated costs, CRAFT determines that no other (equal-area or adjacent) two-way or three-way exchange can further reduce the cost of the layout and it terminates with the final solution shown in Figure 8.5.

```
     1  2  3  4  5  6  7  8  9 10 11 12 13 14 15 16 17 18
 1 | A  A  A  A  A  A  A  A  A  A  G  G  G  G  G  G  G  G |
 2 | A                          A  G                    G |
 3 | A  A  A  A  A  A  A  A  A  A  G  G  G              G |
 4 | B  B  B  B  B  C  C  C  C  C  F  F  G  G  G  G  G  G |
 5 | B           B  C              C  F  F  F  F  F  F  F  F |
 6 | B           B  C  C  C  C  C  C  F  F  F  F  F  F     F |
 7 | B  B  B  B  B  D  D  D  D  D  E  E  E  E  E  E ¦ F     F |
 8 | D  D  D  D  D  D           D  E              E ¦ F     F |
 9 | D                    D  D  E  E  E  E  E  E ¦ F        F |
10 | D  D  D  D  D  D  D  D  H  H  H  H  H  H  E  E ¦ F  F  F |
```

**Figure 8.4**   Intermediate CRAFT layout obtained after exchanging departments E and F ($z = 2953 \times 20 = 59{,}060$ units).

|    | 1 | 2 | 3 | 4 | 5 | 6 | 7 | 8 | 9 | 10 | 11 | 12 | 13 | 14 | 15 | 16 | 17 | 18 |
|----|---|---|---|---|---|---|---|---|---|----|----|----|----|----|----|----|----|----|
| 1  | A | A | A | A | A | A | A | A | A | A  | G  | G  | G  | G  | G  | G  | G  | G  |
| 2  | A |   |   |   |   |   |   |   | A | G  |    |    |    |    |    |    |    | G  |
| 3  | A | A | A | A | A | A | A | A | A | A  | G  | G  | G  |    |    |    |    | G  |
| 4  | C | C | C | B | B | B | B | B | B | B  | F  | F  | G  | G  | G  | G  | G  | G  |
| 5  | C |   | C | C | B |   |   |   | B | F  | F  | F  | F  | F  | F  | F  | F  | F  |
| 6  | C |   |   | C | B | B | B | B | B | B  | F  | F  | F  | F  | F  | F  |    | F  |
| 7  | C | C | C | C | B | D | D | D | D | E  | E  | E  | E  | E  | E  | F  |    | F  |
| 8  | D | D | D | D | D | D |   |   | D | E  |    |    |    |    | E  | F  |    | F  |
| 9  | D |   |   |   |   |   | D | D | E | E  | E  | E  | E  | E  | F  |    |    | F  |
| 10 | D | D | D | D | D | D | D | D | H | H  | H  | H  | H  | E  | E  | F  | F  | F  |

**Figure 8.5**  Final CRAFT layout ($z = 2833.50 \times 20 = 56{,}670$ units).

Recall that a computer-generated layout should not be presented to the decision maker before the analyst molds or massages it into a practical layout. In massaging a layout, the analyst may generally disregard the grids and use a continuous representation. In so doing, he or she may smooth the department borders and slightly change their areas or orientation, if necessary. After massaging the layout shown in Figure 8.5, we obtained the layout shown in Figure 8.6. Note that, other than department H (which is a dummy department), we made no significant changes to any of the departments; yet, the layout shown in Figure 8.6 is more reasonable and perhaps more practical than the one shown in Figure 8.5.

Once it is massaged in the above manner, a layout cannot be generally reevaluated via a computer-based layout algorithm unless one is willing to redefine the grid size and repeat the process using the new grid size. Also, we need to stress that in many real-world problems, layout massaging usually goes beyond adjusting department shapes or areas. In massaging a layout, the analyst must often take into account certain qualitative factors or constraints that may not have been considered by the algorithm.

Earlier we remarked that if two departments are *not equal* in area, then adjacency is a *necessary but not sufficient* condition for being able to exchange them without shifting the other departments. Obviously, adjacency is necessary since it is otherwise physically impossible to exchange two unequal-area departments without shifting other departments. (Recall that extra space is also modeled as a "department.") The fact that adjacency is not sufficient, on the other hand, can be shown through the following example [31].

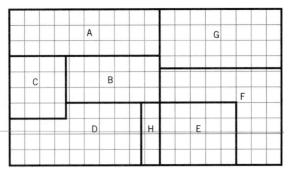

**Figure 8.6**  Final "massaged" layout obtained with CRAFT.

Consider a 7×5 layout with six departments as shown in Figure 8.7. Note that departments 2 and 4 are not equal in area but they are adjacent, which implies that we can draw a single box around them. Yet, one cannot exchange departments 2 and 4 without splitting department 2. In fact, if we give the above layout to CRAFT, and fix the locations of all the departments except 2 and 4, CRAFT does not exchange departments 2 and 4 even if $f_{1,4}$ is set equal to a large value [31]. The above example is, of course, carefully constructed to show that adjacency is not sufficient. In most cases, two unequal-area departments that are adjacent can be exchanged without splitting either one.

Specifying the conditions required for a three-way exchange is somewhat more complicated. Suppose departments $i$, $j$, and $k$ are considered for a three-way exchange such that department $i$ "replaces" $j$, department $j$ "replaces" $k$, and department $k$ "replaces" $i$. If a single "box" can be drawn around departments $i$, $j$, and $k$, and this "box" does not contain any other department, then (except for cases such as the one shown above for departments 2 and 4) we can perform a three-way exchange without shifting any of the other departments. Note that, in order to draw such a box, each of the three departments need not share a border with the other two; one may still draw the above box if department $i$ is adjacent to department $j$ (but not $k$), and department $k$ is adjacent to department $j$ (but not $i$).

If it is not possible to draw such a box, equal-area departments may still permit some three-way exchanges. Suppose departments $i$ and $j$ are adjacent but department $k$ is separated from both. For the above exchange involving departments $i$, $j$, and $k$, for example, one may perform the exchange without shifting any of the other departments if departments $j$ and $k$ are equal in area. Of course, other combinations (including the case where all three departments are nonadjacent but equal in area) are also possible. Computer implementation of three-way exchanges (i.e., deciding which department to move first and how to reassign the grids) is not straightforward. Furthermore, the number of possible three-way exchanges increases quite rapidly with the number of departments (which may result in long execution times). Since more recent (single-floor) facility layout algorithms focus on two-way exchanges only, and since the adjacency or equal-area requirement has been relaxed with various layout formation techniques, we will not present in detail three-way exchanges performed by CRAFT.

A more recent personal-computer implementation of CRAFT by Hosni, Whitehouse, and Atkins [21] is distributed by the Institute of Industrial Engineers (IIE) under the name MICRO-CRAFT. (Since IIE plans to discontinue software distribution, the reader may refer to the original authors for a copy of the code.) MICRO-CRAFT (or

**Figure 8.7** Example to show that CRAFT may not be able to exchange two adjacent departments that are not equal in area.

MCRAFT) is similar to CRAFT except that the above constraint is relaxed; i.e., MCRAFT can exchange any two departments whether they are adjacent or not. Such an improvement is obtained by using a layout formation technique which "automatically" shifts other departments when two unequal-area, nonadjacent departments are exchanged. Instead of assigning each grid to a particular department, MCRAFT first divides the facility into "bands" and the grids in each band are then assigned to one or more departments. The number of bands in the layout is specified by the user. MCRAFT's layout formation technique—which was originally used in an earlier algorithm, namely, ALDEP *(Automated Layout DEsign Program)* [41]—is described through the following example.

## *Example 8.2*

Consider the same data given for Example 8.1. Unlike CRAFT (where the user selects the grid size), MCRAFT asks for the length and width of the building and the number of bands. The program then computes the appropriate grid size and the resulting number of rows and columns in the layout. Thus, setting the building length (width) equal to 360 ft (200 ft), and the number of bands equal to, say, three, we obtain the initial layout shown in Figure 8.8, where each band is six rows wide. MCRAFT forms a layout by starting at the upper-left-hand corner of the building and "sweeping" the bands in a serpentine fashion. In doing so, it follows a particular sequence of department numbers which we will refer to as the *layout vector* or the *fill sequence.* The layout shown in Figure 8.8 was obtained from the layout vector 1-7-5-3-2-4-8-6, which is supplied by the user as the initial layout vector. (Note that MCRAFT designates the departments by numbers instead of letters; department 1 represents department A, department 2 represents department B, etc.)

MCRAFT performs four iterations (i.e., four two-way exchanges) before terminating with the two-opt layout. The departments exchanged at each iteration and the corresponding layout cost are given as follows: first iteration—departments C and E (59,611.11 units); second iteration—departments C and H (58,083.34 units); third iteration—departments C and D (57,483.34 units); and fourth iteration—departments B and C (57,333,34 units). The resulting three-band layout is shown in Figure 8.9, for which the layout vector is given by 1-7-8-5-3-2-

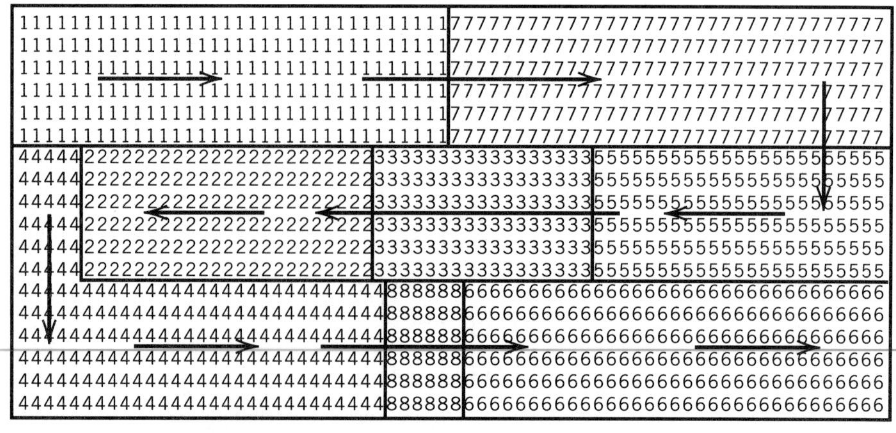

**Figure 8.8**   Initial MCRAFT layout for Example 8.2 ($z = 60,661.11$ units).

```
11111111111111111111111111111111111177777777777777777777777777777777777
11111111111111111111111111111111111177777777777777777777777777777777777
11111111111111111111111111111111111177777777777777777777777777777777777
11111111111111111111111111111111111177777777777777777777777777777777777
11111111111111111111111111111111111177777777777777777777777777777777777
11111111111111111111111111111111111177777777777777777777777777777777777
222222222222222222222222333333333333333333555555555555555555555555888888
222222222222222222222222333333333333333333555555555555555555555555888888
222222222222222222222222333333333333333333555555555555555555555555888888
222222222222222222222222333333333333333333555555555555555555555555888888
222222222222222222222222333333333333333333555555555555555555555555888888
222222222222222222222222333333333333333333555555555555555555555555888888
244444444444444444444444444444444444444466666666666666666666666666666666
244444444444444444444444444444444444444466666666666666666666666666666666
244444444444444444444444444444444444444466666666666666666666666666666666
244444444444444444444444444444444444444466666666666666666666666666666666
244444444444444444444444444444444444444466666666666666666666666666666666
244444444444444444444444444444444444444466666666666666666666666666666666
```

**Figure 8.9** Final MCRAFT layout ($z = 57,333.34$ units).

4-6. Except for department 2, the departments in Figure 8.9 appear to have reasonable shapes. Unlike CRAFT, the department shapes obtained from the sweep method tend to be reasonable if an appropriate number of bands is selected. Of course, alternative initial and final layouts may be generated by varying the number of bands and the initial layout vector.

Due to the above layout formation technique, MCRAFT will not be able to capture the initial layout accurately unless the departments are already arranged in bands. As a result, one may have to massage the initial layout to make it compatible with MCRAFT. (The reader may compare the original initial layout shown in Figure 8.3 with MCRAFT's initial layout shown in Figure 8.8.) The actual cost of the initial layout as computed by CRAFT is equal to $2974 \times 20 = 59,480$ units, whereas MCRAFT's initial layout cost is equal to 60,661.11 units—a relatively small difference considering the fact that the "actual" layout used by CRAFT is itself an approximation of reality. Based on the number of departments and the actual initial layout, however, the above difference may be significant for some problems.

Another limitation of the sweep method used in MCRAFT is that the band width is assumed to be the same for all the bands. As a result, MCRAFT is generally not as effective as CRAFT in treating obstacles and fixed departments. If there is an obstacle in the building, the analyst must ensure that its width does not exceed the band width. Otherwise, the obstacle must be divided into two or more pieces which is likely to further complicate matters. In addition, while it is straightforward in MCRAFT to fix the location of any department, a fixed department may still "shift" or "float" when certain non-equal-area departments are exchanged.

In the above example, the two fixed departments (1 and 7) remained at their current locations since they are the first two departments in the initial layout vector 1-7-5-3-2-4-8-6. However, if we fix, say, department 2, and the algorithm exchanges departments 3 and 4, the location of department 2 will shift when the layout is formed with the new layout vector. (Why?) In general, when two unequal-area departments are exchanged, the location of a fixed department will shift if the two departments fall on either side of the fixed department in the layout vector. Considering that obstacles are also modeled as fixed departments (with zero flow), the above limitation implies that obstacles may shift as well. It is instructive to note that MCRAFT's primary strength (i.e., being able to "automatically" shift other departments as necessary) is also its primary weakness (i.e., fixed departments and obstacles may also be shifted).

## 8.4 BLOCPLAN

BLOCPLAN, which was developed by Donaghey and Pire ([10] and [39]), is similar to MCRAFT in that departments are arranged in bands. However, there are certain differences between the two algorithms. BLOCPLAN uses a relationship chart as well as a from-to chart as input data for the "flow." Layout "cost" can be measured either by the distance-based objective (see Equation 8.1) or the adjacency-based objective (see Equation 8.2). Furthermore, in BLOCPLAN, the number of bands is determined by the program and limited to two or three bands. However, the band widths are allowed to vary. Also, in BLOCPLAN since each department occupies exactly one band, all the departments are rectangular in shape. Lastly, unlike MCRAFT, BLOCPLAN uses the continuous representation.

BLOCPLAN may be used both as a construction algorithm and an improvement algorithm. In the latter case, as with MCRAFT, it may not be possible to capture the initial layout accurately. Nevertheless, improvements in the layout are sought through (two-way) department exchanges. Although the program accepts both a relationship chart and a from-to chart as input, the two charts can be used only one at a time when evaluating a layout. That is, a layout is not evaluated according to some "combination" of the two charts.

BLOCPLAN first assigns each department to one of the two (or three) bands. Given all the departments assigned to a particular band, BLOCPLAN computes the appropriate band width by dividing the total area of the departments in that band by the building length. The complete layout is formed by computing the appropriate width for each band as described above and arranging the departments in each band according to a particular sequence.

The above layout formation procedure and the layout scores computed by BLOCPLAN are explained through the following example.

## *Example 8.3*

Consider the same data given for Example 8.1. As in MCRAFT, BLOCPLAN will not be able to capture the initial layout accurately unless the departments are already arranged in bands. As a result, the analyst may have to massage the initial layout to make it compatible with BLOC-PLAN. (The reader may compare the actual initial layout shown in Figure 8.3 with BLOC-PLAN's initial layout shown in Figure 8.10.) The actual cost of the initial layout as computed by CRAFT is equal to 59,480 units, whereas BLOCPLAN's initial layout cost is equal to 61,061.70 units—a relatively small difference of less than 3%.

BLOCPLAN offers the analyst a variety of options for improving the layout. The analyst may try some two-way exchanges simply by typing the department indices to be exchanged, or he or she may select the "automatic search" option to have the algorithm generate a pre-specified number of layouts. (According to [10], the automatic search option is based on the

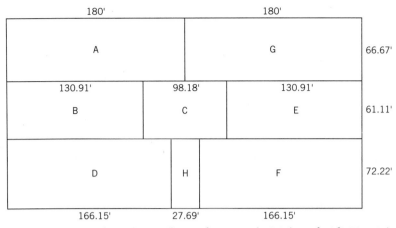

**Figure 8.10**   Initial BLOCPLAN layout for Example 8.3 ($z = 61,061.70$ units).

"procedures that an experienced BLOCPLAN user used in obtaining a 'good' layout;" no details are provided on what these procedures are.) Using BLOCPLAN's improvement algorithm, which interactively considers all possible two-way exchanges, we obtain the layout shown in Figure 8.11, where departments C and H have been exchanged. Since no other two-way exchange leads to a reduction in layout cost, BLOCPLAN was terminated with the final layout as shown in Figure 8.11.

The "cost" of the final layout, as measured by the distance-based objective and the from-to chart given for the example problem—is equal to 58,133.34 units. BLOCPLAN implicitly assumes that all the $c_{ij}$'s are equal to 1.0; that is, the user cannot enter a cost matrix. (Recall that in the example problem all the $c_{ij}$'s are assumed to be equal to 1.0.) The final layout shown in Figure 8.11 can also be evaluated with respect to the adjacency-based objective. Using Equation 8.2, we compute $z = 235$ units, which is obtained by adding the $f_{ij}$ values between all department pairs that are adjacent in Figure 8.11. The normalized adjacency score (or the efficiency rating) given by Equation 8.3 is equal to $235/(235 + 200) = 0.54$.

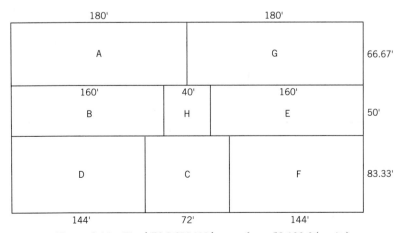

**Figure 8.11**   Final BLOCPLAN layout ($z = 58,133.34$ units).

If the input data is given as a relationship chart instead of a from-to chart, the adjacency score may still be computed provided that a numerical value is assigned to each closeness rating. More specifically, in place of $f_{ij}$ shown in Equation 8.2, one may use the numerical value of the closeness rating assigned to departments $i$ and $j$. The following default values are used in BLOCPLAN: $A = 10$, $E = 5$, $I = 2$, $O = 1$, $U = 0$, and $X = -10$. The user may of course specify different numerical values. Although the values $A = 6$, $E = 5$, $I = 4$, $O = 3$, $U = 2$, and $X = 1$ have been used in some algorithms or texts, for the purposes of computing the adjacency score, using a nonzero value for U and a nonnegative value for X would not be appropriate. Department pairs with an U relationship should not affect the score whether they are adjacent or not, while those pairs with an X relationship should adversely affect the score if they are adjacent.

Even if a from-to chart is provided, BLOCPLAN computes the adjacency score based on the relationship chart. This is accomplished by converting the from-to chart into a relationship chart and using the above numerical values assigned to the closeness ratings. Although one may develop alternate schemes for such a conversion, the one used in BLOCPLAN is straightforward; it is explained through the following example.

## *Example 8.4*

Consider the from-to chart given earlier for Example 8.1 (see Table 8.1). Since BLOCPLAN implicitly assumes that all the unit costs are equal to 1.0, and all the distances are symmetric by definition (i.e., $d_{ij} = d_{ji}$ for all $i$ and $j$), it first converts the from-to chart into a flow-between chart by adding $f_{ij}$ to $f_{ji}$ for all department pairs. That is, the from-to chart shown in Table 8.1 is converted into the flow-between chart shown in Table 8.2a. A flow-between chart is symmetric, by definition.

The maximum flow value in the flow-between chart is equal to 90 units. Dividing the above maximum by 5 we obtain 18, which leads to the following intervals and corresponding closeness ratings: 73 to 90 units (A), 55 to 72 units (E), 37 to 54 units (I), 19 to 36 units (O), and 0 to 18 units (U). That is, BLOCPLAN divides the flow values into five intervals of 18 units each. Any flow value between 73 and 90 units is assigned an A relationship, and so on. Applying this conversion scheme to the above flow-between chart yields the relationship chart shown in Table 8.2b.

**Table 8.2**  *Flow Between Chart and Relationship Chart for Example 8.4*

|   | A | B | C | D | E | F | G | H |   | A | B | C | D | E | F | G | H |
|---|---|---|---|---|---|---|---|---|---|---|---|---|---|---|---|---|---|
| A | 0 | 45 | 15 | 25 | 10 | 5 | 0 | 0 | A | — | I | U | O | U | U | U | U |
| B |   | 0 | 0 | 50 | 25 | 20 | 0 | 0 | B |   | — | U | I | O | O | U | U |
| C |   |   | 0 | 0 | 5 | 10 | 0 | 0 | C |   |   | — | U | U | U | U | U |
| D |   |   |   | 0 | 35 | 0 | 0 | 0 | D |   |   |   | — | O | U | U | U |
| E |   |   |   |   | 0 | 90 | 35 | 0 | E |   |   |   |   | — | A | O | U |
| F |   |   |   |   |   | 0 | 65 | 0 | F |   |   |   |   |   | — | E | U |
| G |   |   |   |   |   |   | 0 | 0 | G |   |   |   |   |   |   | — | U |
| H |   |   |   |   |   |   |   | 0 | H |   |   |   |   |   |   |   | — |
|   | *(a)* Flow-between chart |   |   |   |   |   |   |   |   | *(b)* Relationship chart |   |   |   |   |   |   |   |

Given the default numerical values assigned to the closeness ratings (i.e., $A = 10$, $E = 5$, $I = 2$, $O = 1$, $U = 0$, $X = -10$) and the relationship chart shown in Table 8.2$b$, we compute a normalized adjacency score of $30/48 = 0.63$ for the final layout obtained in Example 8.3 (see Figure 8.11). It is instructive to note that the normalized adjacency score for the initial layout in Example 8.3 is also equal to 0.63, while the distance-based objective for the initial and final layouts in Example 8.3 are equal to 61,061.70 and 58,133.34 units, respectively. The above result illustrates the concern we expressed earlier for the adjacency score; that is, it is possible for two layouts to have virtually the same (normalized) adjacency score but different travel distances for parts flow. BLOCPLAN reports both the normalized adjacency score and the distance-based objective for each layout it generates.

In addition to the distance-based layout cost (using the flow-between chart) and the normalized adjacency score (based on the relationship chart), BLOCPLAN computes a "REL-DIST" score, which is a distance-based layout cost that uses the numerical closeness ratings instead of the flow values. That is, the numerical values of the closeness ratings are multiplied by the rectilinear distances between department centroids. For example, for the initial (final) layout shown in Figure 8.10 (Figure 8.11), the REL-DIST score is equal to 2887.01 (2708.33) units. The REL-DIST score is useful when a from-to chart is not available and the layout must be evaluated with respect to a relationship chart only.

BLOCPLAN also computes a normalized REL-DIST score. However, the upper and lower bounds used in normalizing the REL-DIST score depend on the particular layout being evaluated (we refer the reader to [10]). As a result, caution must be exercised when comparing the normalized REL-DIST scores of two competing layouts.

We remarked earlier that BLOCPLAN implicitly assumes that all the unit costs are equal to 1.0. This assumption is not as restrictive as it may first seem. If the unit costs are not equal to 1.0, then provided that they are symmetric, that is, $c_{ij} = c_{ji}$ for all $i$ and $j$, the user may first multiply each flow value with the corresponding unit cost and then enter the result as the "flow." Note that unit costs must be symmetric since BLOCPLAN works with a flow-between chart as opposed to a from-to chart as shown earlier in Example 8.4.

BLOCPLAN may also be used as a construction algorithm. In doing so, the layout planner may indicate the location of certain departments a priori. As we remarked earlier, whether one inputs the entire initial layout or a few departments that must be located prior to layout construction, it may not be possible to accurately locate all the departments. BLOCPLAN uses the following scheme to specify the location of one or more departments: the entire building is divided into nine cells (labeled A through I) that are arranged in three bands. The top band contains cells A, B, and C (from left to right), while the middle and bottom bands contain cells D, E, F and G, H, I (from left to right), respectively. Each cell is further divided into two halves; a right half and a left half. Hence, there are a total of 18 possible locations, which is also equal to the maximum number of departments that BLOCPLAN will accept.

The location of a department is designated by indicating the appropriate side of a cell. For example, to locate a department on the northeast corner of the building, one would use C - R to indicate the right side of cell C. To locate a department, say, on the west side of the building, one would use D - L, and so on. Of course, regardless of the cell they are assigned to, all departments will be arranged in two or three bands when BLOCPLAN constructs a layout. Each department may be assigned to only one of the 18 cells.

# 8.5   MIP[1]

Using the continuous representation, the facility layout problem may be formulated as a mixed integer programming (MIP) problem if all the departments are assumed to be rectangular. Recall that, with rectangular departments, just the centroid and the length (or width) of a department fully define its location and shape. Although an alternate objective may be used, the model we show here assumes a distance-based objective (given by Equation 8.1) with $d_{ij}$ defined as the rectilinear distance between the centroids of departments $i$ and $j$. Generally speaking, models based on mathematical programming are regarded as *construction-type* layout models since there is no need to enter an initial layout. However, as we discuss later, such models may also be used to improve a given layout.

The model we show here is based on the one presented by Montreuil [35]. A similar model is also presented by Heragu and Kusiak [18], where the department dimensions are treated as parameters with known values instead of decision variables. Assuming that all department dimensions are given and fixed is appropriate in "machine layout" problems, where each "department" represents the rectangular "footprint" of a machine. Note that, with such an approach, the layout model is actually used as a two-dimensional "packing" algorithm to determine the locations of rectangular objects with known shapes. Aside from possible differences in the objective function, such a packing problem is also known as the "two-dimensional bin packing" problem [6].

Treating the department dimensions as decision variables, the facility layout problem may be formulated as follows. Consider first the problem parameters. Let:

$B_x$ be the building length (measured along the $x$-coordinate),

$B_y$ be the building width (measured along the $y$-coordinate),

$A_i$ be the area of department $i$,

$L_i^\ell$ be the lower limit on the length of department $i$,

$L_i^u$ be the upper limit on the length of department $i$,

$W_i^\ell$ be the lower limit on the width of department $i$,

$W_i^u$ be the upper limit on the width of department $i$,

$M$ be a large number.

Consider next the decision variables. Let:

$\alpha_i$ be the $x$-coordinate of the centroid of department $i$,

$\beta_i$ be the $y$-coordinate of the centroid of department $i$,

$x_i'$ be the $x$-coordinate of the left (or west) side of department $i$,

---

[1]The model presented in this section requires a basic knowledge of linear and integer programming.

$x_i''$ be the $x$-coordinate of the right (or east) side of department $i$,

$y_i'$ be the $y$-coordinate of the bottom (or south side) of department $i$,

$y_i''$ be the $y$-coordinate of the top (or north side) of department $i$,

$z_{ij}^x$ be equal to 1 if department $i$ strictly to the east of department $j$, and 0 otherwise,

$z_{ij}^y$ be equal to 1 if department $i$ is strictly to the north of department $j$, and 0 otherwise.

Note that department $i$ would be strictly to the east of department $j$ if and only if $x_j'' \leq x_i'$. Likewise, department $i$ would be strictly to the north of department $j$ if and only if $y_j'' \leq y_i'$. Two departments are guaranteed not to overlap if they are "separated" along either the $x$-coordinate (i.e., one of the departments is strictly to the east of the other) or the $y$-coordinate (i.e., one of the departments is strictly to the north of the other). Of course, it is possible that two departments are separated along both the $x$ and $y$-coordinates.

The above parameter and variable definitions lead to the following model:

$$\text{Minimize } z = \sum_i \sum_j f_{ij} c_{ij} \left( |\alpha_i - \alpha_j| + |\beta_i - \beta_j| \right) \tag{8.5}$$

$$\text{Subject to: } L_i^\ell \leq (x_i'' - x_i') \leq L_i^u \quad \text{for all } i \tag{8.6}$$

$$W_i^\ell \leq (y_i'' - y_i') \leq W_i^u \quad \text{for all } i \tag{8.7}$$

$$(x_i'' - x_i')(y_i'' - y_i') = A_i \quad \text{for all } i \tag{8.8}$$

$$0 \leq x_i' \leq x_i'' \leq B_x \quad \text{for all } i \tag{8.9}$$

$$0 \leq y_i' \leq y_i'' \leq B_y \quad \text{for all } i \tag{8.10}$$

$$\alpha_i = 0.5 x_i' + 0.5 x_i'' \quad \text{for all } i \tag{8.11}$$

$$\beta_i = 0.5 y_i' + 0.5 y_i'' \quad \text{for all } i \tag{8.12}$$

$$x_j'' \leq x_i' + M(1 - z_{ij}^x) \quad \text{for all } i \text{ and } j, \, i \neq j \tag{8.13}$$

$$y_j'' \leq y_i' + M(1 - z_{ij}^y) \quad \text{for all } i \text{ and } j, \, i \neq j \tag{8.14}$$

$$z_{ij}^x + z_{ji}^x + z_{ij}^y + z_{ji}^y \geq 1 \quad \text{for all } i \text{ and } j, \, i < j \tag{8.15}$$

$$\alpha_i, \beta_i \geq 0 \quad \text{for all } i \tag{8.16}$$

$$x_i', x_i'', y_i', y_i'' \geq 0 \quad \text{for all } i \tag{8.17}$$

$$z_{ij}^x, z_{ij}^y \, 0/1 \text{ integer} \quad \text{for all } i \text{ and } j, \, i \neq j \tag{8.18}$$

The objective function given by Equation 8.5 is the distance-based objective shown earlier as Equation 8.1. Constraint sets 8.6 and 8.7, respectively, ensure that the length and width of each department are within the specified bounds. The area requirement of each department is expressed through constraint set 8.8, which are the only non-linear constraints in the model. Constraint sets 8.9 and 8.10 ensure that the department sides are defined properly and that each department is located within the building in the $x$ and $y$ directions, respectively. Constraint sets 8.11 and 8.12, respectively, define the $x$ and $y$ coordinates of the centroid of each department.

Constraint set 8.13 ensures that $x_j'' \leq x_i'$ (i.e., department $i$ is strictly to the east of department $j$) if $z_{ij}^x = 1$. Note that, if $z_{ij}^x = 0$, constraint set 8.13 is satisfied whether $x_j''$ is less than or equal to $x_i'$ or not. In other words, constraint set 8.13 becomes "active" only when $z_{ij}^x = 1$. Constraint set 8.14 serves the same purpose as constraint set 8.13 but in the $y$ direction. Constraint set 8.15 ensures that no two departments overlap by forcing a separation at least in the east–west or north–south direction.[2] Lastly, constraint sets 8.16 and 8.17 represent the nonnegativity constraints while constraint set 8.18 designates the binary variables.

Due to constraint set 8.8, the above model is nonlinearly constrained. Furthermore, the objective function contains the absolute value operator. There are alternative schemes one may employ to approximate constraint 8.8 via a linear function. In [35], the department area is controlled through its perimeter (which is a linear function of the length and width of the department) and constraint sets 8.6 and 8.7.

Provided that $f_{ij} \geq 0$ for all $i$ and $j$, the absolute values in the objective function may be removed by introducing a "positive part" and a "negative part" for each term [37]. That is, if we set

$$\alpha_i - \alpha_j = \alpha_{ij}^+ - \alpha_{ij}^- \quad \text{and} \quad \beta_i - \beta_j = \beta_{ij}^+ - \beta_{ij}^-, \tag{8.19}$$

$$\text{then} \quad |\alpha_i - \alpha_j| = \alpha_{ij}^+ + \alpha_{ij}^- \quad \text{and} \quad |\beta_i - \beta_j| = \beta_{ij}^+ + \beta_{ij}^-, \tag{8.20}$$

where $\alpha_{ij}^+$, $\alpha_{ij}^-$, $\beta_{ij}^+$, and $\beta_{ij}^-$ are all nonnegative variables.

Hence, the nonlinear model given by Equations 8.5 through 8.18 can be transformed into the following linear MIP problem:

$$\text{Minimize } z = \sum_i \sum_j f_{ij} c_{ij} (\alpha_{ij}^+ + \alpha_{ij}^- + \beta_{ij}^+ + \beta_{ij}^-) \tag{8.21}$$

$$\text{Subject to: } L_i^\ell \leq (x_i'' - x_i') \leq L_i^u \quad \text{for all } i \tag{8.22}$$

$$W_i^\ell \leq (y_i'' - y_i') \leq W_i^u \quad \text{for all } i \tag{8.23}$$

$$P_i^\ell \leq 2(x_i'' - x_i' + y_i'' - y_i') \leq P_i^u \quad \text{for all } i \tag{8.24}$$

$$0 \leq x_i' \leq x_i'' \leq B_x \quad \text{for all } i \tag{8.25}$$

$$0 \leq y_i' \leq y_i'' \leq B_y \quad \text{for all } i \tag{8.26}$$

$$\alpha_i = 0.5x_i' + 0.5x_i'' \quad \text{for all } i \tag{8.27}$$

$$\beta_i = 0.5y_i' + 0.5y_i'' \quad \text{for all } i \tag{8.28}$$

$$\alpha_i - \alpha_j = \alpha_{ij}^+ - \alpha_{ij}^- \quad \text{for all } i \text{ and } j, \ i \neq j \tag{8.29}$$

$$\beta_i - \beta_j = \beta_{ij}^+ - \beta_{ij}^- \quad \text{for all } i \text{ and } j, \ i \neq j \tag{8.30}$$

$$x_j'' \leq x_i' + M(1 - z_{ij}^x) \quad \text{for all } i \text{ and } j, \ i \neq j \tag{8.31}$$

$$y_j'' \leq y_i' + M(1 - z_{ij}^y) \quad \text{for all } i \text{ and } j, \ i \neq j \tag{8.32}$$

$$z_{ij}^x + z_{ji}^x + z_{ij}^y + z_{ji}^y \geq 1 \quad \text{for all } i \text{ and } j, \ i < j \tag{8.33}$$

$$\alpha_i, \beta_i \geq 0 \quad \text{for all } i \tag{8.34}$$

$$x_i', x_i'', y_i', y_i'' \geq 0 \quad \text{for all } i \tag{8.35}$$

$$\alpha_{ij}^+, \alpha_{ij}^-, \beta_{ij}^+, \beta_{ij}^- \geq 0 \quad \text{for all } i \text{ and } j, \ i \neq j \tag{8.36}$$

$$z_{ij}^x, z_{ij}^y \ 0/1 \text{ integer} \quad \text{for all } i \text{ and } j, \ i \neq j \tag{8.37}$$

where $P_i^\ell$ and $P_i^u$ that appear in constraint set 8.24 represent the lower and upper limits imposed on the perimeter of department $i$, respectively.

---

[2] Actually, given the nature of constraints 8.13 and 8.14, one may rewrite constraint 8.15 as $z_{ij}^x + z_{ji}^x + z_{ij}^y + z_{ji}^y = 1$. Such constraints are known as "multiple-choice" constraints in integer programming; i.e., only one of the binary variables in the constraint may be set equal to one. Some branch-and-bound algorithms aim at reducing the computational effort by taking advantage of multiple-choice constraints.

In the above formulation, department shapes as well as their areas are indirectly controlled through constraint sets 8.22 through 8.24. In some cases, one may wish to explicitly control the shape of a department by placing an upper limit on the ratio of its longer side to its shorter side. If we designate the upper limit by $R_i$ ($\geq 1.0$), such a ratio may be controlled simply by adding the following (linear) constraints to the model for department $i$:

$$(x''_i - x'_i) \leq R_i(y''_i - y'_i) \tag{8.38}$$

$$(y''_i - y'_i) \leq R_i(x''_i - x'_i). \tag{8.39}$$

Note that, depending on which side is the longer one, at most one of the above two constraints will be "binding" or hold as an equality for non-square departments.

We remarked in Section 8.1 that the layout planner may represent an X relationship between departments $i$ and $j$ by assigning a negative value to $f_{ij}$. However, the transformation we used earlier to linearize the model (see Equation 8.19) requires that $f_{ij} \geq 0$ for all $i$ and $j$. While this requirement rules out the use of "negative flow" values, it does not imply that the above model cannot effectively capture an X relationship between two departments. One may insert a minimum required distance between two departments $i$ and $j$ by modifying constraints 8.31 and 8.32, as follows:

$$x''_j + \Delta^x_{ij} \leq x'_i + M(1 - z^x_{ij}) \qquad \text{for all } i \text{ and } j, \ i \neq j \tag{8.40}$$

$$y''_j + \Delta^y_{ij} \leq y'_i + M(1 - z^y_{ij}) \qquad \text{for all } i \text{ and } j, \ i \neq j \tag{8.41}$$

where $\Delta^x_{ij}$ and $\Delta^y_{ij}$ denote the minimum required distance (or clearance) between departments $i$ and $j$ along the $x$ and $y$-coordinates, respectively. Such clearances should be used sparingly; otherwise, the number of feasible solutions can be very limited. Also, note that if departments $i$ and $j$ are separated along, say, only the $x$-coordinate, then the resulting clearance between them along the $y$-coordinate may be less than $\Delta^y_{ij}$.

An alternate approach to impose an X relationship between departments $i$ and $j$ is to add a constraint of the following type:

$$\alpha^+_{ij} + \alpha^-_{ij} + \beta^+_{ij} + \beta^-_{ij} \geq \Delta_{ij}, \tag{8.42}$$

where the left-hand side is the rectilinear distance between the centroids of departments $i$ and $j$, and $\Delta_{ij}$ is the lower bound.

Hence, in theory, provided that the departmental area requirements need not be satisfied with precision, the optimum layout with rectangular departments can be obtained by solving the MIP model given by Equations 8.21 through 8.37. In practice, however, obtaining an exact solution to the above problem is not straightforward due to the large number of binary variables involved. At present, using "generic" optimization codes, the largest-size problem that one can solve exactly and with reasonable computational effort is approximately a seven- or eight-department problem in general.

Consequently, to use the MIP model in practice (where the number of departments can easily exceed 15 or 18 departments), one may aim for a heuristic solution rather than the optimal solution. One obvious heuristic approach is to terminate the branch-and-bound search when the difference between the least lower bound and the incumbent solution is less than, say, 5% or 10%. The advantage of such an approach is that the resulting solution is guaranteed to be within 5% or 10% of the optimum; the primary drawback is that it may still take a substantial amount of computer time to identify such a solution.

Another possible heuristic approach [36] is to determine the north–south and east–west relationships between the departments a priori by constructing a "design skeleton." With such an approach, first the $z_{ij}^x$ and the $z_{ij}^y$ values are determined heuristically, and the MIP model shown by Equations 8.21 through 8.37 can then be solved as a linear programming problem to "pack" the departments within the building.

## Example 8.5

Consider the same data given for Example 8.1. Assuming that departments A and G are fixed on the north side of the building (see the initial layout shown in Figure 8.3), we use the MIP model to construct the optimal layout. The solution is shown in Figure 8.12. The layout cost is equal to 53,501.17 units. Although some department areas in Figure 8.12 are not exactly equal to the values specified in Table 8.1, the reader may verify that the maximum deviation is only 0.25% (due to department D). Of course, some shape constraints were imposed on the departments to avoid long-and-narrow departments.

There are many real-world applications where the ideal shape of a department is a rectangle. However, there are also cases where L-shaped or U-shaped departments may be more appropriate or necessary. Furthermore, if there are fixed departments or obstacles in the building, enforcing rectangular department shapes with exact area values may make it impossible to find a feasible solution (even if the continuous representation is used and the fixed departments and obstacles are rectangular).

For example, consider a 10×10 building and four departments with the following area requirements: $A_1 = 18$, $A_2 = 20$, $A_3 = 27$, and $A_4 = 33$. Note that the total area requirement is actually less than 100 units. If none of the departments are fixed, then the above problem obviously has many feasible solutions. However, if the first

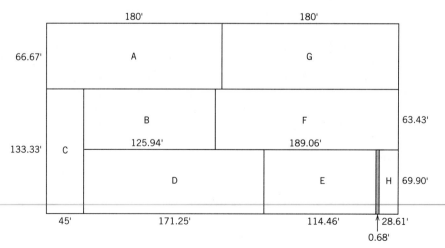

**Figure 8.12**  The optimal layout obtained from the MIP model ($z^* = 53,501.17$ units).

department is fixed, say, at the northeast corner of the building (see Figure 8.13*a*), then there is no feasible solution to the problem. In contrast, if the area of, say, department 3 or 4 can be slightly adjusted, then three feasible solutions can be constructed as shown in Figures 8.13*a*, 8.13*b*, and 8.13*c*.

Thus, retaining some flexibility in departmental area requirements is necessary for the effective use of the MIP model. In fact, approximating the departmental areas through some linear relaxation of constraint 8.8 serves a dual purpose: It allows one to remove a nonlinear constraint and at the same time it significantly reduces the likelihood of having no feasible solutions when obstacles and/or fixed departments are present. Of course, generally speaking, it also increases the number of feasible solutions.

The MIP model may also be used to improve a given layout after specifying the dimensions and location of fixed departments in the initial layout. This is accomplished simply by setting $\alpha_i$, $\beta_i$, $x'_i$, $x''_i$, $y'_i$, and $y''_i$ equal to their appropriate values for all the fixed departments and/or obstacles. (Nonrectangular obstacles may be represented as a collection of rectangular dummy departments.) Note that, with such an approach, the initial layout is "improved" not by exchanging department locations

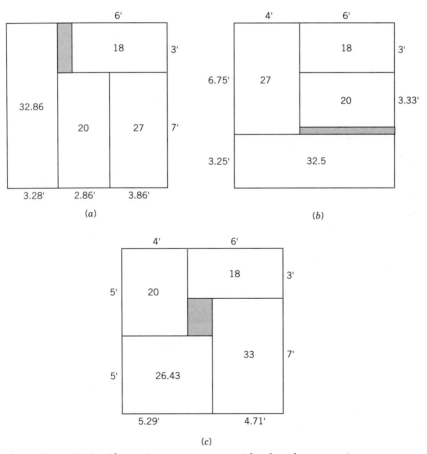

**Figure 8.13**  Alternative arrangements with relaxed area requirements.

but rather determining the new locations of all the (nonfixed) departments at once. Since such an approach does not explicitly consider "incremental" improvements to the layout and the resulting relocation costs versus material handling savings, it may not be a practical approach.

If relocation costs are significant, the layout planner may systematically fix and unfix certain subsets of the (nonfixed) departments and solve the MIP model several times to affect incremental changes in the layout. Alternately, the layout planner may first determine what the north–south/east–west relationships are going to be for two departments after their exchange, and subsequently use these relationships and the above MIP model to "repack" the departments with two of them exchanged.

## 8.6   LOGIC

LOGIC *(Layout Optimization with Guillotine Induced Cuts)* was developed by Tam [42]. In describing LOGIC, we assume that a from–to chart is given as input data for the flow. We also assume that layout "cost" is measured by the distance-based objective function shown in Equation 8.1. The departments generated by LOGIC are rectangular provided that the building is rectangular. The layout is represented in a continuous fashion.

Although LOGIC can be used as a layout improvement algorithm, we will first present it as a construction algorithm. LOGIC is based on dividing the building into smaller and smaller portions by executing successive "guillotine" cuts; that is, straight lines that run from one end of the building to the other. Each cut is either a vertical cut or a horizontal cut. If a cut is vertical, a department is assigned either to the east side of the cut or to the west. Given the building width (or a portion of it), and the total area of all the departments assigned, say, to the east side of a vertical cut, one may compute the exact *x*-coordinate of the cut. Likewise, given the building length (or a portion of it), and the total area of all the departments assigned, say, to the north side of a horizontal cut, one may compute the exact *y*-coordinate of the cut.

LOGIC executes a series of horizontal and vertical cuts. With each cut, an appropriate subset of the departments are assigned to the east-west or north-south side of the cut. In order to systematically execute the cuts and keep track of the departments, LOGIC constructs a tree as described in the following example, where we assume that the cuts and the department assignments are made randomly.

## *Example 8.6*

Consider the same data given for Example 8.1, except that we assume none of the departments (including A and G) are fixed. That is, we assume a vacant building which measures 360 ft×200 ft in length and width, respectively. Suppose the first cut is a vertical cut and departments D, F, and G are assigned to its east while the remaining departments are assigned to its west (see Figure 8.14*a*). Since the total area required by departments D, F, and G is equal to 36,000 ft² and the building width is equal to 200 ft, the above vertical cut divides the building into two pieces that are 36,000/200 = 180 ft long each. The first cut is also

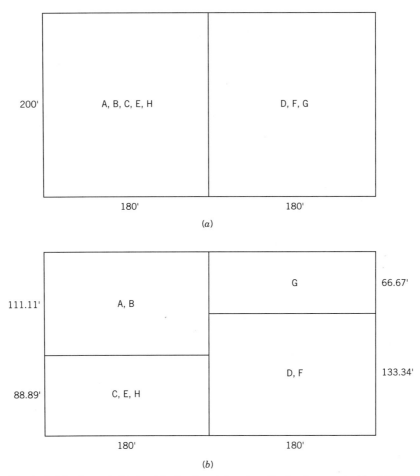

**Figure 8.14**  Layout obtained by horizontal and vertical cuts executed by LOGIC.

shown in Figure 8.15, where each node is labeled with a *v* for a vertical cut and a *h* for a horizontal cut.

LOGIC next treats each portion of the building as a "building" by itself and repeats the above procedure until each "building" contains only one department. Consider first the "building" which contains departments A, B, C, E, and H. Suppose the next cut is a horizontal cut and that departments A and B are assigned to the north of the cut while C, E, and H are assigned to its south. Since the total area required by departments A and B is equal to 20,000 ft² and the "building" length is equal to 180 ft, the width of the "building" which contains departments A and B is equal to 20,000/180 = 111.11 ft (see Figure 8.14*b*).

Consider next the "building" which contains departments D, F, and G. Suppose the third cut is again a horizontal cut and that department G assigned to the north of the cut while D and F are assigned to its south. Since department G requires an area of 12,000 ft² and the "building" length is equal to 180 ft, the width of department G is set equal to 12,000/180 = 66.67 ft as shown in Figure 8.14*b*. The second and third cuts as described above are also shown in Figure 8.15.

Suppose the fourth cut horizontally divides departments A and B, while the fifth cut vertically divides departments {C,H} and E. Further suppose the sixth cut horizontally divides

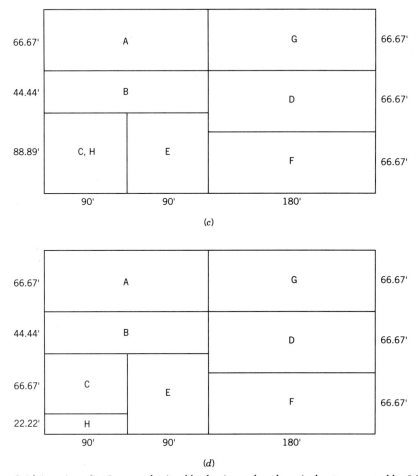

**Figure 8.14 (continued)**   Layout obtained by horizontal and vertical cuts executed by LOGIC.

departments D and F. The layout that results from the above cuts is shown in Figure 8.14c. Assuming that the seventh and final cut horizontally divides departments C and H, we obtain the final layout shown in Figure 8.14d. (The above cuts are also shown in Figure 8.15.)

LOGIC may also be used as an improvement algorithm in a variety of ways. Here we will show how it can be used to exchange two departments *given that the cut-tree remains the same*. Consider the layout shown in Figure 8.14d. Suppose we would like to exchange departments D and E, which are not equal in area. If the cuts remain the same (see Figure 8.15), we simply replace all the D's in the tree with E's and vice versa, and compute the new $x$ and $y$-coordinates of the cuts. The resulting layout is shown in Figure 8.16.

It is instructive to note that, since departments D and E are adjacent, CRAFT would have taken a fundamentally different approach to exchange them (see Section 8.3) and the resulting shapes of the two departments would have been different from those shown in Figure 8.16. As the above example illustrates, LOGIC can exchange two unequal-area departments (whether they are adjacent or not). Naturally, other

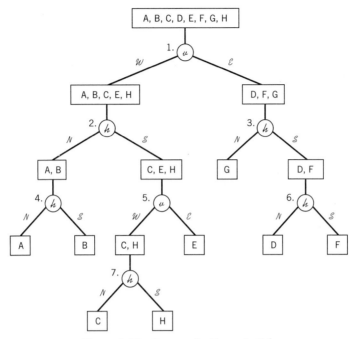

**Figure 8.15**   Cut-tree for Example 8.6.

departments have to shift to accommodate such an exchange. (We encourage the reader to compare the layouts shown in Figures 8.14d and 8.16.) Hence, like MCRAFT, LOGIC can "automatically" shift other departments, when necessary. However, again like MCRAFT, this may pose a problem if a fixed department is shifted in the process. One may try excluding all fixed departments from the tree (to retain their current positions). With such an approach, if a cut goes through one or more fixed departments or obstacles, it will complicate the calculation of its $x$ or $y$ coordinate. With LOGIC, it is generally not straightforward to model fixed departments or obstacles relative to CRAFT or MIP. We refer the reader to [42] for details.

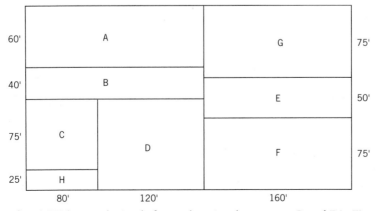

**Figure 8.16**   LOGIC layout obtained after exchanging departments D and E in Figure 8.14*d*.

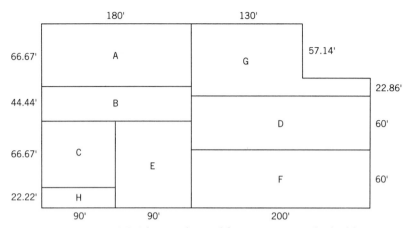

**Figure 8.17**   LOGIC layout obtained for a nonrectangular building.

LOGIC can be applied in nonrectangular buildings provided that the building shape is "reasonable." If a cut intersects a portion of the building where the length or width changes, then LOGIC uses a simple search strategy to compute the exact location of the cut. For example, consider the layout shown in Figure 8.17. The layout and the building are identical to the one shown earlier in Figure 8.14*d* except for the change we made to the east side of the building. The third cut shown earlier in Figure 8.15 (i.e., the cut that horizontally divided departments G and {D,F}) has to be computed differently since the "building" length increases from 130 ft to 200 ft within department G.

In presenting LOGIC, we assumed that the above cuts—and the assignment of departments to either side of each cut—are made "randomly." Actually, in order to use LOGIC in a practical application, such decisions are made within the framework of a pseudorandom search strategy such as "simulated annealing," which we discuss in Section 8.9.

## 8.7   MULTIPLE

MULTIPLE *(MULTI-floor Plant Layout Evaluation)* was developed by Bozer, Meller, and Erlebacher [5]. As the name suggests, MULTIPLE was originally developed for multiple floor facilities. However, it can also be used in single-floor facilities simply by setting the number of floors equal to one and disregarding all the data requirements associated with the lifts.

Except for the exchange procedure and layout formation, MULTIPLE is similar to CRAFT. It uses a from-to chart as input data for the flow, and the objective function is identical to that of CRAFT (i.e., a distance-based objective with distances measured rectilinearly between department centroids). Departments are not restricted to rectangular shapes, and the layout is represented in a discrete fashion. Also, like CRAFT,

MULTIPLE is an improvement-type layout algorithm that starts with an initial layout specified by the layout planner. Improvements to the layout are sought through two-way exchanges, and at each iteration the exchange that leads to the largest reduction in layout cost is selected; that is, MULTIPLE is a steepest-descent procedure.

The fundamental difference between CRAFT and MULTIPLE is that MULTIPLE can exchange any two departments whether they are adjacent or not. Recall that MCRAFT and BLOCPLAN can also exchange any two departments; however, since both algorithms are based on "bands," fixed departments may either shift or change shape. Also, the band approach imposes certain restrictions on the initial layout as well as fixed departments and obstacles as discussed earlier. In essence, MULTIPLE retains the flexibility of CRAFT while relaxing CRAFT's constraint imposed on department exchanges.

MULTIPLE achieves the above task through the use of "spacefilling curves," (SFCs) which were originally developed by the Italian mathematician Peano. Although SFCs (which were considered "mathematical oddities" at the time of their introduction) initially had nothing to do with optimization and industrial engineering, they have been used to construct a heuristic procedure for routing and partitioning problems [2] and for determining efficient locations for items in a storage rack [3]. In MULTIPLE, SFCs are used to reconstruct a new layout when any two departments are exchanged.

MULTIPLE's use of SFCs for the above purpose can perhaps be best described through an example. Consider the SFC shown in Figure 8.18a, which is known as the Hilbert curve [19]. (The procedure to generate such a curve is shown in [5]. The interested reader may also refer to [2].) Note that the curve connects each grid such that a "dot" traveling along the curve will always visit a grid that is adjacent to its current grid. Also note that each grid is visited exactly once. Suppose the following area values (expressed in grids) are given for six departments: $A_1 = 16$, $A_2 = 8$, $A_3 = 4$, $A_4 = 16$, $A_5 = 8$, and $A_6 = 12$. If the *layout vector* or the *fill sequence* is given by 1-2-3-4-5-6, we obtain the layout shown in Figure 8.18b by starting from grid 1 and assigning the first 16 grids (along the SFC) to department 1, the next eight grids (along the SFC) to department 2, and so on. In other words, the SFC allows MULTIPLE to map a unidimensional vector (i.e., the layout vector) into a two-dimensional layout. The above mapping can be performed rapidly since the grids are sorted a priori according to their sequence on the SFC.

Given such a mapping, it is now straightforward to exchange any two departments by exchanging their locations in the layout vector. For example, to exchange departments 1 and 5, which are neither adjacent nor equal in area, we first switch their positions in the layout vector to obtain 5-2-3-4-1-6 and then simply reassign the grids (following the SFC). The resulting layout is shown in Figure 8.18c. Note that all the departments, except for department 6, have "shifted" to accommodate the above exchange. (The shift is fairly significant since department 5 is only half as large as department 1.) In general, whenever two unequal-area departments are exchanged, all the departments that fall between the two departments on the layout vector will be shifted; the other departments will remain in their original locations.

In order to avoid shifting fixed departments or obstacles, the SFC "bypasses" all the grids assigned to such departments. Also, in those cases where the building shape is irregular or there are numerous obstacles (including walls), MULTIPLE can

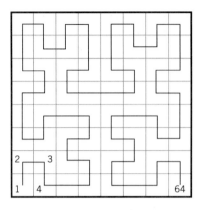

(*a*) Hilbert curve [19]

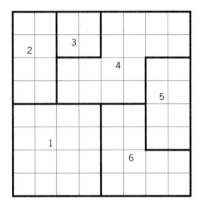

(*b*) Layout vector: 1-2-3-4-5-6

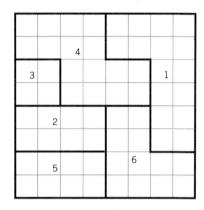

(*c*) Layout vector: 5-2-3-4-1-6

**Figure 8.18**   MULTIPLE's use of spacefilling curves in layout formation; departments 1 and 5 have been exchanged.

be used with a "hand-generated" curve, which may start at any grid, and end at any grid, but must visit all the grids exactly once by taking only horizontal or vertical steps (from one grid to an adjacent grid). Diagonal steps are allowed but generally not recommended since they may split a department. (Recall that two grids that "touch" each other only at the corners are not considered adjacent for facility layout purposes.) Unless a fixed department or wall physically separates the building into two disjoint segments, it is always possible to construct such a curve by reducing the grid size. (Why?)

Of course, such hand-generated curves are mathematically no longer SFCs, but they serve the same function. In [5], the authors report that while using MULTIPLE in a "large, four-floor production facility," they opted for hand-generated curves to capture the "exact building shape, the current layout, and all the obstacles." In the following example, we illustrate how a hand-generated curve can be used to capture the current layout. Such curves are also referred to as "conforming curves" since they fully conform to the current layout.

# *Example 8.7*

Consider the same data given for Example 8.1, where departments A and G are assumed to be fixed. The initial layout (which has a cost of 59,480 units) was shown earlier in Figure 8.3. A hand-generated, conforming curve for the initial layout is shown in Figure 8.19a. Note that the curve does not visit departments A and G. Also note that the curve visits all the grids assigned to a particular department before visiting any other department. The initial layout vector is given by C-B-D-H-F-E.

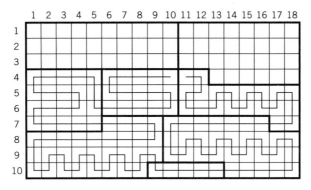

(a) Conforming "hand-generated" curve

| | 1 | 2 | 3 | 4 | 5 | 6 | 7 | 8 | 9 | 10 | 11 | 12 | 13 | 14 | 15 | 16 | 17 | 18 |
|---|---|---|---|---|---|---|---|---|---|---|---|---|---|---|---|---|---|---|
| 1 | A | A | A | A | A | A | A | A | A | A | G | G | G | G | G | G | G | G |
| 2 | A | | | | | | | | | A | G | | | | | | | G |
| 3 | A | A | A | A | A | A | A | A | A | A | G | G | G | | | | | G |
| 4 | D | D | D | D | D | D | D | D | D | D | E | E | G | G | G | G | G | G |
| 5 | D | | | | | | | | | D | E | E | E | E | E | E | E | E |
| 6 | D | D | D | D | D | D | D | D | D | D | E | E | E | E | E | E | E | E |
| 7 | B | B | B | B | B | B | B | B | B | F | F | F | F | F | F | F | E | E |
| 8 | B | B | B | B | B | B | B | B | B | F | F | F | F | F | F | F | F | F |
| 9 | B | H | H | C | C | C | C | C | C | F | F | F | F | F | F | F | F | F |
| 10 | B | H | H | H | C | C | C | C | C | C | C | C | C | F | F | F | F | F |

(b) Final MULTIPLE layout (z = 54,200 units).

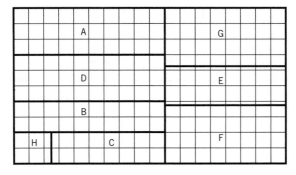

(c) Final "massaged" layout obtained with MULTIPLE.

**Figure 8.19** Layouts obtained with MULTIPLE for Example 8.7.

Given the above initial layout vector and the curve, in the first iteration MULTIPLE exchanges departments C and D, which reduces the layout cost to 54,260 units. In the second and last iteration, MULTIPLE exchanges departments C and H to obtain the layout vector D-B-H-C-F-E and the final layout shown in Figure 8.19*b*. The cost of the final layout is equal to 54,200 units, which is less than the final layout cost obtained by CRAFT (2833.50 × 20 = 56,670 units). In general, MULTIPLE is very likely to obtain lower-cost solutions than CRAFT because it considers a larger set of possible exchanges at each iteration. However, even if both algorithms are started from the same initial layout, MULTIPLE is not guaranteed to find a lower-cost layout than CRAFT. (Why?)

As we indicated for CRAFT, the final layout generated by MULTIPLE may also require massaging to smooth the department borders. Retaining the relative locations of the departments and slightly adjusting some department areas, we obtain the final layout shown in Figure 8.19*c*. Also, note that the particular curve used not only determines the final layout cost, but it also determines the department shapes. Hence, alternative layouts can be generated by trying different curves.

MULTIPLE may also be used as a construction procedure. In such cases, the layout planner may use any SFC or hand-generated curve that best conforms to the (vacant) building and possible obstacles; of course, there is no initial layout that the curve needs to conform to. Generally speaking, curves such as the one shown in Figure 8.18*a* seem to generate more reasonable department shapes than those which

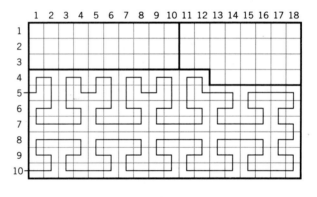

(*a*)

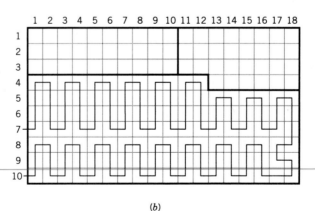

(*b*)

**Figure 8.20**  Alternative curves for MULTIPLE.

"run" in straight lines from one end of the building to the other. Any layout vector may be used as the "initial" layout. It is generally recommended to try alternative layout vectors as the starting point.

It is also instructive to note that, by considering all possible layout vectors for small problems, one may obtain the "optimal" layout *with respect to a given curve*. For example, for the curve we used in Example 8.7, the layout vector D-B-H-C-F-E is optimal and the final layout shown in Figure 8.19*b* (at 54,200 units) is the "optimal" layout. Two alternative curves are shown in Figures 8.20*a* and *b*, for which the optimal layout vector is given by D-E-F-B-C-H (at 54,920 units) and D-E-F-H-B-C (at 54,540 units), respectively. The above results suggest that, while alternative layouts may be obtained from alternative curves, the cost of these layouts may not be very sensitive to the curve, provided that the curves are not dramatically different. As an exercise, the reader may construct the optimal layouts by using the above two layout vectors and the corresponding curves shown in Figures 8.20*a* and *b*.

# *8.8*  DEPARTMENT SHAPES AND MAIN AISLES

In developing alternative block layouts, the main aisles are typically not represented explicitly until the block layout is finalized. As the layout planner "massages" the final block layout, he or she attempts to correct irregular department shapes and aims for smooth department borders primarily for two reasons: first, a poor department shape may make it virtually impossible to develop an efficient and effective detailed layout for that department; second, since main aisles connect all the departments, by definition, irregular department shapes would lead to irregular main aisles. Generally speaking, for efficient material handling and other reasons (including safety, unobstructed travel, evacuation in an emergency), main aisles should connect all the departments in a facility with minimum travel, minimum number of turns, and minimum "jog overs" (i.e., they should run in straight lines as much as possible). Hence, attaining "good" department shapes is an important consideration in finalizing a block layout.

As we showed in Section 8.5 (see Equations 8.38 and 8.39), controlling department shapes is relatively straightforward for rectangular departments (such as those obtained with BLOCPLAN and MIP). This is due to the fact that it is straightforward to define and measure the shape of a rectangle: it is simply the ratio of its longer side to its shorter side (or vice versa). However, if the layout planner needs or wishes to consider nonrectangular department shapes (such as those obtained with CRAFT and MULTIPLE), then shape measurement and control is not straightforward. In fact, given two alternative but "similar" shapes for the same department, one of the alternatives may be regarded as acceptable while the other one is regarded as poor. Although humans are good at making such (subjective) judgments with respect to department shapes, computer-based algorithms require formal and objective measures.

A few alternative measures have been suggested in the literature to control the shape of nonrectangular departments. As we shall see, such measures are intended only to detect and avoid irregular department shapes; they cannot be used to guarantee or prescribe specific department shapes. Although the shape measures we describe below are more suitable for the discrete layout representation, they can be implemented with the continuous representation as well.

Two measures to control department shapes are presented in [29]. Both measures are based on first identifying the smallest rectangle that fully encloses the department. The first measure is obtained by dividing the area of the smallest-enclosing-rectangle (SER) by the area of the department. Given that the department area is fixed, one would expect the above ratio to increase as the department shape becomes more irregular since a larger rectangle would be required to enclose the department. The second measure is obtained by dividing the longer side of the SER by its shorter side. As before, as the department shape becomes more irregular, one would expect the above ratio to increase.

A third measure, presented in [5], originally appeared in papers concerned with geometric modeling (see [15], among others). It is based on the observation that, given an object with a fixed area, the perimeter of the object generally increases as its shape becomes more irregular. Hence, one may measure the shape of a department by dividing its perimeter by its area. However, unless the layout planner generates alternative shapes for each department "by hand" and computes the above ratio for each shape a priori, it is difficult to predict reasonable values for it. To address this difficulty, in [5] the above ratio is normalized as follows: If the "ideal" shape for a department is a square, then the "ideal" shape factor, say, $S^*$, is equal to $(P/A)^* = (4\sqrt{A})/A = 4/\sqrt{A}$, where $P$ denotes the perimeter and $A$ denotes the area of the department. The normalized shape factor, say, $F$, is equal to $S/S^* = (P/A)/(4/\sqrt{A}) = P/(4\sqrt{A})$. Hence, if a department is square shaped, we obtain $F = 1.0$; otherwise, we obtain $F > 1.0$. Generally speaking, reasonable shapes are obtained if $1.0 \le F \le 1.4$. If a square is not the "ideal" shape for a department, the analyst may impose a lower bound greater than 1.0 on $F$.

In [5] the above three shape factors are compared via an example; we present it here with a minor change. Suppose the department area is equal to 16 units. Four alternative shapes, including the SERs, are shown in Figure 8.21. For Figure 8.21$a$, the first measure is equal to 1.0, and it increases to $25/16 = 1.5625$ for Figure 8.21$b$. However, it remains at 1.5625 for Figures 8.21$c$ and 8.21$d$. The second measure, on the other hand, is equal to 4.0 for Figure 8.21$a$, and it decreases to 1.0 for the remaining three department shapes. The third measure, $F$, is equal to 1.25, 1.25, 1.50, and 1.625 for Figures 8.21$a$ through 8.21$d$, respectively. (The reader may verify that $P$ is equal to 20, 20, 24, and 26 for Figures 8.21 $a$ through $d$, respectively.)

As the above example illustrates, no shape measure is perfect, although $F$ seems to generate more accurate results. Furthermore, as we remarked earlier, none of the shape measures can be used to prescribe a particular shape for a department. Rather, they can be used to avoid irregular department shapes by checking the shape factor for each department following an exchange. Since a typical computer run considers many possible department exchanges, any shape measure must be straightforward to compute. The three shape measures we showed above are straightforward to implement within a computer-based layout algorithm that uses the discrete representation; the computational burden they generate is minimal.

For example, consider the third measure. To compute the perimeter of department $i$, the computer first examines each grid assigned to department $i$, one at a time, and simply counts the number of adjacent grids which have *not* been assigned to department $i$. (Unless it is located along the building perimeter, each grid has exactly

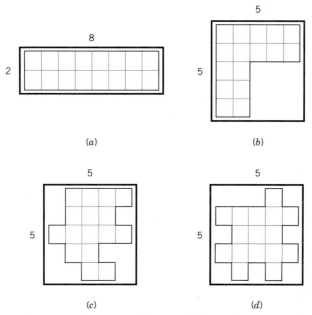

**Figure 8.21** Alternative department shapes and the smallest-enclosing-rectangle.

four adjacent grids. Grids that are located along the perimeter of the building must be treated slightly differently.) Once the perimeter of each department is computed, the computer determines the normalized shape factor for each department. As a result of exchanging two departments, if any department violates a shape constraint (i.e., the normalized shape factor for the department falls outside a user-defined range, which may be department-specific), then the computer simply "rejects" the exchange.

Note that, with the above scheme to compute department perimeters, those with "enclosed voids" will be correctly identified as possibly irregular in shape. For example, for the departments shown in Figures 8.2c, d and e, the normalized shape factor is given by 1.061, 1.591, and 1.414, respectively. Had we not included the "inside perimeter" in Figure 8.2e, the shape factor would have been 1.061, which is a misleading figure.

Although shape measures are useful in avoiding irregular department shapes, they should be used with care for three reasons: First, if strict shape constraints are imposed, many (if not all) possible department exchanges may be rejected regardless of how much they reduce the layout cost. Second, some departments may take irregular shapes only temporarily; i.e., a department which is irregularly shaped in the current iteration may assume a reasonable shape in the next iteration—note that this may occur both with CRAFT and MULTIPLE. Third, the analyst has the option to correct the department shapes by massaging the final layout generated by the computer. If strict shape constraints are imposed, the computer will "automatically" discard lower-cost solutions with irregular department shapes even if the analyst could have corrected the problem through massaging. Hence, we generally recommend imposing no shape constraints until the layout planner makes a few runs and obtains preliminary results. In subsequent runs, a shape constraint may be imposed only on those departments that have unreasonable shapes that cannot be corrected through massaging.

## 8.9   SIMULATED ANNEALING AND GENETIC ALGORITHMS[3]

Simulated Annealing (SA) and Genetic Algorithms (GAs) represent relatively new concepts in optimization. Although both SA and GAs can be used for layout construction, we will limit their presentation to *layout improvement*. Since a formal and thorough treatment of either subject is beyond the scope of this book, we will present only some of the basic concepts of SA and GAs, and show only one SA-based layout algorithm.

The most significant drawback of the steepest descent approach (which we described in Section 8.3) is that it forces the algorithm to terminate the search at the first two-opt or three-opt solution it encounters. As we remarked earlier, such a solution is very likely to be only locally optimal. (Note that, when they consider two-way or three-way department exchanges, CRAFT, BLOCPLAN, and MULTIPLE are examining the local "neighborhood" of the current solution.) As a result, any algorithm that uses the steepest descent technique becomes highly "path dependent" in that the initial solution and the specific department exchanges made by the algorithm play a significant role in determining the cost of the final solution. (The impact of the initial solution is also known as the *initial layout bias*.) Ideally, a heuristic procedure should consistently identify a low-cost, that is, an optimal or near optimal, solution regardless of the starting point.

One of the primary strengths of SA is that, while trying to improve a layout, it may "occasionally" accept nonimproving solutions to allow, in effect, the algorithm to explore other regions of the solution space (instead of stopping at the first "seemingly good" solution it encounters). In fact, a SA-based procedure may accept nonimproving solutions several times during the search in order to "push" the algorithm out of a solution which may be only locally optimal. As a result, the objective function value may actually increase more than once. However, the amount of increase in the objective function that the algorithm will "tolerate" is carefully controlled throughout the search. Also, the algorithm always "remembers" the best solution; that is, the best solution identified since the start of the search is never discarded even as new regions of the solution space are explored.

The fundamental concepts behind SA are based on an interesting analogy between statistical mechanics and combinatorial optimization problems [26]. Statistical mechanics is the "central discipline of condensed matter physics, a body of methods for analyzing aggregate properties of the large numbers of atoms to be found in

[3]This section presents recent research results, rather than detailed algorithms and examples. Intended for first-year graduate students and advanced undergraduate students, it provides insights into the direction the field might take during the next few years.

samples of liquid or solid matter" [26]. One of the key issues in statistical mechanics is the state of the matter (or the arrangement of its atoms) as its temperature is gradually reduced until it reaches the "ground state" (which is also referred to as the "lowest energy state" or the "freezing point").

According to [48], "In practice, experiments designed to find the (lowest energy) states are performed by careful *annealing,* that is, by first melting the (material) at a high temperature, then lowering the temperature slowly (according to an *annealing schedule*), finally spending a long time at temperatures in the vicinity of freezing, or solidification, point. The amount of time spent at each temperature during the annealing process must be sufficiently long to allow the system to reach thermal equilibrium (steady state). If care is not taken in adhering to the annealing schedule (the combination of a set of temperatures and length of time to maintain the system at each temperature), undesirable random fluctuations may be frozen into the material thereby making the attainment of the ground state impossible."

Hence, the analogy between combinatorial optimization problems and statistical mechanics is that each solution in the former corresponds to a particular arrangement of the atoms in the latter. The objective function value is viewed as the energy of the material, which implies that finding a low-cost solution is analogous to achieving the lowest energy state through an effective annealing schedule. Also, since we are working with mathematical optimization models rather than "condensed matters," the annealing schedule applied to a combinatorial optimization problem is not "real" annealing, it is "simulated" annealing. As we shall see shortly, the annealing schedule, that is, the set of temperatures used (including the initial temperature) and the time spent at each temperature, plays an important role in algorithm development and the cost of the final solution.

The concept of "occasionally" accepting nonimproving solutions goes back to a simple Monte Carlo experiment developed in [33]: Given a current arrangement of the atoms (or "elements"), we randomly make incremental changes to the current arrangement to obtain a new arrangement, and we measure the *decrease* in energy, say, $\Delta E$. If $\Delta E > 0$ (i.e., the energy decreases), we accept the new arrangement as the current one and use it to make subsequent changes. However, if $\Delta E \leq 0$, the new arrangement is accepted with probability $P(\Delta E) = \exp(\Delta E/k_b T)$, where $T$ is the temperature and $k_b$ is Boltzmann's constant. (If we do not accept the new arrangement, we generate another one.)

To apply the above procedure to optimization problems we simply treat the "current arrangement" as the *current solution,* the "new arrangement" as the *candidate solution,* and the "energy" as the *objective function value.* A random incremental change is made, for example, by exchanging two randomly picked departments in the current layout. Last, we set $k_b = 1$ since it has no known significance in optimization problems. Note that, as the increase in the objective function gets larger, the probability of accepting the candidate solution gets smaller. For example, if $T = 150$ and the increase in the objective function is equal to 250 units (i.e., $\Delta E = -250$), the probability of accepting the candidate solution is equal to 0.1889. At $T = 150$, if the increase in the objective function is equal to 400 units, however, the above probability decreases to 0.0695. Also note that the probability of accepting a nonimproving solution decreases as the temperature decreases. For example, for $\Delta E = -250$, if we decreased the temperature from 150 to 100 units, the above probability would decrease from 0.1889 to 0.0821. In other words, we are more likely to accept nonimproving solutions early in the annealing process (due to the relatively high temperatures).

It is instructive to further note that the probability in question ultimately depends on the change in the objective function *relative to* the temperature. Therefore, in a SA-based algorithm, we are not only concerned with how fast we "cool" the system but also with the initial temperature we select to start the annealing process. One possible approach is to set the initial temperature according to the objective function value of the starting solution (see, for example, [24]). In the algorithm that follows, the initial temperature is set equal to $z_0/40$, where $z_0$ is the cost of the initial layout. Of course, the layout planner should experiment with different initial temperature settings.

The above process of generating candidate solutions and appropriately updating the current solution continues until the system reaches steady state *at the current temperature*. As shown in the following algorithm, the percent change in the *mean* objective function value is used to determine whether steady state has been reached. According to [48], "at each temperature, the annealing schedule *must* allow the simulation to proceed long enough for the system to reach steady state." Once steady state is reached, the temperature is reduced according to a predetermined temperature reduction factor and we continue to generate and evaluate candidate solutions with the new temperature setting.

Typically, the search is terminated either when a user-specified final temperature is reached (which can also be expressed as the maximum number of temperature reductions to be considered) or a user-specified number of *successive* temperature reductions does not produce an improvement in the best solution identified since the start of the annealing process (i.e, the "current best" solution). The algorithm we present below uses both the latter stopping criterion and one that is based on the maximum number of "epochs." An epoch corresponds to a particular set of candidate solutions accepted by the algorithm. The epoch length is expressed as the number of candidate solutions in the set.

To develop a SA-based layout algorithm, we will apply an annealing schedule to MULTIPLE, which was described in Section 8.7. Recall that, in MULTIPLE, given a particular spacefilling curve and the departmental area requirements, the layout vector (i.e., a sequence of department numbers) fully defines a layout. Therefore, any solution is represented only as a layout vector. Except for the technique used in generating the candidate layout vectors, the algorithm we show here is identical to SABLE *(Simulated Annealing-Based Layout Evaluation)*, which was developed by Meller and Bozer [32]. (SABLE uses a more general technique for generating candidate layout vectors.) When appropriate, variables were used in a manner similar to their use in [48]. Let

$s^0$ be the initial layout vector,

$s^*$ be the "current best" layout vector which corresponds to the lowest cost layout identified by the algorithm,

$s$ be the current layout vector,

$s'$ be the candidate layout vector,

$\alpha$ be the temperature reduction factor (which controls how fast the system is "cooled down"),

$T$ be a set of annealing schedule temperatures $\{t_1, t_2, t_3, \ldots\}$, where $t_i = t_0(\alpha)^i$ for all $i > 1$,

$t_0$ be the initial temperature,

$e$ be the (fixed) epoch length,

$f_j(s)$ be the objective value of the $j$th accepted candidate layout vector, $s$, in an epoch.

$\bar{f}_e$ be the mean objective function value of an epoch, i.e., $\bar{f}_e = [\Sigma^e_{j=1} f_j(s)]/e$,

$\bar{f}'_e$ be the overall mean objective function value of all the layout vectors accepted during the epochs previous to the current one (for a given temperature),

$\epsilon_i$ be a threshold value used to determine whether the system is in equilibrium at temperature $i$,

$M$ be the maximum total number of epochs to be considered (across all temperatures),

$I$ be a counter to record the last temperature setting that produced the "current best" layout vector, $s^*$,

$N$ be the maximum number of *successive* temperature reductions that will be performed with no improvement in $s^*$.

The initial layout vector $s^0$ as well as the values of $\alpha$, $t_0$, $e$, $\epsilon$, $M$, and $N$ are specified by the user a priori. Using the above notation (except for the subscript $j$, which we will omit for brevity), a simple version of SABLE [32] is presented as follows:

**Step 1.**   Set $s = s^0$, $I = 1$, and $i = 1$. Compute the initial layout cost, $z_0$; set $t_0 = (z_0)/40$ and $t_1 = \alpha t_0$.

**Step 2.**   For the current layout vector, compute the layout cost $f(s)$.

**Step 3a.**   Randomly pick two departments in $s$, exchange their locations and store the resulting layout vector (i.e., the candidate layout vector) in $s'$.

**Step 3b.**   Compute the decrease in the layout cost, i.e., set $\Delta f = f(s) - f(s')$. If $\Delta f > 0$, go to step 3d; otherwise, go to step 3c.

**Step 3c.**   Sample a random variable $x \sim U(0, 1)$. If $x < \exp(\Delta f/t_i)$, go to step 3d; otherwise, go to step 3a.

**Step 3d.**   Accept the candidate sequence; i.e., set $s = s'$ and $f(s) = f(s')$. If $f(s) < f(s^*)$, then update the "current best" solution; i.e., set $s^* = s$, $f(s^*) f(s)$, and $I = i$. If $e$ candidate sequences have been accepted, go to step 4; otherwise, go to step 3a.

**Step 4.**   If equilibrium has not been reached at temperature $t_i$, that is, if $|\bar{f}_e - \bar{f}'_e|/\bar{f}'_e \geq \epsilon_i$, reset the counter for accepted candidate solutions and go to step 3a; otherwise, set $i = i + 1$ and $t_i = t_0(\alpha)^i$. If $(i - I) < N$, go to step 5; otherwise, **STOP**—the maximum number of successive nonimproving temperatures has been reached.

**Step 5.**   If the total number of epochs is less than $M$, go to step 3a; otherwise, **STOP.**

The parameter values selected by the user are very likely to have a more-than-minor impact on the cost of the final solution obtained by the above algorithm as well as its execution time. Setting the initial temperature with respect to the initial layout cost and ensuring that the system is not "cooled down" too rapidly generally improves the performance of the algorithm. In [32], for example, the following parameter values were observed empirically to substantially reduce the initial layout bias and yield generally "good" solutions for all the test problems evaluated: $\alpha = 0.80$, $t_0 = (z_0)/40$, $e = 30$, $\epsilon_1 = 0.25$, $\epsilon_i = 0.05$ for all $i \geq 2$, $M = 37$, and $N = 5$. We caution the reader that the above parameter settings are shown only as an example. (Also, recall that in [32] a more general candidate solution generation procedure is used.) Once a SA-based algorithm is developed, it is generally good practice to experiment with a variety of parameter settings.

The above algorithm is certainly not the only possible application of SA to facility layout problems. In fact, as we remarked at the end of Section 8.6, LOGIC [42] is actually a SA-based layout algorithm. Recall that LOGIC develops a layout by executing a series of horizontal and vertical cuts, and by assigning (an appropriate subset of) the departments to one or the other side of each cut. Given a current solution, which is represented by a cut-tree (see Section 8.6), a candidate solution may be obtained, for example, by randomly changing the orientation of one (or more) of the cuts in the tree and/or by randomly changing the partition of a set of departments over one (or more) of the cuts. (We refer the reader to [42] for further details.) Candidate solutions obtained in the above manner can be evaluated within the framework of an annealing schedule as we showed in the above algorithm. Another application of SA in facility layout is presented in [22]. The algorithm in [22] is intended primarily for the special case where all the departments have equal area requirements; using it with general departmental area requirements is possible but it may lead to numerical problems or split departments.

The "biased sampling technique" (BST) [38], which was applied to CRAFT, has certain similarities to SA. With the BST, at each iteration, given a set of exchanges (that are estimated to reduce the layout cost), instead of always selecting the exchange which is estimated to reduce the layout cost the most, the algorithm assigns a nonzero probability to each exchange in the set and randomly selects one of them. By solving the same problem several times with different random number streams and different starting points, the BST was shown to reduce the initial layout bias [38]. However, the BST is not as "formal" a concept as SA and, more importantly, the BST will still terminate at the first locally optimal solution it encounters (i.e., the BST does not accept an exchange that may increase the objective function value).

It is instructive to note that developing effective and efficient SA-based layout algorithms is not only a matter of finding a "good" annealing schedule but also a matter of finding a "good" representation that makes it possible to *rapidly generate and evaluate* a variety of candidate solutions. Note that we are referring to "solution" representation and not "layout" representation (as in discrete versus continuous layout representation). Naturally, the two are not independent. For example, in MULTIPLE [5] and SABLE [32], the solution is represented as a layout vector and the layout is constructed through the use of spacefilling curves (which work well with the discrete layout representation). In LOGIC [42] the solution is represented as a cut-tree and the layout is constructed by applying vertical and horizontal cuts (which work well with the continuous layout representation).

The application of GAs, on the other hand, to facility layout problems (and other optimization problems) is relatively recent, but it has been gaining momentum.

The basic concept behind GAs was developed by Holland [20] who observed that the "survival of the fittest" (SOF) principle in nature may be used in solving decision-making, optimization, and machine learning problems. Although the SOF principle may first appear to have no relation to optimization problems, the "relationship" between the two is as fascinating as the "relationship" we described between statistical mechanics and optimization problems.

Algorithms based on SA "occasionally" accept nonimproving solutions; however, they still work with only one solution at a time. That is, there is only one current solution, from which we generate only one candidate solution. In contrast, GAs work with a family of solutions (known as the "current population") from which we obtain the "next generation" of solutions. When the algorithm is used properly, we obtain progressively better solutions from one generation to the next. That is, good solutions (or actually "parts of good solutions") propagate from one generation to the next and lead to better solutions as we produce more generations.

The basic GA, at least on the surface, is admirably simple. Given a current population of, say, $N$ solutions, we generate the next population as follows: We randomly pick two "parents" (i.e., two solutions) from the current population and "randomly cross over" the two parents to obtain two "offsprings" (i.e., two "new" solutions) for the next population. We repeat the above process until we have a new generation with $N$ solutions. (Note that, unlike most natural systems, the population size is fixed at $N$.) Each time we generate two offsprings, their two parents are picked randomly; however, the probability of picking a particular solution as a parent is proportional to the "fitness" of the parent. For minimization problems, the above implies that the probability of picking a solution as a parent is inversely proportional to the objective function value of that parent (i.e., a "fit" parent with a small objective function value is more likely to be selected). The first generation (i.e., the starting generation) is usually created randomly. Typically, the process of creating new generations continues until a prespecified number of generations have been produced or no noticeable reduction in the average population objective function value is detected for a number of successive generations.

In most GAs, the above "basic" algorithm is modified in a number of ways. One common addition is "mutation," which essentially means altering (in some arbitrary fashion) one or more solutions picked at random within a given population. Another addition is the concept of "elitist reproduction," which basically implies that when a new population is created, the best 10% or 20% of the solutions in the current population are "automatically" copied over to the next population. The remaining 90% or 80% of the solutions in the next population are generated by the "two parents–two offsprings" method we described above. Solutions that were copied over are still eligible to act as parents for creating the remainder of the next population.

The population size, the mutation rate, the rate of elitist reproduction, and the number of generations to create are all user-specified parameters that affect the performance and execution time of a GA. Also, there are alternative cross-over operators that specify how two offsprings are "randomly" generated from the two parents. In fact, the type of solution representation used for the problem is critical in terms of defining appropriate cross-over operators (so that we do not create "infeasible offsprings" from "feasible parents"). That is, even though the two parents represent feasible solutions, if we do not use or develop an appropriate cross-over operator, their offsprings may not. A discussion of alternative cross-over operators and appropriate values to assign to the above parameters are beyond the scope of this book. The reader may refer to [16] for an excellent introduction to basic and some advanced concepts in GAs and their use in optimization problems and machine learning. Also, [8] is a good reference book for GAs; it includes comprehensive examples for GA applications. Furthermore, the reader will find GAs applied to the Quadratic Assignment Problem and the facility layout problem in [43], [44], and [45].

In concluding this section, we note that, generally speaking, if the computational effort required to evaluate a single solution is high, SA-based algorithms are likely to outperform GAs. This stems primarily from the fact that a GA is more likely to evaluate a larger number of solutions than a SA-based algorithm. It may also be the case that, in early generations, GAs spend computer time evaluating many poor solutions before they are removed from the "gene pool." On the other hand, there is a "natural fit" between GAs and parallel computers, where each processor in the computer can independently and concurrently generate and evaluate a pair of off-springs for the new population. This way, a large number of new population members can be rapidly generated and evaluated in parallel. Nevertheless, with or without parallel computers, it is too early to make conclusive statements about SA versus GAs in facility layout; further research is needed to fully develop both techniques and compare their performance on various facility layout problems.

# *8.10* COMMERCIAL FACILITY LAYOUT PACKAGES

As we remarked in Section 8.1, there are a few commercial packages available for facility layout and related topics. FactoryCAD, FactoryPLAN, and FactoryFLOW are offered by CIMTECHNOLOGIES CORPORATION, ISU Research Park, Suite 700, 2501 North Loop Drive, Ames, Iowa, 50010-8285 (Phone: 515-296-9914; Fax: 515-296-9909). FactoryCAD allows the user to customize AutoCAD to facilitate the drawing of layouts. FactoryPLAN generates numeric and graphical output to evaluate a *user-specified* layout. FactoryFLOW incorporates material handling data into a given layout to compute material handling distances and generate material handling equipment utilization reports. In its current form, FactoryPLAN is a descriptive tool; that is, it will help the user evaluate a given layout but it will not prescribe specific changes or show how to improve it.

Another commercial facility layout package, namely, LayOPT, is offered by PRODUCTION MODELING CORPORATION, Three Parklane Boulevard, Suite 910 West, Dearborn, Michigan, 48126 (Phone: 313-441-4460; Fax: 313-441-6098). LayOPT is a commercial implementation of MULTIPLE (see Section 8.7) with an optional AutoCAD interface. The user enters the data, the spacefilling curve, and the current/proposed layout, and LayOPT returns with the "top 10" two-way department exchanges that reduce the layout cost. The user may also interactively perform "layout massaging" to improve the department shapes or make other changes to the layout. In addition to single-floor facilities, LayOPT is designed to work with mezzanines and multifloor facilities with vertical handling equipment.

FactoryModeler is a commercial package that provides an environment for factory planning and design; it is offered by Systemes Espace Temps, Inc., 1043 Gustave Langelier, Cap Rouge, Quebec, Canada (Phone: 418-654-9456; Fax: 418-656-7746). FactoryModeler enables the user to develop a model of the factory. This model can be used for manufacturing process engineering, capacity planning, factory configuration, material handling system requirements planning, and layout design. It supports three-dimensions and a dynamic planning horizon. Based on the object-agent oriented paradigm, FactoryModeler permits navigation from aggregate to detailed designs. It also supports multiple design methodologies and models.

We expect to see more activity and new developments in the area of commercial facility planning and layout packages. Not only are the above packages likely to evolve into more comprehensive and capable products, but we will probably witness the entry of new companies with products based on powerful graphical displays, possibly with multimedia and three-dimensional "rendering." Some of these products may also offer integrated design, analysis, simulation, and animation environments. For student versions or demonstration disks for the above products, the reader should inquire with the companies.

While there are some other packages which may be used to design the detailed layout, say, of offices (such as office furniture), and there are some packages that may be used to plan plant services, the above three packages are, to our knowledge, currently the only ones that can be used to design and/or evaluate the facility (block) layout in the manner addressed in this chapter. We regret any omissions we may have made.

## *8.11* SUMMARY

In this chapter, we presented a number of computer-based facility layout algorithms; namely, CRAFT, MCRAFT, BLOCPLAN, MIP, LOGIC, and MULTIPLE. Among the algorithms we covered, CRAFT, MCRAFT and MULTIPLE use the discrete layout representation, and with the possible exception of MCRAFT, they place no restrictions on department shapes. Also, CRAFT and MULTIPLE can capture the initial layout, the building shape, and fixed departments/obstacles with fairly high accuracy. However, with CRAFT and MULTIPLE it is difficult to generate departments with prescribed shapes (such as rectangular or L-shaped departments). Although the spacefilling curve in MULTIPLE may be revised to correct some department shapes, both algorithms are likely to generate layouts that require considerable massaging.

BLOCPLAN, MIP, and LOGIC, on the other hand, use the continuous layout representation and they work only with rectangular department shapes. (LOGIC may generate one or more nonrectangular departments if the building is nonrectangular or there are obstacles present.) Rectangular departments considerably increase our control over department shapes. Also, in practice, a rectangular shape is very likely to be acceptable (if not desirable) for many departments in a facility. However, if fixed departments or obstacles are present, maintaining rectangular departments may increase the cost of the layout. Furthermore, capturing the initial layout, which may contain one or more nonrectangular departments, may not be possible or straight-forward (especially with BLOCPLAN and LOGIC).

We also demonstrated the fundamental "idea" behind each algorithm. For example, MCRAFT relies on a sweep technique, while BLOCPLAN uses two or three (horizontal) bands. LOGIC, on the other hand, divides the building progressively into smaller portions by performing vertical and horizontal cuts and by assigning one or more departments to each portion of the building. MULTIPLE uses a spacefilling curve (or a hand-generated curve), which "maps" the two-dimensional layout into a single dimension (i.e., the layout vector). Last, MIP is based on mathematical programming and it uses 0/1 integer variables to ensure that departments do not overlap.

As far as the objective function is concerned, all the algorithms we showed use either the distance-based objective or the adjacency-based objective, or both. In reality, almost any layout algorithm can be adapted to accept one objective function or another. For example, the objective functions of CRAFT, LOGIC, and MULTIPLE can be changed from distance based to adjacency based with no fundamental changes to the algorithm itself. In fact, except for MIP, any one of the above algorithms can be used to generate a layout and evaluate it with respect to a user-specified objective function. One can also use an alternate objective function with MIP; however, care must be exercised to maintain a "reasonable" objective function (or at least a linear objective function) since the objective function, unlike the other algorithms, is an inherent part of MIP.

For those readers who might, at least from a theoretical standpoint, be concerned with layout optimality, we note that only two of the algorithms we presented can be used to identify the optimal layout: MIP will find the optimal layout only if all the departments are required to be rectangular; MULTIPLE will find the optimal layout *only with respect to a particular spacefilling curve* by enumerating over all-possible department sequences in the layout vector. LOGIC may also find the "optimal" layout (by enumerating over all-possible cut-trees); however, if all the departments are rectangular, the cost of any optimal solution obtained by LOGIC will be greater than or equal to the cost of an optimal solution obtained by MIP. (Why? Hint: consider the impact of the "guillotine cuts" in LOGIC.)

There are a number of early computer-based layout algorithms that we did not describe in this chapter. Such algorithms include ALDEP [41], COFAD [46], CORELAP [28], and PLANET [9]. Since their introduction, these algorithms have served as the cornerstones of facility layout algorithms, and they fueled interest in computer-based facility layout.[4] We did not present a number of relatively recent computer-based layout algorithms; see, for example, DISCON [11], FLAC [40], and SHAPE [17], and [47]. The reader may refer to [27] for a survey and discussion of these and other single-floor facility layout algorithms. Last, we did not present any results on multifloor facility layout problems. With advances in vertical material handling technology (see, for example, [49] and [50]), there appears to be renewed interest in multifloor facilities. Generally speaking, multifloor facility layout problems are more difficult than their single-floor counterparts. For more information on multifloor facility layout and computer-based algorithms that are developed for solving such problems, the reader may refer to BLOCPLAN [10], MSLP [25], MULTIPLE [5], SPACECRAFT [23], SPS [29], and Meller [30].

---

[4]Due to their historical significance, ALDEP and CORELAP are described in the appendix to the chapter.

Some readers, from a theoretical or practical standpoint, might wonder which layout algorithm is the best? From a theoretical standpoint, it is valid to compare certain layout algorithms. For example, since they use the same objective function and the same representation, one may compare the performance of CRAFT, MCRAFT, and MULTIPLE. In fact, in [5], numerical results are shown to compare CRAFT with MULTIPLE. Likewise, as long as all the departments are rectangular, one may perform a meaningful comparison between BLOCPLAN, LOGIC, and MIP (assuming that MIP is solved through a heuristic procedure such as the one shown in [36]). In contrast, comparing, say, MULTIPLE with MIP, requires a more careful approach; the two algorithms use different representations and the resulting department shapes are likely to be different. Hence, any comparison must include the department shapes. However, if the purpose of the comparison is to simply measure the layout cost impact of allowing nonrectangular departments (at the risk of obtaining some irregularly shaped departments), then it may make sense to compare just the layout costs of MULTIPLE and MIP.

From a practical standpoint, we hope that the reader is now in a position to appreciate that each layout algorithm has certain strengths and weaknesses. Particular constraints imposed by a problem (such as the number and location of fixed departments and/or obstacles, department shapes, the building shape, moving into a vacant building versus improving a given layout), coupled with our description of each algorithm, should help the reader (or the layout planner) select not the "best" algorithm but the "most appropriate" one. Although "dummy or negative flows" can be used in a from-to chart, and some qualitative interactions among the departments may be captured in a relationship chart, we also need to remind the practitioner that no computer-based layout algorithm will capture all the significant aspects of a facility layout problem.

In fact, we do not know if computers will ever be able to fully capture and use human experience and judgment, which play a critical role in almost any facility layout problem. In that sense, a computer-based layout algorithm should not be viewed as an "automated design tool," which can functionally replace the layout planner. To the contrary, we believe the (human) layout planner will continue to play a key role in developing and evaluating the facility layout. The computer-based layout algorithms presented in this chapter (and other such algorithms in the literature) are intended as "design aids" that, when used properly, significantly enhance the productivity of the layout planner. Therefore, for non-trivial problems, we believe that, given two layout planners with comparable experience and skills, the one "armed" with an appropriate computer-based layout algorithm is far more likely to generate a demonstrably "better" solution in a shorter time.

# BIBLIOGRAPHY

1. Armour, G. C., and Buffa, E. S., "A Heuristic Algorithm and Simulation Approach to Relative Location of Facilities," *Management Science,* vol. 9, no. 2, 1963, pp. 294–309.
2. Bartholdi, J. J., and Platzman, L. K., "Heuristics Based on Spacefilling Curves for Combinatorial Problems in Euclidean Space," *Management Science,* vol. 34, no. 3, 1988, pp. 291–305.

3. Bartholdi, J. J., and Platzman, L. K., "Design of Efficient Bin-Numbering Schemes for Warehouses," *Material Flow,* vol. 4, 1988, pp. 247–254.

4. Bozer, Y. A., and Meller, R. D., "A Reexamination of the General Facility Layout Problem," Technical Report 93-01, Auburn University, Auburn, AL, 1993.

5. Bozer, Y. A., Meller, R. D., and Erlebacher, S. J., "An Improvement-Type Layout Algorithm for Single and Multiple Floor Facilities," *Management Science,* vol. 40, no. 7, 1994, pp. 918–932.

6. Brown, A. R., *Optimum Packing and Depletion,* Macdonald and Co., American Elsevier Publishing Co., New York, 1971.

7. Buffa, E. S., Armour, G. C., and Vollman, T. E., "Allocating Facilities with CRAFT," *Harvard Business Review,* vol. 42, 1964, pp. 136–158.

8. Davis, L. (Ed.), *Handbook of Genetic Algorithms,* Van Nostrand Reinhold, New York, 1991.

9. Deisenroth, M. P., and Apple, J. M., "A Computerized Plant Layout Analysis and Evaluation Technique (PLANET)," *Technical Papers 1962,* American Institute of Industrial Engineers, Norcross, GA, 1972.

10. Donaghey, C. E., and Pire, V. F., "Solving the Facility Layout Problem with BLOCPLAN," Industrial Engineering Department, University of Houston, TX, 1990.

11. Drezner, Z., "DISCON: A New Method for the Layout Problem," *Operations Research,* vol. 20, 1980, pp. 1375–1384.

12. Foulds, L. R., "Techniques for Facilities Layout: Deciding Which Pairs of Activities Should be Adjacent," *Management Science,* vol. 29, no. 12, 1983, pp. 1414–1426.

13. Foulds, L. R., and Robinson, D. F., "Graph Theoretic Heuristics for the Plant Layout Problem," *International Journal of Production Research,* vol. 16, 1963, pp. 27–37.

14. Francis, R. L., McGinnis, L. F., and White, J. A., *Facility Layout and Location: An Analytical Approach,* Prentice Hall, Englewood Cliff, NJ, 1992.

15. Freeman, H., "Computer Processing of Line-Drawing Images," *Computing Surveys,* vol. 6, 1974, pp. 57–97.

16. Goldberg, D. E., *Genetic Algorithms in Search, Optimization, and Machine Learning,* Addison-Wesley, Boston 1989.

17. Hassan, M. M. D., Hogg, G. L., and Smith, D. R., "SHAPE: A Construction Algorithm for Area Placement Evaluation," *International Journal of Production Research,* vol. 24, 1986, pp. 1283–1295.

18. Heragu, S. S., and Kusiak, A., "Efficient Models for the Facility Layout Problem," *European Journal of Operational Research,* vol. 53, 1991, pp. 1–13.

19. Hobson, E. W., *The Theory of Functions of a Real Variable and the Theory of Fourier's Series, Volume I,* 3rd ed., Cambridge University Press and Harren Press, Washington D.C., 1950.

20. Holland, J. H., *Adaptation in Natural and Artificial Systems,* The University of Michigan Press, Ann Arbor, MI, 1975.

21. Hosni, Y. A., Whitehouse, G. E., and Atkins, T. S., *MICRO-CRAFT Program Documentation,* Institute of Industrial Engineers, Norcross, GA.

22. Jajodia, S., Minis, I., Harhalakis, G., and Proth, J., "CLASS: Computerized LAyout Solutions using Simulated annealing," *International Journal of Production Research,* vol. 30, no. 1, 1992, pp. 95–108.

23. Johnson, R. V., "SPACECRAFT for Multi-Floor Layout Planning," *Management Science,* vol. 28, no. 4, 1982, pp. 407–417.

24. Johnson, D. S., Aragon, C. R., McGeoch, L. A., and Schevon, C., "Optimization by Simulated Annealing: An Experimental Evaluation; Part I, Graph Partitioning," *Operations Research,* vol. 37, 1989, pp. 865–892.

25. Kaku, K., Thompson, G. L., and Baybars, I., "A Heuristic Method for the Multi-Story Layout Problem," *European Journal of Operational Research,* vol. 37, 1988, pp. 384–397.

26. Kirkpatrick, S., Gelatt, C. D., and Vecchi, M. P., "Optimization by Simulated Annealing," *Science,* vol. 220, 1983, pp. 671–680.

27. Kusiak, A., and Heragu, S. S., "The Facility Layout Problem," *European Journal of Operational Research,* vol. 29, 1987, pp. 229–251.

28. Lee, R. C., and Moore, J. M., "CORELAP—Computerized Relationship Layout Planning," *Journal of Industrial Engineering,* vol. 18, no. 3, 1967, pp. 194–200.

29. Liggett, R. S., and Mitchell, W. J., "Optimal Space Planning in Practice," *Computer Aided Design,* Vol. 13, 1981, pp. 277–288.

30. Meller, R. D., "Layout Algorithms for Single and Multiple Floor Facilities," Ph.D. Dissertation, Department of Industrial and Operations Engineering, The University of Michigan, 1992.

31. Meller, R. D., Personal Communication, 1993.

32. Meller, R. D., and Bozer, Y. A., "Solving the Facility Layout Problem with Simulated Annealing," Technical Report 91-20, The University of Michigan, Ann Arbor, MI, 1991.

33. Metropolis, N., Rosenbluth, A. W., Rosenbluth, M. N., Teller, A. H., "Equation of State Calculation by Fast Computing Machines," *Journal of Chemical Physics,* vol. 21, 1953, pp. 1087–1092.

34. Montreuil, B., Ratliff, H. D., and Goetschalckx, M., "Matching Based Interactive Facility Layout," *IIE Transactions,* vol. 19, no. 3, 1987, pp. 271–279.

35. Montreuil, B., "A Modeling Framework for Integrating Layout Design and Flow Network Design," *Proceedings of the Material Handling Research Colloquium,* Hebron, KY, 1990, pp. 43–58.

36. Montreuil, B., and Ratliff, H. D., "Utilizing Cut Trees as Design Skeletons for Facility Layout," *IIE Transactions,* vol. 21, no. 2, 1989, pp. 136–143.

37. Murty, K., *Linear and Combinatorial Programming,* John Wiley, 1976.

38. Nugent, C. E., Vollman, R. E., and Ruml, J., "An Experimental Comparison of Techniques for the Assignment of Facilities to Locations," *Operations Research,* vol. 16, 1968, pp. 150–173.

39. Pire, V. F., "Automated Multistory Layout System," Unpublished Master's thesis, Industrial Engineering Department, University of Houston, Houston, TX, 1987.

40. Scriabin, M., and Vergin, R. C., "A Cluster-Analytic Approach to Facility Layout," *Management Science,* vol. 31, no. 1, 1985, pp. 33–49.

41. Seehof, J. M., and Evans, W. O., "Automated Layout Design Program," *Journal of Industrial Engineering,* vol. 18, 1967, pp. 690–695.

42. Tam, K. Y., "A Simulated Annealing Algorithm for Allocating Space to Manufacturing Cells," *International Journal of Production Research,* vol. 30, 1991, pp. 63–87.

43. Tanaka, H., and Yoshimoto, K., "Genetic Algorithm Applied to the Facility Layout Problem," Department of Industrial Engineering and Management, Waseda University, Tokyo, 1993.

44. Tate, D. M., and Smith, A. E., "A Genetic Approach to the Quadratic Assignment Problem," *Computers and Operations Research,* vol. 22, no. 1, 1995, pp. 73–83.

45. Tate, D. M., and Smith, A. E., "Genetic Algorithm Optimization Applied to Variations of the Unequal Area Facilities Layout Problem," Proceedings of the 2nd Industrial Engineering Research Conference, 1993, Los Angeles, CA, pp. 335–339.

46. Tompkins, J. A., and Reed, R. Jr., "An Applied Model for the Facilities Design Problem," *International Journal of Production Research,* vol. 14, no. 5, 1976, pp. 583–595.

47. Tretheway, S. J., and Foote, B. L., "Automatic Computation and Drawing of Facility Layouts with Logical Aisle Structures," *International Journal of Production Research,* vol. 32, no. 7, 1994, pp. 1545–1555.

48. Wilhelm, M. R., and Ward, T. L., "Solving Quadratic Assignment Problems by Simulated Annealing," *IIE Transactions,* Vol. 19, 1987, pp. 107–119.

49. *Material Handling Engineering,* "Vertical Reciprocating Conveyors: Flexible Handling Devices," vol. 45, no. 9, 1990, pp. 81–85.

50. *Modern Materials Handling,* "Vertical Conveyors Increase Plant's Efficiency 33%," vol. 40, Casebook Directory, 1985, p. 113.

# APPENDIX 8.A
# ALDEP and CORELAP

## 8.A1 INTRODUCTION

Because of their historical significance, in the appendix we consider briefly two additional computer-aided layout algorithms, ALDEP and CORELAP. Following the development of CRAFT in 1963, ALDEP and CORELAP were developed in 1967. For two decades, the three computer-aided layout algorithms served as the benchmarks for all other such algorithms.

Developed for main frame computers, we are not aware of microcomputer versions of ALDEP and CORELAP currently being supported commercially. For this reason, we did not choose to include them in the main body of the chapter. However, for those who want to understand the evolution of computer-aided layout algorithms to the present, it is useful to consider the earliest such algorithms.

In the appendix, we consider first CORELAP, since ALDEP is but a slight variation of CORELAP. Our reason for including both CORELAP and ALDEP is the randomness aspect of ALDEP and its multifloor capability. After learning about CORELAP and ALDEP, we suspect you will recognize the hereditary roots of several algorithms presented in the chapter.

## 8.A2 CORELAP

CORELAP, an acronym representing **computerized relationship layout planning,** constructs a layout for a facility by calculating the total closeness rating (TCR) for each department where the TCR is the sum of the numerical values assigned to the closeness relationships (A = 6, E = 5, I = 4, O = 3, U = 2, X = 1) between a department and all other departments [28]. The department having the highest TCR is then placed in the center of the layout. If there is a tie for the highest TCR, the following tie-breaking rule is applied: the department having the largest area and then the department having the lowest department number. Next, the relationship chart is scanned and if a department is found having an "A" relationship with the *selected department,* it is brought into the layout. If none exists, the relationship chart is scanned for an "E" relationship, then an "I", and so on. If two or more departments are found having the same relationship with the selected department, the department having the highest TCR is selected; if a tie still exists, the tie-breaking rule is utilized. The third department to enter the layout is determined by scanning the relationship chart to see if an unassigned department exists that has an "A" relationship with the first department selected. If so, this department is brought into the layout. If a tie exists, the TCR and then the tie-breaking rule are utilized. If no unassigned department exists that has an "A" relationship with the second department, the procedure is repeated considering "E" relationships, then "I" relationships, and so on. If a tie occurs, the TCR and then the tie-breaking hierarchy are utilized. The same procedure is repeated for the fourth department to enter the layout, except the three departments previously selected are included in the search. The procedure continues until all departments have been selected to enter the layout.

Once a department is selected to enter the layout, a placement decision must be made. The placement decision is made by calculating the placing rating for the available locations of the department, where the placing rating is the sum of the weighted closeness ratings between the department to enter the layout and its neighbors. For example, consider the layout consisting of departments 1 and 7 in Figure 8.A1a. Suppose department 2 is next to enter the layout and it has an "A" relationship with department 1 and an "E" relationship with department 7. If an "A" relationship is weighted 64 and an "E" is weighted 16 by the user, the placing ratings are as shown in 8.A1*b*, 8.A1*c*, and 8.A1*d*.

If a tie occurs for the placing rating, the boundary lengths of the tied locations are compared. The boundary length is the number of unit square sides that the department to enter the layout has in common with its neighbors. In Figures 8.A1*b* and 8.A1*c*, the boundary lengths are 2, and in 8.A1*d* the boundary length is 3.

After the final CORELAP layout has been prepared, CORELAP evaluates the layout by calculating the layout score. The layout score is defined as:

$$\text{Layout score} = \sum_{\text{all departments}} \begin{array}{c} \text{numerical} \\ \text{closeness} \\ \text{rating} \end{array} \times \begin{array}{c} \text{length of} \\ \text{shortest} \\ \text{path} \end{array} \qquad (8.\text{A}1)$$

It should be noted that CORELAP utilizes the shortest rectilinear path between departments as opposed to the rectilinear distance between department centroids as in CRAFT. The shortest rectilinear path is used as it is assumed that each department will have a dispatch area and a receiving area on the side of its layout nearest its neighbor.

CORELAP-generated layouts often result in irregular building shapes and will need manual adjustment. Care must also be exhibited with the interpretation of the layout scores, as the shortest rectilinear path between departments may not always be a realistic measure.


## *Example 8.A1*

The data requirements for CORELAP are the department areas and the relationship chart. Recall the relationship diagramming example in Chapter 7 with the activity relationship chart

```
                        2 2
    1  1  1           1  1  1              Placing
    1  1  1  7  7  7  1  1  1  7  7  7    rating = 64
          (a)                (b)


    1  1  1     2  2  Placing       1  1  1  2  2     Placing
    1  1  1  7  7  7  rating = 16   1  1  1  7  7  7  rating = 80
          (c)                            (d)
```

**Figure 8.A1**  Illustration of CORELAP placing rating.

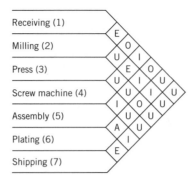

**Figure 8.A2**   Activity relationship chart.

given in Figure 8.A2 and space requirements given in Table 8.A1. CORELAP begins by calculating the TCR for each department. Table 8.A2 illustrates these calculations. The first department selected by CORELAP to enter the layout is department 5, as it has the highest TCR. Scanning the relationship chart indicates that only department 6 has an "A" relationship with department 5, so it will be the second department selected to enter the layout. The selection of the third department to enter the layout begins by scanning the relationship chart and determining that there are no additional "A" relationships with department 5 and that there are no unselected departments having an "A" relationship with department 6. "E" relationships with department 5 also result in no selection. An "E" relationship does exist between departments 6 and 7; therefore, department 7 will be the third department to enter the layout. In an effort to select the fourth department, it is noted that no additional departments have "A" or

**Table 8.A1**   *Department Areas and Number of Unit Area Templates for Example 8.A1*

| Code | Function | Area (Square feet) | Number of Unit Area Templates |
|------|----------|--------------------|-------------------------------|
| 1 | Receiving | 12,000 | 6 |
| 2 | Milling | 8000 | 4 |
| 3 | Press | 6000 | 3 |
| 4 | Screw machine | 12,000 | 6 |
| 5 | Assembly | 8000 | 4 |
| 6 | Plating | 12,000 | 6 |
| 7 | Shipping | 12,000 | 6 |

**Table 8.A2**   *Calculation of the Total Closeness Ratings (TCRs) for Example 8.A1*

| Department | Department Number | Relationships | TCR |
|------------|-------------------|---------------|-----|
| Receiving | 1 | E, O, I,  O, U, U | 19 |
| Milling | 2 | E, U, E, I,  I,  U | 22 |
| Press | 3 | O, U, U, U, U, U | 14 |
| Screw machine | 4 | I, E, U, I,  U, U | 19 |
| Assembly | 5 | O, I,  U, I,  A, I | 23 |
| Plating | 6 | U, I,  O, U, A, E | 22 |
| Shipping | 7 | U, U, U, U, I, E | 17 |

**Table 8.A3   *Weights Assigned to Relationships for Example 8.A1***

| Relationship | Weight |
|:---:|:---:|
| A | 243 |
| E | 81 |
| I | 27 |
| O | 9 |
| U | 1 |

"E" relationships with department 5, 6, or 7. Department 5 has an "I" relationship with departments 2 and 4. By checking the TCR, we see department 2 has the highest value and will be the fourth department to enter the layout. After determining that no "A" relationships exist with the selected departments, "E" relationships are scanned; it may be seen that departments 1 and 4 have an "E" relationship with department 2. Both departments 1 and 4 have a TCR of 19 and an area of 12,000 ft$^2$, so that lowest numbered department, department 1, will be the fifth department to enter the layout. Department 4 will enter the layout next, followed by department 3. The overall order of departments entering the layout is 5-6-7-2-1-4-3.

The weighted values for placing departments for this example are as assigned in Table 8.A3. Department 5 is the first department to enter the layout and is placed in the center. Department 6 is then positioned adjacent to department 5, as shown in Figure 8.A3$a$. It should be noted that CORELAP adds 10 to the department number prior to it being printed, that is, department 1 will be printed as department 11, department 5 as 15, department 18 as 28, and so on. Department 7 will maximize its placing rating by next entering the layout, as shown in Figure 8.A3$b$. Department 2 has an area of 1 unit square and relationships, "I," "I," and "U" with departments 5, 6, and 7, respectively. Therefore, the highest placing rating would be obtained if department 2 could be adjacent to departments 5 and 6. Unfortunately, as noted in Figure 8.A3$b$, no such location is available. In fact, the only location that allows department 2 to be adjacent to more than one department is the location shown in Figure 8.A3$c$. The placing rating for this location is 28 (27 for department 2 being adjacent to department 6, and 1 for department 2 being adjacent to department 7). This is the highest placing rating and is the location CORELAP selects for department 2. By pursuing this same course of action, CORELAP produces the final layout shown in Figure 8.A3$d$. The last CORELAP function would be the calculation of the layout score shown in Table 8.A4.

```
0 0 0  0  0  0 0 0       0 0 0   0   0  0 0 0
0 0 0 16 16 0 0 0        0 0 0  16 16 0 0 0
0 0 0 0  15 0 0 0        0 0 17 17 15 0 0 0
0 0 0 0  0  0 0 0        0 0 0   0   0  0 0 0
0 0 0 0  0  0 0 0        0 0 0   0   0  0 0 0
        (a)                      (b)


0 0 0  0  0  0 0 0       0 14 11 11 13 0 0 0
0 0 12 16 16 0 0 0       0 14 12 16 16 0 0 0
0 0 17 17 15 0 0 0       0 0  17 17 15 0 0 0
0 0 0  0  0  0 0 0       0 0  0   0   0  0 0 0
0 0 0  0  0  0 0 0       0 0  0   0   0  0 0 0
        (c)                      (d)
```

**Figure 8.A3**   CORELAP layout construction for the example problem. ($a$) The first two adjacent departments. ($b$) Department 7 added to departments 5 and 6. ($c$) The layout consisting of departments 5, 6, 7, and 2. ($d$) The final layout.

**Table 8.A4** *Calculation of the Layout Score*

| Relationship | Preset Value | From | To | Distance | Product of Preset Value and Distance |
|:---:|:---:|:---:|:---:|:---:|:---:|
| A | 6 | 15 | 16 | 0 | 0 |
| E | 5 | 11 | 12 | 0 | 0 |
| E | 5 | 12 | 14 | 0 | 0 |
| E | 5 | 16 | 17 | 0 | 0 |
| I | 4 | 11 | 14 | 0 | 0 |
| I | 4 | 12 | 15 | 2 | 8 |
| I | 4 | 12 | 16 | 0 | 0 |
| I | 4 | 14 | 15 | 3 | 12 |
| I | 4 | 15 | 17 | 0 | 0 |
| O | 3 | 11 | 13 | 0 | 0 |
| O | 3 | 13 | 16 | 0 | 0 |
| O | 3 | 11 | 15 | 2 | 6 |
| U | 2 | 11 | 16 | 0 | 0 |
| U | 2 | 11 | 17 | 1 | 2 |
| U | 2 | 12 | 13 | 2 | 4 |
| U | 2 | 12 | 17 | 0 | 0 |
| U | 2 | 13 | 14 | 2 | 4 |
| U | 2 | 13 | 15 | 1 | 2 |
| U | 2 | 13 | 17 | 2 | 4 |
| U | 2 | 14 | 16 | 1 | 2 |
| U | 2 | 14 | 17 | 1 | 2 |
| | | | | Layout score | 46 |

# 8.A3 ALDEP

ALDEP, an acronym representing **automated layout design program,** has the same basic data input requirements and objectives as CORELAP. The basic procedural difference between CORELAP and ALDEP is that CORELAP selects the first department to enter the layout and breaks ties with the total closeness rating, where ALDEP selects the first department and breaks ties randomly. The basic philosophical difference between CORELAP and ALDEP is that CORELAP attempts to produce the one best layout, whereas ALDEP produces many layouts, rates each layout, and leaves the evaluation of the layouts to the facilities designer. ALDEP is the first model having provisions for multiple floors; it can handle up to three floors. Additionally, ALDEP allows departments to be placed in specific locations and is able to include dummy departments [41]. However, ALDEP does not accommodate flow between floors as effectively as MULTIPLE. Also, if there is a fixed department, ALDEP often splits departments.

As was previously mentioned, the first department ALDEP selects to enter the layout is selected randomly. The relationship chart is then scanned to determine if there is a department having an "A" relationship with the randomly selected first department. If one exists, it is selected to enter the layout. If more than one exists, one is randomly selected to enter the layout. If no departments have a relationship at least equal to the minimum acceptable closeness rating specified by the user, the second department to enter the layout will be selected randomly. Once the second department to enter the layout is selected, the selection procedure is repeated for the second department selected and all unselected departments. Once the third department is selected, the next department to enter the layout is determined by repeating the selection procedure. This process continues until all departments have been selected to enter the layout.

The placement routine within ALDEP begins by placing the first department in the upper left corner of the layout and extends it downward. The width of the downward extension of the department entering the layout is input by the user and is termed the **sweep width.** The department is placed in the layout using the sweep pattern illustrated in Figure 8.A4*f.* Each additional department added to the layout begins where the previous department ends and continues to follow the serpentine path. When all departments have entered the layout, ALDEP rates the layout by assigning values to the relationships among adjacent departments. If a department is adjacent to a department with which it has an "A" relationship, a value of 64 is added to the rating of the layout. An "E" relationship adds 16, "I" adds 4, and an "O" relationship adds 1 to the rating of the layout. A "U" relationship has no effect on the rating of the layout and if two adjacent departments have an "X" relationship, 1024

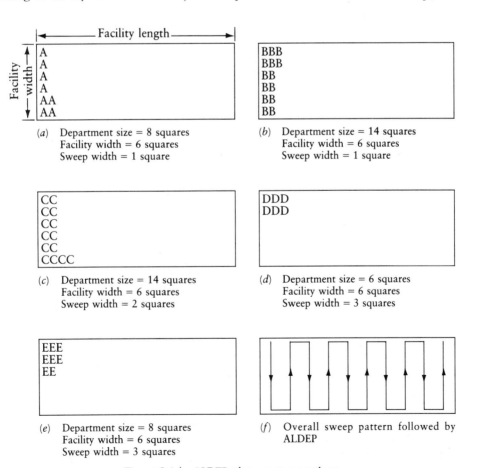

(*a*) Department size = 8 squares
Facility width = 6 squares
Sweep width = 1 square

(*b*) Department size = 14 squares
Facility width = 6 squares
Sweep width = 1 square

(*c*) Department size = 14 squares
Facility width = 6 squares
Sweep width = 2 squares

(*d*) Department size = 6 squares
Facility width = 6 squares
Sweep width = 3 squares

(*e*) Department size = 8 squares
Facility width = 6 squares
Sweep width = 3 squares

(*f*) Overall sweep pattern followed by ALDEP

**Figure 8.A4**   ALDEP placement procedure.

is subtracted from the rating of the layout. ALDEP prints the layout and the rating and then returns to randomly generate the first department to be selected for the next layout. The entire procedure is repeated. ALDEP can be used in this manner to generate up to 20 layouts and ratings per run.

ALDEP will not print layouts whose rating is less than an initially input minimal score. For this reason, users of ALDEP should be aware of the very high penalty that is built into ALDEP for the "X" relationship. Relationship charts having more than a few "X" relations often will not result in layouts being printed. Additionally, users should not accept ALDEP answers without experimenting with the input values for the specified level of importance and the sweep width. The specified level of importance basically segregates all relationships into an important or an unimportant classification. The decision as to which relationships should be considered important and which should be unimportant must be made for each problem and can only be determined by experimentation. The sweep width that provides the best layout is problem dependent and must be determined via experimentation.

## Example 8.A2

In addition to the input data for Example 8.A1, ALDEP requires the specification of the sweep width and the specified level of importance. For the iteration to be described, the sweep width will be assigned a value of 2 and the specified level of importance will be input

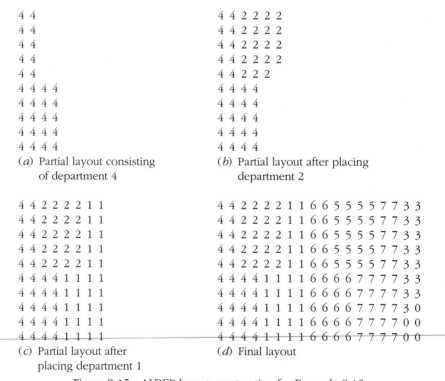

```
4 4
4 4
4 4
4 4
4 4
4 4 4 4
4 4 4 4
4 4 4 4
4 4 4 4
4 4 4 4
```
(*a*) Partial layout consisting
     of department 4

```
4 4 2 2 2 2
4 4 2 2 2 2
4 4 2 2 2 2
4 4 2 2 2 2
4 4 2 2 2
4 4 4 4
4 4 4 4
4 4 4 4
4 4 4 4
4 4 4 4
```
(*b*) Partial layout after placing
     department 2

```
4 4 2 2 2 2 1 1
4 4 2 2 2 2 1 1
4 4 2 2 2 2 1 1
4 4 2 2 2 2 1 1
4 4 2 2 2 2 1 1
4 4 4 4 1 1 1 1
4 4 4 4 1 1 1 1
4 4 4 4 1 1 1 1
4 4 4 4 1 1 1 1
4 4 4 4 1 1 1 1
```
(*c*) Partial layout after
     placing department 1

```
4 4 2 2 2 2 1 1 6 6 5 5 5 5 7 7 3 3
4 4 2 2 2 2 1 1 6 6 5 5 5 5 7 7 3 3
4 4 2 2 2 2 1 1 6 6 5 5 5 5 7 7 3 3
4 4 2 2 2 2 1 1 6 6 5 5 5 5 7 7 3 3
4 4 2 2 2 2 1 1 6 6 5 5 5 5 7 7 3 3
4 4 4 4 1 1 1 1 6 6 6 6 7 7 7 7 3 3
4 4 4 4 1 1 1 1 6 6 6 6 7 7 7 7 3 3
4 4 4 4 1 1 1 1 6 6 6 6 7 7 7 7 3 0
4 4 4 4 1 1 1 1 6 6 6 6 7 7 7 0 0
4 4 4 4 1 1 1 1 6 6 6 6 7 7 7 0 0
```
(*d*) Final layout

**Figure 8.A5** ALDEP layout construction for Example 8.A2

**Table 8.A5**  *ALDEP Scoring (Rating) Procedure Applied to Example 8.A2*

| Adjacent Departments | Relationship | Value | Rating |
|---|---|---|---|
| 4-2 and 2-4 | E | 16 | 32 |
| 4-1 and 1-4 | I | 4 | 8 |
| 2-1 and 1-2 | E | 16 | 32 |
| 1-6 and 6-1 | U | 0 | 0 |
| 6-5 and 5-6 | A | 64 | 128 |
| 6-7 and 7-6 | E | 16 | 32 |
| 5-7 and 7-5 | I | 4 | 8 |
| 7-3 and 3-7 | U | 0 | 0 |
|  | Total |  | 240 |

as an "E" relationship. Assume that ALDEP randomly selects department 4 to be the first department to enter the layout. The relationship chart is scanned to determine if a department has either an "A" or "E" relationship with department 4. Department 2 has an "E" relationship with department 4 and is selected to be the second department to enter the layout. The unselected department having the highest relationship with department 2 is department 1; therefore, department 1 will enter the layout next. Of the remaining unselected departments, none have either an "A" or an "E" relationship with departments 1, 2, or 4, so the next department to enter the layout is selected randomly. Suppose department 6 is selected; it has an "A" relationship with department 5, so department 5 is the fifth department to enter the layout. Next, department 7 is selected to enter the layout, because of its "E" relationship with department 6. Department 7 is followed by the only remaining department, department 3. The order of departments to enter the layout for this iteration is 4-2-1-6-5-7-3.

Given the overall layout is to be 10 units wide and 18 units in length, the placement of department 4 in the layout is shown in Figure 8.A5*a*. The addition of department 2 is given in Figure 8.A5*b*. The inclusion of department 1 in the layout is depicted in Figure 8.A5*c*. The final layout is shown in Figure 8.A5*d*. The rating for the final layout given in Figure 8.A5*d* is 240. The calculation procedure followed by ALDEP to determine this rating is shown in Table 8.A5. Once the layout and rating are printed, ALDEP begins the next layout by once again randomly selecting the first department to enter the layout.

# PROBLEMS

**8.1**  Suppose five departments labeled A through E are located as shown in the layout below. Given the corresponding flow-between chart, compute the efficiency rating for the layout.

| A | B |
|---|---|
| C || 
| D | E |

|  | A | B | C | D | E |
|---|---|---|---|---|---|
| A | — | 5 | 0 | 4 | −3 |
| B |  | — | 6 | −1 | 2 |
| C |  |  | — | −6 | 0 |
| D |  |  |  | — | 3 |
| E |  |  |  |  | — |

**8.2**   Suppose the efficiency rating is designated by $E$ and the cost computed by CRAFT is designated by $K$. Further suppose that four departments (labeled A, B, C, and D) are given for a layout problem. Each department is assumed to be of *equal area* and each department is represented by a unit square. The unit cost used in computing $K$ is assumed to be equal to 1.0 for all department pairs. (That is, $c_{ij} = 1.0$ for all $i, j$.) The flow-between the above departments is given as follows:

|   | A | B | C | D |
|---|---|---|---|---|
| A | — | 10 | 10 | 0 |
| B |   | — | 0 | 4 |
| C |   |   | — | 4 |
| D |   |   |   | — |

**a.**  Compute the values of $E$ and $K$ for the following layout.

**b.**  Compute the values of $E$ and $K$ for the following layout.

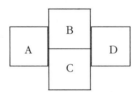

**c.**  Compute the values of $E$ and $K$ for the following layout.

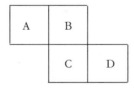

**d.** What can you say regarding the **consistency** of the two measures. That is, if a layout is "good" when measured by its $E$ value, would it still be "good" when it is measured by its $K$ value or vice versa? Justify your answer.

**8.3**   Consider four departments labeled A, B, C, and D. Each department is represented by a $1 \times 1$ square. The following data are given:

Flow-Between Matrix     Unit Cost Matrix

Initial Layout

| A | B |
|---|---|
| C | D |

|   | A | B | C | D |   |   |   | A | B | C | D |
|---|---|---|---|---|---|---|---|---|---|---|---|
| A | — | 6 | 0 | 3 |   |   | A | — | 2 | 0 | 3 |
| B |   | — | 5 | 0 |   |   | B | 2 | — | 1 | 0 |
| C |   |   | — | 0 |   |   | C | 0 | 1 | — | 0 |
| D |   |   |   | — |   |   | D | 3 | 0 | 0 | — |

Location of department A is **fixed.** Answer the following questions using CRAFT with **two-way** exchanges only.

a.  List all the department pairs that CRAFT would consider exchanging. (Do **not** compute their associated cost.)

b.  Compute the **actual** cost of exchanging departments C and D.

c.  Given that department A is fixed and that each department must remain as a $1 \times 1$ square, is the layout obtained by exchanging departments C and D optimum? Why or why not? (Hint: examine the properties of the resulting layout and consider the objective function of CRAFT.)

8.4  The following layout is an illegal CRAFT layout. Nevertheless, given that the volume of flow from A to B is 4, A to C is 3, and B to C is 9, and that all move costs are 1, what is the layout cost?

```
C  C  C  C  C  C
C  B  B  B  B  C
C  B  A  A  B  C
C  B  B  B  B  C
C  C  C  C  C  C
```

8.5  When CRAFT evaluates the exchange of departments, instead of actually exchanging the departments, it only exchanges the centroids of departments.

a.  What is the impact of this method of exchanging if all departments are the same size?

b.  Given the following from-to-chart and scaled layout (each square is $1 \times 1$), what does the evaluation of the exchange of departments B and C indicate should be saved over the existing layout and what is actually saved once this exchange is made?

| To<br>From | A | B | C |
|---|---|---|---|
| A | — | 10 | 6 |
| B | 2 | — | 7 |
| C | | | — |

From-to chart

```
A  A  A  C  B  B  B  B
A  A  A  C  B  B  B  B
A  A  A  C  B  B  B  B
```

Initial layout

8.6  Explain the steps CRAFT would take with the following problem and determine the final layout. Only two-way exchanges are to be considered.

| To<br>From | A | B | C | D | E |
|---|---|---|---|---|---|
| A | — | 3 | 2 | 1 | |
| B | | — | 1 | 3 | |
| C | 1 | | — | 4 | |
| D | | | | — | |
| E | | | | | — |

From-to chart

```
A  A  A  B  B  B
A  A  A  C  C  C
A  A  A  C  C  C
D  D  D  E  E  E
D  D  D  E  E  E
```

Initial layout

**8.7** A manufacturing concern has five departments (labeled A through E) located in a rectangular building as shown below:

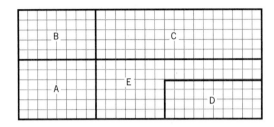

Suppose the flow data, the unit cost data, and the distance matrix are given as follows:

| | FLOW-BETWEEN MATRIX | | | | | | | UNIT COST ($/UNIT DIST.) | | | | | | | DISTANCE MATRIX | | | | |
|---|---|---|---|---|---|---|---|---|---|---|---|---|---|---|---|---|---|---|---|
| | A | B | C | D | E | | | A | B | C | D | E | | | A | B | C | D | E |
| A | — | 0 | 5 | 0 | 5 | | A | — | 0 | 1 | 0 | 1 | | A | — | 6 | 20 | 18 | 11 |
| B | | — | 6 | 2 | 0 | | B | | — | 1 | 4 | 0 | | B | | — | 12 | 22 | 15 |
| C | | | — | 3 | 0 | | C | | | — | 3 | 0 | | C | | | — | 10 | 8 |
| D | | | | — | 7 | | D | | | | — | 1 | | D | | | | — | 7 |
| E | | | | | — | | E | | | | | — | | E | | | | | — |

    **a.** Using the CRAFT two-way exchange procedure, indicate all the department pairs CRAFT would consider exchanging in the above layout.

    **b.** Compute the *estimated* cost of exchanging departments A and E.

**8.8** Suppose the following layout is provided as the initial layout to **CRAFT.** The flow-between matrix and the distance matrix are given as follows. (All the $c_{ij}$ values are equal to 1.0.)

| | Flow-Between Matrix | | | | | | | | Distance Matrix | | | | | |
|---|---|---|---|---|---|---|---|---|---|---|---|---|---|---|
| | A | B | C | D | E | F | | | A | B | C | D | E | F |
| A | — | 0 | 8 | 0 | 4 | 0 | | A | — | 30 | 25 | 55 | 50 | 80 |
| B | | — | 0 | 5 | 0 | 2 | | B | | — | 45 | 25 | 60 | 50 |
| C | | | — | 0 | 1 | 0 | | C | | | — | 30 | 25 | 55 |
| D | | | | — | 6 | 0 | | D | | | | — | 45 | 25 |
| E | | | | | — | 4 | | E | | | | | — | 30 |
| F | | | | | | — | | F | | | | | | — |

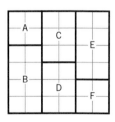

    **a.** Given the above data and initial layout, which department pairs will **not** be considered for exchange.

    **b.** Compute the cost of the initial layout.

    **c.** Compute the *estimated* layout cost assuming that departments E and F are exchanged.

    **d.** In general, when the same data and initial layout are supplied to CRAFT and MULTIPLE, why would one expect MULTIPLE to often (but not always) outperform CRAFT? Also, what type of data and/or initial layout would allow CRAFT to consistently generate layout costs that are comparable to those obtained from MULTIPLE?

**8.9** Answer the following questions for CRAFT.

    **a.** State two principal weaknesses and two principal strengths of CRAFT.

    **b.** CRAFT uses the estimated cost in evaluating the potential impact of exchanging two or three department locations. Suppose we modify the original computer code to obtain a new code, say, NEWCRAFT, where we *always* use the *actual cost* in evaluating the potential impact of exchanging two or three departments. Assuming that we start with the *same initial layout* and that we use the *same exchange option* (such as two-way exchanges only, or any other exchange option), the cost of the final layout obtained from NEWCRAFT will *not necessarily be less than* the cost of the final layout obtained from CRAFT. True or false? Why?

**8.10** Using BLOCPLAN's procedure, convert the following from-to chart to a relationship chart.

| FROM \ TO | A | B | C | D | E | F | G | H |
|---|---|---|---|---|---|---|---|---|
| A | — | 8 | | 3 | | 6 | | |
| B | 1 | — | | | 5 | | | |
| C | | | — | | | | 4 | |
| D | | 9 | | — | | | 18 | |
| E | | | 4 | 1 | — | | | |
| F | 4 | | | 4 | | — | | |
| G | | | | 2 | | | — | 20 |
| H | | | | 7 | | | | — |

**8.11** Consider BLOCPLAN. Suppose the following REL-chart and layout are given for a five-department problem. (It is assumed that each grid in the layout represents a unit square.) Further suppose that the following scoring vector is being used:

$A = 10, E = 5, I = 2, O = 1, U = 0,$ and $X = -10.$

| | 1 | 2 | 3 | 4 | 5 |
|---|---|---|---|---|---|
| 1 | — | A | U | E | U |
| 2 | | — | U | U | I |
| 3 | | | — | U | X |
| 4 | | | | — | A |
| 5 | | | | | — |

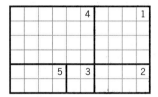

a. Compute the "efficiency rating."
b. Compute the REL-DIST score.
c. In improving a layout, BLOCPLAN can exchange only those departments that are either adjacent or equal in area. True or false? Why? (If true, then explain why there is such a limitation. If false, then explain how two departments that do not meet the above constraint are exchanged.)
d. In constructing and improving a layout, BLOCPLAN maintains rectangular department shapes. Discuss the advantages and limitations of maintaining such department shapes in facility layout.

**8.12** Use the MIP model to obtain an optimal layout with the data given for Problem 8.7. (You may assume that each grid measures $20' \times 20'$.)

**8.13** Use the MIP model to obtain an optimal layout with the data given for Problem 8.8. (You may assume that each grid measures $10' \times 10'$.)

**8.14** Consider the layout shown in Figure 8.14*d*. Use LOGIC and the cut-tree shown in Figure 8.15 to exchange departments B and F.

**8.15** Re-solve Problem 8.14 by exchanging departments D and H in the layout shown in Figure 8.14*d.*

**8.16** Re-solve Problem 8.14 by exchanging departments G and H in the layout shown in Figure 8.17.

**8.17** Answer the following questions for **MULTIPLE.**

    **a.** Consider four departments (labeled A through D) with the following area requirements: A = 7 grids, B = 3 grids, C = 4 grids, and D = 2 grids. Using the spacefilling curve shown below, show the layout that would be obtained from the sequence A-B-C-D.

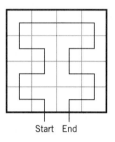

Start  End

    **b.** Using the data given for part (a) and the spacefilling curve shown above, show the layout that would be obtained by exchanging departments B and D.

    **c.** Discuss the advantages and limitations of using spacefilling curves in the manner they have been used in MULTIPLE. Your discussion should include the treatment of fixed departments, dummy departments, unusable floor space (i.e., obstacles), and extra (i.e., empty) floor space.

**8.18** Consider the initial layout and flow/cost data given for Problem 8.7.

    **a.** Using the following conforming curve and MULTIPLE, improve the initial layout via two-way department exchanges.

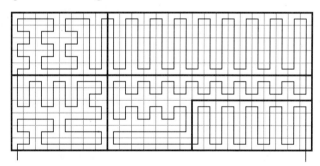

    **b.** In general, what is the disadvantage of using conforming curves?

**8.19** Consider the initial layout and flow data given for Problem 8.8. Use MULTIPLE and the following conforming curve to obtain a two-opt layout.

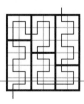

# Part Three

# DEVELOPING ALTERNATIVES: FUNCTIONS

# 9

# *WAREHOUSE OPERATIONS*

## *9.1*  INTRODUCTION

At the mercy of myriad business, logistics, and government initiatives including just-in-time production, quick response, the quest for quality, enhanced customer service, operator safety, and environmental protection, warehouse operations have been and are being revolutionized.

The *Just-In-Time* operating philosophy has moved from manufacturing to distribution. Where it used to be sufficient to ship 100 units of a product on a Monday to complete a week's supply, today we must ship 20 units on each day of the week. The same amount of material is shipped, but there are 5 times as many transactions.

*Quick response* programs have reduced the amount of time allowed to respond to customer demands. Very quickly the standard for order cycle time is becoming same-day or overnight shipment. The compressed time schedules limit the available strategies for productivity improvement and place increased importance on the functionality and capacity of warehouse control and material handling systems.

The *quest for quality* has also moved from manufacturing into warehousing and distribution. As a result, the standards for accuracy performance have increased dramatically. Today, the average shipping accuracy in U.S. warehouses is about 99%. However, the Japanese standard of one error per 10,000 shipments is rapidly becoming the acceptable standard.

A renewed emphasis on *customer service* has increased the number and variety of value-added services in the warehouse. The extra services may include kitting, special packaging, label application, etc. For example, a large fine paper distributor counts and packages individual sheets of paper for overnight shipment. A large discount retailer requires vendors to provide slipsheets between each layer of cases on a pallet to facilitate internal distribution.

Increased emphasis on customer service and evolving consumer demand patterns in the United States have also increased the number of unique items in a typical warehouse or distribution center. The result, *sku proliferation,* is perhaps best illustrated in the beverage industry. Not many years ago the beverage aisle in a typical grocery store was populated with two or three flavors in 12 ounce bottles in six packs. Today the typical beverage aisle is populated with colas (regular and diet caffeinated and non), clear drinks, and fruit flavored drinks in 6-, 12-, and 24-pack glass and plastic bottles and cans and 1-, 2-, and 3-liter bottles.

Finally, an increased concern with the preservation of the *environment,* the *conservation of natural resources,* and *human safety* have brought more stringent government regulations into the design and management of warehousing operations.

The traditional response to increasing demands is to acquire additional resources. In the warehouse those resources include people, equipment, and space. Unfortunately, those resources may be difficult to obtain and maintain. Before the recent recession, economic forecasts identified a coming labor shortage. As the economy recovers, we will again begin to experience the effects of the labor shortage. In addition, we will have to adjust to a workforce characterized by advancing age, minority and non-English speaking demographics, and declining technical skills. In addition, increased safety standards brought on by OSHA and the Americans with Disabilities Act make managing large workforces increasingly difficult. New standards for workforce safety and composition through OSHA's lifting standards and the Americans with Disabilities Act also make it difficult to rely on an increased workforce as a way to address the increased demands on warehousing operations.

When labor is not the answer, we typically turn to mechanization and automation as a means to address increasing demands. Unfortunately, our history of applying high technology in warehousing operations as a substitute for labor has not been distinguished. In many cases we have over-relied on high technology as a substitute for labor problems. In addition, high levels of technology are becoming more difficult to justify as capital becomes more restrictive, as down-sizing becomes more prevalent, as mergers, acquisitions, and the introduction of new competitors make it increasingly difficult to forecast the future.

In the face of rapidly increasing demands on warehouse operations and without a reliable pool of additional resources to turn to, the planning and management of today's warehousing operations is very difficult. To cope, we must turn to simplification and process-improvement as a means of managing warehouses and distribution centers. Toward that end, this chapter is meant to serve as a guide for warehouse operations improvement through the application of best-practice procedures and available material handling systems for warehousing operations. We begin with an introduction to the missions of the warehouse. One way to re-engineer warehousing operations is to justify each warehouse function and handling step relative to the mission of the warehouse. If a function is not clearly serving the warehouse mission, it should be eliminated. Likewise, one or more functions may need to be added to bring the warehouse operations more closely in line with the mission of the warehouse. For example, to improve response time within the warehouse, a cross-docking function may need to be included. We then turn our attention to individual functions and activities within the warehouse. In the introduction we introduce each function and describe its objective. Then, each function is described in detail and best-practice principles and systems for executing each function are defined.

## *9.2* MISSIONS OF A WAREHOUSE

In a distribution network, a warehouse may serve any of the following requirements:

1. It may hold inventory that is used to balance and *buffer* the variation between production schedules and demand. For this purpose, the warehouse is usually located near the point of manufacture and may be characterized by the flow of full pallets in and full pallets out assuming that product size and volume warrant pallet-sized loads. A warehouse serving only this function may have demands ranging from monthly to quarterly replenishment of stock to the next level of distribution.

2. A warehouse may be used to accumulate and *consolidate* products from various points of manufacture within a single firm, or from several firms, for combined shipment to common customers. Such a warehouse may be located central to either the production locations or the customer base. Product movement may be typified by full pallets in and full cases out. The facility is typically responding to regular weekly or monthly orders.

3. Warehouses may be distributed in the field in order to shorten transportation distances to permit *rapid response* to customer demand. Frequently, single items are picked, and the same item may be shipped to the customer every day.

Figure 9.1 illustrates warehouses performing these functions in a typical distribution network. Unfortunately, in many of today's networks, a single item will pass in and out of a warehouse serving each of these functions between the point of manufacture and the customer. When feasible, two or more missions should be combined in the same warehousing operation. Current changes in the availability and cost of trans-

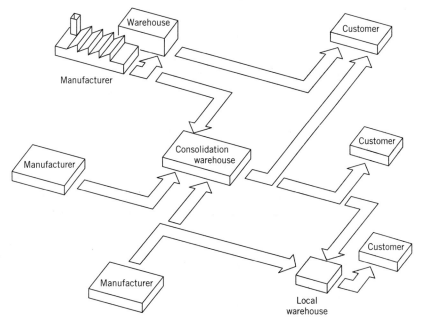

**Figure 9.1.** Warehouse roles within the distribution network.

portation options make the combination possible for many products. In particular, small high-value items with unpredictable demand are frequently shipped worldwide from a single source using overnight delivery services.

## 9.3   FUNCTIONS IN THE WAREHOUSE

Although it is easy to think of a warehouse as being dominated by product storage, there are many activities that occur as part of the process of getting material into and out of the warehouse. The following list includes the activities found in most warehouses. These tasks, or functions, are also indicated on a flow line in Figure 9.2 to make it easier to visualize them in actual operation.

1. Receiving
2. Prepackaging (optional)
3. Put-away
4. Storage
5. Order picking (pallets, cases, loose items)
6. Packaging and/or pricing (optional)
7. Sortation and/or accumulation
8. Packing and shipping
9. Cross-docking (optional)
10. Replenishment (optional)

The functions may be defined briefly as follows:

1. *Receiving* is the collection of activities involved in (a) the orderly receipt of all materials coming into the warehouse, in (b) providing the assurance that the quantity and quality of such materials are as ordered, and in (c) disbursing materials to storage or to other organizational functions requiring them.

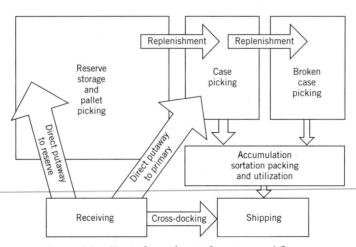

**Figure 9.2.**   Typical warehouse functions and flows.

2. *Prepackaging* is performed in a warehouse when products are received in bulk from a supplier and subsequently packaged singly, in merchandisable quantities, or in combinations with other parts to form kits or assortments. An entire receipt of merchandise may be processed at once, or a portion may be held in bulk form to be processed later. This may be done when packaging greatly increases the storage-cube requirements or when a part is common to several kits or assortments.

3. *Put-away* is the act of placing merchandise in storage. It includes transportation and placement.

4. *Storage* is the physical containment of merchandise while it is awaiting a demand. The form of storage will depend on the size and quantity of the items in inventory and the handling characteristics of the product or its container.

5. *Order picking* is the process of removing items from storage to meet a specific demand. It represents the basic service that the warehouse provides for the customer and is the function around which most warehouse designs are based.

6. *Packaging and/or pricing* may be done as an optional step after the picking process. As in the prepackaging function, individual items or assortments are boxed for more convenient use. Waiting until after picking to perform these functions has the advantage of providing more flexibility in the use of on-hand inventory. Individual items are available for use in any of the packaging configurations right up to the time of need. Pricing is current at the time of sale. Prepricing at manufacture or receipt into the warehouse inevitably leads to some repricing activity as price lists are changed while merchandise sits in inventory. Picking tickets and price stickers are sometimes combined into a single document.

7. *Sortation* of batch picks into individual orders and *accumulation* of distributed picks into orders must be done when an order has more than one item and the accumulation is not done as the picks are made.

8. *Packing and shipping* may include the following tasks:
   - Checking orders for completeness.
   - Packaging merchandise in an appropriate shipping container.
   - Preparing shipping documents, including packing list, address label and bill of lading.
   - Weighing orders to determine shipping charges.
   - Accumulating orders by outbound carrier.
   - Loading trucks (in many instances, this is a carrier's responsibility).

9. *Cross-docking* inbound receipts from the receiving dock directly to the shipping dock.

10. *Replenishing* primary picking locations from reserve storage locations.

For discussion purposes, this chapter includes in *receiving* those activities described above as receiving, prepackaging, and putaway; in *order picking* those activities described above as order picking, packaging, and sortation/accumulation; and in *shipping* those activities described as packing and shipping. The activities, best-practice operating principles, and space planning methodologies for each of the functional areas are described in subsequent sections.

# *9.4*  RECEIVING AND SHIPPING OPERATIONS

Problems can occur in planning receiving and shipping facilities if the carriers that interface with receiving and shipping activities are not properly considered. Positioning of the carriers and their characteristics are important to the operations of receiving and shipping activities. It is useful to think of the carriers that interface with the receiving and shipping functions as a portion of the receiving and shipping facility. Hence, all carrier activities on the site are included in receiving and shipping facility planning. Receiving and shipping functions will be defined to begin and end when carriers cross the property line.

The activities required to receive goods include:

1. Spot the carrier.
2. Chock the carrier wheels.
3. Check the seal.
4. Position and secure dockboard.
5. Palletize or containerize as appropriate.
6. Unload the carrier.
7. Prepare receiving tally.
8. Reconcile receiving tally, packing list, and purchase order.
9. Segregate goods into saleable and nonsaleable categories.
10. Dispatch the carrier.
11. Prepare a receiving report.
12. Dispatch goods.

The facility requirements to perform these receiving activities include:

1. Sufficient area to stage and spot carriers.
2. Dockboards to facilitate carrier unloading.
3. Sufficient area to palletize or containerize goods.
4. Sufficient area to place goods prior to dispatching.
5. An office to house information on purchase orders and allow for report generation.

The activities required to ship goods include:

1. Accumulate and pack the order.
2. Stage and check the order.
3. Reconcile shipping release and customer order.
4. Spot the carrier.
5. Chock the carrier wheels.
6. Position and secure dockboard.
7. Load the carrier.
8. Dispatch the carrier.

The facility requirements to perform these shipping activities include:

1.  Sufficient area to stage orders.
2.  An office to house information on shipping releases and customer orders.
3.  Sufficient area to stage and spot carriers.
4.  Dockboard to facilitate carrier loading.

Some desirable attributes of receiving and shipping facilities plans include:

1.  Directed flow paths among carriers, buffer, or staging areas and storage areas.
2.  A continuous flow without excessive congestion or idleness.
3.  A concentrated area of operation that minimizes material handling and increases the effectiveness of supervision.
4.  Efficient material handling.
5.  Safe operation.
6.  Minimizing damage.
7.  Good housekeeping.

Requirements for people, equipment, and space in receiving and shipping depend on the effectiveness of programs to incorporate prereceiving and postshipping considerations. For example, by working with vendors and suppliers, peak loads at receiving can be reduced. Scheduling inbound shipments is one method of reducing the impact of randomness on the receiving workload.

Another reason for being concerned with prereceiving activities is the opportunity to influence the unit load configurations of inbound material. If, for example, cases are hand-stacked in the trailer by the vendor, they will probably have to be unloaded by hand. Likewise, if either slipsheets or clamp trucks were used by the vendor to load the carrier and receiving does not have similar material handling equipment, the shipment probably will have to be unloaded by hand. Finally, if materials are received in unit loads not compatible with the material handling system, then additional loading, unloading, or both might be necessary.

A third reason for trying to influence prereceiving activities is to provide a smooth interface between the vendor's and receiver's information system. Where automatic identification systems are in use in receiving, some firms supply their vendors with the appropriate labels to be placed on inbound material to facilitate the receiving activity. The faster and more accurately receiving is performed, both vendor and buyer benefit. Information is provided to accounting and payments are made sooner; and materials are available for use sooner, rather than sitting idle.

Just as the receiver wishes to influence the vendor, the customer wishes to influence the shipper. Hence, postshipping activities must be considered. In addition to the reasons cited for considering prereceiving activities, the following are issues in postshipping activities: returnable containers, returned goods, returning carriers, and shipping schedules.

If goods are shipped to customers in returnable containers, a system must be developed to keep track of the containers and to ensure they are returned. Additionally, whether or not the shipping container or support is returnable, there will occur a natural attrition for which replacements must be planned.

Goods are returned because the goods failed to meet the customer's quality specifications, mistakes were made in the type and amount of material shipped, or the customer just simply deciding not to accept the material. Regardless of the reason, returned goods must be *handled* and an appropriate system for handling the material must be designed.

If supplier-owned equipment is used to deliver materials to customers, consideration should be given to the utilization of the capacity of the carrier on the return trip. The "backhaul" of the carrier could be used for returning the returnable containers; additionally, it could be used for other transportation purposes.

Shipping schedules can have a significant impact on the resource requirements for shipping. Hence, close coordination is required between the shipper and the shipping department. The shipper might be the customer's carrier, its own carrier, a contract hauler, or a commercial carrier. If shipping activities are to be planned, then shipping schedules must be accurate and reliable.

In addition to the need for closely coordinating vendor and receiving activities and shipping and customer activities, it is equally important to coordinate the activities of receiving and production, production and shipping, and receiving and shipping.

The natural sequence for the flow of materials is vendor, receiving, storage, production, warehousing, shipping, customer. However, in some cases materials might go directly from receiving to production and from production to shipping. Hence, such possibilities should be included in the system design.

Why should receiving and shipping be coordinated? Common space, equipment, and/or personnel might be used to perform receiving and shipping. Additionally, when slave pallets are used in a manufacturing or warehousing activity, empty pallets will accumulate at shipping and must be returned to the loading point at either receiving or production.

A key decision in designing and receiving and shipping functions is whether or not to centralize the two functions. As depicted in Figure 9.3, the location of receiving and shipping depends on access to transportation facilities.

The decision to centralize receiving and shipping depends on many factors, including the nature of the activity being performed. For example, if receiving is restricted to occur in the morning and shipping is restricted to occur in the afternoon, then it would be appropriate to utilize the same docks, personnel, material handling equipment, and staging space for both receiving and shipping. Alternately, if both occur simultaneously, closer supervision is required to ensure that received goods and goods to be shipped are not mixed.

The opportunity to centralize receiving and shipping should be investigated carefully. The pros and cons of centralization should be enumerated and considered before making a decision.

## Receiving and Shipping Principles

### Receiving Principles

The following principles serve as guidelines for streamlining receiving operations. They are intended to simplify the flow of material through the receiving process and to insure the minimum work content is required. In order they are:

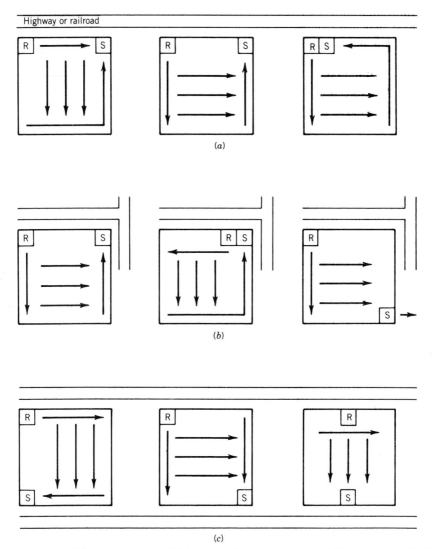

**Figure 9.3.**  Possible arrangements of shipping and receiving areas. (*a*) Transportation facilities on one side of building. (*b*) Transportation facilities on two adjacent sides of building. (*c*) Transportation facilities on two opposite sides of building. *(From Apple [5] with permission.)*

1.  *Don't receive.* For some materials, the best receiving is no receiving. Often, drop shipping, having the vendor ship to the customer directly, can save the time and labor associated with receiving and shipping. Large, bulky items lend themselves to drop shipping. An example is a large camp and sportswear mail order distributor drop shipping canoes and large tents.

2.  *Pre-receive.* The rationale for staging at the receiving dock, the most time and space intensive activity in the receiving function, is often the need to hold the material for location assignment, product identification, and so on. This information can often be captured ahead of time by having the information communicated by the vendor at the time of shipment via EDI link or via fax

Company _____ Date _____ Raw Materials _____ Finished Goods _____
Prepared by _____ Sheet _____ of _____ Plant Supplies _____

| 1 | 2 | 3 | 4 | 5 | 6 | 7 | 8 | 9 | 10 | 11 |
|---|---|---|---|---|---|---|---|---|---|---|
| | UNIT LOADS | | | | Size of Shipment (unit loads) | Frequency of Shipment | TRANSPORTATION | | MATERIAL HANDLING | |
| Description | Type | Capacity | Size | Weight | | | Mode | Specifications | Method | Time |
| | | | | | | | | | | |
| | | | | | | | | | | |
| | | | | | | | | | | |
| | | | | | | | | | | |
| | | | | | | | | | | |

**Figure 9.4.** Receiving and shipping analysis chart.

notification. In some cases, the information describing an inbound load can be captured on a smart card, allowing immediate input of the information at the receiving dock or can be communicated in RF tags readable by transponders located along the road.

3. *Cross-dock "cross-dockable" material.* Since the ultimate objective of the receiving activity is to prepare material for the shipment of orders, the fastest, most productive receiving process is cross-docking—essentially shipping directly from the receiving dock. Palletized material with a single sku per pallet, floor stacked loose cases, and backordered merchandise are excellent candidates for cross-docking.

4. *Put-away directly to primary or reserve locations.* When material cannot be cross-docked, material handling steps can be minimized by bypassing receiving staging and putting material away directly to primary picking locations, essentially replenishing those locations from receiving. When there are no severe constraints on product rotation this may be feasible. Otherwise, material should be directly putaway to reserve locations. In direct putaway systems, the staging and inspection activities are eliminated. Hence, the time, space, and labor associated with those operations is eliminated. In either case, vehicles that serve the dual purpose of truck unloading and product putaway facilitate direct putaway. For example, counterbalanced lift trucks can be equipped with scales, cubing devices, and on-line RF terminals to streamline the unloading and putaway function. The most advanced logistics operations are characterized by automated, direct putaway to storage. The material handling technologies that facilitate direct putaway include roller-bed trailers and extendable conveyors. In addition to pre-receiving, prequalifying vendors helps eliminate the need for receiving staging.

5. *Stage in storage locations.* If material has to be staged, the floorspace required for staging can be minimized by providing storage locations for receiving staging. Often, storage locations may be live storage locations with locations blocked until the unit is officially received. Storage spaces are provided over dock doors.

6. *Complete all necessary steps for efficient load decomposition and movement at receiving.* The most time we will ever have available to prepare a product for shipment is at receiving. Once the demand for the product has been received there is precious little time available for any preparation of the material that needs to be done prior to shipment. Hence any material processing that can be accomplished ahead of time should be accomplished. Those activities include:

   a. *Prepackage in issue increments.* At a large office supplies distributor, quarter and half-pallet loads are built at receiving in anticipation of orders being received in those quantities. Customers are encouraged to order in those quantities by quantity discounts. A large distributor of automotive aftermarket parts conducted an extensive analysis of likely order quantities. Based on that analysis, the company is now prepackaging in those popular issue increments.

   b. *Apply necessary labelling and tags.*

   c. *Cube and weigh for storage and transport planning.*

7. *Sort inbound materials for efficient putaway.* Just as zone picking and location sequencing are effective strategies for improving order picking productivity, inbound materials can be sorted for putaway by warehouse zone and by location sequence. At the U.S. Defense Logistics Agency's Integrated Materiel Complex, receipts are staged in a rotary rack carousel and released and sorted by warehouse aisle and location within the aisle to streamline the putaway activity. Automated guided vehicles take the accumulated loads to the correct aisle for putaway.

8. *Combine putaways and retrievals when possible.* To further streamline the putaway and retrieval process, put-away and retrieval transactions can be combined in a dual command to reduce the amount of empty travel on industrial vehicles. This technique is especially geared for pallet storage and retrieval operations. Again, counterbalance lift trucks, which can unload, putaway, retrieve, and load, are a flexible means for executing dual commands.

9. *Balance the use of resources at receiving by scheduling carriers and shifting time consuming receipts to off-peak hours.* Through computer-to-computer (electronic data interchange) links and fax machines companies have improved access to schedule information on inbound and outbound loads. This information can be used to proactively schedule receipts and to provide advance shipping notice information.

10. *Minimize or eliminate walking by flowing inbound material past workstations.* An effective strategy for enhancing order picking productivity, especially when a variety of tasks must be performed on the retrieved material (i.e., packaging, counting, labeling), is to bring the stock to order picking station equipped with the aids and information to perform the necessary tasks. The same strategy should be emplyed for receipts, which by their nature require special handling. At the flagship distribution center of a large retailer, receipts flow out of inbound trailers on skatewheel conveyor by a stationary receiving station equipped with a touchscreen terminal. At the stationary receiving stations, inbound cases are weighed, cubed, and tagged with a bar code label describing all necessary product and warehouse location information. In addition, if the material is needed for an outbound order, a receiving operator diverts the inbound case down a separate conveyor line for cross-docking.

## Shipping Principles

Many of the best-practice rceiving principles also apply in reverse in shipping including direct loading (the reverse of direct unloading), advanced shipping notice preparation (pre-receiving), and staging in racks. In addition to those principles best-practice principles for unitizing and securing loads, automated loading, and dock management are introduced.

1. **Select cost and space effective handling units.**

   *a. For loose cases.* The options for unitizing loose cases include wood (disposable, returnable, and rentable) pallets, plastic, metal, and nestable pallets. The advantage of plastic pallets over wood pallets include durability, cleanliness, and colorability. The Japanese make excellent use of colored plastic pallets and totes in creating appealing work environments in factories and warehouses. Metal pallets are primarily designed for durability and weight capacity. Nestable pallets offer good space utilization during pallet storage and return but are not very durable or weight capacity. Nestable pallets offer good space utilization during pallet storage and return but are not very durable or weight capacity. Other options for unitizing loose cases include slip sheets and roll carts. Slip sheets improve space utilization in storage systems and in trailers but require special lift truck attachments. Roll carts facilitate the containerization of multiple items in case quantities and facilitate material handling throughout the shipping process from order picking to packaging to checking to trailer loading. The selection factors for unitizing loose cases include initial purchase cost, maintenance costs and requirements, ease of handling, impact on the environment, durability, and product protection.

b. *For loose items.* The options for unitizing loose items include totes (nestable and collapsible) and cardboard containers. As was the case with unitizing loose cases, the selction factors include impact on the environment, initial purchase cost, life cycle cost, cleanliness, and product protection. An excellent review and comparison of carton and tote performance is provided in [36].

2. **To minimize product damage:**

   a. *Unitize and secure loose items in cartons or totes.* In addition to providing a unit load to facilitate material handling, a means must be provided to secure material within the unit load. For loose items in totes or cartons those means include foam, peanuts, popcorn, bubblewrap, newsprint, and airpacks. The selection factors include initial and life cycle cost, impact on the environment, product protection, and reusability. An excellent review of alternative packaging methods is provided in [36].

   b. *Unitize and secure loose cases on pallets.* Though the most popular alternative is stretchwrapping, wrapping loose cases with velcro belts and adhesive tacking are gaining in popularity as environmentally safe means for securing loose cases on pallets.

   c. *Unitize and secure loose pallets in outbound trailers.* The most common methods are foam pads and plywood.

3. **Eliminate shipping staging, and direct-load outbound trailers.** As was the case in receiving, the most space and labor intensive activity in shipping is the staging activity. To facilitate the direct loading of pallets onto outbound trailers, pallet jacks and counterbalance lift trucks can serve as picking and loading vehicles allowing a bypassing of staging. To go one step further, automating pallet loading can be accomplished with pallet conveyor interfacing with specially designed trailer beds to allow pallets to be automatically conveyed onto outbound trailers, with automated fork trucks, and/or automated guided vehicles. Direct, automated loading of loose cases is facilitated with extendable conveyor.

4. **Use storage racks to minimize floorspace requirements for shipping staging.** If shipping staging is required, the floorspace requirements for staging can be minimized by staging in storage racks. A large automotive aftermarkets

supplier places racks along the shipping wall and above dock doors to achieve this objective.

5. **Route on-site drivers through the site with minimum paperwork and time.** A variety of systems are now in place to improve the management of shipping and receiving docks and trailer drivers. At one brewery trailer drivers use a smart card to gain access throughout the distribution center site, to expedite on-site processing, and to insure shipping accuracy. At another brewery terminal stands are provided throughout the site to allow drivers on-line access to load status and dock schedules.

## Receiving and Shipping Space Planning

The steps required to determine the total space requirements for receiving and shipping areas are as follows:

1. Determine what is to be received and shipped.
2. Determine the number and type of docks.
3. Determine the space requirements for the receiving and shipping area within the facility.

Each of these steps is the subject of one of the following sections.

### Determining What Is to Be Received and Shipped

A chart that is useful in defining what is to be received and shipped is a receiving and shipping analysis chart, as shown in Figure 9.4 on page 398. The first seven columns of the receiving and shipping analysis chart include information on the *what, how much,* and *when* of items received or shipped. For an existing receiving or shipping operation or one that is to have similar objectives to an existing operation, this information may be obtained from past receiving reports or shipping releases. For a new receiving or shipping operation, the parts lists and the market analysis information for all products must be analyzed to determine reasonable unit loads and order quantities. Once this information is obtained, the first seven columns of the receiving and shipping analysis chart may be completed.

The eighth and ninth columns of the receiving and shipping analysis chart identify the types of carriers to be utilized for receiving and shipping materials. At a minimum, the type carrier, the overall length, width, and height, and the height to the carrier's dock should be specified. The information can be obtained by investigating the types of carriers utilized for the materials to be received or shipped.

### Determining the Number and Type of Docks

As shown in Chapter 16, waiting line analysis may be utilized to determine the number of docks that will provide the required service if arrivals and/or services are Poisson-distributed and the arrival and service distributions do not vary significantly over time. If arrival and service distributions vary with the time of day, day of week, or number of trucks waiting at the dock, simulation may be used.

The last two columns on the receiving and shipping analysis chart concern the handling of materials on and off a carrier. If the chart is being completed for an existing receiving or shipping area, the current methods of unloading or loading materials should be evaluated and, if acceptable, recorded on the chart. If the chart is being completed for a new operation, methods of handling materials should be determined by following the procedure described in Chapter six. The time required to unload or load a carrier may be determined for an existing operation from historical data, work sampling, or time study. The accuracy and effort required to determine time standards using these aproaches typically results in the facility planner using predetermined time elements. Predetermined time elements are also used for new receiving or shipping operations.

Predetermined time elements vary from very general (macro) standards to very detailed (micro) standards. Probably the most general-type standard is one such as an individual utilizing the proper equipment can unload or load 7500 lb/hr. Much more detailed standards have been compiled by the U.S. Department of Agriculture (USDA) [27]. An application of the standard developed via USDA standards is given in Figure 9.5. Even more detailed are standards developed by material handling

**TIME STANDARD WORK SHEET**

Company BCD Prepared by  JA
Process Unload cartons, palletize, and store
Starting Point Spot carrier  Ending point Close truck door and dispatch truck
Date _____Sheet  1  of  1

| Step | Description | Crew Size | USDA Reference Table Number | Productive Labor (Hours) |
|------|-------------|-----------|------------------------------|--------------------------|
| 1 | Spot carrier. | 1 | XII | .1666 |
| 2 | Open truck door. | 1 | XII | .0163 |
| 3 | Remove bracing. | 1 | XII | .1101 |
| 4 | Get bridge plate. | 1 | XII | .0310 |
| 5 | Position bridge plate. | 1 | XII | .0168 |
| 6 | Place empty pallet at truck tailgate. Repeat 40 times. | 1 | XX | .2333 |
| 7 | Unload cartons onto pallet. Repeat 640 times. | 1 | XIV A | 1.5360 |
| 8 | Pick up loaded pallet and move out of truck. Repeat 40 times. | 1 | VI | .3800 |
| 9 | Transport pallet to storage area (100 ft) and return to truck. Repeat 40 times. | 1 | XI | .6040 |
| 10 | Store pallet. Repeat 40 times. | 1 | VI | .1760 |
| 11 | Remove bridge plate. | 1 | XII | .0048 |
| 12 | Close truck door and dispatch truck | 1 | XII | .0068 |
| | Total | | | 3.2817 hr |

**Figure 9.5.**   Example of use of U.S. Department of Agriculture (USDA) predetermined time elements to determine truck unloading standard time.

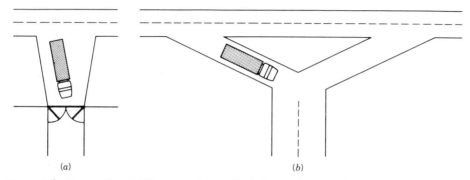

(a)                                                    (b)

**Figure 9.6**  Recessed and "Y" approaches to facilitate truck access to property. (*a*) Recessed truck entrance. (*b*) "Y" truck entrance.

equipment manufacturers dealing with specific types of material handling equipment.

Once all the information for the receiving and shipping analysis chart has been obtained for all materials to be received or shipped, the receiving or shipping function is fully defined. Although considerable effort is required to obtain the information, it must be obtained if receiving and shipping facilities are to be properly planned.

Once the number of docks is determined, the dock configuration must be designed. The first consideration in designing the proper dock configuration is the flow of carriers about the facility. For rail docks, location and configuration of the railroad spur dictate the flow of railroad cars and the configuration of the rail dock. For truck docks, truck traffic patterns must be analyzed. Truck access to the property should be planned so that trucks need not back onto the property. Recessed or "Y" approaches, as shown in Figure 9.6, should be utilized if trucks turn into the facility from a narrow street. Other truck guidelines to be taken into consideration are:

1.   Two-directional service roads should be at least 24 ft wide.

2.   One-way service roads should be at least 12 ft wide.

3.   If pedestrian travel is to be along service roads, a 4-ft wide walk physically separated from the service road should be included.

4.   Gate openings for two-directional travel should be at least 28 ft wide.

5.   Gate openings for one-way travel should be at least 16 ft wide.

6.   Gate openings should be 6 ft wider if pedestrians will also use the gate.

7.   All right-angle intersections must have a minimum of a 50-ft radius.

8.   If possible, all traffic should circulate counterclockwise because left turns are easier and safer to make than right turns (given that steering is on the left of the vehicle).

9.   Truck waiting areas should be allocated adjacent to the dock apron of sufficient magnitude to hold the maximum expected number of trucks waiting at any given time.

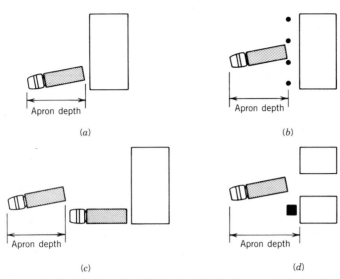

**Figure 9.7** Apron depth definition for 90° docks. (*a*) Unobstructed dock. (*b*) Postsupported canopy. (*c*) Alongside other trucks. (*d*) Driveways and stalls.

After considering the guidelines given above, the overall flow of trucks about a facility may be determined. Care must be taken to ensure that adequate space exists for 90° docks. Space requirements for 90° docks are illustrated in Figure 9.7 and given in Table 9.1. If adequate apron depth does not exist for a 90° dock, a finger dock must be utilized. As can be seen in Figure 9.8, 90° docks require greater apron depth but less bay width. Drive-in docks are typically not economical; 90° docks require a

**Table 9.1**  *Space Requirements for 90° Docks*

| Truck Length (feet) | Dock Width (feet) | Apron Depth (feet) |
|---|---|---|
| 40 | 10 | 46 |
|    | 12 | 43 |
|    | 14 | 39 |
| 45 | 10 | 52 |
|    | 12 | 49 |
|    | 14 | 46 |
| 50 | 10 | 60 |
|    | 12 | 57 |
|    | 14 | 54 |
| 55 | 10 | 65 |
|    | 12 | 63 |
|    | 14 | 58 |
| 60 | 10 | 72 |
|    | 12 | 63 |
|    | 14 | 60 |

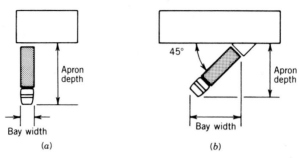

**Figure 9.8** 90° dock and finger dock bay width and apron depth trade-offs. (*a*) 90° dock. (*b*) 45° finger dock.

greater *outside* turning area and finger docks require greater *inside* maneuvering area. Because outside space costs considerably less to construct and maintain than inside space, 90° docks are used whenever space exists. Furthermore, when finger docks must be utilized, the largest-angle finger dock should be utilized. Space requirements for finger docks are given in Table 9.2.

Even though dock widths of 10 ft are adequate for spotting trucks, the potential for accidents, scratches, and increased maneuvering time has resulted in the accepted width being 12 ft. An exception exists for extremely busy docks, where 14-ft dock widths are recommended.

The overall space requirements outside a facility for truck maneuvering may be determined by the following procedure:

1. Determine the required number of docks.
2. Determine truck flow patterns.
3. Determine whether 90° docks may be utilized, and if not, select the largest-angle finger dock for which space is available.
4. Specify a dock width.
5. Determine apron depth and dock width.
6. Establish the overall outside space requirement by allocating the space determined in item 5 for the number of docks determined in item 1.

**Table 9.2** *Finger Dock Space Requirements for a 65-ft Trailer*

| Dock Width (feet) | Finger Angle (degrees) | Apron Depth (feet) | Bay Width (feet) |
|---|---|---|---|
| 10 | 10° | 50 | 65 |
| 12 | 10° | 49 | 66 |
| 14 | 10° | 47 | 67 |
| 10 | 30° | 76 | 61 |
| 12 | 30° | 74 | 62 |
| 14 | 30° | 70 | 64 |
| 10 | 45° | 95 | 53 |
| 12 | 45° | 92 | 54 |
| 14 | 45° | 87 | 56 |

## Determining Internal Receiving and Shipping Area Requirements

Receiving and shipping department area requirements within a facility may include space allocations for the following:

1.   Personnel convenience.
2.   Offices.
3.   Material handling equipment maintenance.
4.   Trash disposal.
5.   Pallet and packaging material storage.
6.   Trucker's lounge.
7.   Buffer or staging area.
8.   Material handling equipment maneuvering.

Space requirements for personnel convenience and office areas for receiving and shipping departments should be determined. Material handling equipment maintenance area requirements should be planned. Trash disposal, pallet storage, and packaging materials storage are functions that must be planned.

A truckers' lounge is space for the drivers while truck unloading or loading is taking place. The objective is to provide a place for the drivers to relax while not involved in receiving or shipping operations. However, the drivers should not be allowed to wander around within a facility. A truckers' lounge should be provided with wash room facilities. The lounge should not be less than 150 ft$^2$. For facilities having more than six docks, 25 ft$^2$ should be planned for the lounge for each dock. The truckers' lounge should be easily accessible from the docks and should be positioned such that receiving and/or shipping supervision may observe all activities of the drivers.

Buffer areas are areas within receiving departments where materials removed from carriers may be placed until dispatched. If the operating procedure is to deliver all merchandise to either stores, inspection, or the department requesting the merchandise immediately upon receipt, a buffer area is not required. If the operating procedure is to remove materials from the carrier and place it into holding area prior to dispatching, space must be allocated to store the merchandise. In a similar manner, staging areas are areas within shipping departments where merchandise is placed and checked prior to being loaded into a carrier. If merchandise is loaded into carriers directly after being withdrawn from the warehouse, no staging area is required.

The space required for buffer or staging areas may be determined by considering the number of carriers for which merchandise is to be stored in these areas and the space required to store the merchandise for each carrier. Typically, when buffer and staging areas are utilized, sufficient space should be allocated for one full carrier for each dock. When fluctuations in hourly unloading or loading rates become pronounced, space for storing two or more carriers of merchandise in buffer or storage areas should be considered. The cost of the buffer or staging area should be compared to the truck unloading and loading costs to determine the proper amount of storage space. If a simulation model is used to determine the number of docks, it may be used to determine the cost trade-off.

The space required to store the merchandise either unloaded from or to be loaded on to a carrier depends on the size of the carrier, the cube utilization in the

**Table 9.3**  *Minimum Maneuvering Allowances for Receiving and Shipping Areas*

| Material Handling Equipment Utilized | Minimum Maneuvering Allowance (ft) |
|---|---|
| Tractor | 14 |
| Platform truck | 12 |
| Forklift | 12 |
| Narrow-aisle truck | 10 |
| Handlift (Jack) | 8 |
| Four-wheel hand truck | 8 |
| Two-wheel hand truck | 6 |
| Manual | 5 |

carrier, and the cube utilization in the buffer or staging area. Merchandise can typically be stored at least carrier height and often much higher in a buffer area. The inverse is true for staging areas where order checking must take place. Storage equipment, as described in Chapter 6, may be utilized in buffer or staging areas, but typically lacks the versatility required for temporary storage in these areas. Aisle spacing between buffer areas for various docks and between staging areas for various docks should follow the guidelines given in Chapter 4.

Maneuvering space required for material handling equipment is provided between the backside of the dockboard and the beginning of the buffer or staging areas. Maneuvering space is dependent on the type of material handling equipment, as indicated in Table 9.3.

## Example 9.1

A simulation study has been completed for a new facility, and four docks have been specified. (See Figure 9.9.) Two docks are to be allocated to receiving and two to shipping. Sixty-foot tractor trailers are to be the largest carriers serving the facility. The new plant is to be positioned to the north of a road that is oriented in an east-west direction. All trucks are to be unloaded and loaded with lift trucks. The docks are located at the northwest corner of the facility. Buffer areas are not required, but one staging area must be included for each dock. Each staging area must be capable of holding fifty-two 48 × 40 × 42-in. pallet loads of merchandise. What are the roadway requirements to the east, north, and west of the facility and what space requirements are needed within the facility?

All trucks should enter the property from the east of the facility and exit to the west. Service roads to the east and west of the facility should be 12 ft wide. The area to the north of the facility on the eastern side should be used for truck waiting and should be at least 24 ft wide. All docks should be 90°, 12-ft docks. A dock face of 48 ft with an apron depth of 63 ft is required. The apron depth of 63 ft is in addition to the 60 ft length of the trucks; therefore, a minimum of 123' should be allowed between the northern extreme of the building and the property line. The dockboards will extend 6 ft into the building; therefore, a maneuvering distance of 18 ft (6 + 12 ft) is required for the entire 48-ft dockface for a total area of 864 ft$^2$. Two staging areas are required, assuming loads are stacked two high and are stored back to back for 13 rows, and each staging area will require an area that is 7 ft wide and 52 ft long. Ten-foot aisles should be on each side of the staged material. A total staging area of

$$[3(10) + 2(7)] \times 52 = 2288 \text{ ft}^2 \qquad (9.1)$$

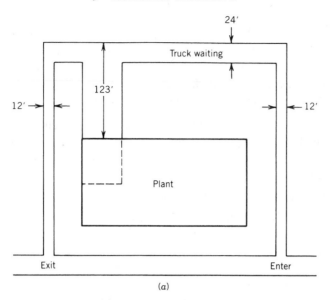

(a)

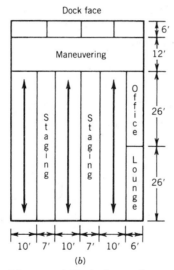

(b)

**Figure 9.9**  (*a*) Site and (*b*) dock area for Example 9.1.

is required. A truckers' lounge of 150 ft² should be included, as well as 150 ft² for a receiving and shipping office. Personnel convenience, material handling equipment maintenance, trash disposal, pallet storage, and packaging material storage areas will be integrated with similar areas within the facility. The total space requirement within the facility is:

| | |
|---|---|
| Maneuvering area | 864 ft² |
| Staging area | 2288 ft² |
| Truckers' lounge | 150 ft² |
| Office | 150 ft² |
| | 3452 ft² |
| or more realistically, | 3500 ft² |

## Dock Operations Planning

Equipment requirements for receiving and shipping areas consist of the equipment required to properly interface between the carriers and the docks. The methods of handling materials through this interface are described in Chapter 6. The dock equipment needed to perform this interface includes:

1. Dock levelers as the interface between a dock at a given height and variable height carriers.
2. Bumper pads as the interface between a fixed dock and a moveable carrier.
3. Dock shelters as the interface between a heated/air-conditioned dock and an unheated/non-air-conditioned carrier.

The following sections describe the three equipment types.

## Dock Levelers

Viewing the carrier as a portion of the receiving and shipping functions, as described in Section 9.1, leads to an interesting question. If a temporary storage area (the carrier) is to be used in conjunction with the receiving and shipping functions and the storage area is at a different height than the dock, how should this difference in height be treated? Five possible answers are:

1. Walking up or down a step to accommodate the difference in height.
2. A portable "ramp" between the dock and carrier.
3. A permanent adjustable "ramp" between the dock and carrier.
4. Raising the carrier to the height of the dock.
5. Raising the dock to the height of the carrier.

The first option of walking up or down a step to accommodate the difference in carrier and dock height is acceptable for any type load not requiring an industrial truck to enter the carrier. Stepping into a carrier to load or unload small packages is acceptable. If a large quantity of small packages are to be loaded or unloaded in this manner, telescoping belt, skatewheel, or roller conveyors may be helpful. If materials are to be loaded or unloaded via crane, the difference in height need only be negotiated by the personnel aligning the materials and step up or down to the carrier is acceptable. If materials are to be lifted off either side of a carrier using an industrial truck, the difference in carrier and dock height is also of little consequence. Unfortunately, most materials are loaded and unloaded with an industrial truck that enters the carrier; therefore, the step up or down is usually unacceptable.

The second option of a portable "ramp" suggests the use of dockboards (also referred to as dock plates or bridge plates) or yard ramps. Dockboards are typically made of aluminum or magnesium and are no more than ramps that can be placed between a dock and a carrier so that an industrial truck may drive into and out of the carrier. In an attempt to minimize dockboard weight, they are often less than 4 ft long; therefore, for even small height differentials, inclines can result that are unsafe for industrial truck travel. The major use of dockboards is for loading or unloading rail cars where rail car spotting locations are variable. Even for rail car applications, unless volume of movement is very low, dock boards should be replaced with a more permanent device to adjust for height differentials.

Yard ramps or portable docks are ramps that are sometimes used so that carriers can be unloaded using an industrial truck at a facility without a permanent dock. Yard ramps are useful for unloading carriers when inadequate dockspace exists, when carriers are to be unloaded in the yard, or for ground-level plants. Figure 9.10 illustrates yard ramps.

The third option, the permanent adjustable dockboard, is the most frequently used approach to compensating for the height differential between carriers and

**Figure 9.10** Yard ramps. (Part *a courtesy of Brooks and Perkins, Inc.;* part *b courtesy of MagLine, Inc.*)

docks. Permanent adjustable dockboards may be manually, mechanically, or hydraulically activated. Because permanent adjustable dockboards are fastened to the dock and not moved from position to position, the dock board may be longer and wider than portable dockboards. The extra length results in a smaller incline between the dock and the carrier. This allows easier and safer handling of hand carts, reduced power drain on electrically powered trucks, and less of a problem with fork and undercarriage fouling on the dockboard. The greater width allows for safer and more efficient carrier loading and unloading. Permanent adjustable dock boards also eliminate the safety, pilferage, and alignment problems associated with dockboards. For these reasons, permanent adjustable dockboards, as shown in Figure 9.11, should always be given serious consideration.

The fourth option of raising the carrier to the height of the dock may be utilized when the variation in height of the trucks to be loaded or unloaded is very large. Truck levelers, as shown in Figure 9.12, are platforms installed under the rear wheels of a truck that lift the rear of the truck to the height of the dock. Typical truck levelers can raise a truck bed up to 3 ft. Although the flexibility and functional aspects of truck levelers are desirable, the installation cost of truck levelers limits their use.

The last approach to matching the height of the dock to the carrier is raising or lowering the level of the dock to that of the carrier. Scissors-type dock elevators, as show in Figure 9.13, are the most common method of raising or lowering the entire dock. Dock elevators may be permanently installed or may be mobile. Dock elevators are typically used when the plant is at ground level and room does not exist for a

(a)

**Figure 9.11** Permanent adjustable dock boards. (*a*) Hydraulic. (*Courtesy of W. B. McGuire C., Inc.*) (*b*) Face-of-dock. (*Courtesy of Brooks and Perkins, Inc.*)

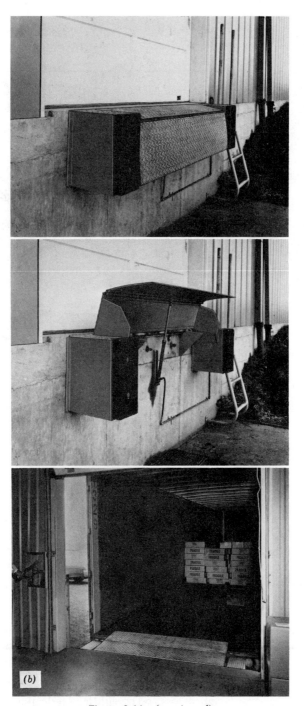

**Figure 9.11**   (continued)

**Figure 9.12** Truck leveler. *(Courtesy of Autoquip Corporations.)*

yard ramp or when the existing dock is too low or too high and sufficient space does not exist for a dockboard.

## Bumper Pads

A trailer carrying a load of 40,000 lb traveling at 4 mph when hitting a dock transmits 150,000 lb of force to the building. Reinforced concrete docks will disintegrate under this type of constant impact. The addition of 1-in. of cushioning to the front of the dock will reduce to 15,000 lb the force transmitted to the building from the same 40,000-lb truck at the same speed. Bumper pads are no more than molded or laminated rubber cushions that; when fastened to a dock, allow for the force to be absorbed from trucks hitting it.

## Dock Shelters

A dock shelter is a flexible shield that, when engaged by a carrier, forms a hermetic seal between the dock and the carrier. Figure 9.14 illustrates several types of dock shelters. The advantages of dock shelters include:

1.  *Energy savings:* Often, the heat loss and cooling load reductions that result from enclosing the docks more than pay for the cost of installing and maintaining dock shelters.
2.  *Increased safety:* Dock shelters protect dock areas from rain, ice, snow, dirt, and debris that enter through open, unprotected dock doors.

**Figure 9.13** Scissors-type lifting docks. (*Part a courtesy of Southworth Inc.; part b courtesy of Autoquip Corporation.*)

3.  *Improved product protection:* Materials entering and leaving a facility are better protected. For foodstuffs and electronic equipment, dock shelters are often required.

4.  *Better security:* Because of the elimination of openings around the carriers, the opportunity for pilferage is reduced.

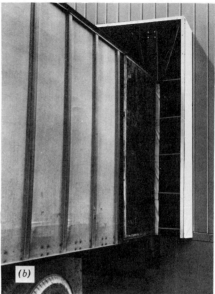

**Figure 9.14**   Dock shelters. (*Parts a and b courtesy of W. B. McGuire Co., Inc.; part c courtesy of Brooks and Perkins, Inc.*)

5.   *Reduced maintenance:* Dock shelters prevent leaves, dirt, debris, and water from entering the facility; therefore, less housekeeping is required.

6.   *Reduced spotting time:* Most dock shelters have guidance stripes to allow quicker and better truck positioning.

## *9.5* DOCK LOCATIONS

Not only are the number and design of docks important considerations for both inbound and outbound shipments, but the locations of the docks at the facilities must be carefully considered as well. One important factor in locating the docks will be the location of receiving and shipping. A discussion of receiving and shipping locations for warehouses is included in Section 10.4, Layout Planning, Popularity (p. 351). An important overall facilities consideration is the centralization or decentralization of receiving and shipping. No general guidelines are presented, but the following factors should be considered:

1. Labor requirements for both functions.
2. Material flow patterns.
3. Operating capacity.
4. Energy considerations.
5. Shipping and receiving schedules.

As has been noted, docks are among the first requirements at a site and are vital for smoothly functioning operations. Additionally, plans for facility expansion must incorporate receiving and shipping. An important rule-of-thumb concerning expansion is to expand the warehouse without disrupting dock operations. As shown in Figure 9.15, numerous expansion alternatives exist for a distribution center. The small arrows

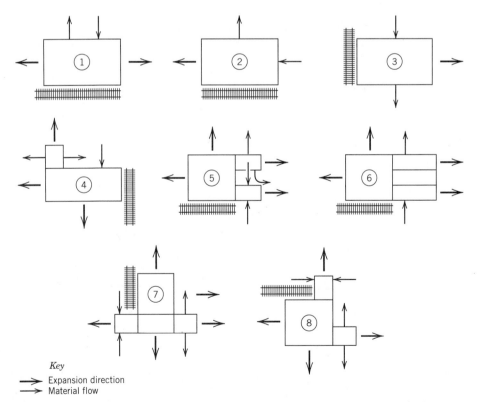

**Figure 9.15** Modular designs of storage/warehouse facilities with expansion alternatives shown.

represent material flows and large, open arrows represent directions for expansion that will minimize disruptions to receiving and shipping activities.

# *9.6*   STORAGE OPERATIONS

The objective of storage and warehousing functions is either to maximize resource utilization while satisfying customer requirements or to maximize customer service subject to a resource constraint. Storage and warehousing resources are space, equipment, and personnel. Customer requirements for storage and warehousing functions are to be able to obtain the desired goods quickly and in good condition. Therefore, in designing storage and warehousing systems it is desirable to:

1. Maximize space utilization.
2. Maximize equipment utilization.
3. Maximize labor utilization.
4. Maximize accessibility of all materials.
5. Maximize protection of all materials.

Planning storage and warehousing facilities follows directly from these objectives. Planning for maximum equipment utilization requires the selection of the correct equipment. The third objective, the maximum utilization of labor, involves providing the needed personnel services and offices. Planning for the maximum accessibility of all materials is a layout issue. Planning for the maximum protection of items follows directly from storing goods in adequate *space* with the proper *equipment* by well-trained *personnel* in a properly planned *layout.*

## Storage Space Planning

A chart that can be used to facilitate the calculation of space requirements for storage and warehousing is the storage analysis chart given in Figure 9.16. The first five columns of the storage analysis chart are identical to the first five columns of the shipping and receiving analysis chart given in Figure 9.4. If materials are stored in the same method whether received or shipped, data may be extracted directly from the shipping and receiving analysis chart. If changes are made in the unit loads after receipt or before shipping, the changes should be reflected in the storage analysis chart. If, for an existing facility, a shipping and receiving analysis chart has not been completed, a physical survey of the items stored will provide the data required in the first five columns.

The maximum and average quantities of unit loads stored (the sixth and seventh columns on the storage analysis chart) are directly related to the method of controlling inventory and inventory control objectives. These quantities should be provided as inputs to the facilities planner by the inventory control function.

Company ________ Date ________ ________ Raw Materials ________ In-Process Goods

Prepared by ________ Sheet ____ of ________ ________ Plant Supplies ________ Finished Goods

| | UNIT LOADS | | | | QUANTITY OF UNIT LOADS STORED | | | STORAGE SPACE | | | |
|---|---|---|---|---|---|---|---|---|---|---|---|
| Description | Type | Capacity | Size | Weight | Maximum | Average | Planned | Method | Space Standard | Area (sq. ft) | Ceiling Height Req'd |
| | | | | | | | | | | | |

Figure 9.16 Storage Analysis Chart.

## *Example 9.2*

At the ABC manufacturing firm, the average daily withdrawal rate for product A is 20 cartons per day, safety stock is 5 days, order lead time is 10 days, and the order quantity is 45 days. What are the maximum and average quantities of unit loads to be stored?

Reorder point $= \dfrac{\text{safety}}{\text{stock}} + \dfrac{\text{demand over}}{\text{lead time}}$

$= (5 \text{ days})(20 \text{ cartons per day}) + (10 \text{ days}) (20 \text{ cartons per day})$

$= 300 \text{ cartons}$

Maximum quantity to be stored $= \dfrac{\text{safety}}{\text{stock}} + \dfrac{\text{order}}{\text{quantity}}$

$= 100 \text{ cartons} + (45 \text{ days})(20 \text{ cartons per day})$

$= 1000 \text{ cartons}$

Average quantity to be stored $= \frac{1}{2} \text{ (order quantity)} + \text{safety stock}$

$= \frac{1}{2} (900) + 100$

$= 550 \text{ cartons}$

The planned number of unit loads for each material to be stored may be determined by considering the receiving schedule and the method of assigning materials to storage locations. If all materials to be stored in a particular manner are to be received together, the planned quantity of unit loads stored must be equated to the maximum quantity of unit loads. If materials to be stored in a particular manner are to arrive over time, then the method of assigning materials to storage locations will determine the planned quantity of unit loads stored.

The two storage location methods that in some sense represent extreme points of view are **randomized storage** and **dedicated storage.** Randomized storage is used when an individual stock-keeping unit (SKU) can be stored in any available storage location. The most common operational definition of randomized storage appears to be the following: When an inbound load arrives for storage, the closest-available-slot is designated as the storage location; retrievals are performed on a first-in, first-out (FIFO) basis. The FIFO retrieval policy provides for a uniform rotation of stocks, and in the long run, is equivalent to "pure" randomized retrieval. The closest-available-slot storage policy can yield results that are similar to the "pure" randomized storage policy *if* the storage level remains fairly constant and at a high level of utilization; otherwise, there will exist differences in the throughput rates obtained.

Dedicated storage is used when an SKU is assigned to a specific storage location or set of locations. The term **fixed slot** is used to describe dedicated storage. Two variations of dedicated storage are commonly used. Storage of items in part number sequence is one form of dedicated storage; another approach is to determine the storage location for an SKU based on its level of activity and inventory level. The latter method is preferred when there are significant differences in either the activity level or the inventory level for SKUs.

The number of openings assigned to an SKU must accommodate its maximum inventory level. Hence, the planned quantity of unit loads stored required for dedicated storage is equal to the sum of the openings required for each SKU. With randomized storage, however, the planned quantity of unit loads stored required in the system is the number of openings required to store *all* SKUs. Since typically all SKUs will not be at their maximum inventory levels at the same time, randomized storage will generally require fewer openings than dedicated storage.

There are two reasons for randomized storage resulting in less storage space than required for dedicated storage. First, if an "out-of-stock" condition exists for a given SKU the empty slot continues to remain "active" with dedicated storage; whereas, it would not with randomized storage. Second, if there exist multiple slots for a given SKU, then as the inventory level decreases empty slots will develop.

## Example 9.3

In order to illustrate the effect of the storage method on the storage space required, suppose six products are received by a warehouse according to the schedule depicted in Table 9.4. By summing the inventory levels of the six products, the **aggregate inventory** level is obtained.

With dedicated storage the required amount of space, as given in Table 9.4 equals the sum of the maximum inventory levels for the six products or 140 pallet positions. With ran-

Table 9.4 *Inventory Levels for Six Products in a Warehouse. Expressed in Pallet Loads of Product*

| Period | PRODUCTS | | | | | | Aggregate |
|---|---|---|---|---|---|---|---|
| | 1 | 2 | 3 | 4 | 5 | 6 | |
| 1 | 24 | 12 | 2 | 12 | 11 | 12 | 73 |
| 2 | 22 | 9 | 8 | 8 | 10 | 9 | 66 |
| 3 | 20 | 6 | 6 | 4 | 9 | 6 | 51 |
| 4 | 18 | 3 | 4 | 24 | 8 | 3 | 60 |
| 5 | 16 | 36 | 2 | 20 | 7 | 24 | 105 |
| 6 | 14 | 33 | 8 | 16 | 6 | 21 | 98 |
| 7 | 12 | 30 | 6 | 12 | 5 | 18 | 83 |
| 8 | 10 | 27 | 4 | 8 | 4 | 15 | 68 |
| 9 | 8 | 24 | 2 | 4 | 3 | 12 | 53 |
| 10 | 6 | 21 | 8 | 24 | 2 | 9 | 70 |
| 11 | 4 | 18 | 6 | 20 | 1 | 6 | 55 |
| 12 | 2 | 15 | 4 | 16 | 24 | 3 | 64 |
| 13 | 24 | 12 | 2 | 12 | 23 | 24 | 97 |
| 14 | 22 | 9 | 8 | 8 | 22 | 21 | 90 |
| 15 | 20 | 6 | 6 | 4 | 21 | 13 | 75 |
| 16 | 13 | 3 | 4 | 24 | 20 | 15 | 84 |
| 17 | 16 | 36 | 2 | 20 | 19 | 12 | 105 |
| 18 | 14 | 33 | 8 | 16 | 13 | 9 | 98 |
| 19 | 12 | 30 | 6 | 12 | 17 | 6 | 83 |
| 20 | 10 | 27 | 4 | 8 | 16 | 3 | 68 |
| 21 | 8 | 24 | 2 | 4 | 15 | 24 | 77 |
| 22 | 6 | 21 | 8 | 24 | 14 | 21 | 94 |
| 23 | 4 | 18 | 6 | 20 | 13 | 18 | 79 |
| 24 | 2 | 15 | 4 | 16 | 12 | 15 | 64 |

Maximum of aggregate inventory level = 105 pallet loads
Sum of individual maximum inventory levels = 140
Average inventory level = 77.5
Minimum of aggregate inventory level = 51

domized storage the required amount of space equals the maximum of the aggregate inventory level or 105 pallet positions. In this example dedicated storage requires one-third more pallet positions than does randomized storage.

When inventory shortages seldom occur and single slots are assigned to SKUs then there are no differences in the storage space requirements for randomized and dedicated storage. Interestingly, many carousel and miniload systems meet these conditions.

To maximize throughput when using dedicated storage, it is established in Chapter 15 that SKUs should be assigned to storage locations based on the ratio of their activity to the number of openings or slots assigned to the SKU. The SKU having the highest ranking is assigned to the preferred openings, and so on, with the lowest-ranking SKU assigned to the least-preferred openings. Because "fast movers" are up front and "slow movers" are in back, throughput is maximized.

In ranking SKUs it is important to define *activity* as the number of storages/retrievals per unit time, not the quantity of materials moved. Also, it is important to think of "part families" as well. "Items that are ordered together should be stored together" is a maxim of activity-based storage.

Despite the greater throughput of dedicated storage, it is not used as often as it should be. One reason is that it requires more *information* to plan the system for maximum efficiency. Very careful estimates of activity levels and space requirements must be made. Also more *management* is required in order to continue to realize the benefits of dedicated storage after the system is installed. When conditions change significantly, items must be relocated to achieve the benefits of dedicated storage. Hence, randomized storage is more appropriately used under highly seasonal and dynamic conditions.

When there exist many SKUs, dedicated storage based on each SKU may not be practical. Instead, SKUs can be assigned to classes based on their activity-to-space ratios. Class-based dedicated storage, with randomized storage within the class, can yield the throughput benefits of dedicated SKU storage and the space benefits of randomized SKU storage. Depending on the activity-to-space ratios, three to five classes might be defined.

## *Example 9.4*

To illustrate the effect on space and throughput of the storage method used, suppose the storage area for the warehouse is designed as shown in Figure 9.17. A single input/output (I/O) point serves the storage area. All movement is in full pallet-load quantities. The storage area is subdivided into 10 × 10 ft storage bays. Three classes of products (A, B, and C) are to be stored. Class A items represent 80% of the input/output activity and have a dedicated storage requirement of 40 storage bays, or 20% of the total storage; class B items generate 15% of the I/O activity and have a dedicated storage requirement of 30% of the total, or 60 storage bays; and class C items account for only 5% of the throughput for the system, but represent 50% of the storage requirement.

Assuming lift truck travel between the I/O point and individual storage bays can be approximated by rectilinear distances between the dock and the centroid of each bay, the distances are as shown in Figure 9.18. Based on the ratio of I/O activity to dedicated storage requirement, the product classes will be placed in the layout in the rank order A, B, and C to obtain the product layout given in Figure 10.4.

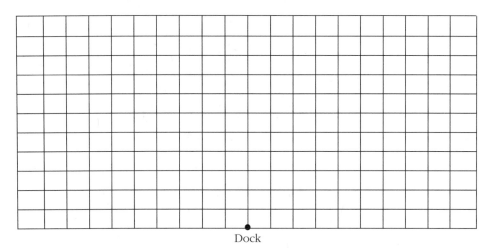

**Figure 9.17** Warehouse layout example.

The expected distance traveled for the dedicated storage layout shown in Figure 9.19 is 53.15 ft. If randomized storage is used, such that each bay is equally likely to be used for storage, then the expected distance traveled will be 100 ft. However, with randomized storage a total storage requirement less than 200 bays is anticipated for the reasons discussed previously. The exact storage requirement will depend on the demand and replenishment patterns for the three product classes.

Even though the storage requirement for randomized storage is not known, it is possible to compute an upper bound for storage that will yield an expected distance traveled equal to or less than that for dedicated storage. To do this, storage bays are eliminated from consideration in reverse order of their distances from the I/O point and expected distance values are computed. The process continues until a sufficient number have been eliminated. For the example, 138 storage bays, or 69% must be eliminated for randomized storage to

| 190 | 180 | 170 | 160 | 150 | 140 | 130 | 120 | 110 | 100 | 100 | 110 | 120 | 130 | 140 | 150 | 160 | 170 | 180 | 190 |
| 180 | 170 | 160 | 150 | 140 | 130 | 120 | 110 | 100 | 90 | 90 | 100 | 110 | 120 | 130 | 140 | 150 | 160 | 170 | 180 |
| 170 | 160 | 150 | 140 | 130 | 120 | 110 | 100 | 90 | 80 | 80 | 90 | 100 | 110 | 120 | 130 | 140 | 150 | 160 | 170 |
| 160 | 150 | 140 | 130 | 120 | 110 | 100 | 90 | 80 | 70 | 70 | 80 | 90 | 100 | 110 | 120 | 130 | 140 | 150 | 160 |
| 150 | 140 | 130 | 120 | 110 | 100 | 90 | 80 | 70 | 60 | 60 | 70 | 80 | 90 | 100 | 110 | 120 | 130 | 140 | 150 |
| 140 | 130 | 120 | 110 | 100 | 90 | 80 | 70 | 60 | 50 | 50 | 60 | 70 | 80 | 90 | 100 | 110 | 120 | 130 | 140 |
| 130 | 120 | 110 | 100 | 90 | 80 | 70 | 60 | 50 | 40 | 40 | 50 | 60 | 70 | 80 | 90 | 100 | 110 | 120 | 130 |
| 120 | 110 | 100 | 90 | 80 | 70 | 60 | 50 | 40 | 30 | 30 | 40 | 50 | 60 | 70 | 80 | 90 | 100 | 110 | 120 |
| 110 | 100 | 90 | 80 | 70 | 60 | 50 | 40 | 30 | 20 | 20 | 30 | 40 | 50 | 60 | 70 | 80 | 90 | 100 | 110 |
| 100 | 90 | 80 | 70 | 60 | 50 | 40 | 30 | 20 | 10 | 10 | 20 | 30 | 40 | 50 | 60 | 70 | 80 | 90 | 100 |

Dock

**Figure 9.18** Average distances traveled.

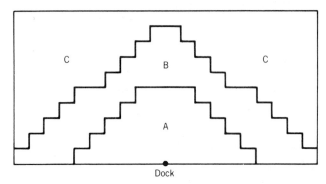

**Figure 9.19**   "Optimum" dedicated storage layout.

yield an expected distance traveled equal to 52.90 ft. The resulting randomized storage layout is shown in Figure 9.20.

If storage space is to be rectangularly shaped and if 10 × 10 ft storage bays are to be used, then it can be shown [15] that the layout having the minimum expected distance will be as given in Figure 9.21. The resulting expected distance traveled for Figure 9.21 is 50 ft. However, only 50 storage bays can be required for storage, rather than the 62 required in Figure 9.20. (The comparison also serves to demonstrate the effect building design can have on expected distance traveled.)

From the bound obtained for randomized storage it is seen that it is not likely that randomized storage will yield a reduction in space for this example sufficient to obtain throughput values comparable to those obtained from dedicated storage. However, if space costs are significantly greater than handling or throughput costs then randomized storage might be preferred, regardless of the impact on throughput.

When a combination of random and assigned storage is utilized, the method of assigning storage is referred to as supermarket storage. This implies that order picking takes place from dedicated storage locations and that backup stocks (reserve storage) are stored randomly. Recall that a supermarket locates merchandise in dedicated or assigned stock location, but backup merchandise is stored randomly in the backroom. When such a storage method is used, the planned quantity of unit loads stored will range from the sum of the individual maximum quantities to the maximum of the aggregate inventory levels, depending on the distribution of goods between reserve storage and active picking (assigned stock).

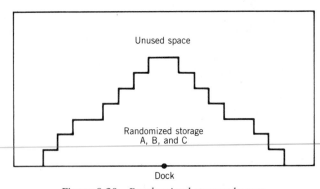

**Figure 9.20**   Randomized storage layout.

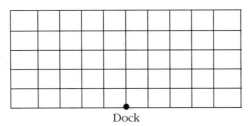

**Figure 9.21**  Rectangularly shaped randomized storage layout.

Once the planned quantity of unit loads to be stored is determined, the method of storing unit loads must be specified. For each method of storage, the loss of cube utilization due to aisles and to honeycombing must be determined. For an existing facility, the loss of cube utilization due to aisles may be determined by drawing various layouts and establishing for each layout the percent loss in cube utilization because of aisles.

## Example 9.5

To demonstrate the calculation of loss in cube utilization because of aisles, suppose a counterbalance lift truck will be used to store and retrieve materials from a pallet rack. A 10-ft service aisle is required. The pallet rack is to be located in an area that is 46 ft deep and has a clear stock height of 18 ft. The warehouse layout is as shown in Figure 9.22. As shown in Figure 9.22, 64% of the total warehouse cube is allocated to aisles.

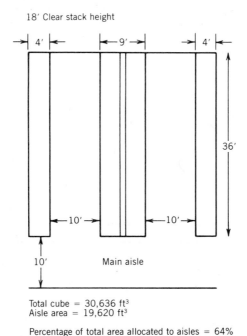

**Figure 9.22**  An example of the calculation of the percentage of the total warehouse area allocated to aisles.

**Honeycombing** is wasted space that results when a partial row or stack cannot be utilized because adding other materials would result in blocked storage. Figure 9.23 demonstrates honeycombing. The percentage of storage space lost because of honeycombing may be calculated using the analytical models given in Chapter 12.

Once the method of storage and the losses in cube utilization due to aisles and honeycombing are determined, space standards for all unit loads to be stored may be calculated. A **space standard** is the volume requirement per unit load stored to include allocated space for aisles and honeycombing. By multiplying the space standard for an item times the planned quantity of unit loads stored, the space requirement for an item may be determined. The sum of the space requirements for all items to be stored is the total storage space requirement. By adding to the total storage space requirement areas for receiving and shipping, offices, maintenance, and plant services, the total area requirement for storage departments or warehouses may be determined.

## Storage Layout Planning

The layout planning objectives for a storage department or a warehouse are:

**1.** To utilize space effectively.
**2.** To provide efficient materials handling.
**3.** To minimize storage cost while providing the required levels of service.
**4.** To provide maximum flexibility.
**5.** To provide good housekeeping.

These objectives are similar to the overall objectives of storage and warehouse planning given in Section 9.6.1. This should not be surprising, as layout planning involves the coordination of labor, equipment, and space. To accomplish the objectives, several storage area principles must be integrated. The principles relate to the following:

**1.** Popularity.
**2.** Similarity.
**3.** Size.
**4.** Characteristics.
**5.** Space utilization.

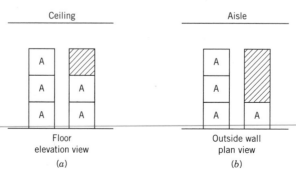

**Figure 9.23**  Examples of honeycombing. The cross-hatched area cannot be used to store other material. (*a*) Vertical honeycombing. (*b*) Horizontal honeycombing.

*Popularity.*  Earlier, Pareto's law was stated as "85% of the wealth of the world is held by 15% of the people." Pareto's law often applies to the popularity of materials stored. Typically, 85% of the turnover will be as a result of 15% of the materials stored. As noted in Section 9.6.1, to maximize throughput the most popular 15% of materials should be stored such that travel distance is minimized. In fact, materials should be stored so that travel distance is inversely related to the popularity of the material. Travel distances may be minimized by storing popular items in deep storage areas

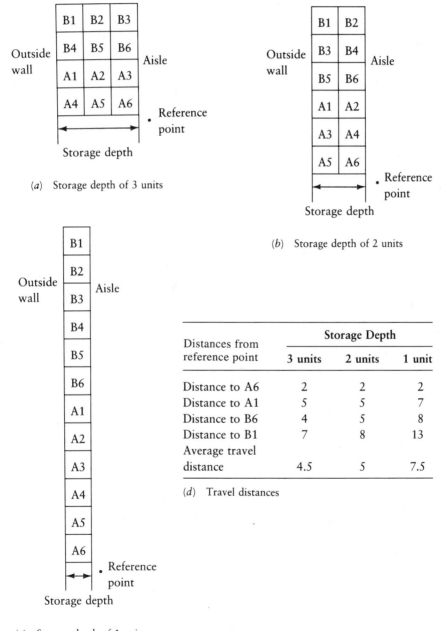

| Distances from reference point | Storage Depth | | |
|---|---|---|---|
| | **3 units** | **2 units** | **1 unit** |
| Distance to A6 | 2 | 2 | 2 |
| Distance to A1 | 5 | 5 | 7 |
| Distance to B6 | 4 | 5 | 8 |
| Distance to B1 | 7 | 8 | 13 |
| Average travel distance | 4.5 | 5 | 7.5 |

(*d*)  Travel distances

Figure 9.24   The impact of storage depth on travel distances.

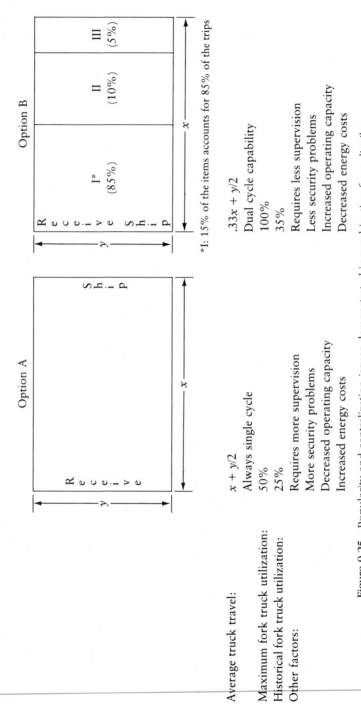

*I: 15% of the items accounts for 85% of the trips

| | Option A | Option B |
|---|---|---|
| Average truck travel: | $x + y/2$ | $.33x + y/2$ |
| Maximum fork truck utilization: | Always single cycle | Dual cycle capability |
| | 50% | 100% |
| Historical fork truck utilization: | 25% | 35% |
| Other factors: | Requires more supervision | Requires less supervision |
| | More security problems | Less security problems |
| | Decreased operating capacity | Increased operating capacity |
| | Increased energy costs | Decreased energy costs |

**Figure 9.25** Popularity and centralization issues demonstrated in combination for a distribution warehouse.

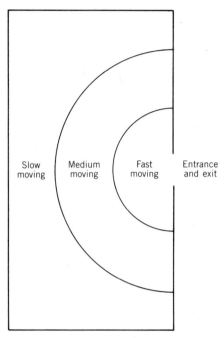

**Figure 9.26**  Material storage by popularity.

and by positioning materials to minimize the total distance traveled. As illustrated in Figure 9.24, by storing popular materials in deep storage areas the travel distance to other materials will be less than if materials were stored in shallow areas. In addition, Figure 9.25 illustrates the popularity issue in combination with the decentralized/centralized shipping and receiving issue (see Section 9.5).

From the previous discussion (and subsequent discussion in Chapter 12), a number of stock location "rules-of-thumb" can be developed. Namely, if materials enter and leave a stores department or warehouse from the same point, popular materials should be positioned as close to this point as possible. Figure 9.26 illustrates how to position materials with respect to such a point. Further, if materials enter and leave a storage area from different points and are received and shipped in the same quantity, the most popular items should be positioned along the most direct route between the entrance and departure points. Also, if materials enter and leave a storage area from different points and are received and shipped in different quantities, the most popular items having the smallest receiving/shipping ratio should be positioned close to the shipping point along the most direct route between the entrance and departure points. Finally, the most popular items having the largest receiving/shipping ratio should be positioned close to the receiving point along the most direct route between the entrance and departure points. (The receiving/shipping ratio is no more than the ratio of the trips to receive and the trips to ship a material.)

## Example 9.6

The products given in Table 9.5 are the most popular in the warehouse shown in Figure 9.27*a*. How should these items be aligned along the main aisle?

**Table 9.5**  *Receiving and Shipping Information on the Most Popular Products*

| Product | Quantity per Receipt | Trips to Receive | Average Customer Order Size | Trips to Ship |
|---------|---------------------|------------------|-----------------------------|---------------|
| A | 40 pallets | 40 | 1.0 pallet | 40 |
| B | 100 pallets | 100 | 0.4 pallets | 250 |
| C | 800 cartons | 200 | 2.0 cartons | 400 |
| D | 30 pallets | 30 | 0.7 pallets | 43 |
| E | 10 pallets | 10 | 0.1 pallets | 100 |
| F | 200 cartons | 67 | 3.0 cartons | 67 |
| G | 1000 cartons | 250 | 8.0 cartons | 125 |
| H | 1000 cartons | 250 | 4.0 cartons | 250 |

The receiving/shipping ratios may be calculated and are given in Table 9.6. A receiving/shipping ratio of 1.0 indicates the same number of trips are required to receive as to ship. Therefore, for products, A, F, and H the travel distance will be the same no matter where along the main aisle the products are stored. A receiving/shipping ratio less than 1.0 indicates that fewer trips are required to receive the product than to ship. Therefore,

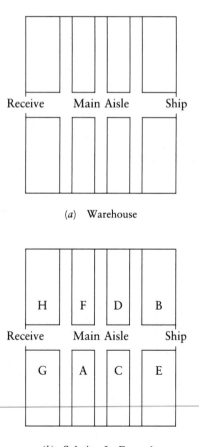

(a)  Warehouse

(b)  Solution for Example

**Figure 9.27**  Warehouse and product assignment for Example 10.5.

**Table 9.6**  *Receiving/Shipping Ratios for Example 10.5*

| Product | Receiving/Shipping Ratio |
|---------|--------------------------|
| A | 1.0 |
| B | 0.4 |
| C | 0.5 |
| D | 0.7 |
| E | 0.1 |
| F | 1.0 |
| G | 2.0 |
| H | 1.0 |

products having ratios less than 1.0 should be located closer to shipping. In order of importance of being close to shipping, the products with ratios less than 1.0 are E, B, C, D. Product G has a ratio greater than 1.0, indicating a need to be close to receiving. One of the assignments of products to locations that will result in minimal travel is given in Figure 9.27b.

*Similarity.*    The second principle for laying out a storage area relates to the similarity of the item stored. Namely, "items that are received and shipped together should be stored together." Even if items are not received together, it is almost always wise to store them together if they are shipped together. For example, in an automotive spare parts warehouse, carburetor parts are often stored together, as are exhaust system parts. A customer is unlikely to request a new carburetor float and a tail pipe bracket. However, it is quite likely that a tail pipe bracket will be included in an order for a new tail pipe gasket when a float is requested. By storing similar items in a common area, travel times for order receipt and order picking may be minimized.

*Size.*    Having small parts stored in spaces designed for large parts wastes storage space. Likewise, it is unfortunate when a large part cannot be stored in the appropriate storage slot (based on similarity and popularity) because it doesn't fit. To alleviate these problems, a variety of storage location sizes should be provided. If uncertainty exists about the size of items to be stored, easily adjustable racks and/or shelves can be utilized to accommodate changing needs.

In general, heavy, bulky, hard-to-handle items should be stored close to their point of use. However, assignment of space should be based on ease of handling and popularity of the item. If two items are equally popular, the bulkiest, most awkward to handle should be given the position closest to the point of use. If one item is more popular than another, but easier to handle, then relative popularities and ease of handling must be traded off.

If the size of items is such that floor loads may become a problem, then heavier items should be stored in areas having the lowest stacking heights. Lighter, easier-to-handle items should be stored where stacking heights are greater.

*Characteristics.*    Characteristics of materials to be stored often require that they be stored and handled contrary to the method indicated by their popularity, similarity, and size. Some important materials characteristics include:

1.  *Perishable materials:* Perishable materials may require that a controlled environment be provided. The shelf life of materials must be considered.

2.  *Oddly shaped and crushable items:* Certain items will not fit the storage areas provided even when various sizes are available. Oddly-shaped items often create significant handling and storage problems; if such items are encountered, open space should be provided for their storage. If items are crushable or become crushable when the humidity is very high, unit load sizes and storage methods must be appropriately adjusted.

4.  *Hazardous materials:* Materials such as paint, varnish, propane, and flammable chemicals require separate storage. Safety codes should be checked and strictly followed for all flammable or explosive materials. Acids, lyes, and other dangerous substances should be segregated to minimize exposure to employees.

5.  *Security items:* Virtually all items are pilferable. However, items with high unit value and/or small size are more often the target of pilferage. These items should be given additional protection within a storage area. With the increasing need to maintain traceability of materials, both pilferage and incorrect stock withdrawal must be prevented. Security of storage areas will be a problem if the layout is not designed to specifically secure the materials stored.

6.  *Compatibility:* Some chemicals are not dangerous when stored alone, but become volatile if allowed to come into contact with other chemicals. Some materials do not require special storage but become easily contaminated if allowed to come in contact with certain other materials. Therefore, the items to be stored in an area must be considered in light of other items to be stored in the same area. For example, butter and fish require refrigeration, but if refrigerated together, the butter quickly absorbs the fish odor.

*Space Utilization.*   As described in Section 9.6.1, space planning includes the determination of space requirements for the storage of materials. However, while considering popularity, similarity, size, and materials characteristics, a layout must be developed that will maximize space utilization as well as maximize the level of service provided. Some factors to be considered while developing the layout are:

1.  *Conservation of space:* Space conservation includes maximizing concentration and cube utilization and minimizing honeycombing. Maximizing space concentration enhances flexibility and the capability of handling large receipts. Cube utilization will increase by storing to a height of 18 in. below the trusses or sprinklers if the ceiling is 15 ft or less and 36 in. below the trusses or sprinklers if the ceiling is greater than 15 ft. Honeycombing is minimized by storing materials at the proper height and depth for the quantity of materials typically stored. Honeycombing also occurs from improper withdrawal of materials from storage.

2.  *Limitations of space:* Space utilization will be limited by the truss, sprinkler and ceiling heights, floor loads, posts and columns, and the safe stacking heights of materials. Floor loading is of particular importance for multistory storage facilities. The negative impact of posts and columns on space utilization should be minimized by storing materials compactly around posts and columns. Safe stacking heights of materials involve the crushability and stability of material

storage, as well as the safe storage and retrieval of materials. In particular, materials to be manually picked should be stacked so that operators can safely pick loads without excessive reaching.

3.  *Accessibility:* An overemphasis on space utilization may result in poor materials accessibility. Aisles must be planned to be wide enough for efficient material handling and be located so that every face of a storage island has aisle access. All main aisles should be straight and lead to doors and should be oriented so that the majority of materials are stored along the long axis of the storage area. Aisles should not be positioned along a wall unless the wall contains doors. These concepts are illustrated in Figure 9.28. Avoid locked stock by planning for the rotation of all inventories. To guarantee stock rotation, use "two-bin" systems in all storage areas. A "two-bin" system consists of two bins, vertical

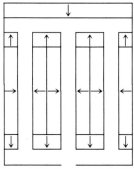

Receiving and shipping

(*a*)  Proper layout. The arrows indicate access to units stored.

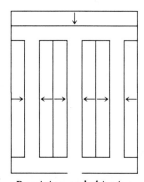

Receiving and shipping

(*b*)  Improper layout as each face of the storage islands does not have access from an aisle.

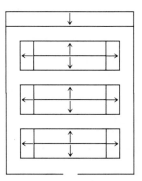

Receiving and shipping

(*c*)  Improper layout as the majority of the stock is not stored along the long axis of the building.

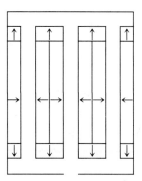

Receiving and shipping

(*d*)  Improper layout as an aisle is positioned in front of a wall that does not contain a door.

**Figure 9.28**  Illustration of storage area accessibility considerations.

stacks, rows, or drums for storage of each material. If only one drum existed, when the reorder quantity was received it would be placed in the drum. Materials would be used until a reorder point was reached, at which point an order would be released and more materials would eventually be received. Materials, upon arrival, would be placed once again into the drum. Thus, the material at the bottom of the drum would be locked into the bottom of the drum and would not be properly rotated. In a "two-bin" system, two drums would be utilized and when materials arrived, they would be placed in a second empty drum. Materials would continue to be used from the first drum until it was empty. When it is empty, the full second drum would be picked from until the reorder point was reached, at which point more would be ordered. When materials arrive, they would be placed into the empty first drum and materials would continue to be withdrawn from the second drum until it was once again empty. By continually repeating this procedure, materials will be properly rotated and no "locked stock" will exist.

## Developing and Maintaining a Stores Department or Warehouse Layout

The easiest method of developing a layout is to develop alternative scaled layouts and to compare these layouts with the principles of popularity, similarity, size, characteristics, and space utilization. The steps required to develop a scaled layout are:

1. Draw the overall area to scale.
2. Include all fixed obstacles, such as columns, elevators, stairs, plant services, and so on.
3. Locate the receiving and shipping areas.
4. Locate various types of storage.
5. Assign materials to storage locations.

Maintenance of the layout requires that materials be stored in an orderly manner and that stock locations be known. Disorderly storage can cause significant losses in cube utilization and accessibility. All materials should be stored in a uniform and neat manner. All materials should be accessible and should be placed properly in the assigned area. Wasted space often results when materials are not properly placed within allotted areas. Aisle marking should be maintained to indicate where loads are to be placed.

All materials within a storage area must be able to be quickly located and picked. If materials are assigned to a fixed location, stock location is easily performed. If materials are assigned to locations randomly, a stock location system must be used to keep track of material storage locations. The stock location system should include the quantity and location of all materials within the storage area. It will serve as the basis for assigning, locating, picking, and changing locations of all materials. A stock location system is essential if a dynamic layout is to be provided in the storage area. It functions as the interface between planning and managing storage facilities and warehouses.

# *9.7*  ORDER PICKING OPERATIONS

A recent survey of warehousing professionals identified order picking as the highest priority activity in the warehouse for productivity improvements [40]. There are several reasons for their concern. First, and foremost, order picking is the most costly activity in a typical warehouse. A recent study in the United Kingdom [12] revealed that 55% of all operating costs in a typical warehouse can be attributed to order picking (Figure 9.29). Second, the order picking activity has become increasingly difficult to manage. The difficulty arises from the introduction of new operating programs such as *Just-In-Time* (JIT), *cycle time reduction, quick response,* and new marketing strategies such as *micromarketing* and *megabrand* strategies. These programs require that (1) smaller orders be delivered to warehouse customers more frequently and more accurately, and that (2) more stock keeping units (SKUs) must be incorporated in the order picking system. As a result, both throughout, storage, and accuracy requirements have increased dramatically. Third, renewed emphasis on quality improvements and customer service have forced warehouse managers to reexamine the order picking activity from the standpoint of minimizing product damage, reducing transaction times and further improving picking accuracy. Finally, the conventional responses to these increased requirements, to hire more people or to invest in more automated equipment have been stymied by labor shortages and high hurdle rates due to uncertain business environments. Fortunately, there are a variety of ways to improve order picking productivity without increasing staffing or making significant investments in highly automated equipment. Twelve ways to improve order picking productivity in light of the increased demands now placed on order picking systems are described below.

## Order Picking Principles

1. *Encourage and design for full-pallet as opposed to loose case picking and full-case as opposed to broken case picking.* By encouraging customers to order in full-pallet quantities, or by creating quarter and/or half pallet loads, much of

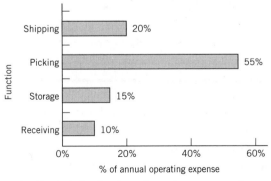

**Figure 9.29**  Typical distribution of warehouse operating expenses.

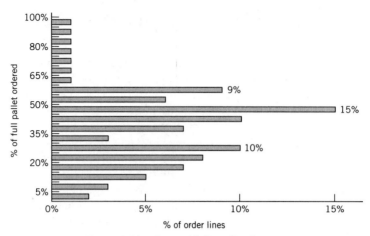

**Figure 9.30**   Case pick line distribution.

the counting and manual physical handling of cases can be avoided both in your warehouse and also in your customer's warehouse. In similar fashion, by encouraging customers to order in full-case quantities, much of the counting and extra packaging associated with loose case picking can be avoided. A pick line profile illustrating the distribution of the portion of a full-pallet or full-case requested by customers frequently reveals an opportunity to reduce the amount of partial pallet and/or partial case picking in the warehouse. An example profile is provide in Figure 9.30.

2. *Pick from storage.* Since a majority of a typical order picker's time is spent travelling and/or searching for pick locations, one of the most effective means for improving picking productivity and accuracy is to bring the storage locations to the picker; preferably reserve storage locations. A large wholesale drug distributor and a large discount retailer have recently installed systems which bring reserve storage locations to stationary order picking stations for batch picking of partial case quantities. In so doing, order picking travel time has been virtually eliminated. In addition, the same system can transfer storage locations to/from receiving, prepackaging, and inspection operations; thus virtually eliminating travel throughout the warehouse. Though expensive, the systems may be justified by increased productivity and accuracy.

3. *Eliminate and combine order picking tasks when possible.* The human work elements involved in order picking may include:
   - **traveling** to, from, and between pick locations,
   - **extracting** items from storage locations,
   - **reaching** and **bending** to access pick locations,
   - **documenting** picking transactions,
   - **sorting** items into orders,
   - **packing** items, and
   - **searching** for pick locations.

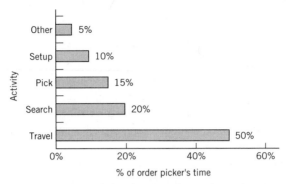

**Figure 9.31** Typical distribution of an order picker's time.

A typical distribution of the order picker's time among these activities is provided in Figure 9.31. Means for eliminating the work elements are outlined in Table 9.7.

When work elements cannot be eliminated, they can often be combined to improve order picking productivity. Some effective combinations of work elements are outlined below.

    a. **Traveling and Extracting Items.** Stock-to-operator (STO) systems such as carousels and the miniload automated storage/retrieval system are designed to keep order pickers extracting while a mechanical device travels to, from, and between storage locations bringing pick locations to the

**Table 9.7** *Order Picking Work Elements and Means for Elimination*

| Work Element | Method of Elimination | Equipment Required |
|---|---|---|
| *Traveling* | Bring pick locations to picker | Stock-to-Picker System<br>Miniload AS/RS<br>Horizontal Carousel<br>Vertical Carousel |
| *Documenting* | Automate information flow | Computer Aided Order Picking<br>Automatic Identification Systems<br>Light-aided Order Picking<br>Radio Frequency Terminals<br>Headsets |
| *Reaching* | Present items at waist level | Vertical Carousels<br>Person-aboard ASRS<br>Miniload ASRS |
| *Sorting* | Assign one picker per order<br>and one order per tour | |
| *Searching* | Bring pick locations to picker<br>Take picker to pick location<br>Illuminate pick locations | Stock-to-Picker Systems<br>Person-aboard AS/RS<br>Pick-to-Light Systems |
| *Extracting* | Automated dispensing | Automatic Item Pickers<br>Robotic Order Pickers |
| *Counting* | Weigh count<br>Prepackage in issue increments | Scales on picking vehicles |

order picker. As a result, a man-machine balancing problem is introduced. If the initial design of stock-to-operator systems is not accurate, a significant portion of the order picker's time may be spent waiting on the storage/retrieval machine to bring pick locations forward.

b. **Traveling and Documenting.** Since a person-aboard storage/retrieval machine is programmed to automatically transport the order picker between successive picking locations, the order picker is free to document picking transactions, sort material, or pack material while the s/r machine is moving.

c. **Picking and Sorting.** If an order picker completes more than one order during a picking tour, picking carts equipped with dividers or totes may be designed to allow the picker to sort material into several orders at a time.

d. **Picking, Sorting, and Packing.** When the cube occupied by a completed order is small, say less than a shoe box, the order picker can sort directly into a packing or shipping container. Packing or shipping containers must be setup ahead of time and placed on picking carts equipped with dividers and/or totes.

4. *Batch orders to reduce total travel time.* By increasing the number of orders (and therefore items) picked by an order picker during a picking tour, the travel time per pick can be reduced. For example, if an order picker picks one order with two items while travelling 100 ft, the distance travelled per pick is 50 ft. If the picker picked two orders with two items on each order, the distance travelled per pick is reduced to 25 ft.

A natural group of orders to batch is single-line orders. Single-line orders can be batched by small zones in the warehouse to further reduce travel time. A profile of the number of lines requested per order helps identify the opportunity for batching single-line orders. An example profile is illustrated in Figure 9.32.

Other order batching strategies are enumerated in Figure 9.33. Note that when an order is assigned to more than one picker, the effort to reestablish order integrity is significantly increased. The additional cost of sortation must be evaluated with respect to the travel time savings generated with batch picking.

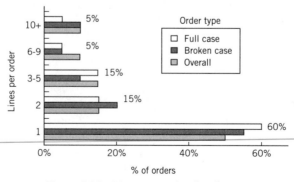

**Figure 9.32** Lines per order distribution.

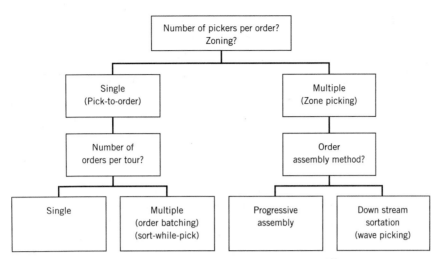

**Figure 9.33** Order picking policies.

a. **Single Order Picking.** In single order picking, each order picker completes one order at a time. For operator-to-stock systems, single order picking is like going through the grocery store and accumulating the items on your grocery list into your cart. Each shopper is concerned with only his or her list.

The major advantage to single order picking is that order integrity is never jeopardized. The major disadvantage is that the order picker is likely to have to travel over a large portion of the warehouse to pick the order. Consequently, the walking time per line item picked is high. However, for large orders (i.e., those greater than 10 line items), a single order may yield an efficient picking tour. In addition, in some systems, response time requirements do not allow orders to build up to create efficient batches for order picking.

b. **Batch Picking.** A second operating strategy for order picking is batch picking. Instead of an order picker working on only one order at a time, orders are batched together. Order pickers take responsibility for retrieving a batch of orders during a picking tour. In the grocery store example, batch picking can be thought of as going to the grocery store with your shopping list and those of some of your neighbors. In one traversal of the grocery store, you will have completed several orders. As a result, the travel time per line item picked will be reduced by approximately the number of orders per batch. The major advantage of batch picking is a reduction in travel time per line item.

The major disadvantages of batch picking are the time required to sort line items into customer orders and the potential for picking errors.

Orders may be sorted in one or two ways. First, the order picker may use separate containers to sort the line items of different orders as he or she traverses the warehouse. Special pick carts and containers are available to facilitate this approach. Second, the line items and quantities of different orders may be grouped together to be sorted later. It is the cost of this sortation process, not required in single order picking, that determines whether batch picking is a cost effective strategy.

Batch picking can also be used in stock-to-operator systems. In those cases, all the line items requested in the batch of orders are picked from a location as it is presented to the picker. Again, the benefits of reduced travel time must be weighed against the cost of sortation and the potential for order filling orders. Batch picking is especially effective for small orders (one to five line items).

c. **Zone Picking.** In zone picking, an order picker is dedicated to pick the line items in his or her assigned zone—one order at a time or in batches. In the grocery store context, zone picking can be thought of as assigning one individual to each aisle in the grocery store. The individual would be responsible for picking all the line items requested in that aisle regardless of the customer order that generated the request.

One advantage of zone picking is travel time savings. Since each picker's coverage has been reduced from the entire warehouse to a smaller area, the travel time per line item should be reduced from that of strict order picking. Again, however, these travel time reductions must be weighed against the costs of sorting and the potential for order filling errors. Additional benefits of zone picking include the order picker's familiarity with the product in his or her zone, reduced interference with other order pickers, and increased accountability for productivity and housekeeping within the zone.

Two methods for establishing order integrity in zone picking systems are progressive assembly and wave picking. In *progressive assembly* (or pick-and-pass) systems, complete orders are established as their components are passed from zone-to-zone in tote pans or cartons along a conveyor or on a cart. In *wave picking,* order pickers truly work in an item picking mode. In a typical wave picking system, an order picker applies a bar code label to each unit picked. Unit-by-unit or in batch, labelled units are placed on a takeaway conveyor for induction into a sortation/accumulation system. As before, the productivity gains in order picking must be weighed against the investment in sortation/accumulation systems.

5. *Establish separate forward and reserve picking areas.* Since a minority of the items in a warehouse generate a majority of pick requests, a condensed picking area, containing some of the inventory of popular items should be established. The smaller the allocation of inventory to the forward area (in terms of the number of SKUs and their inventory allocation), the smaller the forward picking area, the smaller the travel times, and the greater the picking productivity. However, the smaller the allocation, the more frequent the internal replenishment trips between forward and reserve areas, and the greater the staffing requirement for internal replenishments.

Some typical approaches for making forward-reserve decisions include allocating an equal time supply of inventory for all SKUs in the forward area or allocating an equal number of units of all SKUs in the forward area. A near-optimal procedure for solving the forward-reserve problem was recently developed [22]. The procedure makes use of math programming techniques to decide for each SKU whether it should receive a location in the forward area or not, and if it should, its proper allocation. The annual savings in picking and replenishment costs is typically between 20% and 40%.

A simplified approach to the configuration of a forward/reserve system may be determined as follows [21]:

a. **Determine which items should be in the forward picking area.** Because most inventories include many slow-moving items that have relatively small storage cube requirements, a forward picking area may include the entire inventory of each of the slow movers and only a representative quantity of the fast movers. Alternately, to provide the fastest possible picking, very slow-moving items may be stored elsewhere in the warehouse in a less accessible but higher-density storage mode.

b. **Determine the *quantities* of each item to be stored in the forward picking area.** As mentioned, slow movers may have their entire on-hand inventory located in the forward picking area. Storage allocation for other parts may be determined by either (1) an arbitrary allocation of space, as much as one case or one shelf, or (2) space for a quantity sufficient to satisfy the expected weekly or monthly demand. (It is common to select for large quantities from reserve storage and smaller quantities from forward storage.)

c. **Size the total storage cube requirement for items in the forward picking area.** Space planned for each item must be adequate for the expected receipt and/or replenishment quantity, not just for the average balance on-hand.

d. **Identify alternative storage methods that are appropriate for the total forward picking cube and that meet the required throughput.**

e. **Determine the operating methods within each storage alternative in order to project personnel requirements. The description of the operating method must also include consideration of the storage location assignment (i.e., random, dedicated, zoned, or a combination of these), because this will have a significant impact on picking productivity.** It is also necessary to evaluate the opportunity for batch picking (picking multiple orders simultaneously).

f. **Estimate the costs and savings for each alternative system described and implement the preferred system.**

6. *Assign the most popular items to the most easily accessed locations in the warehouse.* Once items have been assigned to storage modes and space has been allocated for their forward and reserve storage locations, the formal assignment of items to warehouse locations can commence. In a typical warehouse, a minority of the items generate a majority of the picking activity. This phenomenon can be used to reduce order picking travel time and reaching and bending. For example, by assigning the most popular items close to the front of the warehouse, an order picker's or s/r machine's average travel time can be significantly reduced. In automated storage/retrieval systems, average dual command travel time can be reduced by as much as 70% over random storage [18]. In miniload

automated storage/retrieval systems and carousel picking systems, order picking productivity can be improved by as much as 50% depending on the number of picks per bin retrieval and the other tasks assigned to the order picker (e.g. packaging, counting, weighing, etc.).

Popularity storage can also be used to reduce stooping and bending, consequently reducing fatigue and improving picking accuracy. Simply, the most popular items should be assigned to the picking locations at or near waist height. In an order picking operation for small vials of radio-pharmaceuticals, a stock location assignment plan was devised which concentrated over 70% of the picks in locations at or near waist level.

The most common mistake in applying this principle in the design of stock location systems is the oversight of the size of the product. The objective in applying the principle is to assign as much picking activity as possible to the locations that are easily accessible. Unfortunately, there are a limited number of picking locations that are easy to access—those near the front of the system and/or at or near waist height. Consequently, the amount of space occupied by an item must be incorporated into the ranking of products for stock assignment. A simple ranking of items based on the ratio of pick frequency (the number of times the item is requested) to shipped cube (the product of unit demand and unit cube) establishes a good baseline for item assignment. Items with high rankings should be assigned to the most easily accessible locations.

Two helpful profiles for stock location assignment are (1) a distribution illustrating items ranked by popularity and the portion of total picking activity they represent (Figure 9.34) and (2) a distribution illustrating items ranked by popularity and the portion of orders those items can complete Figure 9.35. The first profile is the traditional ABC or Pareto plot for item popularity. The second profile may reveal a small grouping of items that complete a large number of orders. Those items become candidates for assignment to a small picking zone dedicated to high density, high throughout order picking.

7.  *Balance picking activity across picking locations to reduce congestion.* In assigning popular items to concentrated areas in operator-to-stock systems, congestion can reduce potential productivity gains. Care must be taken to distribute picking activity over large enough areas to reduce congestion yet, not over so great an area as to significantly increase travel times. This is often achieved in horseshoe configurations of walk-and-pick systems. A typical picking tour will require the picker to traverse the entire horseshoe, however, the most popular

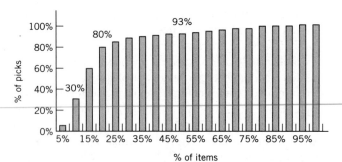

**Figure 9.34**  ABC analysis: Items and picks.

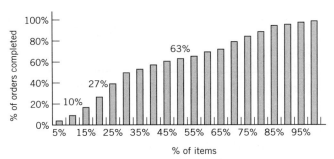

**Figure 9.35** ABC analysis: Items and completed orders.

items are assigned locations on or near the horseshoe. In stock-to-operator systems, system designers must be careful not to overload one carousel unit or one miniload aisle. Balanced systems are more productive.

8.  *Assign items that are likely to be requested together to the same or nearby locations.* Just like a minority of items in a warehouse generate a majority of the picking frequency, there are items in the warehouse that are likely to be requested together. Examples include items in repair kits, items from the same supplier, items in the same subassembly, items of the same size, etc. Correlations can be identified from order profiles [18] and can be capitalized on by storing correlated items in the same or in nearby locations. Travel time is in turn reduced since the distance between pick locations on an order is reduced. In a carousel or miniload AS/RS application, storing items that are likely to be requested together in the same location minimizes the number of locations visits required to complete an order and therefore helps reduce order picker idle time and wear and tear on the system.

    At a major, mail-order apparel distributor, nearly 70% of all orders can be completed from a single size (e.g. small, medium, large, extra large) regardless of the type of item ordered (shirts, pants, belts, etc.). At a major distributor of healthcare products, a majority of the orders can be filled from a single vendor's material. Since material is also received that way, correlated storage by vendor improves productivity in picking and putaway.

    A computerized procedure for jointly considering the popularity and correlation of demand for items in developing intelligent stock assignment plans was recently developed [18]. In the example, order picking productivity was improved by nearly 80%. The procedure suggests that items be clustered into families containing items that are likely to be requested together and then assigning the families to warehouse locations on the basis of the pick frequency and space occupied associated with the cluster. A simple way to begin the process of identifying demand families is to rank pairs of items based on the number of times the pair appears together on an order. The pairs at the top of the list often reveal the rationale for demand family development.

9.  *Sequence pick location visits to reduce travel time.* In both operator-to-stock and stock-to-operator systems, sequencing pick location visits can dramatically reduce travel time. The order picking travel time for a person-aboard AS/RS picking tour can be reduced by 50% by simply dividing the rack into upper and lower halves and visiting pick locations in the lower half in increasing distance from the front of the rack on the outbound leg, and in decreasing distance in

the upper half on the rack during the inbound leg. Location visits should also be sequenced in walk-and-pick systems. In case picking operations, when an order may occupy one or more pallets, the picking tour should be sequenced to allow the picker to build a stable load and to reduce travel distance. A major distributor of photographic supplies uses an expert system to solve this complex problem.

10. *Organize picking documents and displays to minimize search time and errors.* A majority of picking errors are the result of picking documents and displays which are confusing and/or difficult to read. Large, bold characters; color coding; eye-level displays; and floor markings can all be used to minimize confusion. In addition, every attempt should be made to remove similar colors and/or stock numbers from adjacent picking locations. One major distributor of cosmetics uses body parts as opposed to alpha-numerics to name pick locations in bin shelving units. A major distributor of office machines and supplies color codes all put-away tags to simplify product identification.

11. *Design picking vehicles to minimize sorting time and errors and to enhance the picker's comfort.* The order picking vehicle is the order picker's work station. Just as work station design is critical to the productivity and comfort of assembly and office workers, the design of the picking vehicle is critical to the productivity and morale of the order picker. The vehicle should be tailored to the demands of the job. If sorting is required, the vehicle should be equipped with dividers or tote pans. If picking occurs above a comfortable reaching height, the vehicle should be equipped with a ladder. If the picker takes documents on the picking tour, the vehicle should help the picker organize the paperwork. Unfortunately, the design of the picking vehicle is often of secondary concern, yet it is at this work station that order picking really takes place. A major wholesale drug distributor recently installed picking vehicles which are powered and guided by rails running in the ceiling between each picking aisle. The vehicle automatically takes the order picker to the correct location and with an on-board CRT communicates the correct pick location, quantity, and order or container to place the pick quantity into. The vehicle can accommodate multiple containers to allow batch picking and is equipped with on-board scales for on-line weigh counting and pick accuracy verification.

12. *Eliminate paperwork from the order picking activity.* Paperwork is one of the major sources of inaccuracies and productivity losses in the order picking function. Pick-to-light systems, radio frequency data communication, and voice input/output are existing technologies have been successfully used to eliminate paperwork from the order picking function. Each technology was described in Chapter 6.

## *9.8* SUMMARY

Warehousing and distribution center operations are historically one of the most frequently overlooked, underfunded, and inadequately planned corporate functions.

Today, quick response, supply chain integration, electronic data interchange, and the growth of third party logistics have placed warehousing in management's spotlight. As a result, the emphasis on planning and managing these operations has never been greater. In covering the operating principles and space planning methodologies for receiving, storage, picking, shipping, and dock operations, this chapter prepares engineers and managers to professionally plan and re-engineer warehousing and distribution center operations.

# BIBLIOGRAPHY

1.  Ackerman, K. B., Gardner, R. W., and Thomas, L. P., *Understanding Today's Distribution Center,* Traffic Service Corp., Washington, D.C., 1972.
2.  Allred, J. K., "Automated Storage Systems: How They Are Being Justified and Applied Today," paper presented to the Society of Manufacturing Engineers, Moline, Ill., 1975.
3.  Allred, J. K., "Large Automated Warehousing Systems: New Methods of Justifying, Buying, and Applying Them," *Proceedings of the 1975 MHI Material Handling Seminar and MHI Material Handling Symposium,* Material Handling Institute, Pittsburgh, PA, 1975.
4.  Allred, J. K., "How to Plan and Buy a Computer Controlled Storage and Handling System," *Material Handling Engineering,* vol. 29, no. 5, pp. 69–73, May 1974.
5.  Apple, J. M., *Material Handling Systems Design,* Ronald Press, New York, 1972.
6.  Apple, J. M., *Plant Layout and Material Handling,* Ronald Press, 3rd ed., New York, 1977.
7.  Apple, J. M. Jr., and Strahan, B. A., "Proper Planning and Control-The Keys to Effective Storage," *Industrial Engineering,* vol. 13, no. 4, pp. 102–112, April 1981.
8.  Barrett, J. P., "Modern Dock Design Concepts," International Material Management Society, Monroeville, PA. 1976.
9.  Bolz, H. A., and Hageman, G. E., (eds.), *Material Handling Handbook,* Ronald Press, New York, 1958.
10. Bozer, Y. A., and White, J. A., "Optimum Designs of Automated Storage/Retrieval Systems," presentation to the TIMS/ORSA Joint National Meeting, Washington, D.C., May 1980.
11. Briggs, A. J., *Warehouse Operations Planning and Management,* John Wiley, New York, 1960.
12. Drury, J., *Towards More Efficient Order Picking,* IMM Monograph Number 1, The Institute of Materials Management, Cranfield, United Kingdom (1988).
13. Elsayed, E. A., "Algorithms for Optimal Material Handling in Automatic Warehousing Systems," *International Journal of Production Research,* vol. 19, no. 5 (1981), pp. 525–535.
14. Foley, R. D., and Frazelle, E. H., "The Effect of Class-Based Storage on End-of-Aisle Order Picking," Unpublished Working Paper, Material Handling Research Center, Georgia Institute of Technology (1989).
15. Francis, R. L., and White, J. A., *Facility Layout and Location: An Analytical Approach,* Prentice-Hall, Englewood Cliffs, N.J., 1974.
16. Frazelle, E. H., "Automated Storage Retrieval, and Transport Systems in Japan," U.S. Department of Commerce Technical Report, December 1990.
17. Frazelle, E. H., "Small Parts Order Picking: Equipment and Strategy," Material Handling Research Center Technical Report Number 01-88-01, Georgia Institute of Technology, Atlanta, Georgia, 30332-0205.

18. Frazelle, E. H., *Stock Location Assignment and Order Picking Productivity,* Ph.D. Dissertation, Georgia Institute of Technology, December 1989.

19. Frazelle, E. H., and Apple, J. M., Jr., "Warehouse Operations," *The Distribution Management Handbook* (J. A. Tompkins, ed.), McGraw-Hill, New York, 1993.

20. Frazelle, E. H., and Ward, R. E., "Material Handling Technologies in Japan," National Technical Information Service Report #PB93-128197, Washington, D.C., 1992.

21. Frazelle, E. H., Hackman, S. T., and Platzman, L. K., "Intelligent Stock Assignment Planning," *Proceeding of the 1989 Council of Logistics Management's Annual Conference,* St. Louis, MO., October 1989.

22. Frazelle, E. H., Hackman, S. T., Passy, U., and Platzman, L. K., *Solving the Forward-Reserve Problem,* Material Handling Research Center Technical Report, Georgia Institute of Technology, May 1992.

23. Goetschalckx, M., *Storage and Retrieval Policies for Efficient Order Picking Operations,* Ph.D. Dissertation, Georgia Institute of Technology, Atlanta, GA (1983).

24. Goetschalckx, M., and Ratliff, H. D. "Sequencing Picking Operations in a ManAboard Order Picking System," *Material Flow,* vol. 4, no. 4, 1988, pp. 255–263.

25. Graves, S. C., Hausman, W. H., and Schwarz, L. B., "Storage Retrieval Interleaving in Automatic Warehousing Systems," *Management Science,* vol. 23, no. 9 (May 1977), pp. 935–945.

26. Han, M., McGinnis, L. F., Shieh, J. S., and White, J. A., "On Sequencing Retrievals in an Automated Storage/Retrieval System," *IIE Transactions,* vol. 19, no. 2, (March 1987), pp. 56–66.

27. Jenkins, C. H., *Modern Warehouse Management,* McGraw-Hill, New York, 1968.

28. Kinney, H. D., "How to Size the Warehouse," Material Handling Management Course, American Institute of Industrial Engineers, Norcross, GA, June 1979.

29. Kinney, H. D., "A Total Integrated Material Management System," presentation to the 1980 Annual Conference of the International Material Management Society, Cincinnati, OH, June 1980.

30. Mullens, M. A., "Use a Computer to Determine the Size of a New Warehouse, Particularly in Storage and Retrieval Areas," *Industrial Engineering,* vol. 13, no. 6, pp. 24–32, June 1981.

31. Tompkins, J. A., *Facilities Design,* North Carolina State University, Raleigh, NC, 1975.

32. Tompkins, J. A., "Problem Solving Techniques in Material Handling," *Proceedings of the 1976 Material Handling and the Industrial Engineer Seminar,* American Institute of Industrial Engineers, Norcross, GA, 1976.

33. White, J. A., "Justifying Material Handling Expenditures," *Proceedings, 1977 Spring Annual Conference* American Institute of Industrial Engineers, pp. 103–111.

34. White, J. A., "Randomized Storage or Dedicated Storage?," *Modern Materials Handling,* vol. 35, no. 1, p. 19, January 1980.

35. White, J. A., and Kinney, H. D., "Storage and Warehousing," G. Salvendy (ed.) *Handbook of Industrial Engineering,* John Wiley, New York, 1982.

36. Wilde, J. A., "Container Design Issues," *Proceedings of the 1992 Material Handling Management Course,* sponsored by the Institute of Industrial Engineers, Atlanta, GA, June 1992.

37. National Safety Council, Chicago, IL.

38. *Warehousing Education and Research Council's Annual Membership Survey,* Warehousing Education and Research Council, Oak Brook, Illinois, 1988.

39. *Warehouse Modernization and Layout Planning Guide,* U.S. Naval Supply Systems Command Publication 529, U.S. Naval Supply Systems Command, Richmond, VA, 1988.

40. *Warehousing Education and Research Council's Annual Membership Survey,* Warehousing Education and Research Council, Oak Brook, IL, 1988.

# PROBLEMS

9.1  You are employed as an industrial engineer by the Form Utility Co. and feel a project in the dock area would prove beneficial. Before undertaking such a project you must convince your boss, Mr. Save A. Second, that the receiving and shipping areas are worth the study. Prepare a written report justifying your proposed dock study and outlining the types of savings you feel may result.

9.2  If truck arrivals occur in a Poisson fashion at a rate of 20 vehicles per 8-hr day, and truck unloading time is exponentially distributed with a mean of 40 min, how many docks should be planned if it is desired that the average total truck turn around time be less than 50 min?

9.3  A new musical textbook production plant is to be located along a street running north and south. The plant is to be 800 ft long and 500 ft wide. The property is 900 ft long and 700 ft deep. Four receiving docks are to be located in one rear corner of the plant and four shipping docks on the other rear corner. The trucks servicing the plant are all 55 ft. Sufficient truck buffer area must be planned for three trucks for both receiving and shipping. Draw a plot plan for the site.

9.4  A hospital is to have two receiving docks for receipt of foodstuffs, paper forms, and supplies. Forty-foot carriers are to serve the hospital. A narrow-aisle lift truck is to be used to unload the carriers. All foodstuffs, paper forms, and supplies are immediately stored upon receipt. What will be the impact on space requirements if instead of 90° docks, 45° finger docks are utilized?

9.5  A warehousing consultant has recommended that your firm eliminate portable dock boards and install permanent adjustable dock boards to achieve considerable space savings. What is your reaction to this recommendation?

9.6  Perform a survey of the receiving areas on campus and determine how the problem of dock and carrier height differences are handled. Also determine if there exists any potential for dock shelters.

9.7  The legal limits on trucks continue to increase. The length, width, and height have all increased in the last 10 years. What impact does this have on the planning of receiving and shipping areas?

9.8  What fraction of the total cost of operating a warehouse is typically associated with order picking?

9.9  What are four order batching procedures?

9.10  What are five tasks typically performed by an order picker?

9.11  What are three principles of intelligent stock assignment planning?

9.12  List five key data elements to consider in order picking system design.

9.13  Batch picking, as opposed to strict order picking, reduces the _____ between picks, but increases _____ requirements.

9.14  By reducing the amount of stock in the forward picking area, forward picking costs (a) increase, (b) decrease and the cost to replenish the forward picking area (c) increases, (d) decreases.

9.15  What are the three key data elements in determining if an item should be located in the forward picking area?

9.16  What are the two key data elements in determining the location assignment for an item when using popularity-based storage?

9.17  Popularity-based storage generally causes _____ to decrease and _____ to increase.

9.18 List four facets which make the order picking function difficult to manage.

9.19 What are the two principle errors made in order picking?

9.20 In a correlated stock assignment policy, items frequently requested together are stored in _____.

9.21 Visit a warehouse and categorize the type storage used (dedicated vs randomized vs a combination of the two); identify the technologies used to transport material to and from storage, to place material in and retrieve material from storage, to store the material. Assess the age and condition of the technologies identified in the warehouse. Identify the bar code symbologies used, as well as the types of bar code readers used; how are the bar code labels printed and applied?

    Describe the approach used to peform receiving and shipping. Identify the various pallet dimensions and designs found in the warehouse. Assess the quality of house-keeping in the warehouse. Determine the fraction of floor space devoted to the various functions performed in the warehouse; compute the fraction of floor space devoted to aisles. If racked storage is used, determine the fraction of the rack openings that are empty; also, estimate the fraction of cube space within a storage opening that is utilized, i.e., is material, including pallets. Assess the use of cube space within the warehouse.

9.22 Consider three items with the following profiles:

| Item | Annual Demand | Annual Requests | Unit Cube |
|------|---------------|-----------------|-----------|
| A | 100 units | 25 requests | $1 \text{ ft}^3$ |
| B | 50 units | 5 requests | $0.5 \text{ ft}^3$ |
| C | 200 units | 16 requests | $2 \text{ ft}^3$ |

  a. What is the total demand in cube for items A, B, and C?
  b. If $20 \text{ ft}^3$ of space is allocated for item C, how many times will the item be replenished in 1 year?
  c. If each item is to receive a location in the forward area, and an equal-time-supply allocation policy is used, how much space will be allocated for each item assuring a 1-week time supply? Assume 50 weeks.
  d. If the items are to be assigned locations in the forward area on the basis of popularity storage, and they are allocated space as in C above, rank the items so as to maximize the number of picks in the most accessible storage space.

9.23 Under what circumstances or conditions would you prefer to store goods in a warehouse by
  a. placing the product directly on the floor.
  b. palletizing and stacking pallet loads (block storage).
  c. palletizing and storing in conventional pallet rack.
  d. palletizing and storing in pallet flow rack.
  e. palletizing and storing in drive-in rack.
  f. palletizing and storing in drive-through rack.
  g. placing cases directly in flow rack.
  h. storing in cantilever rack.
  i. palletizing and stacking using portable stacking rack.
  j. placing the product directly in a bin.

9.24 Given the assignment to design a new warehouse for storing finished goods involving 1200 different stock-keeping units (SKUs), how would you determine the storage space required? Distinguish between dedicated storage, randomized storage, and class based dedicated storage as they impact space, handling time, and the stock location system.

9.25 Given a warehouse similar to that given in Figure 10.20(a), where would you position the following items?

| Product | Monthly Throughput | Quantity per Receipt | Trips to Receive | Average Customer Order Size | Trips to Ship |
|---|---|---|---|---|---|
| A | High | 300 pallets | 300 | 2.0 pallets | 150 |
| B | Low | 200 cartons | 50 | 4.0 pallets | 50 |
| C | Low | 10 pallets | 10 | 0.2 pallets | 50 |
| D | High | 400 pallets | 400 | 0.5 pallets | 800 |
| E | High | 6000 cartons | 1000 | 10.0 cartons | 600 |
| F | Low | 40 cartons | 40 | 2.0 pallets | 20 |
| G | High | 200 pallets | 200 | 1.0 pallets | 200 |
| H | High | 9000 cartons | 2250 | 5.0 cartons | 1800 |
| I | Low | 50 pallets | 50 | 1.0 pallet | 50 |
| J | High | 500 pallets | 500 | 0.7 pallets | 715 |
| K | Low | 80 pallets | 80 | 2.0 pallets | 40 |
| L | High | 400 pallets | 400 | 1.0 pallets | 400 |
| M | High | 7000 cartons | 1167 | 3.0 cartons | 2334 |
| N | Low | 700 cartons | 140 | 7.0 cartons | 100 |

9.26 All items to be stored in a warehouse are to be block-stacked. All loads are on 48 × 48 × 6-in. pallets. Each load is 4 ft tall. There are 30 different products to be stored and it is planned to store 300 loads of each product. The loads may be stacked four high in a warehouse that has a 22-ft clear ceiling height. The receiving and shipping docks are to be located along one side of the warehouse. A counterbalanced lift truck is to operate in the warehouse; therefore, 13-ft aisles are to be used. Develop a layout for the warehouse.

9.27 What is the relationship between cube utilization and product accessibility?

9.28 Explain the advantages and disadvantages of each of the layouts in Figure 9.36.

9.29 Materials are stored on 42 × 48-in. pallets in selective pallet racks, *double deep*. The pallet has a height of 5 in. The load stored on the pallet is 46 in. high. Two pallet loads are stored side by side in each rack opening. Likewise, two pallet loads are stored back to back in each rack opening. A 5-in. clearance is used (side to side) between loads in an opening and between a load and a vertical rack member. In an opening the load in the first-depth position just touches the load in the second-depth position. Each vertical rack member is 4 in. wide. Pallets are placed on 4-in. load beams, including the bottom load. There is a clearance of 4 in. between the top of each load and the load beam above it. An 8.5-ft aisle (load to load) and a flue spacing (back to back between loads in the second-depth position) of 12 in. is used. The rack has five tiers of storage. Loads do not overhang the pallet. Compute the storage space utilization assuming the loads in the second-depth position are full and the loads in the first-depth position are half full.

9.30 Materials are stored on 42 × 36-in. pallets in selective pallet racks, *one deep*. The pallet has a height of 6 in. The load stored on the pallet is 48 in. high. Three pallets are stored (side by side) in each rack opening. A 4 in. clearance is used between loads in an opening and between a load and a vertical rack member (upright truss). Each vertical rack member is 3 in. wide. Pallets are placed on 4-in. load beams except the bottom load, which is stored on the floor. There is a clearance of 3 in. between the top of each load and the load beam above it. An 8.5 ft. aisle (load to load) and a flue spacing of 15 in. between loads is used. The rack has four tiers of storage. Loads do not overhang the pallet. Compute the storage space utilization.

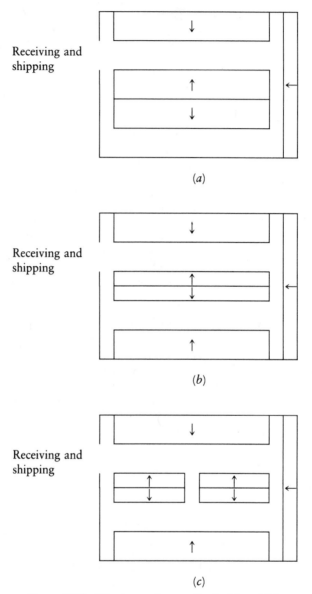

**Figure 9.36**  Warehouse layouts for Problem 9.37.

**9.31**  Materials are stored on 42 × 36 in. pallets in selective pallet racks, *one deep*. The pallet has a height of 6 in. The load stored on the pallet is 40 in. high. Two pallets are stored (side by side) in each rack opening. A 4-in. clearance is used between loads in an opening and between a load and a vertical rack member (upright truss). Each vertical rack member is 3 in. wide. Pallets are placed on 4 in. load beams except the bottom load, which is stored on the floor. There is a clearance of 4 in. between the top of each load and the load beam above it. A 10-ft aisle (load to load) and a back-to-back spacing of 12 in. between loads is used. The rack has six tiers of storage. Loads do not overhang the pallet. Compute the storage space utilization.

# 10

# *MANUFACTURING OPERATIONS*

## *10.1* INTRODUCTION

International competition forced revolutionary changes to occur in manufacturing in the 1980s and 1990s. Traditional paradigms were challenged and new manufacturing principles were developed. Terms such as *lean manufacturing, agile manufacturing,* and *world-class manufacturing* emerged. Regardless of the name given to it, firms gave increased emphasis to delivering products needed by customers faster than the competition, meeting or exceeding "best-in-class" quality requirements, coupled with unparalleled levels of service, and at competitive prices; at the same time, terms such as *breakthrough thinking, enterprise integration, business process reengineering, empowerment, learning organizations,* and *right sizing* became a part of management's vocabulary. Because of their roles in contemporary manufacturing systems, in this chapter we consider concepts, techniques, principles, and technologies from four of the new entries in management's vocabulary: *just-in-time* (JIT), *total quality management* (TQM), *total employee involvement* (TEI), and *computer integrated manufacturing systems* (CIMS).

Companies that have implemented manufacturing systems for the new environment described above have relied on the use of teams in redesigning, improving, and managing the manufacturing system. Some teams challenged and reengineered traditional manufacturing paradigms (e.g., single-receiving docks; centralized storage systems; process layouts; large unit load sizes; a single building for all the products; centralized offices, cafeterias, and parking lots). Other teams pursued continuous improvement of critical processes and took an active role in managing them. The new team environment and the new concepts, techniques, technologies, and ideas dramatically impacted facilities layout; building configuration and size; unit load design; material handling system selection, design, and operation; and storage equipment selection, design, location, and operation.

In this chapter, we consider the impact of agile manufacturing, lean manufacturing, or world-class manufacturing systems on facilities planning; for the purposes of our discussion, we will refer to this new approach to designing and managing manufacturing systems as contemporary manufacturing (CM). In presenting the material, we first contrast CM and traditional manufacturing and consider the impact of CM on facilities planning. Next we present an implementation methodology for CM. Finally, we address a number of facilities planning trends.

## *10.2*  TRADITIONAL MANUFACTURING (TM) VERSUS CONTEMPORARY MANUFACTURING (CM)

We use traditional manufacturing (TM) to identify the way manufacturing organizations were designed, managed, and operated in the 1970s. Specifically, a TM company is characterized by large lot sizes, long setups, many vendors, noncertified vendors, centralized storage systems, corrective quality control, quality control departments in charge of quality assurance, centralized rework departments, centralized offices for support personnel, top-down decision making, vertical organizational structures, specialized personnel, process layout, sequential product-process design, corrective maintenance, inflexible material handling systems, a minimum number of receiving–shipping points, synchronous assembly lines with linear shape arrangements, "push" production strategies, and complex computer-based production scheduling systems. As a consequence, (TM) organizations tend to operate with large inventories, small inventory turnovers, long production lead-times, high manufacturing costs, poor quality, long delivery times, poor customer service, high space and storage requirements, high handling requirements, and low productivity.

The response to the opportunity areas described above, CM, includes all activities required to satisfy customer requirements from product design to delivery and all stages from acquisition and conversion of raw material to delivery of the product. Companies utilizing CM tend to follow a strategy that begins with the development of long term and short term objectives and goals that correlate with corporate, marketing, and product development strategies. Objectives and goals are then translated to tactical and operational strategies that are implemented by teams following an integrated and organized reengineering and/or continuous improvement process (see Section 10.3). The teams generally use techniques and concepts related with JIT, TQM, TEI, and CIMS; the relationships among these concepts and techniques are illustrated in Figures 10.1 and 10.2.

As an example of how one might attempt to transform a TM installation into a CM installation, customer response time can be reduced by using a cellular layout, multiple receiving docks, deliveries to the points of use, shorter setups, smaller production lot sizes, simplified administrative processes, quality at the source, automated and flexible material handling systems, and certified vendors that deliver smaller lot sizes frequently.

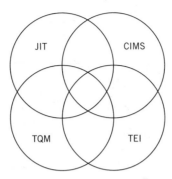

**Figure 10.1**   Relationships among the contemporary manufacturing areas.

Consequently, a typical CM organization could operate with competitive product design and cost, short production lead-times, small inventories, zero defects, small production and purchasing lot sizes, multifunctional workers, certified suppliers, decentralized functions, product and cellular plant layout, decentralized storage systems, kanban control systems with "pulling" procedure, teamwork, efficient material handling, maintenance, and transportation systems, and standardized, simplified, adaptable, flexible, and integrated processes.

As indicated by Gunn [35], "to achieve world-class manufacturing status today, a company needs inventory turnovers in raw materials and work-in-process (WIP) of some 25 to 30 per year to be a Class C world-class manufacturer, about 50 to 60 turns per year for a Class B status, and on the order of 80 to 100 turns per year to be a Class A world-class manufacturer.

As measures of world-class quality, a Class A manufacturer would need to have fewer than 200 defective parts per million of any product that they manufacture. With regard to lead-times, the ratio of value-added lead-time to cumulative manufacturing lead-time must be greater than 0.5 for a world-class manufacturer."

An example of possible stages in progression to CM is presented in Figure 10.3. The most important concepts and techniques of JIT, TQM, TEI, and CIMS and their impact on facilities planning are detailed in the following sections.

## Just-In-Time Production Systems

The JIT system was developed in Japan by Ohno Taiichi at the Toyota Motor Company more than 20 years ago. The JIT production system has been implemented and improved by many companies around the world. The JIT production system has been defined by the American Production and Inventory Control Society (APICS) as a philosophy of manufacturing excellence based on the pursuit of the planned elimination

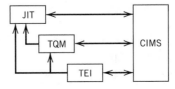

**Figure 10.2**   Relationships among the contemporary manufacturing areas.

Past ———————————→ Present ——————————————————————→ Future

| Traditional Manufacturing | Uncoordinated Incremental Improvements | Focused and Coordinated Product and Process Improvements | Contemporary Manufacturing |
|---|---|---|---|
| • Hierarchical management<br>• Functional focus<br>• Impersonal management<br>• Minimum training<br>• Mass production, inflexible processes<br>• Long production runs<br>• Long lead times<br>• Adversarial relationships with suppliers, customers, and employees<br>• Focus on full machine and labor utilization | • "Waste elimination" focus<br>• Participative management<br>• Management committed to the quality improvement process<br>• Extensive education and training begins<br>• Fewer management layers<br>• Awareness of total cost of quality<br>• "Pilot" projects<br>• Developing supplier relationships<br>• More open communication | • Minimum management layers<br>• Flexible manufacturing<br>• Smaller lots<br>• Quicker response to customer<br>• Detailed process studies<br>• Just-in-time purchasing and manufacturing<br>• Focused plant layouts<br>• Work cells<br>• Statistical control<br>• Reserve capacity<br>• Preventive maintenance<br>• Employee involvement teams<br>• Cross-trained workforce with job rotation | • Customer service excellence<br>• Error-free work<br>• 100% on-time delivery<br>• Paperless business<br>• Short lead-times, quick response<br>• Customer/supplier partnerships<br>• Constant innovation<br>• Produce to customer needs<br>• Low inventories, small lots<br>• Self-managed teams |

**Figure 10.3**   Stages of progression from traditional manufacturing to contemporary manufacturing.

of all waste and consistent improvement of productivity and quality. Possible stages in progression to JIT are presented in Figure 10.4 [35]. According to APICS, the primary elements of JIT may include the following:

- Reduction of work in process, queues, manufacturing and purchasing lead times, lot sizes, transit times, and factory floor space;
- Total productive maintenance;
- Supplier development and certification program;
- Frequent vendor deliveries;
- Focus processing;
- Group technology;
- Cellular manufacturing;
- People involvement;

| Attribute | No JIT Stage 1: Emerging awareness | Beginning JIT Stage 2: Awakening | Intermediate JIT Stage 3: Enlightenment | Full JIT Stage 4: Conviction |
|---|---|---|---|---|
| Production setups | Exceedingly long—many hours | 50% reduction on pilot attempts | 50% reduction in most areas. Another 30% but slower in second attempt | 75% reduction from original in most instances, some as high as 90%—many setups reduced to minutes |
| Annual raw material and work-in-process inventory turns | Less than six | Moving toward 12–15 in pilot attempts | 12–15 overall. 20–25 in pilot areas | 20–25 overall. 40–50 in some lines. 60+ the goal for next year! |
| Waste in operations | Excessive. Signaled by inventory buffers, loosely coupled production, excessive slack everywhere, slow work place | Emerging recognition of opportunities to eliminate waste and how waste comes in many forms | Programs established to attack waste in many forms in all areas | Success encourages further search for waste in more areas. Other company functions encouraged to examine their operations |
| Management attitude and involvement | JIT good for Japanese. Can't work here. Undergoes first exposure to JIT concepts. Perhaps a few customers demanding. JIT means we get stuck with inventory | Must implement JIT! Little or no understanding of implications of JIT for each company function. Implement in 6 months | Growing awareness of benefits possible as well as the enormous scope and complexity of the program. Concerned with value of current (old) accounting system for managing in a JIT environment | Complete dedication to JIT. Change in old accounting system and methods of performance measurement. An advocate of JIT to suppliers. Seek how JIT can be used in other company functions. Growing awareness of relationship of CIM, TQM, and JIT |

**Figure 10.4** Stages in the progression to JIT.

455

| | | | | |
|---|---|---|---|---|
| Worker attitude and involvement | Management doesn't care so why should I? Undergoes first exposure to JIT concepts | Is management serious this time? More exposure to education. Limited training in pilot areas. Maybe we can take pride in our work and company | Benefits for all of us are real. Less fire fighting. Pride, self-reliance, and esprit de corps rising | Pride in company. Knowledge they can compete with anyone. How could we ever have been so bad? |
| Supplier involvement | None | Emerging conceptual education for key suppliers. Beginning attempt at supplier reduction | Key suppliers obtaining benefits from programs. Fewer suppliers means less overhead. Inspectionless receiving started with some qualified suppliers | Number of suppliers reduced by 75%. 90% of suppliers certified for inspectionless receiving. Supplier's JIT programs have lowered their prices, improved their quality and delivery performance |
| Preventive maintenance program | None. "Fix it when it breaks" | May be important after all. Start by allowing worker to maintain own machine and train to do so | Machines run at or below rated speeds. Workers in charge of light machine maintenance. PM a necessity for JIT | A cornerstone of JIT |
| TQM program | None. Can't afford it | Emerging awareness of TQM as a necessity for JIT | Appreciation of close interdependence between JIT and TQM—both receiving equal emphasis | An equal partner with CIM and TQM in WCM |

| | | | | |
|---|---|---|---|---|
| Product and process design involvement | None—JIT an inventory control technique | Preliminary conceptual education. Maybe JIT can be applied to this area | Initial education and training in application of concurrent product and process engineering and designing for manufacturability | Parts counts reduced by 20%–30%. Production complexity reduced. Costs reduced. New product development lead time reduced |
| JIT education and training | Just starting at conceptual level | Companywide program started at all management levels. Detailed training starting on shop floor in pilot areas | Ongoing certification of current workers. Mandatory training of new workers | Ongoing at all levels |
| Standardization theme | Nonexistent | Emerging awareness of benefits and application to JIT | All workers encouraged to use standardized methods | Rigorous use of company standardized procedures and continuous attempts to improve them |
| Continuous improvement theme | Nonexistent | Starting to build on isolated initial success stories | Continuing to build momentum as recognition comes that old goals were limits. A little improvement every day is correct theme | Continual improvement the only way to meet increasingly stringent competitive conditions. "There is no magic wand!" |
| Scheduling system | Push-launch an order and drive it through | As an experiment, eliminate some "fat" out of some lead times | Make only what we need becomes a theme. Eliminate all system slack | Pull-driven by customer demand and service levels. Need to combine best of JIT&CIM (MRP) recognized. |

**Figure 10.4** (continued)

- Point of use storage;
- Level schedules;
- Mixed model scheduling;
- Standard containers;
- Zero defects;
- Quality at the source;
- Flexible manufacturing;
- Minimum bill of materials levels;
- Housekeeping;
- Line balancing; and
- 100% schedule attainment.

In the broad sense, JIT applies to all forms of manufacturing, job shop, process, and repetitive manufacturing. Under JIT, *waste elimination* should be pursued using a philosophy of high *respect for people*. Employee respect means involving the employee in the planning, design, operation, and control of the manufacturing system. It also means recognizing people based on participation, attitude change, knowledge gained, and results.

JIT also pursues *visibility, simplicity, flexibility, organization,* and *standardization.*

*Visibility* can be obtained by the following techniques: electronic boards for quick feedback, pull system with kanbans,[1] problem boards, standard containers, decentralized storage systems, dedicated areas for inventory, tools, etc.

*Simplicity* can be achieved with a pull system with kanbans, simple setup changes, certified processes, small lot sizes, leveled production, simple machines, simple material handling, multi-functional workers, teamwork, etc.

*Flexibility* can be achieved with short setup times, short production times, small lot sizes, kanban carts, flexible material handling equipment, multi-functional and flexible employees, mixed-model sequencing lines, etc.

*Organization* is required for setups, for cleanliness, for work areas, for the kanban system, for the storage areas, for tools, for teamwork activities, etc.

The *Standardization* of tools, equipment, pallets, methods, containers, boxes, materials, and processes is also pursued by JIT.

To eliminate waste, it is important to recognize it first. *Waste* can be defined as any resource that adds cost but does not aggregate value to the product. The most common sources of waste in an organization are equipment, inventories, space, time, labor, handling, transportation, and paperwork.

*Equipment* can be underutilized for many reasons. Some reasons are: poor scheduling approaches, inadequate maintenance programs, absence of feed-

---

[1]A kanban is a card or any signal used to request or authorize production of parts; it contains information on the part, the processes used, identification of the storage area for the part, and the number of parts to produce.

back mechanisms to report maintenance problems, inadequate spare parts inventory, operators not trained on the use of the equipment and on basic preventive maintenance, inefficient product design, long setup times and inefficient setup procedures, and large lot sizes.

*Inventories* for raw materials, components, parts, work-in-process, and finished products are usually wasted for the following reasons: large purchasing and production lot sizes, inappropriate facilities layout, inefficient material handling systems, bad quality, inefficient packaging systems, inadequate training for the workers, poor organization, too many suppliers, centralized warehouses, inefficient product design, lack of standardization, awkward storage and retrieval systems, and poor inventory control systems.

*Space* is another key resource where waste can be identified by excess inventory, inappropriate facilities layout, inadequate building design, unnecessary material handling equipment, inefficient storage systems and administrative policies, and inefficient product design.

*Time* can be a waste due to excessive waiting or delays (due to inefficient scheduling, machine downtimes, long repair times, bad quality, unreliable suppliers, poor facilities layout, inefficient material handling systems, deficient inventory control systems, large lot sizes, and long setup times).

*Labor* is wasted when the employee is assigned to produce unnecessary products, to move and store unnecessary inventory, to rework bad quality products, to wait during machine repairs, and to receive unneeded training. Labor is also wasted when the employee makes mistakes (lack of training, individual responsibility, or mistake-proofing mechanisms).

From the JIT requirements point of view, all the subsystems in the manufacturing system must be improved. For example, in an improved purchasing subsystem, products are purchased from fewer suppliers (located closer) in smaller lot sizes that are delivered frequently to the points of use. Products are transported in standardized, returnable, durable, stackable, and/or collapsible containers. Higher demands are placed on selected suppliers (using EDI or kanbans), but they need to agree to reduce price annually through continuous improvement (long-term contracts with blanket orders are prepared). The customer is willing to develop and eventually certify the supplier (certified suppliers deliver products directly to production areas—no receiving inspection). Suppliers organize themselves to share transportation costs, to have continuous communication during transportation, to consolidate deliveries, and to use side-loading trucks. In some cases, suppliers maintain consignment storage areas inside the customer facility to respond quickly to production requests.

JIT purchasing concepts impact building design and size, plant layout arrangement, unit load design, and material handling and storage equipment type, size, and design. Multiple receiving docks are placed around the facility, decentralized storage areas with simple storage systems are located close to the points of use, and simpler material handling equipment alternatives for lighter loads are used to transport unit loads from receiving docks to decentralized storage areas.

Therefore, there are many concepts and techniques related to the JIT production system that impact building design, plant layout, and the material handling system. As a result, the traditional procedure for the design of manufacturing facilities needs to be revised. JIT concepts that impact building design, plant layout, and the material handling system are examined as follows.

## Reduction of Inventories

One of the main objectives of the JIT production system is the reduction of inventories. Inventories can be reduced if products are produced, purchased, and delivered in small lots; if production schedule is leveled appropriately; if quality control procedures are improved; if production, material handling, and transportation equipment are maintained adequately; if products are "pulled" when needed and in the quantities needed; and so on. Therefore, if inventories can be reduced:

1.  Space requirements are reduced, justifying arranging the machines closer to each other and the construction of smaller buildings. Reducing the distances between machines, reduces the handling requirements.
2.  Smaller loads are moved and stored, justifying the use of material handling and storage equipment alternatives for smaller loads.
3.  Storage requirements are reduced, justifying the use of smaller and simpler storage systems and the reduction of material handling equipment to support the storage facilities.

Consequently, the building could be smaller, a better plant layout could be used, less handling and storage requirements could be needed, and the material handling and storage equipment alternatives could move and store smaller loads.

## Deliveries to Points of Use

If products are purchased and produced in smaller lots, they should be delivered to the points of use to avoid stockouts at the consuming processes. Products can be delivered to the points of use if the building has multiple receiving docks and if a decentralized storage policy is used. If products can be delivered to the points of use, the following could happen:

1.  If parts are coming from many suppliers, several receiving docks surrounding the plant could be required. Receiving docks need extra space for parking of trucks plus equipment for unloading, doors, and so on. Additionally, depending on the number of loads arriving at each receiving dock, in some cases, some storage equipment could be needed at each decentralized storage area. Therefore, if parts are delivered to the points of use, the building could require multiple receiving docks, storage and handling equipment could be needed at each receiving dock, and the parts could be moved shorter distance and less times (promoting a shorter production cycle and improving the inventory turnover, reducing the possibility of having bad quality because of excessive handling, and detecting bad quality parts quicker). The storage and handling equipment alternatives to support multiple receiving docks are usually simple and inexpensive (i.e., pallet racks, pallet jack, pallet truck, walkie stacker). Side loading trailer-trucks could also be used to avoid building expensive receiving docks.
2.  A decentralized storage policy could be required to support the multiple receiving docks and computerized or manual inventory control could be used. If

inventory control is manual (using cards or kanbans), a centralized kanban control system could be used to collect the kanbans and request more deliveries from the suppliers or a decentralized kanban control system could be used with returnable containers and/or withdrawal kanbans being returned to the supplier after every delivery.

3.  If plant layout rearrangements have been performed to support the JIT concepts, internal deliveries to the points of use could be carried out using material handling equipment alternatives for smaller loads and short distances. If the JIT delivery concept is applied without performing layout rearrangements, faster material handling equipment alternatives could be justified depending on the "pulling rate" of the consuming processes.

4.  If a truck is serving different receiving docks, it could be loaded based on the unloading sequence. Sideloading trucks could also be used. If the traffic around the plant is problematic, a receiving terminal could be used to sequence properly the deliveries.

## Quality at the Source

Every supplying process must regard the next consuming process as the ultimate customer and the consuming process must always be able to rely on receiving only good parts from its suppliers. Consequently, the transportation, material handling, and storage processes must deliver to the next process parts with the same level of quality that they received from the preceding process. To achieve the quality at the source concept, the following could be required:

1.  Proper packaging, stacking, and wrapping procedures for parts and boxes on pallets or containers.

2.  Efficient transportation, handling, and storage of parts.

3.  A production system that allows the worker to perform his operation without time pressure and that is supported by team work.

Therefore, if the quality at the source is implemented, better packaging, stacking, and wrapping procedures could be required; transportation, storage, and handling must be carefully done; and nonsynchronous assembly systems, sometimes with U-configurations, could be used. The assembly systems could be supported by the team approach concept and quick feedback procedures supported by electronic boards could be used to achieve line balancing.

## Better Communication, Line Balancing, and Multi-functional Workers

In many JIT manufacturing systems, U-shaped production lines are being used to promote better communication among workers, to use the multi-abilities of the worker allowing him to perform different operations, and to easily balance the production line using visual aids and the team approach. Most of these applications have been very successful and have incorporated "problem boards" to write down the problems that occur during the shift and to analyze them at the end of the shift and to generate solution methods to eliminate the future occurrence of these problems.

Layout arrangements with U-shaped lines minimize handling requirements (in most cases, it eliminates them) and supports the "pulling" procedure with kanbans. In some companies in which this layout arrangement is in operation, inventory buffers between processes of only one part or one container are used.

## Total Employee Involvement (TEI)

TEI is essential for CM implementation. The design, implementation, and operation of most of the JIT, TQM, and CIMS techniques are supported by teams and require employee empowerment, flexibility, and multifunctionality.

Empowerment is needed to reverse the traditional decision making and problem-solving approach (see Figure 10.5 [21]). We want to involve the experts from different departments to improve decision making of complicated issues or involve the experts from a specific department, area, or line to continuously improve their operations.

Three different types of teams are used to involve employees in decision making and problem solving. The *first team type* is usually involved in challenging TM strategies and is often composed of participants from several departments that work for a period of time, achieve drastic changes, and eventually dissolve (known as project specific or reengineering teams). The following list includes typical efforts that can be addressed by project-specific teams:

- Integrated product, process, scheduling, and facilities design;
- Comprehensive maintenance program;
- New plant layout and material handling system;
- Warehousing system redesign;
- Production strategies identification (push, pull, push-pull, MRP role, kanban);
- Reduction of setups;
- Supplier reselection, development, and certification;
- New cost management and performance measurement approaches for CM;
- New training, reward, and recognition systems for CM;
- Plant-wide safety, housekeeping, and organization program (visual management);
- Comprehensive quality assurance program;
- Organizational restructuring;
- Simplification of administrative processes; and
- Customer service improvement.

The *second team type* is involved in continuous improvement and is usually composed of participants from the same department, area, or line that get together (many times in a voluntary basis) to enhance their activities (i.e., communication, layout, paperwork, safety, quality). These teams are commonly known as quality or productivity circles, employee involvement teams, or continuous improvement teams. If the conditions are right, these teams could work permanently in continuous improvement activities.

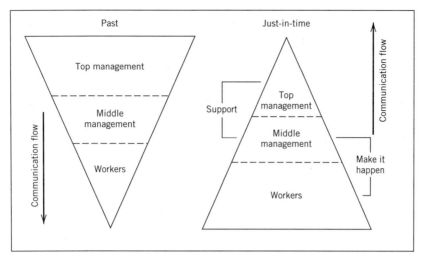

**Figure 10.5**   Decision making and problem solving.

The *third team type* is a combination or expansion of the two other team types. Participants of these teams (known as self-managing, self-control, or self-regulated teams) through education, training, and involvement have mastered knowledge and abilities in several operational, administrative, technical, and human related tasks. This means they can participate in redesign or continuous improvement activities and they also have the capability to get involved in the operation of different machines, and in quality assurance, basic maintenance, housekeeping and organization, material handling, machine setup, training, hiring, production scheduling and inventory control, performance measurement, and so on. The participants are usually from the same department, area, or line and could work together in these activities in a permanent basis.

Obviously, empowerment is not an easy task and requires attitude change, training, involvement, commitment, and support. As illustrated in Figure 10.6, an empowered individual needs to change from dependency to interdependency through development of operational, technical, administrative, and human-related skills and knowledge. On the other hand, managers need to change from "bosses" to leaders, facilitators, educators, technical advisors, and motivators. They need to stop being autocrats and develop themselves through training, education, and involvement to direct, coach, participate, and eventually delegate authority and responsibility.

Demand variability, cellular manufacturing, production leveling, mixed-model sequencing, teamwork, empowerment, and many other issues, concepts, and techniques require flexible and multifunctional employees. The worker may need to operate more than one machine at a time, may need to move continuously to other machining centers or assembly lines, may need to participate in teamwork activities as a member or leader of the team, and/or may need to participate in many other support activities.

The manager and supervisor, on the other hand, may lead a team or group of teams, may coordinate the continuous improvement process, and may participate as

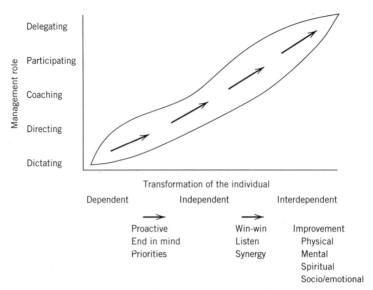

**Figure 10.6**   Empowerment roadmap.

facilitator, technical expert, project manager, technical advisor, instructor, and motivator.

Consequently, facilities are currently designed, redesigned, improved, and managed by teams. Additionally, teams not only require space for meetings and visual management but also they are changing space, storage, and handling requirements and activity relationships through the concepts they are implementing. Teams can reduce lot sizes, decentralize storage areas, reduce the number of components and parts of a product, simplify flow, change the organizational structure, reduce the number of suppliers, and decentralize the organization by product type. Their activities impact building size and configuration, layout arrangement, storage and handling system, and unit load design.

## Quality Practices in Traditional Manufacturing

In TM, the quality department plays the role of a policeman or judge. Quality inspectors sample units at different stages of the production process, especially after these have been completed. Frequently, quality and production employees maintain an adversarial relationship. Defective rates are monitored against pre-established goals, with no concern for customer quality expectations. The emphasis is on complying with established in-house procedures.

Inspection of incoming raw material is another major activity supported by the quality department. Defective raw materials are frequently encountered, since the main driver for supplier selection is part cost. Vendors with a lower price are favored, and issues such as quality and delivery are less important. As a contingency against unreliable suppliers, the purchasing department might buy the same part from various vendors. In such a scenario, additional space is needed for storing raw materials prior to inspection, providing rework areas for defective products, and segregating part numbers that are received from different vendors.

Since the facilities layout is process oriented and setup times are typically large, the manufacturing strategy calls for running large lots of each part number. Such lots move fairly slowly through the different processes, and when completed go to final inspection. If a mistake is made at some operation upstream, there is a good chance for a significant percentage of the lot to have the problem. Once again, such a scenario has a significant impact in space requirements as well as material handling efforts.

In-house quality monitoring includes inspection of the final product, of subassemblies, and raw material. Sophisticated products might require a large investment in test equipment. For example, in printed circuit board (PCB) assembly, tests are performed at the component level (one bad or marginal component might prevent the PCB from working as expected), at the PCB level (to verify that the components are interacting as expected), and afterwards at the system level (to verify that the PCB works fine as part of a computer system). Besides the equipment cost, such inspection strategies will require engineering and technical support, and might demand a significant area in the production floor.

The frequent occurrence of defects requires a fairly visible rework activity, in which components and sometimes complete products need to be scrapped. The rework effort also requires engineering and technical support, as well as space in the production floor.

There is room for improving how many quality departments manage their efforts. Ideally, we want each and every employee to look after quality concerns. The department can move from a policeman to a more participative role. Customer expectations must be understood, and opportunities for customer delight need to be highlighted. Could new products be simplified and process quality improved to require less inspection and test? In addition, emphasis can be placed in supplier quality. If the criterion for vendor selection is modified, the number of vendors can be reduced, and long-term partnerships can be initiated. Basically, all these changes are a minimal requirement for manufacturing companies interested in the global market. The new quality priorities will now be discussed.

## TQM in the CM Scenario

TQM is based on the well articulated concepts pioneered by such visionaries as Deming, Juran, and Crosby, and employs not only the traditional statistically based problem solving techniques, but the more modern approaches of Ishikawa, Taguchi, and others. The operative concept of TQM is "continuous process improvement" involving everyone in the organization, managers and workers alike, in a totally integrated effort toward improving performance of every process at every level. This improved performance is directed toward satisfying such cross-functional goals as quality, cost, schedule, manpower development, and product development. The ultimate focus of every process improvement is increased customer/user satisfaction. Not incidentally, the customer/user's views are actually sought in developing the process improvement methods.

Each organization tailors TQM to their unique needs and environment. Motorola states, "our fundamental objective is total customer satisfaction." Toyota defines it as "building the very best and giving the customer what he wants." Whatever label is applied to a quality management system, all practitioners emphasize common themes such as:

1.   Involvement of all functions,

2.   Involvement of all employees,

3.   Strong customer orientation, and

4.   A philosophy of continuous improvement.

These elements are supported by active vision setting, involvement and commitment of top executives, managerial leadership, use of tools, extensive training, and an environment that fosters teamwork and cultural transformation.

Since customer satisfaction and minimal cost are a must in global competition, we need to be lean in our manufacturing efforts. In this scenario, the Quality Department consists of a small number of employees, who participate in cross-functional teams throughout the organization. They serve as facilitators, providing education and training to all employees so that each department takes charge of continuous improvement, which becomes a way of life in manufacturing as well as administrative and support processes.

Continuous improvement is motivated by viewing the receivers of our work as our customers. *Since external and internal customers* expect perfection, we must improve our processes to comply with and, when possible, exceed their expectations. For example, in the production line, each step plays the role of customer (of the preceding step) and supplier (of the next step). In manufacturing, the customer chain includes (1) vendors, (2) purchasing (3) production, (4) marketing, and (5) the external customer. Other support areas in the organization have customers in this chain. For example, engineering supports purchasing (materials engineering), production (process engineering), and marketing (customer support). Figure 10.7 summarizes this supplier–customer relationship.

The in-house emphasis in manufacturing shifts from reactive (inspection and test) to proactive (quality at source). In labor-intensive operations, *quality at source* is achieved through self- and successive-inspection. *Self-inspection* is performed when an operator verifies if the tasks he or she just performed are correct. *Successive inspection* involves a verification by the next operator of the tasks performed by the preceding operator.

In machine-intensive operations, quality at source is achieved through *in-process verification,* a close monitoring of processes while the product is being built. Two approaches for in-process verification are: mistake proof devices (or "poka-yokes") and source control. *Mistake-proof devices* work in various ways. Some prevent the operator from skipping a process step, others check for missing or faulty materials, while others check if the part just completed is within tolerance. In this way, defects are detected on their first occurrence and defect rates are minimized.

*Source control* is gaining acceptance as a strategy for achieving zero defects. It emphasizes the monitoring and control of key machine parameters, which when out of tolerance, are responsible for the occurrence of defects. Artificial intelligence tools have been developed to improve sensitivity during monitoring. Both mistake-proof-

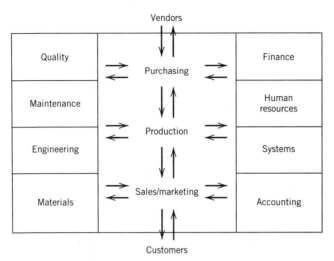

**Figure 10.7**  Customer chain.

ing and source control require stopping the line as soon as a problem arises. The root cause(s) for the problem must be identified and completely resolved.

*Design of Experiments (DOE)* provides efficient procedures for identifying the key process parameters to be monitored through source control techniques. The emphasis on preventing defects through DOE and in-process verification should allow us to simplify inspection and test efforts on finished products.

To minimize the need for incoming inspection, companies work at reducing their vendor base, strengthening ties with those with better performance. *A supplier certification program* provides a formal methodology for monitoring supplier performance. Typical areas for monitoring vendor performance are: technology (processes used), quality results, responsiveness (to changes in customer requirements), on-time delivery performance, product cost, and business stability (financial health of the supplier). Parts coming from certified vendors are sent to the stockroom or directly to the line, without any inspection step at incoming.

A reduction in quality problems helps shorten the manufacturing lead-time. Obviously, much better quality implies happier customers, and an opportunity for increasing business in the future. Since less test and inspection equipment is required, capital can be reallocated for acquiring more "value-adding" equipment (for machining, assembly, etc.). The space needed for new business is obtained from the reduction of space requirements for test and inspection, work-in-process, raw material inspection, and rework activities.

In product design, the effort can be initiated by understanding the customer "wants." *Quality Function Deployment (QFD)* is being used by leading companies to carefully look at product design, part/component needs, manufacturing planning, and production activities. Quality can be built into new products by emphasizing ease of manufacturability, assembly, and test. New products can be developed in a team environment with participants from product and process engineering, production, purchasing, and cost accounting. *Concurrent engineering* practices help teams achieve cost and time goals during new product development efforts.

Many companies interested in the global market have chartered their Quality Department to champion plant-wide efforts looking for recognition as quality leaders. The *Malcolm Baldrige National Quality Award (United States) and the Deming Prize (Japan)* are the best known awards being presented to companies who demonstrate excellence in quality. In addition, companies that want to sell their products in the European Community are improving their quality systems to comply with the international standards in the *ISO 9000* series.

The efforts described as characteristic of a world-class manufacturing environment will allow breakthroughs in the reduction of the *cost of quality*. Figure 10.8 summarizes the changes in quality cost from the traditional to the world class scenario. The increase in prevention costs is offset by reductions in warranty expenses, field failures, rework activities, scrapped parts, and test and inspection efforts.

Two main areas in which manufacturing companies must excel with customers are on-time delivery and quality. *Customer satisfaction* is continuously improved in these two areas by correlating the customer results with the in-house performance. Figure 10.9 suggests the process to follow. Namely, if customer satisfaction is measured by delivery and quality results, production impediments at final assembly (closest production stage to the external customer) must be identified. These results must then be related to problems in subassembly or feeding areas, and possibly to the external suppliers. With the right education and training, line employees can be organized into teams to tackle these improvement opportunities.

*The Deming circle* (attributed to W. Edwards Deming) is a well-known continuous process improvement methodology. It consists of four steps: plan (identify process changes and set goals), do (execute), check (analyze results), and act (identify corrective actions). Figure 10.10 displays how the Deming circle can be put into practice. In the example, the hours lost in a production line due to machine breakdowns are being monitored. The upper-left graph presents a trend chart with the hours lost per period (daily, weekly, or monthly). Note that, on the trend chart, a goal has been established.

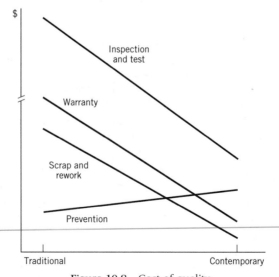

**Figure 10.8**   Cost of quality.

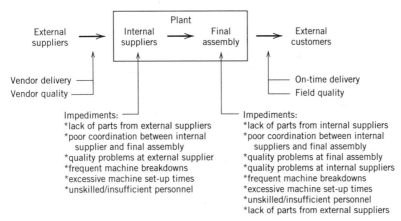

Figure 10.9   Relating customer with in-house performance.

The upper-right graph presents a Pareto chart of the hours lost by the most problematic machines. Another Pareto chart is presented in the lower left corner, which focuses on the actual problems (or "root" causes) and their weight found at the most critical machine. In the lower-right corner, a corrective action plan is developed to tackle the specific issues identified in the previous Pareto. If we are successful in solving the critical problems, we will achieve and possibly exceed our goal. To stress continuous improvement, we can then reset the goal, as the moving goal in the trend chart suggests.

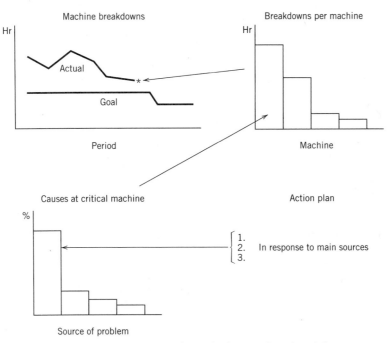

Figure 10.10   Deming circle applied to machine breakdowns.

| Attribute | Stage 1: Traditional "Quality Control" | Stage 2: Beginning TQC Awakening | Stage 3: Intermediate TQM Enlightenment | Stage 4: Full TQM Conviction |
|---|---|---|---|---|
| Quality costs as a % of sales | Perceived 3%–5%. Probable 15–25%. All quality costs really not measured | Perceived 10–15%. Probable 15–25%. Need now recognized to measure all four costs—prevention, appraisal, internal, and external | Perceived 15%–20%. Probable 15%–20% | Actual less than or equal to 1% |
| Management attitude and involvement | More quality equals higher cost. Our QC people do a good job of inspection and quality control—see them if you have any questions on quality—I don't have time to get involved | Growing awareness of the real cost of quality and the customers' demand for higher quality. Exposure to concepts of TQM | Awareness of strategic importance of high quality and the fact that quality is a management of responsibility. Working knowledge of all TQM tools | CEO responsible for quality. Quality built into everyone's performance measures. Only way to be low-cost manufacturer is to be high-quality manufacturer. Quality is an attitude. Quality is everyone's job! Awareness of interrelationship of TQM, CIM, JIT |
| Worker attitude and involvement | Management doesn't care so why should we? Higher quality means more inspectors. Meet the schedule is the only concern. What can we do about quality? | Maybe quality is broader than we thought. We might be able to obtain higher quality if given the tools | Management does care about quality! Better quality means a better work environment—less fire fighting, more predictable, less hassle. | Quality is our first job! We can stop the process to avoid poor quality. More pride, esprit de corps |
| Supplier involvement | Limited incoming sampling inspection. Material review boards | TQM education and training program for suppliers | Supplier certification for quality permits inspectionless receiving | Long-term partnerships (sole source) with fewer high quality suppliers |

| | Stage 1 | Stage 2 | Stage 3 | Stage 4 |
|---|---|---|---|---|
| Involvement with customer | Little or none—sales does that! How does the customer know what good quality is? | More complaint feedback procedures setup and made easier—this information is fed back to design, manufacturing, and top management | Customers have more influence on product design. Our customers really appreciate quality! | The customer is the sole arbiter of quality |
| Preventive maintenance | Nonexistent | PM education started. Worker involvement started | PM program firmly established. Machines run at or below rated speeds. Workers responsible for light maintenance | TPM, the cornerstone of TQM and JIT |
| Product and process design involvement and knowledge of TQM | Little. The inspectors aren't tough enough. Inspectors lack education in SQC, Taguchi methods, quality function deployment. Walls between product and process design | Design plays a key role in meeting quality goals | Must design quality into the product and process. Teamwork between product and process design | Full understanding of manufacturing as a science |
| Definition of quality | Quality = conformance to specification | Initial exposure to Garvin's broad definition—8 attributes of quality—and to the idea that each action has a customer | Quality's broad definition accepted. Concept of the customer being the next person in the process accepted | Quality defined broadly, and above all as total satisfaction of each customer, internally and externally |

**Figure 10.11**  Stages in progression to TQM.

| | | | |
|---|---|---|---|
| Quality education and training program | Nonexistent | Initial TQM education and awareness-building. SQC training in pilot areas | All employees trained in SQC. Further TQM education program established for design for quality, Taguchi methods, quality function deployment | Ongoing selection of TQM education and training programs mandatory for current and all new employees |
| Quality organization | Has an internal focus. Not a part of top management. Low in organization structure. Disliked by production workers. Not enough people to improve quality. "A necessary evil" | Emerging appreciation of true role of quality organization | Quality's external role with customers and suppliers strengthened. Quality organization now seen as a provider of education and tools to allow everyone to perform their quality job | Quality function represented by a corporate office reporting to CEO. Quality head spends a lot of time with customers and design function |
| Role of technology in attaining quality | Limited to mechanical gauges, blocks, etc. for inspection | Initial use of computers and software-driven process control/quality equipment in pilot areas. Use of software to capture quality data for analysis | Use of CAD/CAE technology in design. Use of CIM technology and software to compare manufacturing process results with product and process design data | Technology (CIM) plays an inseparable role with people in promoting, measuring, and ensuring quality |

**Figure 10.11** (continued)

The Deming circle and a graph similar to Figure 10.10 can be used in any process for monitoring, analysis, and corrective action. For example, if a production line is suffering from an inadequate service by the material handling vehicles, we can develop a trend chart for the weekly time spent waiting for vehicles. An initial Pareto can be prepared to identify in which locations waiting times are longer. Then, the second Pareto can pinpoint from where the vehicles are coming, identifying the "root causes" for the slow vehicle response at this specific location. And last the action plan will propose approaches for solving the issues of vehicles coming to the troubled location.

In conclusion, the application of the quality approaches discussed in this section will help reduce material handling complexity, will allow for significant space savings, will free valuable resources (e.g., engineers and technicians) for new business, will make manufacturing activities more predictable, will shorten manufacturing lead-times, will have a major impact in the cost of quality, and more importantly, will delight our customers. This opens the possibility for more business, increased market share, and recognition as a "world-class" competitor. Stages in progression to TQM are presented in Figure 10.11 [35].

# Computer Integrated Manufacturing Systems (CIMS)

The term computer integrated manufacturing systems (CIMS) was coined by Joseph Harrington in his book by the same title, written in 1973 [35]. CIMS brings people, technology, products, and processes together into a single integrated system.

CIMS is a concept for integrating all components involved in the design, planning, production, and control of a product as shown in Figure 10.12 [35]. Note that the integrator of product and process design, manufacturing planning and control, and production process is information technology.

CIMS has a great potential to reduce the economic waste resulting from huge quantities of work in process in conventional factories and very low percentage utilization of capital equipment in those factories. CIMS also provides an opportunity for faster identification of quality problems, reliable equipment for consistent manufacturing processes, reliable software and tools, and enormous information related to manufacturing operations. These features of CIMS enable us to improve and control the plant and manufacture good quality products.

The four CIM functional areas and CIMS and facilities design are described in the following sections.

## Functional Area 1: Product and Process Design

Product and process design (PPD) encompasses four major concepts. Obviously, product design and process design are two of these concepts. Group technology and engineering control are the two other areas that apply to both product and process design as illustrated in Figure 10.13.

The two major aspects of product design are computer aided design (CAD) and computer aided engineering (CAE). CAD captures the geometry of the part in an electronic (computer-based) engineering database. This geometric data serve as the basis for most manufacturing activities. Since every product has to start with design, drawing, and part details, CAD helps in every phase of these operations to form an

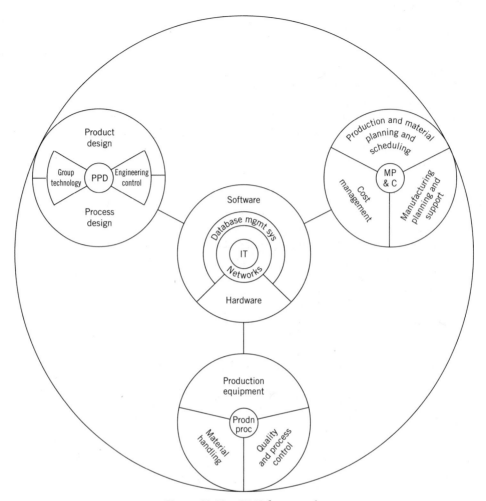

**Figure 10.12**  CIMS framework.

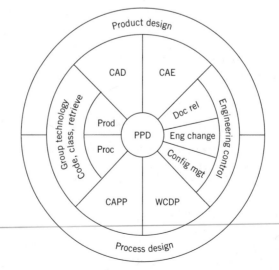

**Figure 10.13**  CIM product and process design.

interaction between designers and production engineers during the design stage. Further integration of the CAD data to production is achieved through computer aided manufacturing (CAM).

CAE uses analytical software to analyze part designs. This could involve the use of a finite-element modeling package to calculate and show the stress or strain in a mechanical part, or it could involve the use of simulation software to verify the logic and timing functions in a semiconductor chip. CAE also helps to ensure compliance with professional standards and codes, and helps to make cost analysis.

Process design uses computer aided process planning (CAPP) and work cell device programming (WCDP). CAPP is used to plan each operation a part will go through on the shop floor. A well-designed CAPP system helps manufacturers by making better use of capital equipment and reducing manufacturing cost and lead-times. In a full CAM environment, feedback from a flexible manufacturing system (FMS) can be sent directly to the CAPP system to further enhance its performance. Therefore, CAPP is an important interface between CAD and CAM together with group technology.

WCDP is also referred to as computer numerically controlled (CNC) or numer-ically controlled (NC) part programming. Devices in a work cell today could be a CNC machine, a robot, a vision system, or a coordinate measuring machine.

Group technology (GT) and engineering control are the two other areas that apply to both product and process design. GT is a technique to group parts with similar geometric and processing requirements to improve production efficiency. GT is the glue that can provide the integration of CAD/CAM into CIM, because it provides a common vocabulary for users and facilitates the creation of a database, which includes design and manufacturing attributes, processes, tooling, and machine ca-pabilities. A GT based classification and coding system also provides the means for extracting data in usable formats for a variety of different users.

Engineering control includes (1) documentation and release of the initial part design, (2) engineering change control for future modifications to existing part de-signs, and (3) configuration management to document the components of the final product that goes to the customer.

## Functional Area 2: Manufacturing Planning and Control

Manufacturing planning and control includes three major areas as shown in Figure 10.14. The first area is production and material planning and scheduling, the second is cost management systems, and the third is manufacturing planning and support.

Manufacturing resource planning (MRP II), distribution resource planning (DRP), and kanban are the types of systems that are used in the production and material planning and scheduling area. Kanban represents the information system side of JIT production. The three tools support the purchasing function by means of MRP II software.

Artificial intelligence (AI), expert systems (ES), optimization packages, and sim-ulation/animation are some of the tools used in the manufacturing planning and control area. Cost management and preventive maintenance systems are central to any manufacturing function.

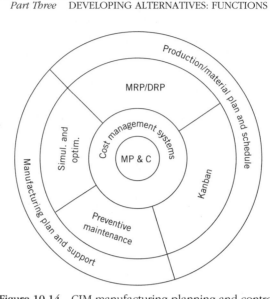

**Figure 10.14**   CIM manufacturing planning and control.

Simulation and animation are widely used tools in a CIMS environment. Since CIMS include advanced and expensive systems such as robots, AGVS, AS/RS, and FMS, it becomes more and more important to be able to simulate their behavior. For example, the ability to react quickly to market changes can be improved through the use of simulation. When simulation/animation is integrated with AI/ES, it becomes an even more powerful tool for CIMS applications. For example, one can use the rules of the expert system for selection of material handling equipment alternatives and then use simulation/animation for design and control.

## *Functional Area 3: The Production Process*

The production process includes three major areas: production equipment, material handling equipment, and quality and process control devices. The center circle in Figure 10.15 includes the concepts of flexible manufacturing systems (FMS), flexible assembly systems, and flexible packaging systems, which are crucial to CIM based manufacturing.

A FMS is a reprogrammable manufacturing system capable of producing a variety of products automatically. The incorporation of robots and numerical control machines provides reprogramming capabilities at the machine level with minimum setup time. FMS are the basic building blocks for most of the CIM systems. The degree of flexibility of the CIM system is a strategic issue that should be decided in the planning stage of the facility. After the decision on flexibility, the design consideration of layout and building should be addressed by using appropriate tools [58].

Production equipment includes robots, machines tools driven by NC, CNC, or DNC, and group technology cells. Quality and process control devices include programmable logic controllers and microprocessors, statistical quality control and re-

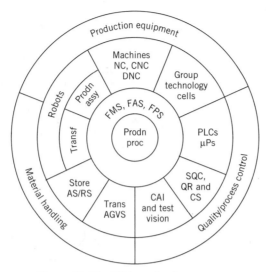

**Figure 10.15** CIM production process.

porting systems, and computer-aided inspection and test equipment (i.e., vision systems).

On the material handling side, robots can be used for transferring materials. Automated storage and retrieval systems (AS/RS), automated guided vehicle systems (AGVS), intelligent monorails, automated palletizers, and many other material handling equipment automated alternatives can be used to support CIMS.

In a typical company, automation starts with the creation of automated islands. Establishing a CIM system requires integration of these islands of automation. Material handling systems should deliver and remove work pieces from machine tools, and eliminate manual handling to reduce work in process between workstations. Moreover, material handling should be under computer control to integrate the whole system and interface the other components of the CIMS. In this sense, automated material handling (AMH) plays the most important role for this kind of integration. AMH provides higher degree of flexibility, reliability, and faster response and makes possible a computer controlled material movement between work stations.

Although certain types of material handling equipment alternatives such as conventional forklift trucks, towline, roller, and belt conveyors do not look as CIMS material handling equipment alternatives, with the increased use of computers, sensors, PLC's, servo mechanisms, and electronic control, it is possible to convert them to automated material handling systems. For example, roller and belt conveyors can be automated if we incorporate computer control by using sensors with or without PLC's. In a typical CIMS environment, conveyors provide movement over a short fixed path within the manufacturing cells, while the AGVS provide interdepartmental transport. All these activities are controlled by the computer by using database management systems and other necessary software. Since AMH provides less people and less handling, it reduces space requirements, but it may require additional room for control and loading/unloading operations [43], [86].

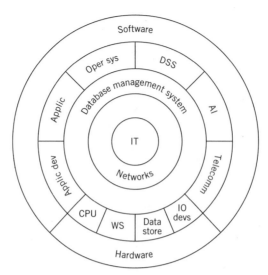

**Figure 10.16**   CIM information technology.

## *Functional Area 4: Information Technology*

In a CIM system, data can be gathered, defined, stored, manipulated, and communicated to form the information by using computers which allow us to integrate the entire manufacturing activities as shown in Figure 10.16. All these functions can be achieved by using a common database management system (DBMS) together with other support software such as application development tools, program generators, and programming productivity devices. CAD, MRP II, and accounting or finance packages are examples of application software. AI/ES, decision support systems (DSS), and simulation/animation are also important software packages in the information technology area.

Operating systems such as DOS/VSE, OS/2, VM, MVS/XA, and UNIX are main components of the information system. Moreover, local area networks or wide area networks are used for communication needs for local and global purposes. On the other hand, necessary hardware consists of CPU, workstations, data storage devices, and I/O devices.

## *CIMS and Facilities Planning*

The integration of the manufacturing system under CIMS concepts has many important effects on the facilities planning process. Since a full CIMS provides less inventory, less paperwork, better quality, shorter production lead-times, faster design processes, less personnel, fewer offices, and reliable equipment, it requires less space, less handling, less storage, and smaller buildings. Stages in progression to CIM are included in Figure 10.17 [35].

| Attribute | Stage 1: Manual | Stage 2: Some Computerization and Beginning Coordination | Stage 3: Much Computerization and Control Some Integration | Stage 4: Full Integration Maturity and Controlled Refinement |
|---|---|---|---|---|
| Computing environment | Batch. Little timely operations feedback. Paper reports only | Some on-line inquiry. Weekly batch update | On-line inquiry. Daily batch update | On-line interactive. "Real time" update |
| Computing equipment | None | A proliferation of equipment, languages, operating systems, protocols | Emerging corporate standards for hardware, application software, database management systems, data dictionaries, and telecommunication networks | Required use of corporate systems. Controlled innovation and enhancement |
| Corporate data telecommunications networks | None. All communication between plants, headquarters, offices, customers, and suppliers by voice or paper | Limited computer communications between plants and corporate headquarters. Alphanumeric data only. Low-speed public phone lines | More and faster computer communications between plants, headquarters, offices. Emerging supplier/customer electronic links. Still primarily alphanumeric data. Lease, private, high-speed phone lines | Full satellite- or fiber optic-based worldwide telecommunications between all plants, headquarters, offices, suppliers, and customers. Both alphanumeric and geometric data |
| Data processing capability | None | Corporate mainframe. Remote job entry from dumb terminal(s) in each plant | Hierarchical dispersion of intelligent processing power from mainframes to micros in one or more plants | Geographic and hierarchical dispersion of intelligent processing power to all facilities |
| Database management systems | None | Many flat files, one or more DBMS, often no data dictionary, much redundant data. No data "ownership" | Several DBMS, emerging data dictionary use, emerging use of relational DBMS. Emerging data "ownership" | Several DBMS but one logical integrated data base design. Use of standardized data dictionary |

**Figure 10.17** Stages in progression to CIM.

479

| | | | | |
|---|---|---|---|---|
| CAD/CAE system use | None. All geometric part data on the drawing | One or more turnkey CAD systems used primarily for drafting. Part drawing still master part definition. Not integrated with business systems | Emerging CAD/CAE systems standardization. Fewer CAD/CAE systems used in design and analysis, work cell device programming. Electronic geometric part description now master part definition integration with business systems | One to three corporate standardized CAD/CAE systems. Fully integrated for design and work cell device programming, electronically linked to MRP (bill of material and process/routing) and other business systems |
| User attitude toward computing | Apprehensive, little knowledge or concern. Unaware of benefits | Systems not "user friendly." Computers in MIS run by MIS people who are different. Mistrust of MIS. MIS drives computerization. Emerging awareness of benefits | Systems "user friendly." Users aware of benefits and competitive advantage offered by information technology. Users drive computerization | Information technology an essential and transparent part of daily business life |
| Factory local area networks | None | Shop floor data collection. Some CNC-controlled production equipment not connected | Move to DNC networks. Experiment with work cell integration via MAP or proprietary network | Fully integrated factory LAN using latest MAP specification or proprietary network for scheduling, quality, process control, preventive maintenance, material movement |

**Figure 10.17**  (continued)

# *10.3* A METHODOLOGY FOR CM IMPLEMENTATION

A methodology for CM implementation is needed when new or existing facilities need to be designed or redesigned, respectively. In any case, the owners of the business are trying to increase profits exceeding customer expectations and optimizing company's resources.

Any methodology for CM should answer the following questions:

1. *What* do you want to achieve?
2. *How much* and *when* do you want to achieve it?
3. *How* do you plan to achieve it?

What to achieve is a function of long-term objectives, customer needs, competition analysis, and current company's situation. The answer to this question should provide general and specific objectives and goals for the company. *Benchmarking, quality function deployment,* and *hoshin* planning are some of the techniques that top managers can use to identify what the competition is doing, what the customer wants, and which objectives need to be deployed through the organization (see Figure 10.18).

The establishment of goals for each objective (how much and when to achieve the objective) is critical to measure progress and direction of the effort. The determination and deployment of objectives and goals is the responsibility of top management and is one of the most important factors of success in the implementation of a CM strategy.

Once objectives and goals have been defined and deployed, the next important question is HOW to achieve them in an organized and integrated way. The seven management and planning tools explained in Chapter 3 are often used to translate objectives and goals to specific actions, to create teams to work on these actions, and to develop implementation plans for each team. This is achieved using the following steps:

1. Identification and grouping of important opportunity areas impeding the achievement of objectives and goals (affinity diagram).
2. Identification of critical opportunity areas (interrelationship digraph).
3. Identification of specific actions to solve each critical opportunity area (tree diagram).
4. Evaluation and selection of best specific actions (prioritization matrix).
5. Grouping of best specific actions and creation of project specific teams to work on each group of actions (affinity diagram).
6. The members of each team consider the assigned actions as their objectives. The teams repeat steps 1 to 4 to identify best specific actions to implement.
7. Each team develops an implementation plan (activity network diagram).

This planning phase of the CM implementation methodology is vital and includes identification of objectives and goals and specific actions and teams to achieve them. It also includes the development of implementation plans for each team.

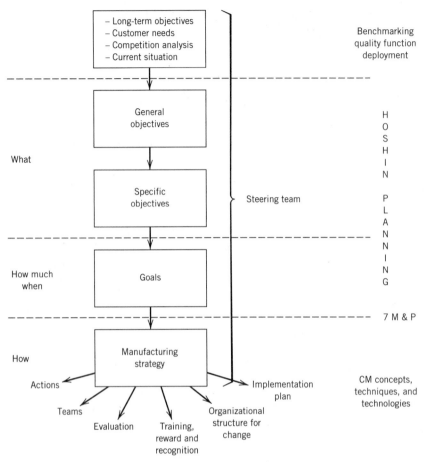

**Figure 10.18**   Business and manufacturing strategy.

The execution phase is as important as the planning phase and includes an organization structure for the team activities, guidance team meetings, education and training, recognition, and evaluation. This is explained as follows:

1.   *Organization structure for team activities.* It is created to provide leadership, coordination, education, training, evaluation, and continuous follow-up to team activities. As illustrated in Figure 10.19, the organizational structure for change identifies the CM implementation leader(s), coordinator(s), production area or product family owner(s), support area owner, and teams (project specific, continuous improvement, and self-managing teams). This organizational structure for change is used in parallel to the functional organizational structure to support the design, redesign, and continuous improvement process.

      As indicated before, CM can be implemented in phases and this is illustrated in Figure 10.19. Some companies start implementation in a specific area of the plant or in a specific product line and then they continue the implementation to the rest of the organization. On the other hand, some companies start implementation in the production area and then they expand the implementation to administrative processes.

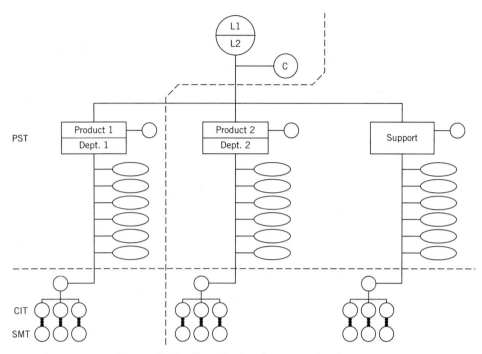

**Figure 10.19** Organizational structure for change.

In any case, most companies start creating project specific teams to perform drastic changes in the organization and challenge the traditional system and then they involve everybody else in continuous improvement teams that eventually become self-managing teams. Project specific teams eliminate many traditional roles (i.e., inspectors, material handlers, dispatchers) and self-managing teams require drastic changes to the traditional organizational structure (i.e., facilitators instead of supervisors, decentralization of technical and support functions, and reduction of organizational levels).

2. *Guidance team.* It is integrated by CM implementation leader(s), coordinator(s), owner(s), project specific team leaders, and facilitator(s). They meet once per week to review status, identify needs, and evaluate team activities.

3. *Education, training, recognition, and evaluation.* Typically, two project specific teams to support these team activities are created. One of them concentrates on education, training, and recognition and the other one on evaluation.

The education, training, and recognition team identifies training needs and satisfy them in a "pull" basis through internal or external instructors (training is delivered when needed, to the people needed, and in the quantities needed). This team also identifies recognition programs to continuously motivate team participants.

The evaluation team works with the other teams to identify performance measures that they can use to evaluate the impact of their actions on company's objectives and goals (see Figure 10.20). They also develop a system to generate the key performance measures and provide performance feedback to the teams

| | | | Objectives and goals | | | | | | | | |
| | | | Objective A | | | Objective B | | | Objective C | | |
| | | | I | A | G | I | A | G | I | A | G |
|---|---|---|---|---|---|---|---|---|---|---|---|
| Area 1 | Team 1-1 | Action 1<br>2<br>3<br>4 | | | | | | | | | |
| | | Metric 1 | | | | | | | | | |
| | | Metric 2 | | | | | | | | | |
| | | | | | | | | | | | |
| Area 2 | Team 2-1 | | | | | | | | | | |
| | | | | | | | | | | | |
| Support | Team 3-1 | | | | | | | | | | |
| | | | | | | | | | | | |

**Figure 10.20** Evaluation of contemporary manufacturing.

I: Initial   A: Actual   G: Goal

so they can redirect their actions. The work of this team is very important because it is continuously focusing team activities on achievement of company's objectives and goals. They also document the CM effort.

In the CM implementation methodology, there are several project specific teams that impact facilities design, namely, the layout and material handling systems, kanban, concurrent engineering, setup time reduction, total productive maintenance, and visual management teams. The typical roles of these teams are explained as follows:

1. *Layout and material handling system team.* This team uses the facilities design process and CM concepts and techniques to generate layout, handling, storage, unit load, and building configuration alternatives and select the best combination. The team is usually integrated with personnel from production, engineering, quality, human resources, maintenance, and materials.

   Some of the CM concepts that need to be considered by this team are: small lot purchasing and production, multiple receiving docks, deliveries to the points of use, decentralized storage areas, kanbans, decentralized functions, open office layouts, focused factories, manufacturing cells, visual management, mixed-model sequencing, and no receiving inspection. The members of this team interface with other teams to identify flow, space, and activity relationship requirements (i.e., kanban, visual management, setup time reduction, total productive maintenance, and concurrent engineering teams).

2. *Kanban team.* This team identifies products to be "pulled" between processes. Additionally, the team determines number of kanbans, kanban quantities, kanban types, kanban boxes, and kanban boards. The members of this team also develop the kanban implementation plan, train the users, and coordinate the kanban system. They need to interface with the layout and material handling system team to consider material handling alternatives, replenishment procedures, decentralized storage areas, and unit loads for the kanban products.

3. *Concurrent engineering team.* This team is in charge of designing new products and redesigning existing ones in shorter periods of time and using an integrated design approach. This team can potentially reduce the number of components in a product, standardize materials and components, design quality into the product, simplify fabrication and assembly processes, and so on. This impacts dramatically space, flow, and activity relationships requirements. For this reason, the concurrent engineering team should work together with the layout and material handling system team to design the best facility, layout, unit loads, and material handling system for new and/or redesigned products.

4. *Setup time reduction team.* This team participates in setup time reduction efforts for critical machines. Typical changes recommended by this team impact: material handling alternatives to move dies, molds, tools, and materials; storage areas for dies, fixtures, molds, and tools; quality assurance approach to verify quality of first parts run; location of auxiliary equipment and materials; tool management approaches; and fastening and adjustment methods. Obviously this impacts space, flow, and activity relationships requirements and the layout and material handling system team should be aware of the setup time reduction team activities.

5.    *Total productive maintenance team.* This team has a tremendous role in CM implementation. Their main objective is to increase machine uptime and product quality. Potentially they can be involved in the implementation of effective corrective, preventive, predictive, and autonomous maintenance programs, setup time reduction, tool management, visual management and housekeeping, and spare parts inventory control. However, some of these activities are divided and assigned to other teams (setup time reduction, visual management, and kanban teams). The implementation of total productive maintenance programs impact material handling and storage alternatives to move and store tools, preventive maintenance materials, testing equipment, and spare parts. Autonomous maintenance programs also impact workstation layout because we need to include space for materials, spare parts, and tools that the operator will need in this new role. All these issues need to be considered by the layout and material handling system team.

6.    *Visual management team.* As described in Chapter 4, visual management approaches can also impact dramatically the effort of the layout and material handling system team. The visual management team is in charge of identification, housekeeping, and organization; visual documentation; visual production, maintenance, inventory, and quality control; performance measurement; and progress status. This will require the efficient use of walls and aisles and dedicated areas for materials, dies, housekeeping and maintenance tools, team meetings, and computer terminals.

# *10.4*  FACILITIES PLANNING TRENDS

Plants are now being designed to provide the smoothest possible flow of materials, achieve flexibility (to adapt to proliferating product lines and volatile market demands), improve quality, increase productivity and space utilization, and simultaneously reduce facilities and operating costs. Based on the concepts and techniques recommended by CM and CM implementations that have been reported in the literature, the following can be considered trends in facilities planning.

1.    Buildings with more than one receiving dock and smaller in size (less space requirements for staging, storage, warehousing, offices, and manufacturing primarily because of JIT purchasing, small lot production, deliveries to the points of use, decentralized storage areas, better production and inventory control, visual management, manufacturing cells, EDI, focused factories, open layout offices, decentralized functions, and simplified organizational structures). Multiple-receiving docks require efficient loading/unloading of side-loading trucks when used.

2.    Smaller centralized storage areas and more decentralized storage areas (supermarkets) for smaller and lighter loads.

3.    Decentralized material handling equipment alternatives at receiving docks.

4.    Material handling equipment alternatives for smaller and lighter loads.

5. Visible, accessible, returnable, durable, collapsible, stackable, and readily transferable containers.

6. Nonsynchronous production lines with mixed-model sequencing for "star" products.

7. Group technology layout arrangements for medium-volume products with similar characteristics to support cellular manufacturing concept.

8. U-shaped assembly lines and fabrication cells.

9. Better handling and transportation part protection.

10. Less material handling requirements at internal processes. More manual handling within manufacturing cells.

11. New fabrics and construction materials for dock seals and shelters.

12. Standard containers, trays, or pallets.

13. Sideloading trucks for easier access and faster loading and unloading operations.

14. The traditional function of storing changes to one of staging.

15. Use of bar codes, laser scanners, EDI, and machine vision to monitor and control the flow of units.

16. Focused factories with product and/or cellular approaches and decentralized support efforts.

17. Lift trucks equipped with radio data terminals and mounted scales to permit on-board weighing.

18. Flow through terminals and/or public warehouses to receive, sort, and route materials.

19. Facilities design process as a coordinated effort between many people.

## *10.5* SUMMARY

In summary, we have described a number of new and emerging concepts that significantly affect the design of manufacturing systems. From a facilities planning perspective, it is important to stay abreast of the changes that are occurring in order to ensure that the facility continues to meet the needs of manufacturing. Further, facilities planning is not exempt from these changes. For example, continuous improvements in the facilities planning process should be sought. The facilities planning process is a candidate for business process reengineering; it, too, needs to be lean and agile. Too often, facilities planners lose sight of the needs of their customers.

The list of references provides a starting point for the reader who is unfamiliar with the concepts and techniques described in the chapter. Obviously, since entire books have been devoted to each of the concepts we described, we cannot provide a comprehensive treatment. However, we would be remiss if we did not bring these concepts to your attention, since they are having a significant impact on the way new manufacturing operations are being designed and the way existing manufacturing operations are being improved.

# BIBLIOGRAPHY

1.  Albano, R. E., Friedman, D. J., and Hendryx, S. A., "Manufacturing Execution: Equipment," *AT&T Technical Journal,* July 1990, pp. 53–63.

2.  Anderson, R. S., Porter, A. L., and Trevino, J., *Automatic Guided Vehicle Systems: Technology Forecast and Evaluation,* Material Handling Research Center, MHRC-OP-87-01, Georgia Institute of Technology, Atlanta, GA, January 1987.

3.  Apple, J. M. Jr., and McGinnis, L. F., "Innovations in Facilities and Material Handling Systems: An Introduction," *Industrial Engineering,* March 1987, vol. 19, no. 3, pp. 33–38.

4.  Auguston, K. A., "For Just $142,000 We Doubled Picking Productivity," *Modern Materials Handling,* February 1992, vol. 47, no. 2, pp. 54–55.

5.  Auguston, K. A., "Automated Handling is Our Key to Quality," *Modern Materials Handling,* May 1991, vol. 46, no. 6, pp. 44–46.

6.  Bagchi, P. K., "Management of Materials Under Just-In Time Inventory System: A New Look," *Journal of Business Logistics,* vol. 9, no. 2, pp. 89–101.

7.  Binning, R. L., "New Uses of Traditional Storage Systems Can Minimize Inventory and Work-In-Process," *Industrial Engineering,* March 1984, vol. 16, no. 3, pp. 81–83.

8.  Black, J. T., *The Design of the Factory with a Future,* McGraw Hill, New York, 1991.

9.  Burgam, Patrick M., "JIT: On the Move and Out of the Aisles," *Manufacturing Engineering,* June 1984, pp. 65–71.

10. Chang, T., Wysk, R. A., and Wang, H., *Computer Aided Manufacturing,* Prentice Hall, NJ, 1991.

11. Costanza, J. R., *The Quantum Leap in Speed to Market,* JIT Institute of Technology, Inc., 1990.

12. Covey, S. R., *The Seven Habits of Highly Effective People,* Simon & Schuster, 1989.

13. Dauch, R. E., *Passion for Manufacturing,* Society of Manufacturing Engineers, Dearborn, MI, 1993.

14. Deming, W. E., *Quality, Productivity, and Competitive Position,* MIT, Cambridge, MA, 1982.

15. Deming, W. E., *The New Economics for Industry, Government, Education,* Massachusetts Institute of Technology, Center for Advanced Engineering Study, Cambridge, MA, 1993.

16. Forger, G., "AS/RS: Changing with the Times," *Modern Materials Handling,* December 1990, vol. 45, no. 14, pp. 50–54.

17. Forger, G., "More Compact Storage Speeds Parts to the Shop Floor," *Modern Materials Handling,* December 1991, vol. 46, no. 14, pp. 54–55.

18. Forger, G., "Breakthrough System Helps The Gap Ship to Hundreds of Stores," *Modern Materials Handling,* May 1992, vol. 47, no. 6, pp. 42–46.

19. Forger, G., "Better Data Cuts Handling Time at McKesson," *Modern Materials Handling,* August 1992, vol. 47, no. 9, pp. 38–40.

20. Goddard, W. E., "Kanban-One Trick in a Magic Act," *Modern Materials Handling,* April 1992, vol. 47, no. 5, p. 31.

21. Goddard, W. E., *Just-In-Time, Surviving by Creating Tradition,* Oliver Wight Limited Publications, Essex Junction, VT, 1986.

22. Gotoh, F., *Equipment Planning for TPM,* Productivity Press, Cambridge, MA, 1991.

23. Gould, L., "Manufacturing Cells Trim Throughout Time 80%," *Modern Materials Handling,* September 1986, vol. 41, no. 10, pp. 107–109.

24. Gould, L., "How Handling and Storage Support a FMS," *Modern Materials Handling,* October 1987, vol. 42, no. 12, pp. 69–71.

25. Gould, L., "Warehouse Renovation Lifts Storage Capacity 40%," *Modern Materials Handling,* November 1987, vol. 42, no. 13, pp. 74–75.

26. Gould, L., "Auto Assembly on AGVs: More Flexibility, Faster Changeovers," *Modern Materials Handling,* May 1990, vol. 45, no. 6, pp. 60–62.

27. Gould, L., "Flow Racks Improve Picking Efficiency by 30%," *Modern Materials Handling,* July 1990, vol. 45, no. 8, pp. 87.

28. Gould, L., "An Automated Monorail System: the Right Choice for AT&T," *Modern Materials Handling,* October 1990, vol. 45, no. 12, p. 85.

29. Gould, L., "Sideloaders: the Best Choice for Long Loads," *Modern Materials Handling,* October 1990, vol. 45, no. 12, pp. 66–67.

30. Gould, L., "JIT: Call Letters for Radio Crane Control," *Modern Materials Handling,* November 1990, vol. 45, no. 13, pp. 70–71.

31. Gould, L., *"Vertical Carousel Reduces Picking Time by 82%," Modern Materials Handling,* June 1991, vol. 46, no. 7, p. 63.

32. Gould, L., "New Handling Systems Help Us Compete," *Modern Materials Handling,* August 1991, vol. 46, no. 9, pp. 40–43.

33. Greif, M., *The Visual Factory,* Productivity Press, Inc. Cambridge, MA, 1991.

34. Grieco, P. L., Gozzo, M. W., and Claonch, J. W., *Just-in-Time Purchasing,* PT Publications, 1988.

35. Gunn, T. G., *Manufacturing for Competitive Advantage,* Ballinger Publishing, 1987.

36. Gupta, Y. P., and Wilborn, W. O., "JIT and Quality Assurance Form a New Partnership in Manufacturing Operations," *Industrial Engineering,* December 1990, vol. 22, no. 12, pp. 34–40.

37. Hall, R. W., *Attaining Manufacturing Excellence,* Dow Jones-Irwin, 1987.

38. Hammer, M., and James, C., *Reengineering the Corporation,* Harper Collins, 1993.

39. Harper, B., and Harper, A., *Succeeding as a Self-Directed Work Team,* MW Corporation, 1990.

40. Hill, J. M., "Part 1 of 2: The Changing Profile of Material Handling Systems and Controls," *Industrial Engineering,* November 1986, vol. 18, no. 11, pp. 68–73.

41. Hill, J. M., "Part 2 of 2: The Changing Profile of Material Handling Systems and Controls," *Industrial Engineering,* December 1986, vol. 18, no. 12, pp. 26–29.

42. Hickman, C. R., and Silva, M. A., *Creating Excellence,* New American Library, New York, 1984.

43. Hunt, D. V., *Computer-Integrated Manufacturing Handbook,* Chapman and Hall, 1989.

44. Imai, M., *Kaizen,* McGraw-Hill, Inc. New York, 1986.

45. Johnson, R. S., "TQM: Leadership for Quality Transformation," *Quality Progress,* January–May 1993, vol. 26, no. 1, 2, 3, 4, 5, pp. 73–76, 55–58, 91–94, 47–50, 83–86.

46. Kenfield, J. E., "A Nine Step Approach to JIT Implementation," *1st International Electronics Assembly Conference,* Santa Clara, CA, October 7–9, 1985.

47. Kenny, A. A., *"A New Paradigm for Quality Assurance," Quality Progress,* June 1988, vol. 21, no. 6, pp. 30–34.

48. Kerr, J., "Priam Stays Onshore with Automated Factory," *Electronic Business,* July 1984.

49. Kinney, H. D., Jr., and McGinnis, L. F., "Manufacturing Cells Solve Material Handling Problems," *Industrial Engineering,* August 1987, vol. 19, no. 8, pp. 54–60.

50. Knight, S. C., "Launching a Program to Gain World Production Competitiveness: One Company's Approach," *Conference Proceedings, American Production and Inventory Control Society,* 1983, pp. 697–699.

51. Knill, B., "Dock Layout: Changing the Way You Do Business," *Material Handling Engineering,* November 1992, vol. 47, no. 11, pp. 85–88.

52. Kusiak, A., *Intelligent Manufacturing System,* Prentice Hall, NJ, 1990.

53. Kulwiec, R., "Toyota's New Parts Center Targets Top Service," *Modern Materials Handling,* November 1992, vol. 47, no. 13, pp. 73–75.

54. Lawler III, E. E., *High Involvement Management,* Jossey-Bass, 1986.

55. Lorincz, J. A., "Machine Cells Meet JIT Need," *Tooling & Production,* May 1992, pp. 50–54.

56. Lubben, R. T., *Just-In-Time Manufacturing,* McGraw-Hill, New York, 1988.

57. Majima, I., *The Shift to JIT,* Productivity, Cambridge, MA, 1982.

58. Maleki, R. A., *FMS: The Technology and Management,* Prentice Hall, Englewood Cliffs, NJ, 1991.

59. McGinnis, L. F., Toro-Ramos, Z., and Trevino, J., *A Review of the Toyota Production System,* Material Handling Research Center, RS-85-01, Georgia Institute of Technology, Atlanta, GA, September 1985.

60. McGinnis, L. F., and Trevino, J., "Integrated Analysis of Nonsynchronous Assembly Systems Used in the Electronics Industry," *1st International Electronics Assembly Conference,* Santa Clara, CA, October 7–9, 1985.

61. McNair, C. J., Mosconi, W. J., and Norris, T. F., *Beyond the Bottom Line,* Richard D. Irwin, 1989.

62. Mohrman, S. A., and Cummings, T. G., *Self Designing Organizations,* Addison Wesley, 1989.

63. Monden, Y., *Toyota Production System,* Industrial Engineering and Management Press, Institute of Industrial Engineers, Norcross, GA, 1983.

64. Nakajima, S., *Introduction to TPM,* Productivity Press, Inc., Cambridge, MA, 1988.

65. Ohno T., and Setsuo, M., *Just-in-Time for Today and Tomorrow,* Productivity Press, Cambridge, MA, 1988.

66. Orr, G. B., Sopher, S. M., and Apple, J. M., Jr., "Material Handling Equipment Alternatives Examined for Progressive Build in Light Assembly Operations," *Industrial Engineering,* April 1985, vol. 17, no. 4, pp. 68–73.

67. Parmelee, R. A., "JIT Implementation by the Numbers," *American Production and Inventory Control Society, Conference Proceedings,* 1985, pp. 457–461.

68. Pasmore, W. A., *Designing Effective Organizations,* John Wiley, New York, 1988.

69. Peeler, D., *QFD: Customer-Driven Engineering,* Hewlett Packard Quality Publishing, Spring 1990.

70. Pierson, R. A., "Adapting Horizontal Material Handling Systems to Flexible Manufacturing Setups," *Industrial Engineering,* March 1984, vol. 16, no. 3, pp. 62–71.

71. Prowell, T., "How Raychem Meets the Challenge of High-Volume, High-Quality Output," *Industrial Engineering,* November 1992, vol. 24, no. 11, pp. 30–33.

72. Quinlan, J., "Carousels Team with Modular Robots," *Material Handling Engineering,* April 1981, vol. 36, no. 5, pp. 78–83.

73. Quinlan, J., "Rotating for Profit," *Material Handling Engineering,* October 1980, vol. 36, no. 10, pp. 88–92.

74. Rhea, N., "Mini-loads, Carousels, Offer Big Benefits to Small-Load Handling," *Material Handling Engineering,* November 1983, vol. 38, no. 11, pp. 42–48.

75. Robinson, A., *Continuous Improvement in Operations,* Productivity, Cambridge, MA, 1991.

76. Rothery, B., *ISO 9000,* Gower Press, 1993.

77. Rygh, O. B., "For AS/R Systems-An Expanding Role in Manufacturing," *Modern Materials Handling,* February 1978, vol. 33, no. 2, pp. 56–60.

78. Sandras, B., and Johnson, T., "Job Shop JIT," *American Production and Inventory Control Society, 1985 Conference Proceedings,* 1988, pp. 448–452.

79. Schonberger, R. J., *Japanese Manufacturing Techniques,* The Free Press, New York, N.Y., 1982.

80. Schonberger, R. J., *World Class Manufacturing,* The Free Press, New York, N.Y., 1986.

81. Schwind, G., "Self-Powered Monorail Carriers: When and Where to Use Them," *Material Handling Engineering,* February 1983, vol. 38, no. 2, pp. 48–53.

82. Schwind, G., "Harley-Davidson Explains the Techniques of Revitalization," *Material Handling Engineering,* February 1985, vol. 40, no. 2, pp. 174–178.

83. Sepehri, M., "Case in Point: Buick City Genuine JIT Delivery," *P & IM Review with APICS News,* 1988, pp. 34–36.

84. Sepehri, M., "A Machine Builds Machines at Apple Computer's Highly Automated Macintosh Manufacturing Facility," *Industrial Engineering,* vol. 17, no. 4, April 1985, pp. 60–67.

85. Sepehri, M., *Just-In-Time, Not Just in Japan,* The Library of American Production, 1986.

86. Sheridan, J. H., "Toward the CIM Solution," *Industry Week,* October 1989, vol. 238, no. 20, pp. 35–80.

87. Shingo, S., *A Revolution in Manufacturing: The SMED System,* Productivity Press, Cambridge, MA, 1985.

88. Shingo, S., *Zero Quality Control,* Productivity Press, Cambridge, MA, 1986.

89. Shingo, S., *Non-Stock Production,* Productivity Press, Inc., Cambridge, MA, 1988.

90. Shingo, S., *Study of Toyota Production System,* Japan Management Association, Tokyo, Japan 1981.

91. Shingo, S., *The Sayings of Shigeo Shingo,* Productivity Press, Cambridge, MA, 1987.

92. Shinohara, I., *New Production System,* Productivity Press, Cambridge, MA, 1988.

93. Sule, D. R., *Manufacturing Facilities,* PWS-KENT Publishing Company, Boston, MA, 1988.

94. Sule, D, R., *Manufacturing Facilities: Location, Planning and Design,* 2nd edition, PWS Publishing Company, Boston, MA, 1994.

95. Tajiri, M., and Gotoh, F., *TPM Implementation,* McGraw-Hill, New York, NY, 1992.

96. Teicholz, E., and Orr, J., *Computer Integrated Manufacturing Handbook,* McGraw-Hill, New York, NY, 1987.

97. Trevino, J., *Next Generation JIT Systems,* Material Handling Focus, Atlanta, GA, September 1989.

98. Trevino, J., *The Impact of Just-In-Time on Material Handling, Facilities Layout, and Building Design,* Material Handling Focus, Atlanta, GA, September 1987.

99. Trevino, J., Irizarry-Resto, M., and Hwang, J., *Just-In-Time Impact on Industrial Logistics and Plant Design,* Working Paper, Department of Industrial Engineering, North Carolina State University, Raleigh, NC, 1993.

100. Wantuck, K. A., *Just-in-Time for America,* The Forum, 1989.

101. White, J. A., "The Changing World of Material Handling," *Material Flow,* June 1987, vol. 4, no. 3, pp. 205–207.

102. "A New Breed of Monorails-Tops for Automation," *Modern Materials Handling,* November 1982, vol. 37, no. 13, pp. 52–59.

103. "AS/RS and Just-In-Time: Partners in Manufacturing," *Modern Materials Handling,* August 1984, vol. 39, no. 9, pp. 56–62.

104. "Automated Storage Systems," *Modern Materials Handling,* October 1983, vol. 38, no. 11, pp. 44–51.

105. "Buick City Heralds a New Era in Automaking," *Modern Materials Handling,* November 1985, vol. 40, no. 13, pp. 58–62.

106. "Containers: Now They're more than Storage Boxes," Modern Materials Handling, January 1986, vol. 41, no. 1, pp. 87–90.

107. "Conveyor Systems for Flexible Assembly Operations," *Modern Materials Handling,* July 1985, vol. 40, no. 8, pp. 66–70.

108. "Flexible Assembly-A Boon for Short Production Runs," *Modern Materials Handling,* June 1984, vol. 39, no. 7, pp. 48–54.

109. "How Chrysler Customizes Racks for Just-In-Time Service," *Modern Materials Handling,* November 1984, vol. 39, no. 13, pp. 48–50.

110. "How We Gained Flexible Flow to 64 Assembly Stations," *Modern Materials Handling,* November 1983, vol. 38, no. 13, pp. 52–55.

111. *"Mini-load AS/RS Trims Inventory, Speeds Assembly,"* *Modern Materials Handling,* September 1984, vol. 39, no. 10, pp. 47–49.

112. "Monorails: A Bigger Role in Factory Automation," *Modern Materials Handling,* July 1984, pp. 52–57.

113. "Our Flexible Systems Help Us Make a Better Product," *Modern Materials Handling,* July 1984, vol. 39, no. 8, pp. 46–49.

114. "The Dock Area—Where Manufacturing Begins and Ends," *Modern Materials Handling,* 1985 Manufacturing Guidebook, vol. 40, pp. 30–31.

# PROBLEMS

**10.1.** The layout procedures described in Chapter 7 have been developed for TM approaches (i.e., process layouts, centralized support and administrative offices, and centralized storage areas). Select one of the layout procedures of Chapter 7 and explain some of the changes required to include decentralized storage areas, product layouts, cellular manufacturing, visual management, and multiple-receiving areas.

**10.2.** Describe and illustrate how to use the Deming Circle to develop an implementation plan to improve the internal layout of a production area.

**10.3.** Identify from papers and books at least five companies that have transformed their organizations from TM to CM. For each company, describe the methodology employed to achieve such transformation, the roadblocks encountered through the process, the main concepts and techniques used, and results achieved. Also describe layout type, material handling and storage alternatives, unit loads, and building arrangements used.

**10.4.** Which JIT, TQM, TEI, and CIMS concepts, techniques, or technologies (CTT) can be used to reduce inventories, to improve quality, and to reduce production lead-time (provide at least three CTT's from each WCW area for each performance measure). Prepare a table summarizing your answers.

**10.5.** Describe the impact of the following CTT's on facilities design:
   a. Bakayokes (fool-proof mechanisms) to prevent overproduction.
   b. Decentralized storage areas.
   c. Supplier development and certification.
   d. Operator performing self-inspection.
   e. No receiving inspection.
   f. Self-managing teams.
   g. Lean or flat organizational structures.
   h. Electronic data interchange.
   i. CAD.
   j. CNC machines.
   k. FMS.
   l. Artificial intelligence/expert systems.

**10.6.** CM organizations need to be flexible to respond to demand variations, competition, employee absenteeism, demand contractions, customer needs, etc. Identify at least ten CM CTT's that companies need to implement to achieve flexibility. Explain why.

**10.7.** Propose at least 20 CM CTT's that can be used in a university to improve any of its activities (i.e., education, research, extension, administration, library, student center). Describe briefly how to implement the CTT's proposed and which facilities design issues will be impacted by these CTT's.

**10.8.** Same as 7, but for a hospital.

**10.9.** Same as 7, but for an airport.

**10.10.** Same as 7, but for a restaurant.

**10.11.** Company ABC is planning to move to a new building and is requesting your help. They would like to have a facility with short production lead-times, excellent quality, low inventory levels, high productivity, high space utilization, minimum handling, and high flexibility. Currently, they are using a process layout, a centralized storage area, long setups, only one receiving/shipping area, specialized operators, a centralized area for offices, a centralized parking lot, a centralized cafeteria, a vertical organizational structure with many levels, centralized functions (i.e., maintenance, quality, purchasing, engineering, materials), product inspection (performed by inspectors), a "push" production strategy, and corrective maintenance.

    **a.** Which CM CTTs can be used to achieve the new facility objectives? Explain why.

    **b.** Describe the characteristics of the new facility (assuming that your recommendations are implemented).

    **c.** Modify the facilities design process described in Chapter 3 to accommodate the implementation of your recommended CM CTTs. Describe the new process.

**10.12.** Which are the most popular material handling alternatives used in CM organizations? Provide examples of at least three CM companies employing these alternatives (include references consulted).

**10.13.** Which CIM CTTs should be used to enhance the facilities design process described in Chapter 3?

**10.14.** Identify from papers and books the new global performance measures used in CM organizations. Using some of these global measures, identify metrics that can be used to evaluate layout, handling, and warehousing strategies.

**10.15.** Describe the impact of JIT on Facilities Design.

**10.16.** Describe the impact of TQM on Facilities Design.

**10.17.** Describe the impact of TEI on Facilities Design.

**10.18.** Describe the impact of CIMS on Facilities Design.

**10.19.** Which CM CTTs can be used to reduce space drastically? Explain why.

**10.20.** Which CM CTTs can be used to reduce handling requirements drastically? Explain why.

**10.21.** Which CM CTTs can be used to reduce storage requirements drastically? Explain why.

**10.22.** Which CM CTTs impact building size and shape? Explain why.

**10.23.** Contrast the differences between traditional facilities and CM facilities (include layout, building, warehousing, material handling alternatives, offices, cafeteria, parking lot, restrooms, unit loads, support departments, and tool room).

**10.24.** Critique the CM methodology proposed in this chapter.

# 11

## *FACILITIES SYSTEMS*

## *11.1* INTRODUCTION

The objective of this chapter is not to prepare the facilities planner to become an architect, mechanical engineer, structural engineer, or a builder. Rather it is to provide the facilities planner with a unified picture of the building technology and the inter-relationship of the facility systems, so that the interior handling system and the layout will not be made without recognizing the reality of the facility constraints. The intent is to provide an understanding of the various system elements within the facility, it is not intended to prepare the facilities planner to design a facilities structure, its heating, ventilation and air conditioning system or other facility systems; the intent is to provide an overview of how the facility system elements impact the overall process of facilities planning.

The planning of facility systems is typically not the responsibility of the facilities planner. However, the specification of *what* systems are required *where* and the integration of the facilities systems into the overall facility are generally the responsibility of the facilities planner. The facilities planner must be well aware that:

1. The cost of constructing a facility is significantly impacted by the facility systems.
2. A critical factor affecting facility flexibility is facilities systems.
3. A facility plan is not complete until all facility systems are specified.
4. Facility systems have an important impact on employee performance, morale, and safety.
5. Facility systems have an important impact on the fire protection, maintenance, and security of a facility.

The facility systems considered in this chapter are:

1. Structural system performance
2. Atmospheric systems
3. Enclosure systems
4. Lighting and electrical systems
5. Life safety systems
6. Sanitation systems

# *11.2* STRUCTURAL SYSTEM PERFORMANCE

The most common structural type for industrial facilities is the steel skeleton frame or reinforced concrete skeleton frame. Many factors will impact the choice of the structural type or the choice of materials: for example, fire protection (steel is notorious for losing its strength when heated above 1000°F), use environment (steel becomes brittle below −20°F and fails easily), and the overall planning grid or module. For facilities planners, the structure and planning needs should inter-mesh, as the column configuration will impact the layout which will impact the structural design. The reverse is also true. The structural design and the grid configuration chosen (e.g., 20 ft × 30 ft or 25 ft × 25 ft) will impact the layout which will impact the planner's options with respect to interior flow. The facilities planner must insist that the structural grid spacing be such that it optimizes the function of the facility. For example, in a warehouse the column spacing should be dictated by the rack dimensions and the access aisles between racks, that is, the clear spacing between columns must be compatible with the storage system, or in a parking garage the grid element should be a multiple of the car bay width and length and the circulation needs. The facilities planner must recognize that the structural dimensions of the building grid are secondary to the optimal layout of the facility without affecting its structural integrity. Given today's mill capability, specific structural members besides 20 ft and 40 ft length can be obtained without a significant penalty cost. Therefore, if a 24 ft × 36 ft bay will optimize the layout this grid configuration should be specified instead of the more traditional 20 ft × 40 ft grid.

A common error in the design of warehousing facilities is the use of building cost criteria to define column spacings and thus determine the grid configuration. Steel mills generally produce structural members in 20- and 40-ft lengths. Architects and structural engineers habitually use these lengths in designing a storage system. The structural dimensions of the steel beams and girders or other structural members should be secondary to the storage system for which the facility is intended. For example, if as shown in Figure 11.1, a 48- by 40-in. pallet is used and a normal 4-in. clearances is allowed between pallets and between uprights and pallets, a minimum of 92 in. clear distance is required between the faces of the uprights in a pallet rack. If the pallet rack is a conventional 4-high unit capable of carrying 2000-lb unit loads, the uprights will normally be 4 in². Thus the overall width of the pallet rack is 100 in., or 96 in. centerline to centerline of the columns.

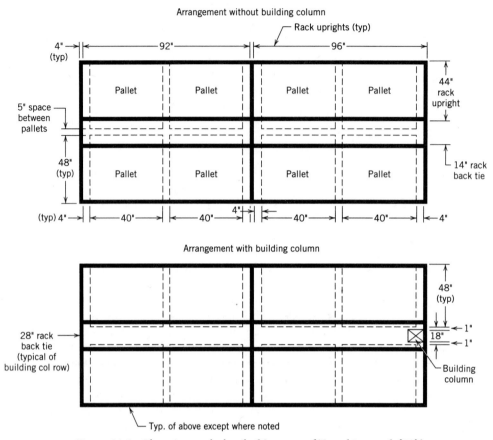

**Figure 11.1** Plan view rack detail. *(Courtesy of Tompkins et al. [10].)*

Carrying this analysis a step further, a typical 26-ft clear storage building will require 12-in. heavy wall pipe columns or 12- by 12-in.-wide flange beam or box columns. If pallet racks are to be used and three sections of pallet racks are to be placed between each pair of columns, the clear space between the faces of the columns must be (96 in. × 3) + 4 in., or 24 ft 4 in. A 4-unit bay would require 32 ft 4 in. (96 in. × 4) + 4 in. Placement of six double-pallet-width units requires a clear space of (96 in. × 3) + 4 in. times 2 plus 12 in. for the column or 49 ft-8 in. Thus, in a typical warehouse design, a 33- by 51-ft center-to-center bay spacing with 12-in. columns would probably accommodate rack storage in either direction and provide optimum flexibility of layout with elimination of column loss.

In such a design pattern, arranging the 33-ft bay parallel with the truck docks provides further advantages. The placement of the truck docks on 15-ft centers two docks in each bay, and the 51-ft span allows a column-free truck service area behind the docks (Figure 11.2). If, as is normally recommended, the truck docks and the rail docks are at right angles on one corner of the building, the 51-ft span along the rail docks would allow optimum flexibility for placement of cars varying in length from 40 to 60 ft or even 90 ft. The 33-ft bay would provide approximately 20 ft from the face of an inside rail dock to the column line, or 33 ft from the outer wall with clear spacing if building face doors were used.

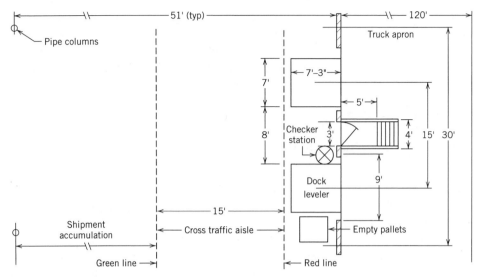

**Figure 11.2**  Basic dock layout. *(Courtesy of Tompkins et al. [10].)*

Needless to say, different pallet sizes affect the dimensions of the racks, and rack dimensions further affect the column spacing dimensions. Thus, before designing a building, it is essential to determine how the facility is to be used. Since the 48- × 40-in. pallet is the Grocery Manufacturers Association's standard and the most commonly used size in consumer and many industrial distribution systems, the dimensional analysis above will probably provide a suitable basis for most warehouse designs where conventional rack systems are to be used. This dimensional pattern eliminates the need to modify the building in accordance with vehicle dimensions, since the racks will fit between the columns in either configuration and aisle spacing is at the option of the layout designer. In the event that a different aisle width is required by the storage machine or forklift truck, a modification in column dimensions may be appropriate.

In this analysis, the size and character of the columns must also be considered. Heavy-wall round or square tubular columns should be used in warehouses. These types of columns eliminate rodent and vermin nesting places, ensure easy maintenance and cleanliness, provide the ability to place downspouts within the columns, and minimize the effect of denting on column strength. Such columns are a little more expensive than H beams, but in the long run provide a much better facility. They also eliminate the corners that can pull packages from pallets when fork lift truck operators move too close to the column. In an all-concrete structure, a square column is normally preferable.

In the final analysis, however, the building's structural integrity is the objective of the structural engineer and the architect. And the building must exhibit stability under the following various load conditions:

- Gravity—continuous (dead loads—roofs, floors, etc.)
  —intermittent (live loads—snow, equipment, people, etc.)
- Wind—hours, days
- Seismic—seconds, minutes

In short, the objective of the structural system is to ensure that the building will be stable. The degree of stability required to be designed in to the building is dictated by the load conditions. The facilities planner must be cognizant of the environment where the facility will be located and the types of load conditions that might render the building unstable.

# *11.3* ENCLOSURE SYSTEMS

The enclosure system provides a barrier against the effects of extreme cold or heat, lateral forces (wind), water, and undesirable entries (human, insects). It is also used as a controlling mechanism. Doors and windows act as filters to provide not only a barrier but controls access to desirables, ventilation, light transmission (privacy) and sound. The facility's enclosure is dictated by different and conflicting influences.

The enclosure elements are floor, walls, and roof. These elements are intended to provide the facility with a specified comfort level. This comfort level is often impacted by the thermal performance of the building enclosure. The tendency, however, is to ignore/underestimate the enclosure's thermal role and rely on environmental systems to make it right. In fact, the following generalization holds true.

Building Enclosure + Services Input = Comfort

Thermal performance is usually required to rectify the heat transmission imbalance between inside and outside the enclosure. Because this transmission is contingent on the temperature differential, climate is therefore the major determinant. One of the major problems therefore in thermal performance is how to make effective use of solar gain. Solar transmission if effectively controlled can significantly reduce the building's dependence on artificial atmospheric systems. Figure 11.3 shows how solar gain can be controlled. In Figure 11.3*a*, 90% of the incident solar rays are transmitted due to the poor absorption and reflection quality of single glazed element. Note in Figure 11.3*b* that a double pane configuration with a heat absorbing element reduces the amount of transmission from 90% to 45%. The solar gain can be reduced additionally from 90% to 25% if a reflective glass member is substituted for a heat-absorbing member in the double pane configuration shown in Figure 11.3*c*.

Typically for manufacturing and warehousing facilities, keeping undesirables out is a key criteria. As such, the material selected should be more impervious to penetration such as metal, plastic, or masonry cladding. Thermal performance coupled with water exclusion forms the backbone of most enclosure systems. There are two areas of considerations; aboveground and below the ground. Aboveground, facilities planners must be aware of the problems that a poorly designed roof can cause. The primary performance need of the roof is water exclusion. However, the importance of thermal comfort necessitates the need for adequate insulation. The three factors listed below are essential to an effective roofing system.

- Vapor migration
- Good insulation
- Water barrier and removal

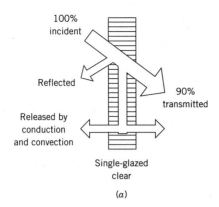

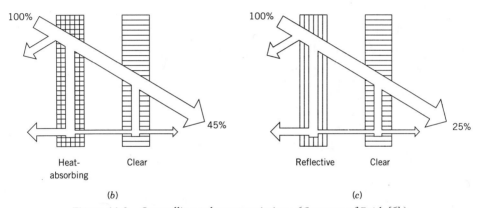

**Figure 11.3** Controlling solar transmission. *(Courtesy of Reid. [6].)*

The built-up roof shown in Figure 11.4 details an effective flat roof design. Note that a membrane layer to prevent water penetration, an insulation layer to assist with thermal comfort and a vapor check to stop vapor migration are integral elements of the system.

The ground and below-ground conditions for industrial facilities is typically a concrete slab sitting directly on the ground. The primary concern with ground conditions are:

- Water penetration
- Vapor migration

As swimming pools must be sealed to keep water in, similarly, the enclosure elements making contact with the ground must be sealed to keep water out. There are two principal ways to accomplish this:

- Integral water-proofing
- Applied membrane

Integral water-proofing consists of an additive to the concrete to make it "water tight." This method, however, is not very effective in controlling vapor migration. Typically, it can take 12 in. or more of concrete to prevent vapor penetration inwards from the surrounding wet soil.

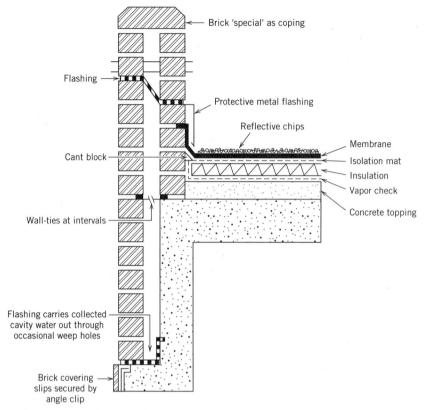

**Figure 11.4**  Typical effective built-up-roof system.

Membranes, on the other hand, have the potential to be 100% effective in handling vapor and water penetration. However, poor designs and shoddy installation can make them ineffective. When membranes are applied to the internal surface, problems may occur with adherence, as the hydrostatic pressure increases. A more effective design method is to apply the membrane on the external surface. This position will allow the membrane to work in conjunction with the hydrostatic pressure of the wet soil surface, and as the hydrostatic pressure increases, the membrane will be pressed more firmly against the structure. However, regardless of the design method, poor installation which results in cracked or punctured membranes will make the application ineffective.

## *11.4*  ATMOSPHERIC SYSTEMS

Atmospheric systems provide for the health and comfort of the building equipment, machinery and occupants. The criteria for equipment and machinery is dictated by manufacturer's specification but the criteria for human comfort for these systems were vague and intuitive until 1900. However, as new technologies were developed for controlling building atmospheres, the criteria became more precisely defined.

The definitive study of human comfort criteria was done in the early 1960s at

Kansas State University using large groups of college students. The data gathered was used to develop equations of the comfort parameters by using the following variables:

A. Temperature
B. Humidity
C. Air speed
D. Clothing
E. Metabolic activity

Atmospheric systems respond to these criteria by heating and cooling building air, and by controling humidity. Air speed is usually held very low, so it is not readily perceived; clothing and metabolic activity are assumed to be constant.

The response to these criteria required by codes is limited to minimum indoor temperatures in winter. More precise responses involving cooling and humidification are provided by market standards for different building types.

One example of such a standard is found in healthcare facilities where humidity is held close to 50%; this humidity level is the optimum for inhibiting the growth of viruses and bacteria. Oxygen in air is essential for building occupants to breathe. But oxygen depletion would occur only at very high population densities, and this is usually not a consideration in building air quality.

A more significant problem is the introduction of inorganic chemical pollutants into building atmospheres from sources in the building. One important pollutant particularly found in single story buildings is radioactive radon gas, which occurs naturally in the earth and ground water, and seeps into these buildings. Such pollutants are best handled by removing the sources from the building, or blocking seepage in the case of radon.

But the major air quality problem in most buildings is odor build-up and air-borne particulate matter. By definition the rate of dilution is expressed as $ft^2/min$— $ft^2$ of floor area. One can see that the way to eliminate unwanted substances is to dilute them by both mixing the stale air with air in the building and adding fresh outdoor air. This air exchange (dilution rate) is typically expressed in air changes per hour as stipulated by local code requirements as it pertains to the building use. Table 11.1 shows typical air change rates for several building use types.

Table 11.1 *Typical Air Changes Rates*

| Use | Supply Air $Ft^3/Min$-$Ft^2$ | Exhause Air $Ft^3/Min$-$Ft^2$ | Air Changes/Hr. |
|---|---|---|---|
| Residence | | | |
| Toilet room | — | 1.5 | 10 |
| Kitchen | — | 1.5 | 10 |
| Other rooms | Natural ventilation—5% of floor area | | — |
| Office/commercial | 0.6 | 0.3 | 4 |
| School | 1.5 | 0.75 | 9 |
| Public assembly | 3.0 | 1.5 | — |
| Hospital | 2.0 | 1.0 | 13 |
| Labs | 1.2 | 1.2 | 8 |
| Animal laboratory | — | 3.0 | 20 |

**Table 11.2**   *Central Equipment Room Area by Use*

| Use | Area of Equipment Rooms |
|---|---|
| Residential | 2% |
| Office/industrial | 5–7% |
| Public assembly | 10–15% |
| Hospital | 25% |
| Laboratory | 25–50% |
| Animal laboratory | 50% |

Because of the importance of eliminating unwanted substances, the equipment to bring in fresh air and remove stale air is a primary consideration in the planning of a facility. This space requirement is governed by the size of the equipment room, the air speed capabilities of main and branch ducts, and the louver size needed to accommodate the required exhaust and intake air needs.

*Equipment Rooms.*   Central equipment rooms can be estimated roughly from Table 11.2. Note that as the air change rate requirement increases, the handling equipment area increases accordingly.

*Duct Sizes.*   Duct sizes can be estimated from air quantities and air speeds in ducts.

Main duct—1800 ft/min
Branch ducts—900 to 1100 ft/min

*Louver Sizes.*   These areas are also estimated from air quantities and air speeds.

Exhaust louvers: Recommended air speed of 2000 ft/min
Intake louvers: Recommended air speed of 1000 ft/min

The following example illustrates how atmospheric space is allocated.

# Example 11.1

Considering the light assembly operation as shown in Figure 11.5, what space requirements should be considered for atmosphere systems?

First, allocate the mechanical equipment room space.

(i)   Total gross area:
240 ft × 120 ft = 28,800 ft$^2$, from Table 13.2 using office/industrial use type, assume 5% of gross area for mechanical equipment space.

(ii)   Assume system will be roof mounted

Second, determine the air handling requirements.

(i)   Minimum air supply rate: 0.6 ft$^3$/min-ft$^2$
Maximum air supply rate: 1.0 ft$^3$/min-ft$^2$
Exhaust 0.3 ft$^3$/min-ft$^2$

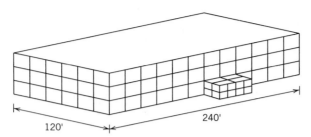

**Figure 11.5** Light assembly plant dimensions for Example 11.1.

(ii)  Air supply: 28,800 ft$^2$ × 1.0 ft$^3$/min-ft$^2$ = 28,800 ft$^3$/min (CFM)
Air Exhaust: 28,800 ft$^2$ × 0.3 ft$^3$/min-ft$^2$ = 8640 ft$^3$/min (CFM)

(iii)  Determine louver cross section using respective CFM requirements and allowable air speeds

Supply: $\dfrac{28,800 \text{ ft}^3/\text{min}}{1000 \text{ ft/min}} = 28.8 \text{ ft}^2$

Exhaust: $\dfrac{8,640 \text{ ft}^3/\text{min}}{2000 \text{ ft/min}} = 4.3 \text{ ft}^2$

Note supply and exhaust louvers must be 15 ft apart to avoid "short circuit" of exhaust air back into building.

(iv)  Main duct calculation and layout
Main duct sizing is determined as follows:

two main ducts, each covering a half of the building's area times the air supply rate

0.5 × 28,800 ft$^2$ × 1 ft$^3$/min—ft$^2$ = 14,400 ft$^3$/min CFM. See Figure 11.6.

Divide this air flow rate by the main duct air speed to obtain the cross-sectional area

$\dfrac{14,400 \text{ CFM}}{1800 \text{ ft/min}} = 8 \text{ ft}^2$

Main duct can be 2 ft × 4 ft. See Figure 11.7.

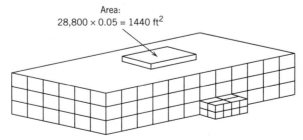

Area:
28,800 × 0.05 = 1440 ft$^2$

**Figure 11.6** Assembly plant with roof mounted air handling unit for Example 11.1.

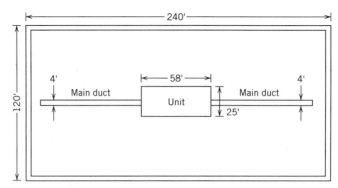

**Figure 11.7**   Main duct cross-sectional area.

**(v)**   Branch duct calculation and layout

For each main duct, assume 10 branch ducts.

Divide flow rate for each branch duct by the number of branches to obtain the branch flow rate.

$$\frac{14{,}400 \text{ ft}^3/\text{min}}{10} = 1440 \text{ ft}^3/\text{min}$$

Divide the branch flow rate by the allowable branch duct air speed.

$$\frac{1440 \text{ ft}^3/\text{min}}{1000 \text{ ft/min}} = 1.4 \text{ ft}^2$$

For the example problem, (2) two main ducts 2 ft-0 in. by 4 ft-0 in. with (10) ten 1 ft-4 in. by 1 ft-0 in. branch ducts would provide the required air dilution to meet the air handling needs (see Figure 11.8).

Although the air dilution method is prevalent and widely accepted, it has some drawbacks. The major drawback to this method is that when you exhaust the air, you are removing heat and/or air conditioning. Therefore, there is a substantial loss

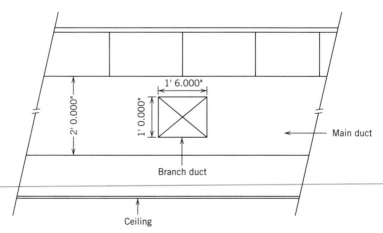

**Figure 11.8**   Cross section of main and branch ducts.

of energy. To reduce the volume of make-up air required, many operations uses mechanical filtration or electrostatic precipitation. Mechanical filtration removes contaminates by passing air through filtering material, sometimes called media or impingement filters to remove particulate matter. Electrostatic precipitation removes particulate by charging the air particles and then collecting them on oppositely charged plates. The air passes through a prefilter and then into the ionizer section of the unit. Here the particles receive a positive charge. These positively charged particles are then attracted to and collected on high-voltage, negative-charged collecting plates. The cleaner air is then returned to the area.

As facilities planners, we are often called upon to design for carefully controlled environments for critical operations in electronics, pharmaceuticals or food processing. These controlled environments are known as clean rooms. There are four official cleanliness room classifications: class 100,000, class 10,000, class 1000, and class 100. These rooms are kept under positive air pressure to prevent dust infiltration. The classification classes refer to the number of particles 0.5 micron and larger in a cubic foot of air. One widely recognized method used to achieve these air quality levels is with the use of HEPA (high efficiency particulate air) filters. These filters are successfully used in a wide variety of industrial, commercial, and institutional installations where the highest possible degree of air cleaning is required. They are also used when it is necessary to remove radioactive or toxic dust, pathogenic organisms, or mold spores. They are available in efficiencies of 95.00%, 99.97%, and 99.999% guaranteed minimum in removing 0.3 micron particles or larger. In actual operation, the HEPA is 100% efficient in removing all known airborne organisms since bacteria, pollens, yeast, and fungi range from 0.5 micron.

The purpose of heating, ventilation, and air conditioning (HVAC) is to control the temperature, humidity, and cleanliness of the environment within a facility. HVAC is important for employee comfort, process control, and/or product control. HVAC is an extremely complex field and is undergoing continuous change. A facilities planner must realize the complex and dynamic nature of HVAC and not make the mistake of attempting to design HVAC for a facility.

The input required by an HVAC expert to design a HVAC system for a facility includes the facility layout and the construction specifications. The facilities planner should be able to provide the HVAC expert with this input as well as understand how the HVAC expert translates this input into HVAC requirements. The heating requirement for a facility is referred to as *heat loss* and is calculated using Equation 11.1.

$$Q_H = Q_F + Q_R + Q_G + Q_D + Q_W + Q_I \qquad (11.1)$$

where

$Q_H$ = total heat loss

$Q_F$ = heat loss through the floor

$Q_R$ = heat loss through the roof

$Q_G$ = heat loss through glass windows

$Q_D$ = heat loss through doors

$Q_W$ = heat loss through walls

$Q_I$ = heat loss due to infiltration

Each of the individual heat losses are calculated using Equation 11.2.

$$Q = AU(t_i - t_o) \tag{11.2}$$

where

$Q$ = heat loss for facility component

$A$ = area of facility component

$U$ = coefficient of transmission of facility component

$t_i$ = temperature inside the facility

$t_o$ = temperature outside the facility

The air conditioning requirement for a facility is referred to as the *cooling load* and is obtained from Equation 11.3.

$$Q_C = Q_F + Q_R + Q_G + Q_D + Q_W + Q_V + Q_S + Q_L + Q_P \tag{11.3}$$

where

$Q_C$ = total cooling load

$Q_F$ = cooling load due to the floor

$Q_R$ = cooling load due to the roof

$Q_G$ = cooling load due to glass windows

$Q_D$ = cooling load due to doors

$Q_W$ = cooling load due to walls

$Q_V$ = cooling load due to ventilation

$Q_S$ = cooling load due to solar radiation

$Q_L$ = cooling load due to lighting

$Q_P$ = cooling load due to personnel

The cooling loads $Q_F$, $Q_R$, $Q_G$, $Q_D$, $Q_W$, and $Q_V$ are determined using Equation 11.4.

$$Q = AU(t_o - t_i) \tag{11.4}$$

where

$Q$ = cooling load for facility component

$A$, $U$, $t_o$ and $t_i$ are defined as for Equation 11.2

The cooling load from solar radiation $Q_S$ is computed using Equation 11.5.

$$Q_S = (A)(H)(S) \tag{11.5}$$

where

$A$ = area of glass surface

$H$ = heat absorption of the building surface

$S$ = shade factor for various types of shading

The cooling load from lighting $Q_L$ may be calculated by using the following conversion factors:

$$Q_L \text{ (incandescent)} = \left(\frac{\text{Incandescent}}{\text{wattage}}\right)\left(3.4 \frac{\text{Btu}}{\text{(hr)(watt)}}\right) \tag{11.6}$$

$$Q_L \text{ (flourescent)} = \left(\frac{\text{Fluorescent}}{\text{wattage}}\right)\left(4.25 \frac{\text{Btu}}{\text{(hr)(watt)}}\right) \tag{11.7}$$

Table 11.3  *Heat Gains for Males and Females Performing Light and Heavy Work*

| Activity | Heat Gains (Btu/hr) |
| --- | --- |
| Male—light work | 800 |
| Male—heavy work | 1500 |
| Female—light work | 700 |
| Female—heavy work | 1300 |

The cooling load from personnel $Q_P$ may be calculated by determining the number of males and females performing light and heavy work and multiplying that times the heat gain factors given in Table 11.3.

## Example 11.2

An 18-ft tall facility having the dimensions 250 ft × 100 ft has eight 4 ft × 8 ft double-pane glass windows and two 3 ft × 8 ft glass doors on both the front and back sides. The facility is made of 8-in. solid cinder block, 1-in. metal-insulated roof, with an insulated ceiling and an uninsulated slab floor. The facility is located in Chicago, Illinois. What is the heat loss?

The following data needed to calculate the heat loss may be obtained from a variety of sources (see [2] and [3]).

$$U_F = 0.81 \text{ Btu/(hr)(ft}^2)(°F) \qquad t_i = 70°F$$
$$U_R = 0.20 \text{ Btu/(hr)(ft}^2)(°F) \qquad t_o = 0°F$$
$$U_G = 0.63 \text{ Btu/(hr)(ft}^2)(°F)$$
$$U_D = 1.13 \text{ Btu/(hr)(ft}^2)(°F)$$
$$U_W = 0.39 \text{ Btu/(hr)(ft}^2)(°F)$$
$$U_I = 0.20 \text{ Btu/(hr)(ft}^2)(°F)$$

Then, using Equation 11.2:

$$Q_F = (25{,}000 \text{ ft}^2)[0.81 \text{ Btu/(hr)(ft}^2)(°F)](70°F - 0°F)$$
$$Q_F = 14.175 \times 10^5 \text{ Btu/hr}$$
$$Q_R = (25{,}000 \text{ ft}^2)[0.20 \text{ Btu/(hr)(ft}^2)(°F)](70°F - 0°F)$$
$$Q_R = 3.5 \times 10^5 \text{ Btu/hr}$$
$$Q_G = (16)(32 \text{ ft}^2)[0.63 \text{ Btu/(hr)(ft}^2)(°F)](70°F - 0°F)$$
$$Q_G = 0.226 \times 10^5 \text{ Btu/hr}$$
$$Q_D = (4)(24 \text{ ft}^2)(1.13 \text{ Btu/(hr)(ft}^2)(°F)](70°F-0°F)$$

$Q_D = 0.076 \times 10^5$ Btu/hr

$Q_W = [(2)(4500 \text{ ft}^2) + (2)(1800 \text{ ft}^2) - (16)(32 \text{ ft}^2)$
$\qquad - (4)(24 \text{ ft}^2)][(0.39 \text{ Btu/(hr)(ft}^2)(°F)](70°F - 0°F)$

$Q_W = 3.274 \times 10^5$ Btu/hr

$Q_I = [(25,000 \text{ ft}^2) + (12,600 \text{ ft}^2)](0.20 \text{ Btu/(hr)(ft}^2)(°F)](70°F - 0°F)$

$Q_I = 0.526 \times 10^5$ Btu/hr

Then, using Equation 11.1, the total heat loss may be calculated as follows:

$Q_H = 14.175 \times 10^5$ Btu/hr $+ 3.5 \times 10^5$ But/hr
$\qquad + 0.226 \times 10^5$ Btu/hr
$\qquad + 0.076 \times 10^5$ Btu/hr $+ 3.274 \times 10^5$ Btu/hr $+ 0.526 \times 10^5$ Btu/hr

$Q_H = 21.777 \times 10^5$ Btu/hr

## *Example 11.3*

For the facility described in Example 11.3, given the following additional information, what is the cooling load? The front of the facility faces west and all windows and doors have awnings. The lighting within the facility consists of 575 fluorescent luminaries, each containing two 60 W lamps. Within the facility, 75 men performing light work and 50 women performing light work will be employed. The outside design temperature is 97°F, and the inside design temperature is 78°F [3].

Using Equation 11.4 results in the following cooling loads for facility components:

$Q_F = (25,000 \text{ ft}^2)[0.81 \text{ Btu/(hr)(ft}^2)(°F)](97°F - 78°F)$

$Q_F = 3.85 \times 10^5$ Btu/hr

$Q_R = (25,000 \text{ ft}^2)[0.20 \text{ Btu/(hr)(ft}^2)(°F)](97°F - 78°F)$

$Q_R = 0.95 \times 10^5$ Btu/hr

$Q_G = (16)(32 \text{ ft}^2)[0.63 \text{ Btu/(hr)(ft}^2)(°F)](97°F - 78°F)$

$Q_G = 0.06 \times 10^5$ Btu/hr

$Q_D = (4)(24 \text{ ft}^2)(1.13 \text{ Btu/(hr)(ft}^2)(°F)](97°F - 78°F)$

$Q_D = 0.02 \times 10^5$ Btu/hr

$Q_W = [(2)(4500 \text{ ft}^2) + (2)(1800 \text{ ft}^2) - (16)(32 \text{ ft}^2)$
$\qquad - (4)(24 \text{ ft}^2)][(0.39 \text{ Btu/(hr)(ft}^2)(°F)](97°F - 78°F)$

$Q_W = 0.88 \times 10^5$ Btu/hr

$Q_V = [(25,000 \text{ ft}^2) + (12,600 \text{ ft}^2)](0.20 \text{ Btu/(hr)(ft}^2)(°F)](97°F - 78°F)$

$Q_V = 0.14 \times 10^5$ Btu/hr

The heat absorption factor for the west side of a facility in Chicago, Illinois is 100 Btu/(hr)(ft$^2$) [3]. The shade factor for awnings is 0.3 [3]. Using Equation 11.5, the cooling load from solar radiation may be calculated as:

$Q_S = [(8)(32 \text{ ft}^2) + (2)(24 \text{ ft}^2)][100 \text{ Btu/(hr)(ft}^2)(°F)(0.3)]$
$\qquad + [(8)(32 \text{ ft}^2) + (2)(24 \text{ ft}^2)][75 \text{ Btu/(hr)(ft}^2)(°F)(0.3)]$

$Q_S = 0.16 \times 10^5$ Btu/hr

The cooling load from lighting may be calculated via Equation 11.7, as follows:

$$Q_L \text{ (fluorescent)} = (575)(2)(60 \text{ W})\left(4.25 \frac{\text{Btu}}{(\text{hr})(\text{watt})}\right)$$

$$Q_L = 2.93 \times 10^5 \text{ Btu/hr}$$

With the aid of Table 11.2, the cooling load from personnel may be calculated as:

$$Q_P = 75(800 \text{ Btu/hr}) + 50(700 \text{ Btu/hr})$$

$$Q_P = 0.95 \times 10^5 \text{ Btu/hr}$$

Then, using Equation 11.3, the total cooling load may be calculated as follows:

$$Q_C = 3.85 \times 10^5 + 0.95 \times 10^5 + 0.06 \times 10^5 + 0.02 \times 10^5 + 0.88 \times 10^5$$
$$+ 0.14 \times 10^5 + 0.16 \times 10^5 + 2.93 \times 10^5 + 0.95 \times 10^5$$

$$Q_C = 9.94 \times 10^5 \text{ Btu/hr}$$

Or, utilizing the conversion factor 12,000 Btu/hr = 1 ton, a total cooling load of 83 tons is required to air condition the total facility.

## *11.5* ELECTRICAL AND LIGHTING SYSTEMS

The electrical load requirement for a facility is needed by the power company well before construction begins. This usually requires that facilities planners have a good preliminary estimate since the design of the facility is often not finalized when this information is required. In general, every electrical system should have sufficient capacity to serve the loads for which it is designed, plus spare capacity to meet anticipated growth in the load on the system.

Allowance for load growth is probably the most neglected consideration in the planning and design of the electrical system. The facilities planner must ensure that major load requirements, both present and future are included in the design data, so that the design engineer can adequately size mains, switchgear, transformers, feeders, panel boards and circuits to effectively handle the growth of the load. The process begins with analyzing the building type and its traditional electrical loads. An idea of the average load requirement for several building types is shown in Table 11.4.

The facilities planner should then determine for the building type under consideration, all unusual and special electrical requirements, current trends and prac-

Table 11.4  *Average Load Requirements*

| Building Type | Load Requirements |
| --- | --- |
| Plants | 20 watts per s.f. |
| Office Buildings | 15 watts per s.f. |
| Hospitals | 3000 watts per bed |
| Schools | 3–7 watts per s.f. |
| Shopping Centers | 3–10 watts per s.f. |

tices. Once this activity is complete, a list of load categories can then be generated. Once the loads are tabulated in each of the electrical load categories:

- Lighting
- Miscellaneous power, which includes convenience outlets and small motors
- HVAC
- Plumbing or sanitary equipment
- Vertical transportation equipment
- Kitchen equipment
- Special equipment

the point of service, service voltage, and metering location should then be estimated with the local utility company. In addition, the facilities planner should determine with the client the type and rating of all equipment to be used in the facility. The space required for electrical equipment and electrical closets can then be determined. Finally, the facilities planner should provide detailed input for the lighting requirements for the facility.

The facilities planner's responsibility with respect to electrical services is to specify *what* level of service is required and *where* it is required. With the exception of facility lighting, the facilities planner's responsibility may be satisfied by giving the individual responsible for planning electrical services a facilities layout and a department service and area requirement sheet (see Figure 4.27) for each department in the facility. Facilities lighting should not be planned by the person planning the electrical services because illumination must be custom designed for the areas within a facility. The facilities planner fully understands the tasks to be performed within a facility and is in the best position to custom design the lighting system. Lighting systems may be designed using the following eight-step procedure:

### Step 1: *Determine the Level of Illumination*

The minimum levels of illumination for specific visual tasks are given in Table 11.5. These levels may be achieved using general lighting or a combination of general lighting and supplementary lighting. When supplementary lighting is utilized, general lighting should contribute at least one-tenth of the total illumination and should be no less than 20 footcandles [1]. Supplementary lighting, particularly when used in conjunction with modular office furniture, has resulted in significant energy savings.

### Step 2: *Determine the Room Cavity Ratio (RCR)*

The RCR is an index of the shape of a room to be lighted. The higher and narrower a room, the larger the RCR and the more illumination is needed to achieve the required level of lighting. The RCR may be calculated using Equation 11.8.

$$\text{RCR} = \frac{(5)\left(\begin{array}{c}\text{Height from the working}\\\text{surface to the luminaries}\end{array}\right)\left(\begin{array}{cc}\text{Room} & \text{Room}\\\text{length} + \text{width}\end{array}\right)}{\left(\dfrac{\text{Room}}{\text{length}}\right)\left(\dfrac{\text{Room}}{\text{width}}\right)} \tag{11.8}$$

### Step 3: *Determine the Ceiling Cavity Ratio (CCR)*

If luminaries are ceiling mounted or recessed into the ceiling, the reflective property of the ceiling will not be impacted by the luminaries' mounting height, so the CCR

**Table 11.5** *Minimum Illumination Levels for Specific Visual Tasks*

| Task | Minimum Illumination Level (footcandles) |
|---|---|
| Assembly | |
|   Rough easy-seeing | 30 |
|   Rough difficult-seeing | 50 |
|   Medium | 100 |
|   Fine | 500 |
|   Extra Fine | 1000 |
| Inspection | |
|   Ordinary | 50 |
|   Difficult | 100 |
|   Highly difficult | 200 |
|   Very difficult | 500 |
|   Most difficult | 1000 |
| Machine Shop | |
|   Rough bench and machine work | 50 |
|   Medium bench and machine work, ordinary automatic machines, rough grinding, medium buffing and polishing | 100 |
|   Fine bench and machine work, fine automatic machine, medium grinding, medium buffing and polishing | 500 |
|   Extra fine bench and machine work, fine grinding | 1000 |
| Material Handling | |
|   Wrapping, packing, labeling | 50 |
|   Picking stock, classifying | 30 |
|   Loading | 20 |
| Offices | |
|   Reading high or well-printed material; tasks not involving critical or prolonged seeing, such as conferring, interviewing, and inactive files | 30 |
|   Reading or transcribing handwriting in ink or medium pencil on good-quality paper; intermittent filing | 70 |
|   Regular office work; reading good reproductions, reading or transcribing handwriting in hard pencil or on poor paper; active filing; index references; mail sorting | 100 |
|   Accounting, auditing, tabulation, bookkeeping, business machine oepration; reading poor reproductions; rough layout drafting | 150 |
|   Cartography, designing, detailed drafting | 200 |
|   Corridors, elevators, escalators, stairways | 20 |
| Paint Shops | |
|   Dipping, spraying, rubbing, firing, and ordinary hand painting | 50 |
|   Fine hand painting and finishing | 100 |
|   Extra fine hand painting and finishing | 300 |
| Storage Rooms and Warehouses | 2 |
|   Inactive | 5 |
|   Active | |
|     Rough bulky | 10 |
|     Medium | 20 |
|     Fine | 50 |
| Welding | |
|   General illumination | 50 |
|   Precision manual arc welding | 1000 |

need not be considered. The greater the distance from the luminaries to the ceiling, the greater will be the CCR and the reduction in the ceiling reflectance. The CCR may be calculated using Equation 11.9.

$$CCR = \frac{\left(\begin{array}{c} \text{Height from} \\ \text{luminaries to ceiling} \end{array}\right)(RCR)}{\begin{array}{c} \text{Height from the working} \\ \text{surface to the luminaries} \end{array}} \tag{11.9}$$

**Step 4:** *Determine the Wall Reflections (WR) and the Effective Ceiling Reflectance (ECR)*

The WR and a base ceiling reflectance (BCR) value can be obtained from Table 11.6. If the luminaries are to be ceiling mounted or recessed into the ceiling, the ECR is equal to the BCR. If luminaries are to be suspended, the ECR is determined from the CCR, WR, and BCR using Table 11.7.

**Step 5:** *Determine the Coefficient of Utilization (CU)*

The CU is the ratio of lumens reaching the work plane to those emitted by the lamp. It is a function of the luminaries utilized, the RCR, the WR, and the ECR. The CU for standard luminaries is given in Table 11.8.

**Step 6:** *Determine the Light Loss Factor (LLF)*

The two most significant light loss factors are lamp lumen depreciation and luminaries dirt depreciation. Lamp lumen depreciation is the gradual reduction in lumen output over the life of the lamp. Typically, lamp lumen depreciation factors are expressed as the ratio of the lumen output of the lamp at 70% of rated life to the initial value. The lumen outputs for various lamps given in Table 11.9 include the lamp lumen depreciation; therefore, no additional consideration is needed. Hence, the LLF may be equated to the lamp luminaries dirt depreciation which, as shown in Table 11.10, varies with the degree of dirty conditions and the length of time between cleaning luminaries.

**Step 7:** *Calculate the Number of Lamps and Luminaries*

Utilize Equation 11.10 to calculate the number of lamps and Equation 11.11 to calculate the number of luminaries.

**Table 11.6**  *Approximate Reflectance for Wall and Ceiling Finishes*

| Materials | Approximate Reflectance (%) |
|---|---|
| White paint, light-colored paint, mirrored glass, and porcelain enamel | 80 |
| Aluminum paint, stainless steel, and polished aluminum | 65 |
| Medium-colored paint | 50 |
| Brick, cement, and concrete | 35 |
| Dark-colored paint, asphalt | 10 |
| Black paint | 5 |

*Source:* From [14].

Table 11.7  The Percent (%) Effective Ceiling Reflectance (ECR) for Various Combinations of Ceiling Cavity Ratios (CCR), Wall Reflectances (WR), and Base Ceiling Reflectances (BCR)

| | BCR | | | | | | | | | | | | | | | | | | | | | | | | | | | | | | | | | | | |
|---|---|---|---|---|---|---|---|---|---|---|---|---|---|---|---|---|---|---|---|---|---|---|---|---|---|---|---|---|---|---|---|---|---|---|---|---|
| | 80 | | | | | | 65 | | | | | | 50 | | | | | | 35 | | | | | | 10 | | | | | | 5 | | | | | |
| | WR | | | | | | WR | | | | | | WR | | | | | | WR | | | | | | WR | | | | | | WR | | | | | |
| CCR | 80 | 65 | 50 | 35 | 10 | 5 | 80 | 65 | 50 | 35 | 10 | 5 | 80 | 65 | 50 | 35 | 10 | 5 | 80 | 65 | 50 | 35 | 10 | 5 | 80 | 65 | 50 | 35 | 10 | 5 | 80 | 65 | 50 | 35 | 10 | 5 |
| .5 | 76 | 74 | 72 | 69 | 67 | 65 | 64 | 60 | 58 | 56 | 54 | 52 | 49 | 47 | 46 | 44 | 42 | 41 | 36 | 34 | 32 | 31 | 29 | 28 | 12 | 12 | 11 | 11 | 9 | 8 | 8 | 8 | 7 | 6 | 5 | 5 |
| 1.0 | 74 | 71 | 67 | 63 | 57 | 56 | 60 | 55 | 53 | 49 | 45 | 43 | 48 | 45 | 43 | 39 | 36 | 35 | 35 | 33 | 31 | 30 | 26 | 25 | 14 | 13 | 12 | 11 | 9 | 8 | 10 | 8 | 8 | 7 | 5 | 4 |
| 1.5 | 72 | 67 | 62 | 55 | 49 | 47 | 58 | 52 | 49 | 44 | 38 | 36 | 47 | 44 | 40 | 35 | 31 | 28 | 35 | 33 | 31 | 26 | 21 | 20 | 16 | 15 | 12 | 11 | 8 | 7 | 14 | 11 | 9 | 7 | 4 | 4 |
| 2.0 | 69 | 63 | 56 | 49 | 41 | 39 | 55 | 49 | 44 | 38 | 32 | 30 | 46 | 42 | 37 | 31 | 26 | 25 | 35 | 32 | 28 | 23 | 18 | 17 | 18 | 17 | 13 | 10 | 8 | 6 | 15 | 12 | 10 | 7 | 4 | 4 |
| 2.5 | 67 | 60 | 51 | 43 | 35 | 33 | 54 | 45 | 40 | 33 | 26 | 25 | 46 | 40 | 35 | 28 | 22 | 21 | 35 | 31 | 26 | 21 | 16 | 13 | 20 | 19 | 13 | 10 | 7 | 5 | 17 | 14 | 10 | 7 | 4 | 3 |
| 3.0 | 65 | 57 | 47 | 38 | 30 | 28 | 53 | 42 | 38 | 29 | 22 | 21 | 45 | 39 | 32 | 25 | 19 | 18 | 35 | 31 | 24 | 20 | 14 | 12 | 21 | 20 | 13 | 10 | 7 | 4 | 19 | 15 | 11 | 8 | 3 | 3 |
| 3.5 | 63 | 54 | 43 | 34 | 26 | 25 | 52 | 39 | 33 | 26 | 18 | 17 | 44 | 38 | 30 | 23 | 17 | 16 | 35 | 31 | 23 | 18 | 12 | 10 | 22 | 21 | 13 | 10 | 7 | 4 | 20 | 16 | 11 | 8 | 3 | 3 |
| 4.0 | 61 | 52 | 40 | 31 | 22 | 21 | 50 | 37 | 31 | 23 | 15 | 14 | 44 | 38 | 28 | 21 | 15 | 13 | 34 | 30 | 23 | 17 | 10 | 8 | 23 | 22 | 14 | 10 | 7 | 3 | 20 | 17 | 12 | 8 | 3 | 2 |
| 5.0 | 58 | 46 | 35 | 26 | 18 | 15 | 48 | 33 | 26 | 18 | 9 | 8 | 42 | 35 | 25 | 18 | 12 | 10 | 34 | 29 | 21 | 16 | 9 | 7 | 25 | 23 | 14 | 10 | 6 | 3 | 23 | 18 | 12 | 8 | 3 | 2 |
| 8.0 | 50 | 36 | 25 | 17 | 11 | 6 | 41 | 24 | 18 | 11 | 5 | 3 | 40 | 30 | 19 | 13 | 7 | 5 | 34 | 28 | 17 | 11 | 5 | 4 | 27 | 24 | 13 | 10 | 5 | 2 | 26 | 19 | 12 | 6 | 3 | 1 |

*Source:* From [14].

**Table 11.8  Coefficients of Utilization for Standard Luminaries**

| Luminaire | Spacing Not to Exceed | RCR | ECR | | | | | | | | | | | | |
|---|---|---|---|---|---|---|---|---|---|---|---|---|---|---|---|
| | | | 80% | | | | 50% | | | | 10% | | | | 0% |
| | | | WR | | | | WR | | | | WR | | | | WR |
| | | | 80% | 50% | 30% | 10% | 80% | 50% | 30% | 10% | 80% | 50% | 30% | 10% | 0% |
| Filament reflector lamps | 1.5 × mounting height | 1 | 1.11 | 1.09 | 1.07 | 1.03 | 1.04 | 1.02 | 1.00 | .98 | .96 | .95 | .94 | .93 | .91 |
| | | 2 | 1.04 | 1.00 | .95 | .92 | .99 | .95 | .92 | .88 | .92 | .90 | .87 | .85 | .83 |
| | | 3 | .95 | .92 | .88 | .82 | .92 | .88 | .84 | .80 | .85 | .83 | .80 | .77 | .75 |
| | | 4 | .90 | .85 | .79 | .73 | .86 | .81 | .76 | .71 | .79 | .77 | .73 | .70 | .68 |
| | | 5 | .82 | .77 | .71 | .65 | .80 | .75 | .69 | .64 | .75 | .71 | .67 | .63 | .61 |
| | | 10 | .58 | .50 | .43 | .38 | .54 | .49 | .43 | .38 | .51 | .47 | .42 | .38 | .36 |
| High-intensity discharge lamps (mercury, metal halide or sodium) | 1.3 × mounting height | 1 | .89 | .87 | .84 | .82 | .81 | .80 | .78 | .77 | .74 | .73 | .72 | .71 | .70 |
| | | 2 | .82 | .79 | .75 | .72 | .77 | .74 | .71 | .68 | .70 | .68 | .66 | .64 | .63 |
| | | 3 | .76 | .72 | .67 | .63 | .72 | .68 | .64 | .61 | .65 | .63 | .60 | .58 | .56 |
| | | 4 | .70 | .66 | .61 | .57 | .67 | .63 | .58 | .55 | .57 | .55 | .54 | .53 | .51 |
| | | 5 | .64 | .60 | .55 | .51 | .62 | .58 | .53 | .49 | .56 | .54 | .52 | .49 | .46 |
| | | 10 | .45 | .40 | .34 | .30 | .42 | .38 | .34 | .30 | .40 | .36 | .31 | .20 | .28 |
| Fluorescent lamps in uncovered fixtures | 1.3 × mounting height | 1 | .88 | .85 | .82 | .79 | .79 | .75 | .72 | .71 | .65 | .64 | .63 | .62 | .59 |
| | | 2 | .78 | .75 | .70 | .65 | .71 | .67 | .63 | .59 | .60 | .57 | .55 | .52 | .50 |
| | | 3 | .69 | .66 | .60 | .55 | .63 | .59 | .54 | .50 | .54 | .51 | .48 | .45 | .42 |
| | | 4 | .61 | .59 | .52 | .46 | .56 | .52 | .47 | .43 | .48 | .45 | .41 | .38 | .36 |
| | | 5 | .53 | .51 | .44 | .39 | .51 | .46 | .40 | .36 | .43 | .40 | .36 | .33 | .30 |
| | | 10 | .35 | .30 | .23 | .19 | .32 | .27 | .21 | .18 | .26 | .23 | .19 | .16 | .14 |
| Fluorescent lamps in prismatic lens fixture | 1.2 × mounting height | 1 | .65 | .63 | .61 | .59 | .60 | .59 | .58 | .56 | .56 | .55 | .54 | .53 | .52 |
| | | 2 | .60 | .57 | .54 | .51 | .56 | .54 | .51 | .49 | .51 | .50 | .49 | .47 | .46 |
| | | 3 | .54 | .51 | .48 | .44 | .51 | .49 | .46 | .43 | .47 | .46 | .44 | .42 | .41 |
| | | 4 | .49 | .46 | .42 | .39 | .48 | .44 | .41 | .38 | .44 | .42 | .39 | .37 | .36 |
| | | 5 | .45 | .42 | .37 | .34 | .44 | .40 | .36 | .34 | .40 | .38 | .35 | .33 | .32 |
| | | 10 | .31 | .26 | .21 | .18 | .29 | .25 | .21 | .18 | .27 | .24 | .20 | .18 | .17 |

*Source:* From [14].

514

**Table 11.9** *Lamp Output at 70% of Rated Life*

| Lamp Type | Watts | Lamp Output at 70% of Rated Life (lumens) |
|---|---|---|
| Filament | 100 | 1,600 |
| | 150 | 2,600 |
| | 300 | 5,000 |
| | 500 | 10,000 |
| | 750 | 15,000 |
| | 1000 | 21,000 |
| High-intensity discharge | 400 | 15,000 |
| | 700 | 28,000 |
| | 1000 | 38,000 |
| Fluorescent | 40 | 2,500 |
| | 60 | 3,300 |
| | 60 | 3,300 |
| | 85 | 5,400 |
| | 110 | 7,500 |

*Source:* From [14].

$$\text{Number of Lamps} = \frac{\left(\begin{array}{c}\text{Required level} \\ \text{of illumination}\end{array}\right)\left(\begin{array}{c}\text{Area to} \\ \text{be lit}\end{array}\right)}{(CU)(LIF)\left(\begin{array}{c}\text{Lamp output at} \\ \text{70\% of rated life}\end{array}\right)} \quad (11.10)$$

$$\text{Number of luminaries} = \frac{(\text{number of lamps})}{(\text{lamps per luminarie})} \quad (11.11)$$

**Step 8**: *Determine the Location of the Luminaries*

The number of luminaries calculated via Equation 11.11 will result in the correct *quantity* of light. In addition to the quantity of light, the *quality* of the light must be considered. The most important factors of quality of light are glare and diffusion. Glare is defined as any brightness that causes discomfort, interference with vision, or eye fatigue. The brighter the luminaries, the greater the potential for glare. Also, the more a luminary is moved from the line of vision, the less the potential for glare. Furthermore, the greater the brightness contrast between a luminary and its surroundings, the greater the potential for glare.

Several design considerations result from observations concerning glare. For example, overly bright luminaries should not be used, light sources should be mounted above the normal line of vision, and ceilings and walls should be painted with light colors to reduce contrast and provide sufficient background lighting if supplementary lighting is used.

Diffusion of light indicates that illumination results from light coming from many directions. The greater the number of luminaries, the more diffuse the light. Fluorescent luminaries provide more diffuse light than incandescent luminaries. Also, light will not diffuse properly if luminaries are spaced too far apart.

Design considerations resulting from a concern for diffusion of light are as follows: Because shadows will produce eye fatigue, many luminaries should be used

Table 11.10 *Lamp Luminary Dirt Depreciation Factors*

DIRT –CONDITION[a]

| Luminarie | Clean—Offices, Light Assembly, or Inspection | | | | | Medium—Mill Offices Paper Processing, or Light Machining | | | | | Dirty—Heat Treating, High-Speed Printing, or Medium Machining | | | | | Very Dirty—Foundary or Heavy Machining | | | | |
|---|---|---|---|---|---|---|---|---|---|---|---|---|---|---|---|---|---|---|---|---|
| | Months Between Cleaning | | | | | Months Between Cleaning | | | | | Months Between Cleaning | | | | | Months Between Cleaning | | | | |
| | 6 | 12 | 24 | 36 | 48 | 6 | 12 | 24 | 36 | 48 | 6 | 12 | 24 | 36 | 48 | 6 | 12 | 24 | 36 | 48 |
| Filament reflector lamps | 0.95 | 0.93 | 0.89 | 0.86 | 0.83 | 0.94 | 0.89 | 0.85 | 0.81 | 0.78 | 0.87 | 0.84 | 0.79 | 0.74 | 0.70 | 0.83 | 0.74 | 0.60 | 0.56 | 0.52 |
| High-intensity discharge lamps | 0.94 | 0.90 | 0.84 | 0.80 | 0.75 | 0.92 | 0.88 | 0.80 | 0.74 | 0.69 | 0.90 | 0.83 | 0.76 | 0.68 | 0.64 | 0.86 | 0.79 | 0.69 | 0.63 | 0.57 |
| Fluorescent lamps in uncovered fixtures | 0.97 | 0.94 | 0.89 | 0.87 | 0.85 | 0.93 | 0.90 | 0.85 | 0.83 | 0.79 | 0.93 | 0.87 | 0.80 | 0.73 | 0.70 | 0.88 | 0.83 | 0.75 | 0.70 | 0.64 |
| Fluorescent lamps in prismatic lens fixtures | 0.92 | 0.88 | 0.83 | 0.80 | 0.78 | 0.88 | 0.84 | 0.77 | 0.73 | 0.71 | 0.82 | 0.78 | 0.71 | 0.67 | 0.62 | 0.78 | 0.72 | 0.64 | 0.60 | 0.57 |

[a]Information under this heading from [14].

to illuminate an area. Also, the selection of fluorescent versus diffusion and the types of available diffusion panels and hoods. To obtain properly diffused light, use the maximum permissible ratio of luminaries spacing to mounting height above the working surface as given in Table 11.8 in the column labeled "Spacing not to exceed."

## *Example 11.4*

A machine shop 100 ft × 40 ft with a 13-ft ceiling is to be illuminated for automatic machining and rough grinding. Uniform lighting is required through the machine shop. If the luminaries are to be ceiling mounted, all ceilings and walls are to be painted white, and all luminaries are to be cleaned every 24 months, what lighting should be specified?

**Step 1.** According to Table 11.5, a minimum illumination level of 100 foot candles is required.

**Step 2.** Assume all working surfaces are to be 3 ft from the floor. Utilizing Equation 11.8 to calculate the RCR results in the following:

$$\text{RCR} = \frac{(5)(13 - 3)(100 + 40)}{(100)(40)} = 1.75$$

**Step 3.** The CCR need not to be considered as the luminaries are ceiling mounted.

**Step 4.** According to Table 11.6, the WR and BCR are 80%. Because the luminaries are to be ceiling mounted, the ECR is also 80%.

**Step 5.** If fluorescent lamps in uncovered fixtures are to be utilized, the CR may be interpolated in Table 11.8 as being between 0.88 and 0.78. A CU of 0.80 will be utilized.

**Step 6.** For fluorescent lamps in uncovered fixtures in a "medium dirty environment" that are cleaned every 24 months, a lamp luminary dirt depreciation factor of 0.85 may be obtained from Table 11.10. Therefore, the LLF is 0.85.

**Step 7.** According the Equation 11.10, the number of lamps required if 60 W fluorescent lamps are utilized is

$$\text{Number of lamps} = \frac{(100)(100)(40)}{(0.8)(0.85)(3300)} = 179$$

If two lamps are placed in each luminary, according to Equation 11.11, 90 luminaries are required.

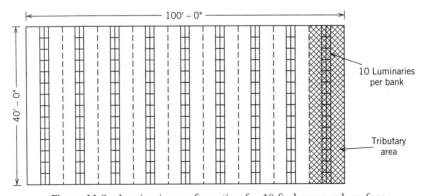

**Figure 11.9** Luminaries configuration for 10 ft above work surface.

**Step 8.** Table 11.8 indicates that for a mounting height 10 ft above the work surface, the luminaries should be spaced no more than 13 ft. Each fluorescent fixture is 4 ft long. By placing nine rows of 10 luminaries across the room with the first and last rows 6 ft from the wall and all other rows 11 ft apart, the illumination level within the room will be evenly distributed and adequate to perform the tasks within the machine shop. This lighting layout is shown in Figure 11.9.

## Example 11.5

Do the lighting requirements change for the machine shop described in Example 11.4 if the luminaries are still to be hung at 13 ft but suspended from a 41 ft ceiling?

**Steps 1 and 2.** The same as Example 11.4.

**Step 3.** According to Equation 11.9, the CCR may be calculated as

$$CCR = \frac{(41 - 13)}{(13 - 3)} \times 1.75 = 4.9$$

**Step 4.** As in Example 11.4, the WR and BCR are 80%. According to Table 11.7, the ECR is 58%.

**Step 5.** Given the fluorescent lamps in uncovered fixtures are still to be utilized, the CU may be interpolated from Table 11.8 as 7.5%.

**Step 6.** Unchanged from Example 11.4.

**Step 7.** According to Equation 11.10, the number of lamps required if 60 W fluorescent lamps are utilized is

$$\text{Number of lamps} = \frac{(100)(100)(40)}{(0.75)(0.85)(3300)} = 191$$

If two lamps are placed in each luminary according to Equation 11.11, 96 luminaries are required.

**Step 8.** Table 11.8 indicates that for a mounting height 10 ft above the work surface, the luminaries should be spaced no more than 13 ft. Each fluorescent fixture is 4 ft long. By placing 4 rows of 24 luminaries along the long axis of the machine shop so that each row begins and ends 2 ft from the wall and the first and last rows are 5 ft from the wall and all other rows are spaced 10 ft apart, the illumination level within the room will be evenly distributed and adequate to perform the required tasks. This lighting layout is shown in Figure 11.10.

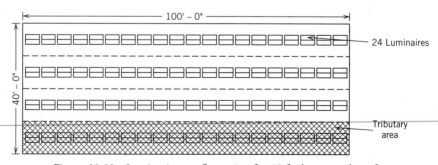

**Figure 11.10** Luminaries configuration for 13 ft above work surface.

The lighting information given to the person planning the electrical services should include a lighting layout and a description of the type of luminaries and lamps to be used. With this information and the location of the equipment required electrical services, as described on the department service and area requirement sheets, the electrical service requirement can be planned.

## *11.6* LIFE SAFETY SYSTEMS

Life safety systems are designed to control emergency situations that will disrupt normal operations. These emergencies are primarily created by:

1. Fire
2. Seismic events
3. Power failure

Fire is the most pervasive of the three and accounts for the majority of costs. Fire resistance is therefore critical in the design of any facility. Fire protection in buildings is governed by the Uniform Building Code (UBC). UBC is a model building code which outlines the protection features which must be included in the building design. The features that are typically covered are fire ratings of walls, floors and roofs as well as egress, sprinklers and standpipe requirements.

The first objective of the facilities planner is to determine the building's function and construction type as defined by the occupancy classification. This occupancy classification consists of over 20 UBC specific classifications. In general, however, the following seven facility types typically account for all occupancy classifications.

| Group | Type |
|-------|------|
| A | Assemblies, theaters additions |
| E | Educational facilities |
| I | Institutional occupancies |
| H | Hazardous occupancies (fuel, paint, chemicals) |
| B | Business, office, factory, commercial facilities |
| R | Residential |
| M | Miscellaneous |

The degree of fire resistance is also governed by the type of structure. These construction types range from Type I (nearly fireproof) to Type D (conventional wood frame construction). Fire resistance, therefore, refers to the ability of a structure to act as a barrier that will not allow the fire to spread from its point of origin. Even with these design efforts, there is no such thing as a fire immune building. Given this fact, facilities planners must be cognizant of the need to provide for safety routes as an integral part of the layout. The Uniform Building Code requires that means of egress be provided in every facility from every part of every floor to a public street or alley. The minimum requirement therefore is at least one exit from every facility and in some cases when the rated number of occupants exceed a certain number two or more exits are needed as stipulated by the building code. In general most buildings require two or more exits. This is done so that other exits are available if one is blocked by fire. In addition, most local building codes will require that maximum allowable times to exit the building not be violated. This time could be the time to reach a protected area, that is, outside the building or to another compartment if the facility has fully fireproof segregated compartments. Also, each exit must be within 150 ft of any point (200 ft in a sprinkled building), and access provided for handicapped persons. This may entail ramps or an exit passage way to a protected compartment. It is important to note that the use of elevators are to be avoided. In general, elevator shafts collect smoke from the floor that is on fire, and their use is strongly discouraged. In any event, facilities planners must recognize that when a fire occurs, the order of priority is:

1.  Life of occupants in the facility
2.  The building
3.  The goods and equipment in the building

As an example, consider the determination of the exits for a warehouse shown in Figure 11.11 below.

## *Example 11.6*

Using BOAC (Building Officials and Code Administrators) code from Table 11.11 (maximum floor area allowances per occupant) for industrial areas 100 SF per occupant is the recommended area.

$$\text{Maximum population} = \frac{120 \text{ ft} \times 200 \text{ ft}}{100 \text{ sf/person}}$$

$$= 240 \text{ people}$$

From Table 11.12 (minimum number of exits for occupant load) given the population, the minimum number of exits required is 2

Number of exits = 2

From Table 11.13, the remoteness requirements can be addressed, that is how far apart these doors should be to safely move occupants out of the facility.

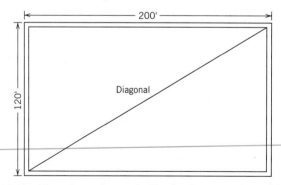

**Figure 11.11**  Plan of warehouse facility for the example problem 13.6.

**Table 11.11** *(BOCA table 806)* **Maximum Floor Area Allowances per Occupant**

| Use | Floor Area per Occupant (ft$^2$) |
|---|:---:|
| Assembly with fixed seats | See Section 806.1.6 |
| Assembly without fixed seats | |
| Concentrated (chairs only—not fixed) | 7 net |
| Standing space | 3 net |
| Unconcentrated (tables and chairs) | 15 net |
| Bowling alleys allow five persons for each alley, including 15 ft of runaway, and for additional areas | 7 net |
| Business areas | 100 gross |
| Court rooms—other than fixed seating areas | 40 net |
| Educational | |
| Classroom area | 20 net |
| Shops and other vocational room areas | 50 net |
| Industrial areas | 100 gross |
| Institutional areas | |
| Inpatient treatment areas | 240 gross |
| Outpatient areas | 100 gross |
| Sleeping areas | 120 gross |
| Library | |
| Reading rooms | 50 net |
| Stack areas | 100 gross |
| Mercantile, basement and grade floor areas | 30 gross |
| Areas on other floors | 60 gross |
| Storage, stock, shipping areas | 300 gross |
| Parking garages | 200 gross |
| Residential | 200 gross |
| Storage areas, mechanical equipment room | 300 gross |

*Note:* 1 ft = 304.8 mm; 1 ft$^2$ = 0.093 m.
*Source:* BOCA [12].

For the building shown in Figure 11.11,

$$\text{The distance between exits} = \frac{\text{the diagonal dimension}}{4} = \frac{d}{4}$$

$$d = \left((200)^2 + (120)^2\right)^{1/2} = 233$$

$$\text{Minimum distance between exits} = \frac{233}{4} = 58$$

Note if the building was not sprinkled, exits should be d/2.

**Table 11.12** *(BOAC table 809.2)* **Minimum Number of Exits for Occupant Load**

| Occupant Load | Minimum Number of Exits |
|:---:|:---:|
| 500 or less | 2 |
| 501–1000 | 3 |
| Over 1000 | 4 |

*Source:* BOCA [12].

**Table 11.13**  *BOCA Remoteness Requirements*

Where two exit access doors are required, they shall be placed a distance apart equal to not less than one-half the length of the maximum overall diagonal dimension of the building or area to be served. Where exit enclosures are provided as a portion of the required means of egress and are interconnected by a corridor conforming to the requirements for corridor construction, the exit separation distance shall be measured along the line of travel within the corridor. In all other cases, the separation distance shall be measured in a straight line between exits or exit access doors.

*Source:* BOCA [12].

Using Table 11.12, we can now determine the allowable minimum width for each door.

$$\text{Capacity per exit} = \frac{240 \text{ people}}{2 \text{ exits}} = 120 \text{ people/exit}$$

$$\text{Minimum width} = 120 \times 0.2 \text{ in} = 24''$$

Since the minimum width is below the 2 ft-8 in. recommended for barrier-free access, each door should be increased to 2 ft-8 in. in width (see Figure 11.12). Finally, check the recommended solution to determine if the allowable length of travel exiting the building is satisfied. From Table 11.14, for the example use group (I) with fire suppression travel distance to exits is okay.

The planner must balance exits with the maximum allowable travel distance required by code. From Table 11.14 the maximum allowable travel distance for industrial use groups with fire a suppression system is 250 ft. This suggests that both doors could be on the same elevation of the warehouse.

Industrial engineers are often concerned with fire protection systems for rack storage facilities. In the 1980s, a great deal of fire protection literature concentrated on the risk associated with rack storage facilities. This resulted in the development of the National Fire Protection Association Section 231C. It is important to note that

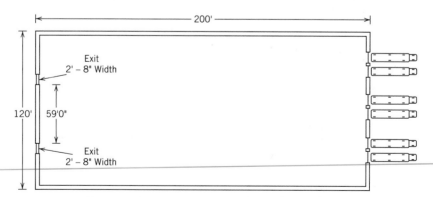

**Figure 11.12**  Exit requirements for 200' × 240' w/h.

Table 11.14  *Length of Exit Access Travel (ft)*

| Use Group | Without Fire Suppression System | With Fire Suppression System |
|---|---|---|
| A, B, E, F-1, I-1, M, R, S-1 | 200 | 250 |
| F-2, S-2 | 300 | 400 |
| H | — | 75 |
| I-2, I-3 | 150 | 200 |

*Source:* BOCA [12].

there are four classes of fires: A, B, C, and D. This classification is determined by the method or equipment used to extinguish the fire. Class A fires are those involving ordinary materials and can be extinguished with water. Class B fires involve flammable gases and liquids and Class C electrical equipment. For both Class B and C water is ineffective. With Class B fires, oil will float on top of water, while with Class C, the extinguishing material must be electrically nonconductive. Class D fires involve combustible metals (sodium, potassium, etc.) and need special extinguishers. Note also that NFPA 231C governs only low or ordinary hazards (Class A fires) and excludes commodities that could be classified as highly hazardous materials such as tires (covered by NFPA 231D), rolled paper stored on end (covered by NFPA 231F), and flammable liquids (covered by NFPA 30). With the storage of these materials, an appropriate extinguishing agent is recommended.

NFPA 231C's hazard categories are established based on the design of automatic sprinkler systems. These automatic sprinkler systems are a widely used and very effective means of extinguishing or controlling fires in their early stage. This effectiveness is due principally to the fact that being automatic they can provide protection during periods when the facility is not occupied. Automatic sprinklers, therefore, because of their effectiveness are mandatory in certain occupancy classifications. They are not mandatory in rack storage systems, but their usefulness is quite evident. Within a rack storage system, sprinklers are generally located along the longitudinal flue spaces (the space between rows of storage perpendicular to the direction of loading), or transverse flue spaces (the space between rows of storage parallel to the direction of loading). If these sprinklers are located along the longitudinal flue passages, they are known generally as in rack sprinklers. However, if they are located in transverse flue spaces along the aisle or in the rack within 18 in. of the aisle face, they are known as face sprinklers. This particular sprinkler location is designed to oppose vertical development of fire on the external face of the rack.

The design requirements of a rack storage sprinkler system is quite involved. The designer must consider a variety of design elements such as height of storage, sprinkler spacing, sprinkler pipe size, water shield location, discharge pressure, water demand requirements, and so on, in order to properly determine the in-rack sprinkler arrangement. One such arrangement is shown in Figure 11.13. The objective here is not to make the facilities planner a fire protection expert but to provide them with a sense of the complex nature of providing an adequate fire protection system.

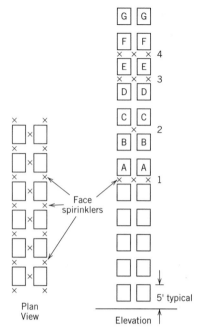

**Figure 11.13** Typical in-rack sprinkler arrangement for height of storage over 25 ft.

Notes:
1. Sprinklers labeled 1 required when loads labeled A or B represent top of storage.
2. Sprinklers labeled 1 and 2 required when loads labeled C or D represent top of storage.
3. Sprinklers labeled 1 and 3 required when loads E represent top of storage.
4. Sprinklers labeled 1 and 4 required when loads F or G represent top of storage.
5. For storage higher than represented by loads labeled G, the cycle defined by notes 2, 3, and 4 is represented.
6. The symbol X indicates face and in-rack sprinklers.

*Source:* NFPA [15].

## 11.7 SANITATION SYSTEMS

Sanitation systems consist of hot and cold water supply, a distribution network to supply potable and cleansing water, as well as a collection network for refuse. Refuse handing goes beyond the collecting and disposing of waste water and "soil solids" (which is known as sewage). It also includes refuse chutes, incinerators, and shredders. The plumbing system, which typically handles the supply of hot and cold water and the sewage system, forms the bulk of the sanitation system. A discharge system typically consists of a series of lateral branches tied to a vertical stack. The system is designed to handle both soil (which contains solid waste material and waste water, i.e., discharge from basins, bath, washing machines). The primary problem with drainage networks is the problem of preventing odor passing back into the facility. The water seal trap shown in Figure 11.14*b* is the most common way of alleviating this problem. This method however is not fool proof and even with design precau-

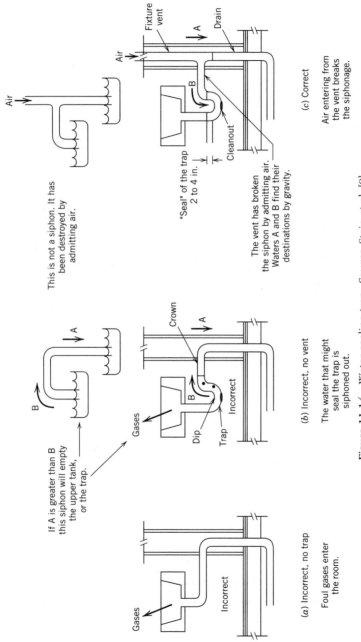

This is not a siphon. It has been destroyed by admitting air.

If A is greater than B this siphon will empty the upper tank, or the trap.

**(a) Incorrect, no trap**

Foul gases enter the room.

**(b) Incorrect, no vent**

The water that might seal the trap is siphoned out.

**(c) Correct**

Air entering from the vent breaks the siphonage.

The vent has broken the siphon by admitting air. Waters A and B find their destinations by gravity.

"Seal" of the trap 2 to 4 in.

**Figure 11.14** Water sealing trap. *Source*: Stein et al. [9].

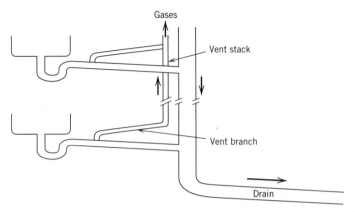

**Figure 11.15**   Vent used to improve water seal integrity. *Source:* Reid [6].

tions such as making the trap deeper it is wise to use venting. Venting is widely used to provide assurance that smells are effectively removed from the facility and vented to the atmosphere. Figure 11.14*c* shows how the trap can be made more effective, by the introduction of an air vent breaking the siphon. A typical trap with air vent arrangement is shown in Figure 11.15.

Plumbing systems within a facility may be required for personnel, processing, and fire protection. Personnel plumbing systems include drinking fountains, toilets, urinals, sinks, and showers. The facilities planner should supply the person planning the overall plumbing systems with the location of all personnel plumbing systems and the number of personnel to utilize these services.

## *Example 11.4*

For the office building with floor area, as shown in Figure 11.16, the facilities planner is required to determine the plumbing requirements for each floor.

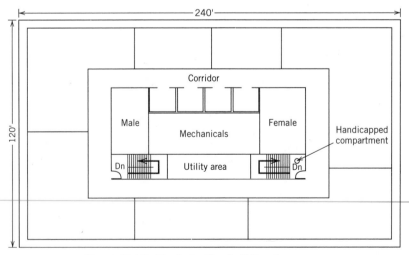

**Figure 11.16**   Typical office building floor area.

Table 11.15 *Plumbing Fixture Requirements for the Example Problem*

| Plumbing Fixture | Women | Men | |
|---|---|---|---|
| | Option 1 | Option 1 | Option 2* |
| Water Closet | 6 | 6 | 4 |
| Lavatories | 7 | 7 | 7 |
| Urinals | — | — | 2 |

*Note up to 1/3 water closets can be replaced by urinals.

Area of floor = 120′ × 240′
= 28,800 sf

Using BOCA code Table 11.11

Minimum recommended space for each occupant is 100 sf/person. For business use type, the maximum allowable floor population therefore is:

$$\frac{28,800 \text{ sf}}{100 \text{ sf/person}} = 288 \text{ persons}$$

Assume 144 men, 144 women

Using Table 5.2, the minimum number of fixtures needed were obtained. The adjusted requirement for urinals is as shown in Table 11.15

For industrial or plant operations, the facilities planner must provide process water and drain requirements to the person planning the plumbing services. In addition, the facilities planner must also provide the composition of all process liquids. The treatment or filtering of these liquids must be carefully planned.

Plumbing services for fire protection require not only that the proper quantity of water is available but also that the proper water pressure is provided. The most common approach to fire protection is automatic sprinkler systems. The facilities planner's input into the planning of fire protection plumbing services is a facility layout and a description of the activities to be performed in various areas within the layout.

## *11.8* SUMMARY

The planning of facility systems is typically not the responsibility of the facilities planner. Nevertheless, the facilities planner will be required to interact with the architects and engineers responsible for planning and designing the facility systems and should therefore be familiar with the approaches to be utilized to plan these systems. This chapter presents an introduction to the approaches used to plan structural, atmospheric, enclosure, lighting and electrical, life safety, and sanitation systems.

# BIBLIOGRAPHY

1. Jennings, B. J., *Environmental Engineering: Analysis and Practice*, International Textbook, Scranton, PA, 1970.
2. Kaufman, J. E., *IES Lighting Handbook*, Illuminating Engineering Society of North America, New York, 1981.
3. McKinnon, G. P., *Fire Protection Handbook*, 15th ed., NFPA, Quincy, MA, 1981.
4. Olivieri, J. B., *How to Design Heating-Cooling Comfort Systems*, Business News Publishing, Birmingham, MI, 1971.
5. Ramsey, C. G., and Sleeper, H. R., *Architectural Graphics Standards*, John Wiley & Sons, New York, 1988.
6. Reid, E., *Understanding Buildings*, MIT Press, Cambridge, MA, 1989.
7. Schiller, M., *Architectural License Seminars*, ALS Publishing, Los Angeles, CA, 1988.
8. Stanford, H. W., *Analysis & Design of Heating, Ventilating and Air Conditioning Systems*, Prentice Hall, Englewood Cliffs, NJ, 1988.
9. Stein, B., Reynolds, J. S., and McGuinness, W. J., *Mechanical And Electrical Equipment For Buildings*, 7th ed., John Wiley, New York, 1986.
10. Tompkins, J. A., and Smith, J. D., *The Warehouse Management Handbook*, McGraw-Hill, New York, 1988.
11. *ASHRAE Handbook of Fundamentals*, American Society of Heating, Refrigerating and Air Conditioning Engineers, New York, 1977.
12. *BOCA, National Building Code*, 11th ed., Building Officials & Code Administrators International, Inc., Country Club Hills, IL, 1990.
13. *Fire Resistance Directory*, Underwriters Laboratories, Northbrook, IL, 1993.
14. *Life Safety Code Handbook*, NFPA, Boston, MA, 1990.
15. *Lighting Handbook*, Westinghouse Lamp Division, Westinghouse Electric Corp., Bloomfield, NJ, 1974.
16. National Fire Protection Association—Section 231C, NFPA Publications, Quincy MA, 1992.

# PROBLEMS

**11.1** A general office that is 250 × 150 ft is to be used for regular office work. The ceiling height is 10 ft and all working surfaces are to be 3 ft from the floor. If the luminaries are to be ceiling mounted, all ceilings and walls are to be painted white, and all luminaries are to be cleaned every 6 months, what lighting should be specified? Utilize fluorescent lamps in prismatic lens fixtures.

**11.2** What would be the changes in lighting specifications for the facility described in Problem 11.1 if the office were to be utilized for accounting and bookkeeping work?

**11.3** What would be the changes in lighting specifications for the facility described in Problem 11.1 if the ceiling height were 15 ft?

**11.4** What would be the changes in lighting specifications for the facility described in Problem 11.1 if the ceiling were 15 ft and the luminaries were to be mounted at 10-ft height?

**11.5** What would be the changes in lighting specifications for the facility described in Problem 11.1, if all ceiling and walls were painted a medium color?

**11.6** What would be the changes in lighting specifications for the facility described in Problem 11.1 if the fixtures were cleaned only once every 36 months?

**11.7** What would be the heat loss of a facility having the following characteristics if the inside design temperature is 72°F and the outside design temperature is 12°F?

> 400 × 300 ft
>
> 20 ft tall
>
> No windows
>
> Six glass doors measuring 3 × 8 ft
>
> 8 ft solid cinder block construction
>
> 1 in. metal insulated roof with an insulated ceiling
>
> Uninsulated slab floor

**11.8** What would be the heat loss of the facility described in Problem 11.7 if 20, 4 × 8 ft double-pane glass windows were installed?

**11.9** What would be the heat loss of the facility described in Problem 11.7 if the roof were replaced with a 2-in. metal roof having an insulated ceiling [$U_R = 0.13$ Btu/(hr)(ft$^2$)(°F)]?

**11.10** What would be the heat loss of the facility described in Problem 11.7 if the slab floor were insulated [$U_F = 0.55$ Btu/(hr)(ft$^2$)(°F)]?

**11.11** Given the following additional information, what is the cooling load for the facility described in Problem 11.7?

> Three doors on 400 ft east side
>
> Three doors on 400 ft west side
>
> Three doors on 400 ft east side = 55 Btu/(hr)(ft$^2$)
>
> Three doors on 400 ft west side = 110 Btu/(hr)(ft$^2$)
>
> Lighting consists of 700 60 W fluorescent lamps
>
> Fifty men perform heavy work within the facility and sixty women perform light work
>
> Outside design temperature = 98°F
>
> Inside design temperature = 78°F

**11.12** If 10, 4 × 8 double-pane glass windows were installed on the east and west sides of the facility described in Problem 11.11, what would be the cooling load of the facility?

**11.13** The shade factor for awnings is 0.3 and for blinds is 0.7. What change in cooling load would occur if awnings were placed on the windows described in Problem 11.12? What if blinds were placed in these windows?

**11.14** On what basis are building occupancy groups determined?

**11.15** For the building shown in Figure 11.11, if the building occupancy was considered to educational classroom use, determine the following:
  a. Maximum population
  b. Minimum number of exits
  c. Minimum distance between exits (assume building has no sprinkler system)
  d. Provide a sketch showing possible location of the exits

**11.16** What is the major benefit of using a sprinkler system?

**11.17** Rank the following design objectives in descending order of priority.
  - Safety of fire fighters
  - Safety of regular building occupants
  - Salvage of building
  - The goods and equipment in the building

**11.18** What is the main purpose of face sprinklers in an in-rack sprinkler system?

**11.19** What is the purpose of a trap in a sewage system?

**11.20** In the design of a plumbing system, what are the two principal reasons, why a vent stack is critical?

**11.21** As a facilities planner, is the following statement true or false, "The grid structure should defer to the function of the facility"? State reasons.

**11.22** How can the thermal performance of a facility be improved? What is the primary purpose of an enclosure system for a manufacturing facility?

**11.23** Using Figure 11.6, if the floor area of the building is 120 ft × 180 ft, provide recommendations for the main duct, louver supply, and exhaust sizes.

Assume
| | |
|---|---|
| Air supply rate | $1.0 \text{ ft}^3/\text{min - ft}^2$ |
| Exhaust rate | $0.3 \text{ ft}^3/\text{min - ft}^2$ |
| Air speed | 1800 ft/min |

Louver air speed
- Exhaust        2000 ft/min
- Intake         1000 ft/min

# Part Four

# DEVELOPING ALTERNATIVES: QUANTITATIVE APPROACHES

# *12*

# *QUANTITATIVE FACILITIES PLANNING MODELS*[1]

## *12.1*  INTRODUCTION

Part 4 of the text consists of a single chapter, which provides a number of quantitative models that can be used to facilitate the development of alternative facilities plans. We make no claim that all relevant models are contained in the chapter. (Recall, computerized layout and graph-based layout models were presented in Chapters 7 and 8.) Instead, we claim that we have found the models presented in this chapter to be useful in both teaching and practicing facilities planning.

In this chapter, we present, first, a number of prescriptive models for solving some relatively simple facilities location and facilities layout problems. Next, we present a number of descriptive and prescriptive models that can be used to design a warehouse or storage system.

Our presentation of warehouse design models begins with a consideration of the classical storage method of block stacking; specifically, prescriptive models are presented for minimizing the average amount of floor space required to store a product. Based on the results obtained for block stacking, similar models are provided for several varieties of racked storage.

The focus on storage systems continues with a consideration of automated storage and retrieval systems. In addition to presenting models for performing rough-cut estimations of the acquisition cost for an AS/RS, a number of models of the operating performance of an AS/RS are presented.

---

[1]This chapter is intended for advanced undergraduate students and first-year-graduate students who are familiar with the concepts of classical optimization, branch and bound, linear programming, and probability theory.

Our consideration of storage systems design concludes with a presentation of prescriptive models for use in designing order picking systems. Both in-the-aisle and end-of-aisle order picking are considered.

Following the storage systems models, we present a number of descriptive models that can be used to design conveyor systems. First, we model a recirculating trolley conveyor; next, we model the horsepower requirements for belt and roller conveyors.

Since the design of conveyor systems is influenced by random variation, following the presentation of deterministic models of recirculating trolley conveyors and horsepower requirements for belt and roller conveyors, we present a number of waiting-line models. The subject of many books and papers, our coverage of waiting lines or queues is limited to relatively simple models that have obvious applications to facilities planning. Although several "expected value" models are presented in the chapter, the section on waiting-line models is the first place where we give explicit consideration to the impact of random phenomena on facilities planning.

Following our presentation of waiting-line models, we consider the use of simulation models to address a broader set of probabilistic phenomena than can be addressed using waiting-line models. As with waiting lines, entire books have been devoted to simulation modeling. Hence, our treatment of the subject is very abbreviated. Beyond the rudiments of simulation, we provide an overview of simulation software that is specifically designed for material handling systems.

As noted above, we have not attempted to provide a comprehensive treatment of quantitative models for use in the facilities planning. In addition to the computerized layout and graph-based layout models cited, we note our omission of assembly line balancing models, models of flexible manufacturing systems and transfer machines, and simulation models of a variety of elements of the facilities planning regime. Further, many more facilities layout and location models are available—the same is true for storage systems, conveyors, and waiting lines.

In summary, there is no shortage of quantitative models that have been applied to gain some insight or facilitate a design decision in facilities planning. However, there is a shortage of space available within this text on the subject. For that reason, we encourage you to maintain a currency in the facilities planning literature, with the references for this chapter being a good starting point.

## *12.2*   FACILITY LOCATION MODELS

In this section, a number of analytical models of facility location problems are presented. The subject of location analysis is sufficiently vast to warrant one or more texts devoted entirely to it. Our objective in this section is to illustrate its applications to facilities planning problems. Hence, the models and solution procedures we present are selected for their illustrative value.

Location analysis can be applied to many problems including locating an airport, a school, a machine tool, a warehouse, a sewage treatment plant, a production facility, a post office, a hospital, a library, and so on. Because of the breath of application of location analysis, there exists a strong interdisciplinary interest in the subject.

There are various objectives used in facility location decisions, for example, minimizing the sum of the weighted distances between the new facility and the other existing facilities (denoted as the *minisum location problem*), and minimizing the maximum distance between the new facility and any existing facility (denoted as the *minimax location problem*).

The distance measures involved in a facility location problem are an important element in formulating an analytical model. Distance measures can be categorized as:

1.  Rectilinear where distances are measured along paths that are orthogonal (or perpendicular) to each other. This measure is also known as Manhattan distance due to the fact that many streets in the city are perpendicular or parallel to each other. An example would be a material transporter moving along rectilinear aisles in a factory.

2.  Euclidean (or straight line) where distances are measured along the straight line path between two points. A straight conveyor segment linking two workstations illustrates Euclidean distance.

3.  Flow path distance where distances are measured along the actual path traversed between two points. For instance, in an automated guided vehicle system a vehicle on a transport mission must follow the directionality of the guide path network. Thus, the flow path distance may be longer compared to rectilinear or Euclidean distance.

We will discuss rectilinear models below and defer the coverage of flow path distance models to Section 12.4. Discussion on Euclidean models can be found in Francis, et al. [29].

## Rectilinear Facility Location Problem

We use the following notation:

$X = (x, y)$         location of the new facility
$P = (a_i, b_i)$       location of existing facility $i$, $i = 1, 2, \ldots m$
$w_i$              proportionality constant
$d(X, P_i)$      distance between the new facility and existing facility $i$

The annual cost of travel between the new facility and existing facility $i$ is assumed to be proportional to the distance between the points $X$ and $P_i$, with $w_i$ denoting the constant of proportionality.

The objective is to

$$\text{Minimize } f(X) = \sum_{i=1}^{m} w_i \, d(X, P_i) \tag{12.1}$$

In a rectilinear model, the distances are measured by the sum of the absolute difference in their coordinates, that is,

$$d(X, P_i) = |x - a_i| + |y - b_i| \tag{12.2}$$

## *Single-Facility Minisum Location Problem*

The minisum location problem is formulated as follows:

$$\text{Minimize } f(X) = \sum_{i=1}^{m} w_i |x - a_i| + \sum_{i=1}^{m} w_i |y - b_i| \qquad (12.3)$$

Since Equation 12.3 is written in such a way that terms involving $x$ are separate from terms involving $y$, the optimum values of $x$ and $y$ can be obtained independently.

In order to find the optimum value of $x$, two mathematical properties of such a solution are employed. Namely, the $x$-coordinate of the new facility will be the same as the $x$-coordinate of some existing facility; and the optimum $x$-coordinate will be such that no more than half the total weight is to the left of $x$ and no more than half the total weight is to the right of $x$. The latter condition is referred to as the *median condition*. Both properties also apply in determining the optimum value of $y$. A justification for these conditions is given in Askin and Standrige [1] and Francis and White [30].

# *Example 12.1*

In order to illustrate how one determines the optimum solution to the minisum problem, consider an example involving the minisum location of a new machine tool in a maintenance department. Suppose there are five existing machines that have a material handling relationship with the new machine. The existing machines are located at the points $P_1 = (1, 1)$, $P_2 = (5, 2)$, $P_3 = (2, 8)$, $P_4 = (4, 4)$, and $P_5 = (8, 6)$. The cost per unit distance traveled is the same between the new machine and each existing machine. The number of trips per day between the new machine and the existing machines are 5, 6, 2, 4, and 8, respectively.

Ordering the $x$-coordinates of the new facilities gives the facilities sequenced 1, 2, 4, 5, and 8, with the corresponding sequence of weights being 5, 2, 4, 6, and 8. The total weight

**Table 12.1  *x-Coordinate Solution for Example 12.1***

| Machine $i$ | Coordinate $a_i$ | Weight $w_i$ | $\sum_{j=1}^{i} w_j$ |
|---|---|---|---|
| 1 | 1 | 5 | 5 |
| 3 | 2 | 2 | 7 |
| 4 | 4 | 4 | $11 < 25/2$ |
| 2 | 5 | 6 | $17 > 25/2$ |
| 5 | 8 | 8 | 25 |
| | $x^* = a_2 = 5$ | | |

**Table 12.2  *y-Coordinate Solution for Example 12.1***

| Machine $i$ | Coordinate $b_i$ | Weight $w_i$ | $\sum_{j=1}^{i} w_j$ |
|---|---|---|---|
| 1 | 1 | 5 | 5 |
| 2 | 2 | 6 | $11 < 25/2$ |
| 4 | 4 | 4 | $15 > 25/2$ |
| 5 | 6 | 8 | 23 |
| 3 | 8 | 2 | 25 |
| | $y^* = b_4 = 4$ | | |

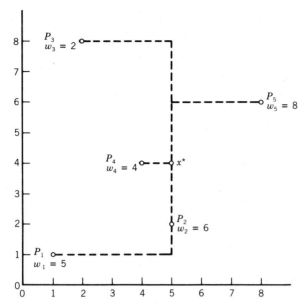

**Figure 12.1**  Facility locations for Example 12.1.

is 25. As shown in Table 12.1, the partial sum first equals or exceeds one-half the total for $i = 2$; hence $x^* = a_2 = 5$. In a similar fashion, as depicted in Table 12.2, the optimum $y$ coordinate is found to be $y^* = b_4 = 4$. Thus, $X^* = (5, 4)$, as depicted in Figure 12.1. Rectilinear paths between the new facility and each existing facility are indicated by dashed lines in Figure 12.1.

The total weighted distance resulting from the location $x = (5, 4)$ is:

$$f(5, 4) = 5(|5 - 1| + |4 - 1|) + 6(|5 - 5| + |4 - 2|) + 2(|5 - 2| + |4 - 8|)$$
$$+ 4(|5 - 4| + |4 - 4|) + 8(|5 - 8| + |4 - 6|)$$
$$= 35 + 12 + 14 + 4 + 40 = 105$$

If one were unable to locate the new machine at the point $(5, 4)$ because the location coincided with, say, a heat treatment furnace, then alternate sites could be evaluated by computing the value of $f(X)$ for each site and selecting the site with the smallest value of $f(X)$.

It should be noted that in the case where the partial sum is equal to one-half of the sum of all the weights, then the solution includes all points between the coordinate where the equality occurred and the next coordinate value.

## Example 12.2

In the previous example suppose it is not possible to place the new machine at a point other than the following candidate sites: $Q_1 = (5, 6)$, $Q_2 = (4, 2)$ and $Q_3 = (8, 4)$. Which would be preferred? Computing the value of $f(X)$ for $X = Q_i$, $k = 1, 2, 3$ yields:

$$f(5, 6) = 45 + 24 + 10 + 12 + 24 = 115$$
$$f(4, 2) = 20 + 6 + 16 + 8 + 64 \quad = 114$$
$$f(8, 4) = 50 + 30 + 20 + 16 + 16 = 132$$

Hence, the best site would be $Q_2$; however, $Q_1$ has very nearly the same value of $f(X)$. Qualitative considerations, as well as quantitative considerations not reflected in $f(X)$, might indicate $Q_1$ is preferred to $Q_2$.

In general, one could construct iso-cost contour lines as an aid in determining an appropriate location for the new facility. Software tools such as Matlab [53], Mathematica, and other similar tools may be used for this purpose. Figure 12.2 shows the iso-cost contour lines for the rectilinear function

$$f(X) = 5(|x - 1| + |y - 1|) + 6(|x - 5| + |y - 2|) + 2(|x - 2| + |y - 8|)$$
$$+ 4(|x - 4| + |y - 4|) + 8(|x - 8| + |y - 6|)$$

## Single-Facility Minimax Location Problem

In the case of rectilinear distances the minimax location problem is given by

$$\text{Minimize } f(X) = \max [(|x - a_i| + |y - b_i|), i = 1, 2, \ldots m] \qquad (12.4)$$

In order to obtain the minimax solution, let

$$c_1 = \text{minimum } (a_i + b_i) \qquad (12.5)$$
$$c_2 = \text{maximum } (a_i + b_i) \qquad (12.6)$$
$$c_3 = \text{minimum } (-a_i + b_i) \qquad (12.7)$$
$$c_4 = \text{maximum } (-a_i + b_i) \qquad (12.8)$$

Optimum solutions to the minimax location problem can be shown to be all points on the line segment connecting the point

$$(x_1^*, y_1^*) = 0.5 \,(c_1 - c_3, c_1 + c_3 + c_5) \qquad (12.9)$$

and the point

$$(x_2^*, y_2^*) = 0.5(c_2 - c_4, c_2 + c_4 - c_5) \qquad (12.10)$$

where $c_5 = \max (c_2 - c_1, c_4 - c_3)$. The maximum distance will be equal to $c_5/2$.

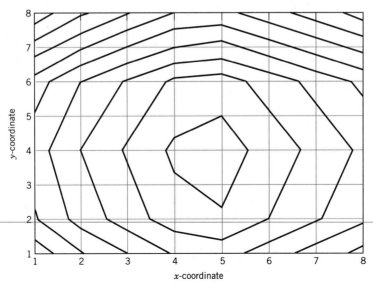

**Figure 12.2**   Contour lines for Example 12.1.

**Table 12.3** *Data for Example 12.3*

| $i$ | $a_i$ | $b_i$ | $a_i + b_i$ | $-a_i + b_i$ |
|---|---|---|---|---|
| 1 | 0 | 0 | 0 | 0 |
| 2 | 4 | 6 | 10 | 2 |
| 3 | 8 | 2 | 10 | -6 |
| 4 | 10 | 4 | 14 | -6 |
| 5 | 4 | 8 | 12 | 4 |
| 6 | 2 | 4 | 6 | 2 |
| 7 | 6 | 4 | 10 | -2 |
| 8 | 8 | 8 | 16 | 0 |
| $c_1 = 0$ | $c_2 = 16$ | $c_3 = -6$ | $c_4 = 4$ | $c_5 = 16$ |

# Example 12.3

In order to illustrate the use of the minimax solution procedure, consider the problem of locating a maintenance department in a production area. It is desirable to locate the maintenance facility as close to each machine as possible, in order to minimize machine downtime.

Eight machines are to be maintained by crews from the central maintenance facility. The coordinate locations of the machines are (0, 0), (4, 6), (8, 2), (10, 4), (4, 8), (2, 4), (6, 4), and (8, 8). From Table 12.3 it is seen that $c_1 = 0$, $c_2 = 16$, $c_3 = -6$, $c_4 = 4$, and $c_5 = 16$. Hence, the optimum solutions lie on the line segment connecting the point.

$$(x_1^*, y_1^*) = \tfrac{1}{2}(6, 10) = (3, 5)$$

and the point

$$(x_2^*, y_2^*) = \tfrac{1}{2}(12, 4) = (6, 2)$$

as shown in Figure 12.3. The point (3, 5) is 8 distance units away from $P_1$, $P_3$, $P_4$, and $P_8$; the point (6, 2) is 8 distance units away from $P_1$, $P_5$, and $P_8$; the remaining points on the line segment are 8 distance units away from $P_1$ and $P_8$.

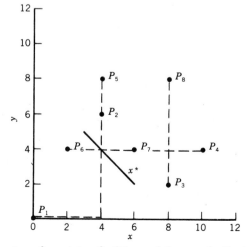

**Figure 12.3** Locations for existing facilities and the new facility for Example 12.3.

## *n-Facility Location Problem*

The general formulation of the *n*-facility location problem is given by

$$\text{Minimum } f(X) = \sum_{j=1}^{n} \sum_{k=1}^{n} v_{jk}\, d(X_j, X_k) + \sum_{j=1}^{n} \sum_{i=1}^{m} w_{ji}\, d(X_j, P_i) \tag{12.11}$$

where $X = (x_j, y_j)$ location of new facility $j$, $j = 1, 2, \ldots, n$

$P = (a_i, b_i)$ location of existing facility $i$, $i = 1, 2, \ldots m$

$v_{jk}$        cost per unit distance travel between new facilities $j$ and $k$

$w_{ji}$        cost per unit distance travel between $j$ and $i$

$d(X_j, X_k)$    distance between $j$ and $k$

$d(X_j, P_i)$    distance between $j$ and $i$

i.     with rectilinear distance

$$d(X_j, X_k) = |x_j - x_k| + |y_j - y_k|, \text{ and} \tag{12.12}$$
$$d(X_j, P_i) = |x_j - a_i| + |y_j - b_i|. \tag{12.13}$$

ii.     with Euclidean distance

$$d(X_j, X_k) = [(x_j - x_k)^2 + (y_j - y_k)^2]^{1/2}, \text{ and} \tag{12.14}$$
$$d(X_j, P_i) = [(x_j - a_i)^2 + (y_j - b_i)^2]^{1/2}. \tag{12.15}$$

With a transformation of terms with absolute values, the *n*-facility rectilinear distance location problem can be solved using linear programming methods. The Euclidean distance problem is more difficult to solve since the function to be minimized is continuous but nondifferentiable over the entire range of both the *x*- and *y*-coordinates.

We refer you to [29] for detailed discussions on solution methods to the *n*-facility location problem.

## Location-Allocation Models

An important class of facility location problems is that involving not only the determination of where new facilities are to be located but also which customers (existing facilities) will be served by each new facility. Such a problem is referred to as a **location-allocation problem**. In its most general form the location-allocation problem also involves a determination of the optimum number of new facilities.

A mathematical formulation of the location-allocation problem is given as follows:

$$\text{Minimize } \Psi = \sum_{j=1}^{n} \sum_{i=1}^{m} z_{ji}\, w_{ji}\, d(X_j, P_i) + g(n) \tag{12.16}$$

subject to

$$\sum_{j=1}^{n} z_{ji} = 1 \qquad i = 1, \ldots, m \tag{12.17}$$
$$n = 1, 2, \ldots, m$$

where

$\Psi$ = total cost per unit time

$n$ = number of new facilities

$X_j = (x_j, y_j)$, the coordinate location of new facility $j$

$w_{ji}$ = cost per unit time per unit distance if new facility $j$ interacts with existing facility $i$

$z_{ji}$ = 1, if new facility $j$ interacts with existing facility $i$; 0, otherwise

$g(n)$ = cost per unit time of providing $n$ new facilities

The decision variables in the location-allocation problem are $n$, the number of new facilities, $Z$, the allocation matrix, and $X_j$, $j = 1, \ldots, n$, the locations of the new facilities. Constraints of Equation 12.17 ensure that each existing facility interacts with only one new facility.

Because of the difficulty in solving the general problem, the location-allocation problem is solved by using an enumeration procedure. Namely, for a given value of $n$ the resulting problem is solved and the value of the objective function, $\Psi_n$, recorded. Then, the optimum value of $n$ is obtained by searching for the minimum value of $\Psi_n$.

For the purposes of this chapter, we will simply enumerate all of the allocation combinations for each value of $n$, determine the optimum location for each new facility for each allocation combination, and specify the minimum cost solution. The number of possible allocations, given $m$ customers and $n$ new facilities is given by

$$S(n, m) = \sum_{k=0}^{n-1} \frac{(-1)^k (n - k)^m}{k! \, (n - k)!}$$

(12.18)

## *Example 12.4*

To illustrate an approach to solving location-allocation problems, consider four machines which are located along a common aisle as follows: $P_1 = (0, 0)$, $P_2 = (3, 0)$, $P_3 = (6, 0)$, and $P_4 = (12, 0)$. For simplicity, we assume that $w_1 = w_2 = w_3 = 1$ and $w_4 = 2$. The cost of setting up $n$ new facilities is $g(n) = 5n$.

For $n = 1$, the problem reduces to a single-facility location problem with $x = 6$. The total cost is computed as $5(1) + [1(6) + 1(3) + 1(0) + 2(6)] = 26$.

For $n = 2$, the following allocations can be made

| New Facility 1 | New Facility 2 |
|---|---|
| $P_1$ | $P_2, P_3, P_4$ |
| $P_1, P_2$ | $P_3, P_4$ |
| $P_1, P_2, P_3$ | $P_4$ |

For $n = 3$,

| New Facility 1 | New Facility 2 | New Facility 3 |
|---|---|---|
| $P_1$ | $P_2$ | $P_3, P_4$ |
| $P_1$ | $P_2, P_3$ | $P_4$ |
| $P_1, P_2$ | $P_3$ | $P_4$ |

For $n = 4$,

| New Facility 1 | New Facility 2 | New Facility 3 | New Facility 4 |
|---|---|---|---|
| $P_1$ | $P_2$ | $P_3$ | $P_4$ |

The number of assignments given above is limited since the $y$-coordinates of all existing facilities are equal to zero. Without this constraint, the total number of possible assignments will be that given by Equation 12.18.

We note that for each assignment, a single-facility location problem is solved. Based on these allocations, the minimum total cost location-allocation can be found.

## Single-Plant Location Model

The plant location problem and/or warehouse location problem has received considerable attention in the research literature. A mathematical formulation of one version of the plant location problem is given as follows:

$$\text{Minimize } z = \sum_{i=1}^{m} \sum_{j=1}^{n} c_{ij} y_{ij} + \sum_{j=1}^{n} f_j x_j \tag{12.19}$$

subject to

$$\sum_{i=1}^{m} y_{ij} \leq m \, x_j \qquad j = 1, \ldots, n \tag{12.20}$$

$$\sum_{j=1}^{m} y_{ij} = 1 \qquad i = 1, \ldots, m \tag{12.21}$$

$$y_{ij} \geq 0 \text{ for all } i, j$$

$$x_j = (0, 1) \text{ for all } j$$

where

$m$ = number of customers

$n$ = number of plant sites

$y_{ij}$ = proportion of customer $i$ demand supplied by a plant at site $j$

$x_j$ = 1, if a plant is located at site $j$; otherwise, 0

$c_{ij}$ = cost of supplying all demand of customer $i$ from a plant located at site $j$

$f_j$ = fixed cost of locating a plant at site $j$

The objective function gives the cost of locating plants at the sites corresponding to the positive-valued $x_j$. The first set of $n$ constraints indicates that the total proportion of customer demand supplied by a plant at site $j$ either equals zero when $x_j$ equals zero or cannot exceed the number of customers when $x_j$ equals one. The second set of $m$ constraints ensures that all demand must be met for customer $i$ by some combination of plants. Non-negativity restrictions on $y_{ij}$ and zero-one restrictions on $x_j$ complete the mixed-integer programming formulation.

A branch and bound solution procedure can be used to solve exactly the plant location problem formulated above, as well as more general formations. Because a consideration of branch and bound methods is beyond the scope of this text, rather than solve the plant location problem exactly, we will use a quick-and-dirty method. In particular, we employ an "add" or "construction" heuristic in which one continues to add new facilities until no additional new facilities are justified. Alternatively, we could use a "drop" or "elimination" heuristic in which one drops new facilities until no additional facilities can be dropped.

## Example 12.5

To motivate the discussion of the plant location problem, suppose there exist five potential sites for a new warehouse A, . . . , E and the firm's major customers are located in five cities 1, . . . , 5. The annual costs of meeting the customer's demands in a city from each site are given in Table 12.4. The annual fixed cost of providing a new warehouse is also given for each site.

If only one warehouse is to be built, which site would you prefer? Selecting site A yields a total annual cost of $19,900, including the cost of the warehouse. The total annual costs for the remaining sites are: $17,700, $9,600, $14,800, and $11,100. Hence, site C would be selected. Notice that if a warehouse is placed at site C, then cities 4 and 5 will be served by the warehouse at site C regardless of any subsequent decisions to locate warehouses at other sites.

Suppose we decide to place a warehouse at site A, in conjunction with the warehouse at site C. What will be the reduction in total annual cost, if any? Cities 1 and 2 would be served from site A at a cost savings of $1,700 + $1,100 = $2,800, which is less than the fixed cost of $3,000 for providing a warehouse at A. Placing a warehouse at site B yields cost savings of $1,300 + $2,400 + $500 = $4,200 for cities 1, 2, and 3; since the cost savings is $2,200 greater than the fixed cost of $2,000, site B is a feasible candidate for an additional warehouse. Placing a warehouse at site D yields annual savings of $3,100 to be offset by the fixed cost of $3,000. Site E yields annual savings of $100 for city 1 at a fixed cost of $4,000. A second warehouse will be placed at site B as the greatest net annual savings occurs for site B. The new total annual cost is ($9,600 − $4,200) + $2,000 = $7,400.

Currently, we have assigned warehouses to sites B and C. If a third one is added, the net annual savings resulting from the placement of the third warehouse at site j, denoted $NAS(j)$, will be

$$NAS(A) = 400 - 3,000 = -2,600$$
$$NAS(D) = 900 - 3,000 = -2,100$$
$$NAS(E) = 0 - 4,000 \quad = -4,000$$

**Table 12.4   Data for Example 12.5**

| Customer Location | Warehouse Sites | | | | |
|---|---|---|---|---|---|
| | A | B | C | D | E |
| 1 | 100 | 500 | 1,800 | 1,300 | 1,700 |
| 2 | 1,500 | 200 | 2,600 | 1,400 | 1,800 |
| 3 | 2,500 | 1,200 | 1,700 | 300 | 1,900 |
| 4 | 2,800 | 1,800 | 700 | 800 | 800 |
| 5 | 10,000 | 12,000 | 800 | 8,000 | 900 |
| Fixed costs | 3,000 | 2,000 | 2,000 | 3,000 | 4,000 |

Hence, no additional warehouses are justified and warehouses will be placed only at sites B and C for a total annual cost of $7,400.

The approach of continuing to add warehouses to sites having the greatest positive net annual savings is not guaranteed to yield an optimum solution. In particular, it may occur that a subsequent addition of warehouses causes some previous decision to add a warehouse to be nonoptimum. One approach that can be used to partially overcome such a possibility is to drop from solution the site that produces the greatest annual savings. A check for feasible "drops" would be made after each check for feasible "adds."

# 12.3  FACILITY LAYOUT MODELS

In this section, we describe several models for facility layout problems. You will recall that in our discussion on facility location problems, we were mostly concerned with finding the optimum values of $x$ and $y$ in continuous space. Here, the solutions will be restricted to discrete locations. In addition, area-consuming facilities will be accounted for and their relative placements considered in attempting to find the best layout for a facility. Examples of area-consuming facilities include the work envelope of a machine tool and the area required in the storage of materials in a warehouse.

## Quadratic Assignment Problem

Location-allocation problems involve a determination of the number and location of new facilities. In some situations the number of new facilities is known a priori and the decision problem involves the location of the new facilities. In this section we consider the problem of locating multiple new facilities in discrete space. Such problems are referred to as assignment problems because the problem reduces to assigning new facilities to sites.

If there exists no interaction between new facilities such that one is concerned only with locating new facilities relative to existing facilities, the problem is a **linear assignment problem**. When interaction exists between new facilities, the problem is a **quadratic assignment problem**. Quadratic assignment problems are treated in this section. A mathematical formulation of the problem is as follows:

$$\text{Minimize } z = \sum_{j=1}^{n} \sum_{k=1}^{n} \sum_{h=1}^{n} \sum_{\ell=1}^{n} c_{jkh\ell} x_{jk} x_{h\ell} \tag{12.22}$$

subject to

$$\sum_{j=1}^{n} x_{jk} = 1 \qquad k = 1, \ldots, n$$

$$\tag{12.23}$$

$$\sum_{k=1}^{n} x_{jk} = 1 \qquad j = 1, \ldots, n$$

$$x_{jk} = (0, 1) \text{ for all } j \text{ and } k \tag{12.24}$$

where $c_{jkh\ell}$ is the cost of assigning new facility $j$ to site $k$ when new facility $h$ is assigned to site $\ell$.

Heuristic procedures are generally used to solve quadratic assignment problems. Two heuristic procedures are considered in this section.

Generally, heuristic solution procedures for the quadratic assignment problem can be categorized as **construction** and **improvement** procedures. Construction procedures develop a solution "from scratch"; namely, facilities are located one at a time until all are located. Improvement procedures begin with all facilities located and seek ways to improve on the solution by interchanging or switching locations for facilities. We consider one construction procedure and one improvement procedure in this chapter.

## Construction Procedure

A quick-and-dirty solution for the quadratic assignment problem can be obtained using a construction procedure based on the work of Conway and Maxwell [18] and Wimmert [85]. The procedure requires that the interaction cost be given by the product of flow and distance. Hence, we will assume that $c_{jkh\ell}$ equals the product of $v_{jh}$, the flow between facilities $j$ and $h$, and $d_{k\ell}$, the distance between sites $k$ and $\ell$. A number of different approaches can be taken; the one we present requires that the $v_{jh}$ values be ordered from largest to smallest and the $d_{k\ell}$ values be ordered from smallest to largest. The approach is as follows:

> Assign pairs of facilities to pairs of sites such that the facilities with the greatest flow are located at the sites having the smallest distance separation: continue the assignment process until a feasible assignment is made.

In order to provide some measure of the effectiveness of the quick-and-dirty solution, a *lower bound* can be calculated for the optimum solution to the quadratic assignment problem. Such a lower bound is determined by multiplying the ordered vector of $v_{jh}$ values and the ordered vector of $d_{k\ell}$ values. The total distance traveled for any assignment will be at least as great as the vector product. If the quick-and-dirty solution results in a total distance equal to the lower bound, then the solution is an optimum solution; otherwise, it is not known if the solution is optimum, as the lower bound may be unattainable.

# Example 12.6

In order to illustrate the quick-and-dirty solution procedure, suppose four machines are to be placed in a job shop. Flow and distance values are given by the symmetric matrices $V$ and $D$,

$$V = \begin{pmatrix} 0 & 2 & 8 & 3 \\ 2 & 0 & 4 & 9 \\ 8 & 4 & 0 & 5 \\ 3 & 9 & 5 & 0 \end{pmatrix} \qquad D = \begin{pmatrix} 0 & 8 & 10 & 2 \\ 8 & 0 & 4 & 7 \\ 10 & 4 & 0 & 9 \\ 2 & 7 & 9 & 0 \end{pmatrix}$$

Ordering the $v_{jh}$ values and the $d_{k\ell}$ values gives the vectors **v** and **d**.

$$\mathbf{v} = (9, 8, 5, 4, 3, 2)$$

and

$$\mathbf{d} = (2, 4, 7, 8, 9, 10)$$

The calculation below is based on the fact that the matrices $V$ and $D$ are symmetric. If they are not symmetric, then the equivalent flow-between matrices must be obtained first.

The lower bound $LB$ is given by

$$LB = \mathbf{vd'} = 9(2) + 8(4) + 5(7) + 4(8) + 3(9) + 2(10)$$

$$= 164$$

The largest flow value ($v_{24} = 9$) is between facilities 2 and 4; the smallest distance ($d_{14} = 2$) is between sites 1 and 4. Hence, facility 2 will be assigned to either site 1 or site 4 and facility 4 will be assigned to the remaining site. The next flow value in $\mathbf{v}$ is $v_{13} = 8$ and the next distance value in $\mathbf{d}$ is $d_{23} = 4$; therefore, it is desired to assign facilities 1 and 3 among sites 2 and 3. The next flow value in $\mathbf{v}$ is $v_{34} = 5$ and the next distance value in $\mathbf{d}$ is $d_{24} = 7$; therefore, it is desired to assign facilities 3 and 4 among sites 2 and 4. Since facility 4 is to be assigned to either site 1 or site 4 and facility 4 is to be assigned to either 2 or site 4, it follows that facility 4 will be assigned to site 4. Thus, facility 2 will be assigned to site 1, facility 3 will be assigned to site 2, and facility 1 will be assigned to site 3. The cost of the assignment is

$$z = v_{12}\, d_{31} + v_{13}\, d_{32} + v_{14}\, d_{34} + v_{23}\, d_{12} + v_{24}\, d_{14} + v_{34}\, d_{24}$$

$$= 2(10) + 8(4) + 3(9) + 4(8) + 9(2) + 5(7)$$

$$= 164$$

Since the objective function value for the assignment equals the lower bound, an optimum solution has been obtained. Such will not always be the case! For larger problems, the details of the quick-and-dirty method become a bit more involved; however, the principle remains the same: *Locate facilities having the greatest interaction (flow) as close together as possible.*

## Improvement Procedures

The improvement procedure we present is based on the CRAFT procedure described in Chapters 7 and 8. It is referred to as a pairwise exchange method. The procedure begins with all facilities assigned to sites. Next, all pairwise interchanges are considered and the interchange that yields the greatest reduction in total cost is performed. The process continues until no pairwise interchange can be found that will yield a reduction in total cost.

# Example 12.7

The same problem solved using the construction procedure will be solved using the improvement procedure. For convenience, label sites numerically and facilities alphabetically. The following flows occur *between* facilities: AB(2), AC(8), AD(3), BC(4), BD(9), and CD(5). Suppose we begin by arbitrarily assigning facilities A, B, C, and D to sites 1, 2, 3, and 4, respectively: Hence, the distances between facilities will be AB(8), AC(10), AD(2), BC(4), BD(7), and CD(9). The resulting total cost for the initial layout is obtained by summing the product of flow and distance for each pair of facilities. Thus, the total cost of the initial solution is

$$2(8) + 8(10) + 3(2) + 4(4) + 9(7) + 5(9)$$

or 226.

Given an initial solution, we now consider switching locations for pairs of departments. Suppose A and B are interchanged. By locating B at site 1 and A at site 2 the new distances will be AB(8), AC(4), AD(7), BC(10), BD(2), and CD(9). The new total cost will be

$$2(8) + 8(4) + 3(7) + 4(10) + 9(2) + 5(9)$$

or 172. We do not make the AB interchange yet, even though a reduction in total cost is possible. We will compute the total cost for all pairwise interchanges and then make the interchange that yields the greatest reduction in the total cost.

Next, we consider the effect of interchanging A and C, but still based on the initial solution. By placing C at site 1 and A at site 3 the following distances result: AB(4), AC(10), AD(9), BC(8), BD(7), and CD(2). The new total cost will be

$$2(4) + 8(10) + 3(9) + 4(8) + 9(7) + 5(2)$$

or 220.

The process continues by considering interchanges of A and D, B and C, B and D, and C and D. The results of the interchanges are given in Table 12.5 for the initial solution. As can be seen, the best interchange is A and B; hence, it will be made to obtain the first improved solution.

Next, we consider the effect on total cost of making pairwise interchanges for the first improved solution. The results of the interchanges are shown in Table 12.6. As can be seen, the greatest reduction in total cost occurs when facilities A and C are interchanged. Hence, the interchange is made and the second improved solution is obtained.

Table 12.5   *Pairwise Exchange Results for the Initial Solution to Example 12.7*

| | | | Distances | | | | | |
|---|---|---|---|---|---|---|---|---|
| | | | | Pairwise Interchanges | | | | |
| | Facility | Initial | | | | | | |
| Flows | Pairs | Solution | AB | AC | AD | BC | BD | CD |
| 2 | AB | 8 | 8 | 4 | 7 | 10 | 2 | 8 |
| 8 | AC | 10 | 4 | 10 | 9 | 8 | 10 | 2 |
| 3 | AD | 2 | 7 | 9 | 2 | 2 | 8 | 10 |
| 4 | BC | 4 | 10 | 8 | 4 | 4 | 9 | 4 |
| 9 | BD | 7 | 2 | 7 | 8 | 9 | 7 | 7 |
| 5 | CD | 9 | 9 | 2 | 10 | 7 | 4 | 9 |
| Total Cost | | 226 | 172 | 220 | 230 | 222 | 227 | 186 |

Table 12.6   *Pairwise Exchange Results for the First Improved Solution to Example 12.7*

| | | | Distances | | | | | |
|---|---|---|---|---|---|---|---|---|
| | | | | Pairwise Interchanges | | | | |
| | Facility | First Improved | | | | | | |
| Flows | Pairs | Solution | AB | AC | AD | BC | BD | CD |
| 2 | AB | 8 | 8 | 10 | 2 | 4 | 7 | 8 |
| 8 | AC | 4 | 10 | 4 | 9 | 8 | 4 | 7 |
| 3 | AD | 7 | 2 | 9 | 7 | 7 | 8 | 4 |
| 4 | BC | 10 | 4 | 8 | 10 | 10 | 9 | 2 |
| 9 | BD | 2 | 7 | 2 | 8 | 9 | 2 | 10 |
| 5 | CD | 9 | 8 | 7 | 4 | 2 | 10 | 9 |
| Total Cost | | 172 | 221 | 164 | 229 | 224 | 174 | 227 |

Normally, the process would continue and pairwise interchanges for the second improved solution would be considered. The search for the best solution ends when either no interchanges are found that yield a reduction in total cost or the lower bound is obtained. In this case, we know it is not necessary to search further, since the second improved solution has a total cost equal to the lower bound computed in Example 12.6. Hence, the search ends with facilities A, B, C, and D located at sites 3, 1, 2, and 4, respectively.

Since the improvement procedure requires an initial solution, the final solution obtained from the construction procedure could be used to initiate the improvement procedure.

## Warehouse Layout Models

In this section, a quantitative warehouse layout model is considered. Specifically, the determination of the location of products for storage in a warehouse is considered. To motivate the discussion, it is necessary to recall the distinction given in Chapter 9 between **dedicated** or **fixed-slow storage** and **randomized** or **floating slot storage**. Recall, with dedicated storage a particular set of storage slots or locations is assigned to a specific product; hence, a number of slots equal to the maximum inventory level for the product must be provided. With a pure randomized storage system each unit of a particular product is equally likely to be retrieved when a retrieval operation is performed; likewise, each empty storage slot is equally likely to be selected for storage when a storage operation is performed.

In this section, an approach is presented for determining the optimum dedicated storage layout; rectilinear travel is assumed. The warehouse layout problem considered involves the assignment of products to storage locations in the warehouse. The following notation is used:

$q$ = number of storage locations

$n$ = number of products

$m$ = number of input/output (I/O) points (docks)

$S_j$ = number of storage locations required for product $j$

$T_j$ = number of trips in/out of storage for product $j$, that is, throughput of product $j$

$p_i$ = percentage of travel in/out of storage to/from I/O point $i$

$d_{ik}$ = distance (or time) required to travel from I/O point $i$ to storage location $k$

$x_{jk}$ = 1 if product $j$ is assigned to storage location $k$; otherwise, 0

$f(\mathbf{x})$ = average distance (or time) traveled

The warehouse layout problem can be formulated as follows:

$$\text{Minimize} \sum_{j=1}^{n} \sum_{k=1}^{q} \frac{T_j}{S_j} \sum_{i=1}^{m} p_i d_{ik} \tag{12.25}$$

subject to

$$\sum_{j=1}^{n} x_{jk} = 1 \qquad k = 1, \ldots, q$$

$$\tag{12.26}$$

$$\sum_{k=1}^{p} x_{jk} = S_j \qquad j = 1, \ldots, n$$

$$x_{jk} = (0, 1) \text{ for all } j \text{ and } k \tag{12.27}$$

It is assumed that each item is equally likely to travel between I/O point or dock $i$ and any storage location assigned to item $j$. Hence, the quantity $(1/S_j)$ is the probability that a particular storage location assigned to product $j$ will be selected for travel to/from a dock. For convenience, let

$$f_k = \sum_{i=1}^{m} p_i d_{ik}$$

In words, $f_k$ is the expected distance traveled between storage location $k$ and the docks.

In order to minimize the total expected distance traveled the following approach is taken.

1.  Number the products according to their $T_j/S_j$ value, such that

$$\frac{T_1}{S_1} \geqslant \frac{T_2}{S_1} \geqslant \cdots \geqslant \frac{T_n}{S_n}$$

2.  Compute the $f_k$ values for all storage locations.
3.  Assign product 1 to the $S_1$ storage locations having the lowest $f_k$ values; assign product 2 to the $S_2$ storage locations having the next lowest $f_k$ values; and so on.

## Example 12.8

As an illustration of the solution procedure for designing a warehouse layout, consider the warehouse given in Figure 12.4$a$. Storage bays are of size 20×20 ft. Docks $P_1$ and $P_2$ are for truck delivery; docks $P_3$ and $P_4$ are for rail delivery. Dedicated storage is used. Sixty percent of all item movement in and out of storage is from/to either $P_1$ or $P_2$, with each dock equally likely to be used. Forty percent of all item movement in and out of storage is equally divided between docks $P_3$ and $P_4$. Three products, A, B, and C, are to be stored in the warehouse with only one-type product stored in a given storage bay. Product A requires 3600 ft$^2$ of storage space and enters and leaves storage at a rate of 750 loads per month; product B requires 6400 ft$^2$ of storage space and enters and leaves storage at a rate of 900 loads per month; product C requires 4000 ft$^2$ of storage space and enters and leaves storage at a rate of 800 loads per month. Rectilinear travel is used and is measured between the centroids of storage bays.

The $f_k$ values are shown in each storage bay in Figure 12.4$b$. As an illustration of the computation of an $f_k$ value, suppose $k = 29$. Measuring the rectilinear distance from the centroid of storage bay 29 and each of the four docks gives $d_{1,29} = 120$, $d_{2,29} = 100$, $d_{3,29} = 100$, and $d_{4,29} = 80$. Hence,

$$f_{29} = 0.3(120) + 0.3(100) + 0.2(100) + 0.2(80)$$
$$= 102$$

The number of storage bays required for each product equals $S_A = 3600/400 = 9$, $S_B = 16$, and $S_C = 10$. The $T_j$ values are $T_A = 750$, $T_B = 900$, and $T_C = 800$. Therefore, the $T_j/S_j$ values are $T_A/S_A = 83.33$, $T_B/S_B = 56.25$, and $T_C/S_C = 80$. Hence the products will be numbered 1(A), 2(C), and 3(B).

Product 1(A) requires 9 storage bays; hence, the storage bays assigned to product A include [17, 18, 19, 9, 25, 20, 10, 26, 21]. Product 2(C) requires 10 storage bays; hence, [11, 27, 22, 12, 28, 23, 13, 29, 24, 1] are assigned to product C. Product 3(B)

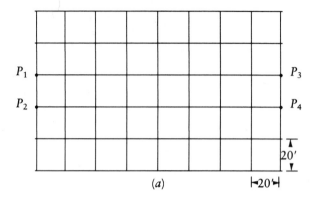

(a)

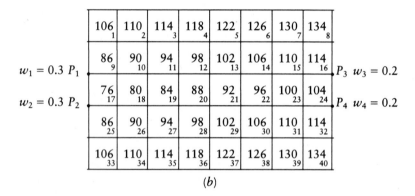

(b)

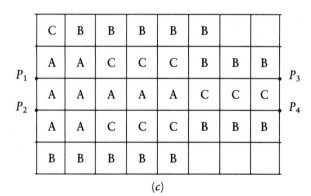

(c)

**Figure 12.4** Warehouse layout for Example 12.8.

requires 16 storage bays; hence, product B is assigned to storage locations [14, 30, 33, 2, 15, 31, 34, 3, 16, 32, 35, 4, 36, 5, 37, 6]. Storage bays 7, 8, 38, 39, and 40 are available for equipment storage, rest rooms, offices, and so on.

The layout design that minimizes expected distance traveled per unit time is given in Figure 12.4c. It is important to point out that the layout design obtained is not necessarily a final layout. Considerations other than expected distance traveled have not been incorporated; the use of the dedicated storage approach must be examined in a particular application. However, the layout does serve as a basis for evaluating other designs.

# *12.4* MACHINE LAYOUT MODELS

Thus far, the layout models we have covered in previous sections are aggregate in nature; that is, we considered rectangular or square blocks and distances are measured from department centroids or from fixed locations of pick-up/delivery stations. However, there are other design issues that affect machine layout problems, which are not addressed in the previous models. In particular, we mention two below.

a.  The interface points for incoming and outgoing parts for individual machines are usually at fixed locations relative to the entire work envelope of the machine.

b.  Minimum spaces between machines must be provided to accommodate access to machines for maintenance and service, and allow enough space for material handling devices and in-process storage areas.

Based on these two factors, we can see that the relative position of two adjacent machines would give two different values for, say, total distance parts are moved between the two machines.

Machine layout problems may be classified as

a.  Circular machine layout
b.  Linear single-row machine layout
c.  Linear double-row machine layout
d.  Cluster machine layout

## Single-Row Machine Layout Model[2]

Let

$x_i^P$ = location of pick-up point of machine $i$, $i = 1, 2, \ldots n$

$x_i^D$ = location of delivery point of machine $i$

$w_i$ = width of machine $i$ oriented parallel to the aisle

$a_{ij}$ = minimum clearance between machines $i$ and $j$

$c_{ij}$ = cost per trip per unit distance from machine $i$ to $j$

$f_{ij}^L$ = total loaded material handling trips from machine $i$ to machine $j$

$f_{ij}^E$ = total empty (deadhead) material handling trips from machine $i$ to machine $j$

---

[2]The presentation in this section is a modified version of the model and algorithm given by Heragu [38] and Heragu and Kusiak [39].

The single-row model formulation is as follows:

$$\text{Minimize TC} = \sum_{i=1}^{n} \sum_{j=1}^{n} c_{ij}(f^L_{ij} d^L_{ij} + f^e_{ij} d^E_{ij}) \tag{12.29}$$

The objective function gives the total cost of material handling where

$$d^L_{ij} = |x^P_i - x^D_j|, \quad \text{loaded travel distance}$$
$$d^E_{ij} = |x^D_i - x^P_j|, \quad \text{empty (deadhead) travel distance}$$

For simplicity, we will assume that each machine $P/D$ point is located at the midpoint along the edge of the machine work area parallel to the aisle. Thus, $d^L_{ij} = d^E_{ij}$. Also, let $f_{ij} = f^L_{ij} + f^E_{ij}$.

The following constraints ensures that the positions of machine work areas satisfy the required clearance between machines

$$|x^P_i - x^D_j| \geq \frac{1}{2}(w_i + w_j) + a_{ij}, \quad i = 1, 2, \ldots, n - 1;$$
$$j = i + 1, \ldots, n \text{ for loaded trips, and} \tag{12.30}$$

$$|x^D_i - x^P_j| \geq \frac{1}{2}(w_i + w_j) + a_{ij}, \quad i = 1, 2, \ldots, n - 1;$$
$$j = i + 1, \ldots, n \text{ for deadhead trips.} \tag{12.31}$$

Direct solution to the above model requires a transformation of the absolute value terms in both the objective function and the constraints. Instead, we describe a construction-type procedure below.

**Step 1.** Determine the first two machines to enter the layout by computing max $\{c_{ij}f_{ij}\}$. The rationale behind the selection of the machine pair $ij$ with the maximum value of $c_{ij}f_{ij}$ is that this value results in the maximum cost increase when the distance between them is set farther apart. The solution is denoted as $\{i^*, j^*\}$.

**Step 2.** Place $i^*$ and $j^*$ adjacent to each other. Since the $P/D$ points are located at the midpoint of the machine edge along its width, there would be no difference in total cost between the placement order $i^* \rightarrow j^*$ and the order $j^* \rightarrow i^*$.

**Step 3.** The next step is to select the next machine, denoted by $k^*$, to place in the layout and to determine where to locate this machine relative to those that are already in the layout. The placement order is either $k^* \rightarrow i^* \rightarrow j^*$ or $i^* \rightarrow j^* \rightarrow k^*$. The selection of $k^*$ is based on evaluating the relative placement cost, RPC, of setting machine $k$ to the left or to the right side of the set of machines already in the layout. The evaluation function is

$$\text{RPC} = \min_{k \in U} \left\{ \sum_{i \in A} c_{ki} f_{ki} d_{ki}, \sum_{j \in A} c_{jk} f_{jk} d_{jk} \right\} \tag{12.32}$$

where $A$ is the set of all machines already assigned specific locations and $U$ is the set of unassigned machines. The first summation gives the cost of placing of machine $k$ to the left and the second summation gives the cost of placing machine $k$ to the right of all assigned machines.

**Step 4.** Continue Step 3 until all machines are assigned.

# Example 12.9

Consider the problem of laying out four machines in a single row. There are a total of 4! = 24 ways of arranging these four machines. Let $c_{ij} = 1$ and $a_{ij} = 1$. The machine dimensions are

| Machine | 1 | 2 | 3 | 4 |
|---|---|---|---|---|
| Dimensions | $2 \times 2$ | $3 \times 3$ | $4 \times 4$ | $5 \times 5$ |

The loaded material handling trips between machines are

| Machine | | 1 | 2 | 3 | 4 |
|---|---|---|---|---|---|
| | 1 | — | 10 | 5 | 6 |
| | 2 | 8 | — | 3 | 8 |
| | 3 | 7 | 9 | — | 4 |
| | 4 | 5 | 11 | 13 | — |

The total flow-between matrix is

| Machine | | 1 | 2 | 3 | 4 |
|---|---|---|---|---|---|
| | 1 | — | 18 | 12 | 11 |
| | 2 | 18 | — | 12 | 19 |
| | 3 | 12 | 12 | — | 17 |
| | 4 | 11 | 19 | 17 | — |

**Steps 1 and 2.**   Since the maximum $\{f_{ij}\}$ value is 19, we select machines 2 and 4 to enter the layout. At this point, the placement order is not important. We arbitrarily select the order 2→4. Thus, $A = \{2, 4\}$ and $U = \{1, 3\}$.

**Step 3.**   For this step, we will evaluate the following placement orders: 1→2→4, 2→4→1, 3→2→4, and 2→4→3.
For placement order 1→2→4,

$$f_{12}d_{12} + f_{14}d_{14} = 18(3.5) + 11(8.5) = 156.5$$

For placement order 2→4→1,

$$f_{21}d_{21} + f_{41}d_{41} = 18(9.5) + 11(4.5) = 220.5$$

For placement order 3→2→4,

$$f_{32}d_{32} + f_{34}d_{34} = 12(4.5) + 17(9.5) = 215.5$$

For placement order 2→4→3,

$$f_{23}d_{23} + f_{43}d_{43} = 12(10.5) + 17(5.5) = 219.5$$

Since the minimum relative placement cost is 156.5, we select the placement order 1→2→4.

**Step 4.**   Since there is only one machine unassigned, $k = 3$, the remaining placement orders to be evaluated are 3→1→2→4 and 1→2→4→3.
For placement order 3→1→2→4,

$$f_{31}d_{31} + f_{32}d_{32} + f_{34}f_{34} = 12(4) + 12(7.5) + 17(12.5) = 350.5$$

For placement order 1→2→4→3,

$$f_{13}d_{13} + f_{23}d_{23} + f_{43}d_{43} = 17(14) + 12(10.5) + 12(5.5) = 430$$

Thus, the final placement order is 3→1→2→4 with a total cost of 602.

# *12.5*   STORAGE MODELS

Recall, in Chapter 1 it was emphasized that in designing storage systems the familiar six-step design process should be used. Namely,

1. Define storage system objectives and scope.
2. Analyze storage system requirements.
3. Generate alternative storage system designs.
4. Evaluate storage system design alternatives.
5. Select the preferred storage system design.
6. Implement the preferred system.

Unfortunately, the "real world" of storage systems design often appears to consist of only two steps:

1. Select the preferred storage system.
2. Implement the preferred system.

While some would take exception to this assertion, most would agree that a number of aspects of the design process do deserve more attention. Specifically, more attention should be paid to:

1. The determination of the objectives and scope for the system.
2. The development of a definitive database.
3. The consideration of a *number* of alternatives, ranging from conventional to automated and including hybrid system combinations.
4. The determination of "optimum" designs for each subsystem.
5. The formal evaluation of alternatives using multiple criteria.
6. The postaudit of the system.

In this section, we concentrate on one aspect of the storage systems design process. Namely, we focus on the determination of "optimum" designs for each subsystem. Specifically, the following unit load storage alternatives are considered:

1. Block stacking.
2. Deep lane storage.
3. Single-deep selective pallet rack.
4. Double-deep selective pallet rack.

Furthermore, the emphasis is on minimizing the space required to meet the storage requirements of the system.

## Block Stacking

As noted in Chapter 9, block stacking involves the storage of unit loads in stacks within storage rows. It is frequently used when large quantities of a few products are to be stored and the product is stackable to some reasonable height without load crushing. Frequently, unit loads are block stacked three high in rows that are 10 or more loads deep. The practice of block stacking is prevalent for food, beverages, appliances, and paper products, among others.

An important design question is how deep should the storage rows be. Block stacking is typically used to achieve a high space utilization at a low investment cost. Hence, it is often the case that storage rows are used with depths of 15, 20, 30, or more.

During the storage and retrieval cycle of a product lot, vacancies can occur in a storage row. To achieve first-in, first-out (FIFO) lot rotation, these vacant storage positions cannot be used for storage of other products or lots until all loads have been withdrawn from the row. The space losses resulting from unusable storage positions are referred to as "honeycomb loss"; block stacking suffers from both vertical and horizontal honeycomb loss. Figure 12.5 depicts the space losses resulting from honeycombing.

The design of the block stacking storage system is characterized by: the depth of the storage row $(x)$, the number of storage rows required for a given product lot $(y)$, and the height of the stack $(z)$, where the decision variables, $x$, $y$, and $z$ must be integer valued. If the height of the stack is fixed, then the key decision variable is the depth of the storage row.

For a single product, factors that may influence the optimum row depth include lot size, load dimensions, aisle widths, row clearances, allowable stacking heights, storage/retrieval times, and storage/retrieval distribution. For multiple products, other decision variables must be considered. For example, the optimum number of unique row depths, row depths, the assignment of products to depths, and aggregate space requirements must be determined.

The following notation is used throughout this section:

$S =$ average amount of floor space required during the life of a storage lot

$S_{BS} =$ average amount of floor space required, with block stacking and no safety stock

$S_{BSSS} =$ average amount of floor space required, with block stacking and safety stock

$S_{BS}^c =$ continuous approximation to the average amount of floor space required, with block stacking and no safety stock

$S_{BSSS}^c =$ continuous approximation to the average amount of floor space required, with block stacking and safety stock

$S_{DL} =$ average amount of floor space required, with deep lane storage and no safety stock

$S_{DLSS} =$ average amount of floor space required, with deep lane storage and safety stock

$S_{DL}^c =$ continuous approximation to the average amount of floor space required, with deep lane storage and no safety stock

$S_{DLSS}^c =$ continuous approximation to the average amount of floor space required, with deep lane storage and safety stock

$S_{DD} =$ average amount of floor space required, with double-deep storage and no safety stock

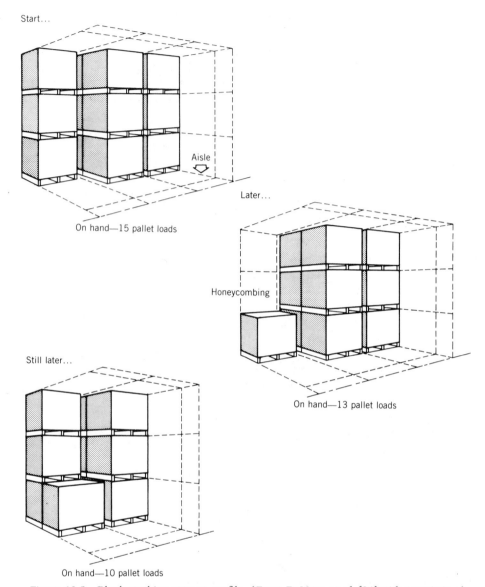

Start...

Aisle

On hand—15 pallet loads

Later...

Honeycombing

On hand—13 pallet loads

Still later...

On hand—10 pallet loads

**Figure 12.5**   Block stacking storage profile. (*From DeMars, et al. [21] with permission.*)

$S_{\mathrm{DDSS}}$ = average amount of floor space required, with double-deep storage and safety stock

$S_{\mathrm{SD}}$ = average amount of floor space required, with single-deep storage and no safety stock

$S_{\mathrm{SDSS}}$ = average amount of floor space required, with single-deep storage and safety stock

$Q$ = size of the storage lot, in unit loads

$s =$ safety stock, in unit loads

$W =$ width of a unit load

$L =$ length or depth of a unit load

$c =$ side-to-side clearance between unit loads and between a unit load and a vertical rack member

$r =$ width of a vertical rack member

$A =$ width of the storage aisle

$f =$ depth of the back-to-back flue space between rack sections

$x =$ depth of a storage row or storage lane, in unit loads

$x_{BS} =$ optimum depth of a storage row, in unit loads, with block stacking

$x^c{}_{BS} =$ continuous approximation of the optimum depth of a storage row, in unit loads, with block stacking

$x_{BSSS} =$ optimum depth of a storage row, in unit loads, with block stacking and safety stock

$x^c_{BSSS} =$ continuous approximation of the optimum depth of a storage row, in unit loads, with block stacking and safety stock

$x_{DL} =$ optimum depth of a storage lane, in unit loads, with deep lane storage

$x^c_{DL} =$ continuous approximation of the optimum depth of a storage lane, in unit loads, with deep lane storage

$x_{DLSS} =$ optimum depth of a storage lane, in unit loads, with deep lane storage and safety stock

$x^c_{DLSS} =$ continuous approximation of the optimum depth of a storage lane, in unit loads, with deep lane storage and safety stock

$z =$ storage height, in unit loads or levels of storage

$y =$ number of storage rows required to accommodate $Q$ unit loads with block stacking

$=$ smallest integer, greater than or equal to $Q/xz$

$v =$ number of storage lanes required to accommodate $Q$ unit loads with deep lane storage

$=$ smallest integer, greater than or equal to $Q/x$

$\eta =$ average number of storage rows required over the life of a storage lot, with block stacking

$\xi =$ average number of storage lanes required over the life of a storage lot, with deep lane storage

To motivate the development of a model of the average amount of floor space required when using block stacking, notice that the average amount of floor space required is equal to the footprint of a storage row (including half the aisle and side-to-side clearances) times the average number of storage rows required during the life of a storage lot of a product. Thus,

$$S_{BS} = \eta(W + c)(xL + 0.5A) \tag{12.33}$$

# Example 12.10

To illustrate the calculation of $\eta$, suppose the inventory level over the life of a particular storage lot is given by

| Period | Inventory Level | Period | Inventory Level | Period | Inventory Level |
|--------|-----------------|--------|-----------------|--------|-----------------|
| 1 | 15 | 6 | 10 | 11 | 5 |
| 2 | 14 | 7 | 9 | 12 | 4 |
| 3 | 13 | 8 | 8 | 13 | 3 |
| 4 | 12 | 9 | 7 | 14 | 2 |
| 5 | 11 | 10 | 6 | 15 | 1 |

If $x = 2$ and $z = 3$, what will be the distribution of the number of storage rows required during the life of the storage lot?

| # Storage Rows | Periods | # Periods |
|----------------|---------|-----------|
| 3 | 1, 2, 3 | 3 |
| 2 | 4, 5, 6, 7, 8, 9 | 6 |
| 1 | 10, 11, 12, 13, 14, 15 | 6 |

Hence, the average number of storage rows will be

$$\eta = 3(3/15) + 2(6/15) + 1(6/15) = 27/15 = 1.8.$$

When the inventory level in period $k$, $I_k$, is equal to $Q + 1 - k$ (i.e., uniform withdrawal, one unit load per period), the value of $\eta$ can be obtained using the following relation:

$$\eta = \{y[Q - (y - 1)xz] + (y - 1)xz + (y - 2)xz + \cdots + 2xz + xz\}/Q \qquad (12.34)$$

Since the sum of the integers 1 through $n - 1$ equals $n(n - 1)/2$, then

$$\eta = [0.5y(y - 1)xz + y(Q - xyz + xz)]/Q$$

which reduces to

$$\eta = y[2Q - xyz + xz]/2Q \qquad (12.35)$$

Note, $y - 1$ storage rows will be full at the time the withdrawal cycle begins. The single partial row will be depleted first; the number of unit loads in the partial row equals $Q - (y - 1)xyz$.

Substituting (12.35) in (12.33) gives

$$S_{BS} = y(W + c)(xL + 0.5A)[2Q - xyz + xz]/2Q \qquad (12.36)$$

Since $S_{BS}$ is not a convex function of $x$, to obtain its minimum it is necessary to enumerate over $x$. Notice that the optimum value of $x$ will not depend on $W$ or $c$.

## Continuous Approximation

For large values of $Q$, a continuous approximation to $S_{BS}$ can be used by letting $Q = xyz$ in (12.36). Replacing $y$ with $Q/xz$ and $xyz$ with $Q$ gives

$$S_{BS}^{c} = (W + c)(xL + 0.5A)(Q + xz)/2xz \qquad (12.37)$$

Taking the derivative of $S_{BS}^{c}$ with respect to $x$, setting the result equal to zero, and solving for $x$ gives a continuous approximation to the optimum value of $x$,

$$x_{BS}^{c} = [AQ/2Lz]^{\frac{1}{2}} \qquad (12.38)$$

# *Example 12.11*

Suppose $Q = 200$, $A = 12'$, $L = 4'$, and $z = 4$. Substituting in (12.38) and solving for $x^c_{BS}$ yields a value of 8.66. Therefore, an approximation of the optimum value of $x$ would be either 8 or 9.

For the example, for what range of values of $Q$ will the continuous approximation of $x$ be equal to 25? Substituting $x$ equal to 24.5 and $x$ equal to 25.5 in (12.38) and solving for $Q$ yields values 1600 and 1734 for $Q$, which suggests the optimum value of $x$ is not especially sensitive to changes in the value of $Q$.

## *Safety Stock*

How might the block stacking model be modified to include safety stock? It is important to recognize what conditions cause safety stock to occur. In particular, safety stock for an individual product is created by having a replacement lot arrive before depleting the inventory of the product in question. Since our optimization is based on the inventory profile for an individual lot, rather than the inventory profile for the product, we note that a safety stock implies that, for the lot that just arrived, no withdrawals occur for some period of time.

Returning to our earlier example, suppose the inventory profile for an item is as follows:

| Period | Inventory Level | Period | Inventory Level | Period | Inventory Level |
|--------|--------|--------|--------|--------|--------|
| 1 | 15 | 8 | 14 | 15 | 7 |
| 2 | 15 | 9 | 13 | 16 | 6 |
| 3 | 15 | 10 | 12 | 17 | 5 |
| 4 | 15 | 11 | 11 | 18 | 4 |
| 5 | 15 | 12 | 10 | 19 | 3 |
| 6 | 15 | 13 | 9 | 20 | 2 |
| 7 | 15 | 14 | 8 | 21 | 1 |

If uniform withdrawals normally occur, then the delay in initiating depletion is due to a safety stock condition. For the example, $s = 6$ and the distribution of storage rows is

| # Storage Rows | # Periods |
|--------|--------|
| 3 | 9 |
| 2 | 6 |
| 1 | 6 |

The average number of storage rows will be

$$\eta = 3(9/21) + 2(6/21) + 1(6/21) = 45/21 = 2.14.$$

To incorporate safety stock in the equation for $\eta$, (2) is modified as follows

$$\eta = \{y[Q + s - (y - 1)xz] + (y - 1)xz + \cdots + 2xz + xz\}/(Q + s) \tag{12.39}$$

which reduces to

$$\eta = y[2(Q + s) - xyz + xz]/2(Q + s) \tag{12.40}$$

Thus, the average amount of space required over the life of a lot with safety stock is given by

$$S_{\text{BSSS}} = y(W + c)(xL + 0.5A)[2(Q + s) - xyz + xz]/2(Q + s) \qquad (12.41)$$

Notice, the denominator is twice the length of the cycle time, not twice the lot size.

## *Example 12.12*

Suppose $L = 48''$, $W = 50''$, $A = 156''$, $c = 10''$, $z = 3$, $Q = 25$, and $s = 10$. What will be the optimum value of $x$? Enumerating over $x$, the values shown below are obtained for $S_{\text{BSSS}}$. Thus, $x_{\text{BSSS}} = 3$, with $x = 5$ being the next best row depth.

| $x$ | $y$ | $S_{\text{BSSS}}$ |
|-----|-----|-------------------|
| 1 | 9 | 310.50 ft$^2$ |
| 2 | 5 | 238.21 |
| 3 | 3 | 206.14* |
| 4 | 3 | 221.79 |
| 5 | 2 | 208.21 |
| 6 | 2 | 226.57 |
| 7 | 2 | 241.50 |
| 8 | 2 | 253.00 |
| 9 | 1 | 212.50 |

### *Continuous Approximation for Safety Stock*

A continuous approximation of the safety stock condition is obtained by substituting $Q$ for $xyz$ and substituting $Q/xz$ for $y$ in (12.41). The resulting expressions for average amount of space and the optimum row depth are

$$S_{\text{BSSS}}^c = Q(W + c)(xL + 0.5A)(Q + 2s + xz]/2(Q + s)xz \qquad (12.42)$$

and

$$x_{\text{BSSS}}^c = [A(Q + 2s)/2Lz]^{\frac{1}{2}} \qquad (12.43)$$

## Deep Lane Storage

Deep lane storage is illustrated in Figure 12.6. Notice that it is very similar to block stacking, except every unit load is individually supported. Hence, there is no vertical honeycomb loss with deep lane storage. The depth of the footprint of a deep lane is equal to half the sum of the storage aisle and the flue space at the back of the deep lane; the width of the footprint is the sum of the width of the unit load, two clearances, and the width of the vertical rack member. With deep lane storage, each deep lane is independent from every other lane, both horizontally and vertically. The square footage allocation for a lane involves a proration of the footprint over the number of levels of storage. Hence, the average amount of floor space required for a deep lane of storage is given by

$$(W + 2c + r)[xL + 0.5(A + f)]/z. \qquad (12.44)$$

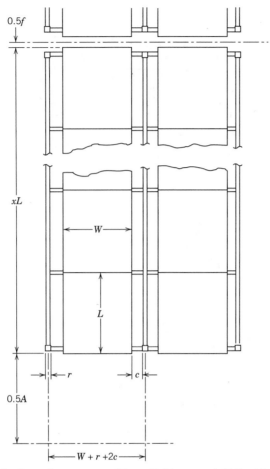

**Figure 12.6**  Deep lane storage. (*From DeMars, et al. [21] with permission*)

The average number of deep lanes of depth $x$ required over the life of a lot is given by

$$\xi = v[2(Q + s) - xv + x]/2(Q + s) \tag{12.45}$$

Therefore,

$$S_{\text{DLSS}} = v(W + 2c + r)[xL + 0.5(A + f)][2(Q + s) \\ - xv + x]/2(Q + s)z \tag{12.46}$$

A continuous approximation for deep lane storage with safety stock yields

$$S^c_{\text{DLSS}} = Q(W + 2c + r)[xL + 0.5(A + f)](Q + 2s + x)/2(Q + s)z \tag{12.47}$$

and

$$x^c_{\text{DLSS}} = [(A + f)(Q + 2s)/2L]^{\frac{1}{2}} \tag{12.48}$$

## *Example 12.13*

Suppose $Q = 147$, $s = 29$, $L = 50''$, $W = 42''$, $c = 10''$, $f = 6''$, $A = 144''$, and $z = 3$. Using deep lane storage, what is the optimum lane depth? Using a continuous approximation yields a value of 17.535. Therefore, it appears that 17 or 18 unit loads is the optimum lane depth. However, when (12.41) is minimized by enumerating over $x$, the following results are obtained:

| $x$ | $v$ | $S_{DLSS}$ (ft$^2$) |
|-----|-----|---------------------|
| 14 | 11 | 12,550.69 |
| 15 | 10 | 12,432.29 |
| 16 | 10 | 12,638.89 |
| 17 | 9 | 12,487.50 |
| 18 | 9 | 12,675.00 |
| 19 | 8 | 12,470.83 |
| 20 | 8 | 12,661.11 |
| 21 | 7 | 12,359.38* |
| 22 | 7 | 12,565.97 |
| 23 | 7 | 12,743.40 |
| 24 | 7 | 12,891.67 |
| 25 | 6 | 12,532.29 |

Thus, if a lane depth of 17 is used, the resulting value of $S$ will be only 1.04% greater than would be obtained by using the optimum lane depth of 21. (If a lane depth of 18 is used, the resulting floor space will be 2.55% greater than the minimum value.)

## Pallet Rack

The single-deep and double-deep pallet racks can be considered special cases of deep lane storage with $x = 2$ and $x = 1$, respectively. The difference between standard pallet rack and deep lane rack is that two loads (not necessarily the same product or from the same lot) can be stored side-by-side using either single-deep or double-deep pallet rack. Hence, the width of a storage footprint for pallet rack will be the width of the load ($W$) plus one full clearance between the load and the vertical rack member ($c$) plus half the side-to-side clearance between loads on a common load beam ($0.5c$) plus half the width of a vertical rack member ($0.5r$), or $W + 1.5c + 0.5r$, as depicted in Figure 12.7.

### *Double-Deep Pallet Storage Rack*

Given the results for deep lane storage, it is relatively easy to compute the value of $S$ for double-deep storage rack installations. Double-deep storage is a special case of deep lane storage with $x = 2$. With double-deep storage, two loads can be stored side-by-side in a pallet opening on a common load beam. Therefore, the width of a double-deep storage footprint will be $W + 1.5c + 0.5r$. The depth of a double-deep footprint will be $2L + 0.5(A + f)$. Again, the area of the footprint is prorated over the number of levels of storage. Hence, the average amount of floor space required for double-deep storage, with safety stock, is given by

$$S_{DDSS} = v(W + 1.5c + 0.5r)[2L + 0.5(A + f)][2(Q + s) \\ - 2v + 2]/2(Q + s)z \tag{12.49}$$

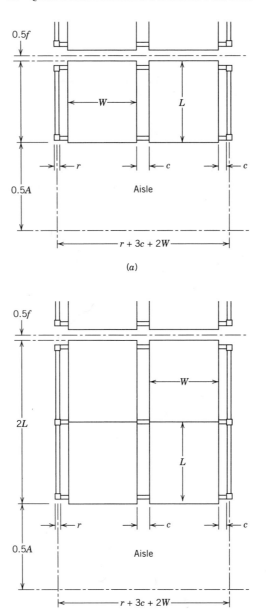

**Figure 12.7** Pallet rack storage. (*a*) Single-deep rack. (*b*) Double-deep rack. (*From DeMars et al. [21] with permission*)

without safety stock,

$$S_{\text{DD}} = v(W + 1.5c + 0.5r)[2L + 0.5(A + f)](Q - v + 1)/Qz \qquad (12.50)$$

Since the storage depth is known, $v$ equal $Q/2$ if $Q$ is even and $(Q + 1)/2$ if $Q$ is odd.

Hence, if $Q$ is even, then

$$S_{DDSS} = Q(W + 1.5c + 0.5r)[2L + 0.5(A + f)](Q \tag{12.51}$$
$$+ 2s + 2)/4(Q + s)z$$

$$S_{DD} = (W + 1.5c + 0.5r)[2L + 0.5(A + f)](Q + 2)/4z \tag{12.52}$$

and, if $Q$ is odd, then

$$S_{DDSS} = (Q + 1)(W + 1.5c + 0.5r)[2L + 0.5(A + f)](Q + \tag{12.53}$$
$$2s + 1)/4(Q + s)z$$

$$S_{DD} = (W + 1.5c + 0.5r)[2L + 0.5(A + f)](Q + 1)^2/4Qz \tag{12.54}$$

### Single-Deep Pallet Storage Rack

To determine the average floor space requirements for single-deep storage rack installations, the depth of the storage footprint will equal $L + 0.5(A + f)$. The width of the footprint is the same as for double-deep storage rack. Letting $x = 1$ and $v = Q$ and modifying appropriately the deep lane storage results, the average amount of floor space required for single-deep storage rack, with safety stock, is given by

$$S_{SDSS} = Q(W + 1.5c + 0.5r)[L + 0.5(A + f)](Q + 2s + 1)/2(Q + s)z \tag{12.55}$$

In the absence of safety stock, (12.55) reduces to

$$S_{SD} = (W + 1.5c + 0.5r)[L + 0.5(A + f)](Q + 1)/2z \tag{12.56}$$

## Example 12.14

Let $Q = 15$
   $L = 50$ in.
   $W = 42$ in.
   $A = 144$ in.
   $r = 3$ in.
   $c = 4$ in.
   $f = 12$ in.
   $z = 3$

Using double-deep storage rack with no safety stock,

$$S_{DD} = [42 + 1.5(4) + 0.5(3)][2(50) + 0.5(144 + 12)](15 + 1)^2/4(15)(3)$$
$$= 12{,}531.2 \text{ in}^2 \text{ or } 87.02 \text{ ft}^2$$

With single-deep storage rack and no safety stock,

$$S_{SD} = [42 + 1.5(4) + 0.5(3)][50 + 0.5(144 + 12)](15 + 1)/2(3)$$
$$= 16{,}896 \text{ in}^2 \text{ or } 117.33 \text{ ft}^2$$

# *12.6*  AUTOMATED STORAGE AND RETRIEVAL SYSTEMS

The automated storage and retrieval system (AS/RS) has had a dramatic impact on manufacturing and warehousing. Through the use of computer control, handling and storage systems have been integrated into manufacturing and distribution processes. While the primary emphasis originally was in storing and retrieving finished goods, more recently the focus has been on work-in-process (WIP), raw materials, and supplies.

The AS/RS has also affected the design of the warehouse. Rack-supported buildings, for example, are frequently constructed to house the AS/RS. A typical rack-supported building is shown in Figure 12.8. Such structures are usually more economical to construct than conventional buildings; additionally, they frequently receive favorable income tax treatment because of their special-purpose construction.

As noted in Chapter 6, the AS/RS consists of storage racks, storage/retrieval (S/R) machines, and input/output (I/O) or pickup/deposit (P/D) stations. A typical AS/RS layout is given in Figure 12.9. Typically, each S/R machine operates in a single aisle and services storage racks on each side of the aisle.

Various types of AS/R systems were presented in Chapter 6. In this chapter, we will first focus on the unit load AS/RS, which is designed to store and retrieve palletized loads. After we present a (deterministic) cost model to design and evaluate unit load AS/R systems, we will present cycle time results, which are based on (non-deterministic) cycles (or trips) performed by the S/R machine. Subsequently, we will consider order picking systems, which include the person-on-board AS/RS and the miniload AS/RS.

In designing an AS/RS, a number of decisions must be made. Depending on the particular situation, some of the decisions will be made by the system supplier and others will be made by the user of the system. Information must be collected on a number of aspects and values must be established for many design parameters. Among some of the most important considerations are the following [88]:

1. Load size(s) and opening size(s).
2. Number and location of P/D stations.
3. Building construction; rack supported or conventional.
4. Land availability, condition, cost, and zoning restrictions.
5. Number, height, and length of storage aisles.
6. Percentage of the operations (storages and retrievals) to be performed on a dual-command basis.
7. Applicability of transfer cars.
8. Randomized or dedicated storage or some combination thereof.
9. Dwell point for the S/R machine when it is idle, in aisle, or at P/D station.
10. Level of automation.
11. Level of computer control.
12. Requirement for physical inventory.

**Figure 12.8**   Rack-supported automated storage/retrieval systems (AS/RS). (Part *a* courtesy of Clark Equipment Co.; part *b courtesy of Hartman Material Handling Systems*)

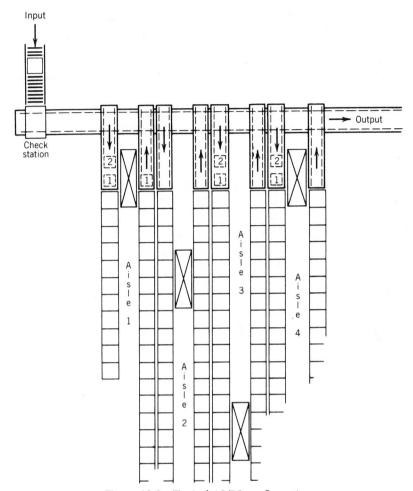

**Figure 12.9** Typical AS/RS configuration.

13. Replenishment requirements.
14. Requirement for maintenance.
15. Use of slave pallets versus vendor pallets.
16. Mode of input/output.
17. Plan for evolution and change.
18. Throughput requirement, peak and average.
19. Level of specifications to be developed for hardware and software.
20. Provision for interrupt and priority storage/retrieval.
21. Storage depth (single, double, or deep lane).
22. Provision for mixed loads on a pallet.
23. Use of automatic identification systems.
24. Use of simulations to support design decisions.

25. Amount of queue space required for inbound and outbound loads.
26. Impact of randomness versus scheduled operation on the requirements for the system.
27. Energy and utilities.
28. Impact of inflation and income taxes.
29. Sprinkler requirements.
30. Plan for training, startup, debugging, and postauditing.

Very early in the AS/RS design process a fixed scope of work should be developed. A schedule of management reviews, a list of deliverables, and a definition of the level of effort should be developed for the project.

The AS/RS Product Section of the Material Handling Institute, Inc. developed a listing of benefit-cost elements to be considered in AS/RS economic justification; included are the following elements [88].

1. Initial investment considerations
   a. Land value
   b. Building structure
   c. Utilities
   d. Machinery and equipment
   e. Investment tax credit
   f. Income tax
   g. Interest expense
2. Annual operating cost considerations
   a. Direct operating personnel
   b. Secondary personnel
   c. Floor space savings and cube utilization
   d. Utilities
   e. Depreciation
   f. Insurance and taxes
   g. Product damage
   h. Pilferage
   i. Inventory carrying costs
   j. Job enrichment
   k. Maintenance
   l. Improved service levels
   m. Improved materials control

For some of these factors, reasonably accurate estimates can be obtained using existing accounting data and work measurement data. Land, building, equipment, and labor costs are relatively easy to estimate. On the other hand, such items as heat, light, power, insurance, pilferage, product damage, employee satisfaction, reduced inventory levels and production delays, poor management decisions, and customer service are not easily traced and measured. However, it is readily agreed that all of these factors, as well as others, are important and do contribute to the firm's overall economic performance.

## Sizing and Cost Estimation[3]

In order to develop a basis for estimating AS/RS investment costs, Zollinger [87] compiled detailed cost information for more than 60 AS/RS installations. He found that a reasonably accurate estimate of the acquisition cost of an AS/RS could be obtained by summing the costs of the storage rack, the S/R machines, and the building used to house the AS/RS.

Further, he found that the cost of the storage rack, including installation and S/R machine support rail, was a function of the number of unit loads stored in the rack, the cubic size of the unit load, the weight of the unit load, and the height of the rack. The cost of an S/R machine was found to be a function of the height of the AS/RS, the weight of the unit load, and the type and location of the S/R machine controls. The cost of the building was found to be a function of its height and the cost per square foot to construct a 25-ft-high building.

To facilitate the presentation, the following notation and estimates are used:

$x$ = depth of the unit load (i.e., stringer dimension) in inches

$y$ = width of the unit load in inches

$z$ = height of the unit load in inches

$v$ = volume of unit load in cubic feet = $xyz/1728$

$w$ = weight of unit load in pounds

$W$ = width of an aisle of AS/RS storage in inches

$L$ = length of an aisle of AS/RS storage in inches

$H$ = height of an aisle of AS/RS storage in inches

$n$ = number of tiers or levels of storage

$m$ = number of columns of storage per aisle side

$a$ = number of storage aisles

$BH$ = building height

$BW$ = building width

$BL$ = building length

$\lambda$ = allowance, measured in feet

$\alpha$ = parameter for computing rack cost, expressed in dollars

$\beta$ = height parameter for computing S/R machine cost, expressed in dollars

$\gamma$ = weight parameter for computing S/R machine cost, expressed in dollars

$\phi$ = control parameter for computing S/R machine cost, expressed in dollars

---

[3]The cost-estimating procedure presented is based on one developed by Zollinger [87]. To simplify the presentation, it is assumed that all unit loads stored in the AS/RS are the same size and weight. It should be noted that the cost estimates are "quick-and-dirty" values and are presented for expository purposes only. Quotations from AS/RS suppliers should not be evaluated on the basis of the "rough cut" cost estimates provided in this section.

$\delta$ = cost per square foot to construct a 25-ft-tall building

CF = conversion factor for converting cost per sq ft to construct a 25-ft-tall building to the cost per sq ft to construct a building of height, BH

The dimensions of a storage aisle can be estimated as follows:

$$W = \begin{cases} 3(x + 6'') & \text{(with in-rack sprinklers)} \\ 3(x + 4'') & \text{(without in-rack sprinklers)} \end{cases} \tag{12.57}$$

$$L = m(y + 8'') \tag{12.58}$$

$$H = n(z + 10'') \tag{12.59}$$

The width of a storage aisle is determined by measuring from the center-line of the flue space to the right of the aisle to the center-line of the flue space to the left of the aisle. Hence, the width of a storage aisle includes the width of the storage spaces on both sides of the aisle and the width of the aisle. The width of the aisle is assumed to equal the depth of the load, plus 4 in. for clearance between the load and the rack. The depth of a storage space is equal to the depth of the load, plus 4 in. The 4-in. addition to the load depth accounts for half the flue space behind the load. In the case of in-rack sprinklers, an additional 6 in. is added to the flue space dimension.

In calculating the length of the storage aisle, it is assumed that a 4 in. upright truss is used to support the loads. Further, a clearance of 2 in. is assumed between an upright truss and the load. Hence, with a 2-in. clearance on each side of the load and a 4-in. upright truss, the width of a storage opening equals the width of the load plus 8 in.

In calculating the height of an aisle of storage, it is assumed that each load rests on cantilever arms that are 4 in. thick. Likewise, it is assumed that a clearance of 6 in. is required between the top of a load and the cantilever arm above it to allow the S/R's shuttle to retrieve the load.

In converting the dimensions of a storage aisle into building dimensions, it should be noted that the lowest storage level in an aisle cannot be at floor level. (The shuttle for the S/R machine cannot retrieve the load unless it is approximately 28 in. above floor level.) Likewise, approximately 20 in. of space must be provided above the top load for installation of the power conductor bar for the S/R machine and clearance for the S/R machine carriage in positioning the shuttle to retrieve the top-most load. Hence, the building height will be approximately 4 ft greater than the height of an aisle of storage.

In estimating the building width, one can multiply the width of a storage aisle by the number of storage aisles. However, since the outermost racks should not be installed "snug" with the wall, an addition of 2 ft should be provided. (If food is being stored, regulations might require a larger clearance in order to prevent rodent infestation. Sufficient space might be required to allow removal of trash between the wall and the rack.)

In determining the building length, space must be provided for pickup and deposit (P/D) stations, as well as S/R extension or runout at the end of the rack for maintenance access. The amount of the allowance ($\lambda$) is dependent on the width of the unit load.

Based on the above, the dimensions of the building can be estimated as follows:

$$BH = H + 48 \text{ in.} \tag{12.60}$$

$$BW = aW + 24 \text{ in.} \tag{12.61}$$

$$BL = L + \lambda \tag{12.62}$$

where $\lambda$, for values of $y$ between 24 in. and 54 in., can be estimated by

$$\lambda = \begin{cases} 12.5 + 0.45y & \text{(without transfer cars)} \\ 29.5 + 0.45y & \text{(with transfer cars)} \end{cases} \tag{12.63}$$

## Example 12.15

Suppose an AS/RS is to be designed for 40 in.$\times$48 in. (depth$\times$width) unit loads that are 48 in. tall. There are to be eight aisles; each aisle is 12 loads high and 80 loads long. There are to be sprinklers and no transfer cars. What will be the minimum dimensions for the building?

The building height will be 12(48 in. + 10 in.)/12 + 48 in./12 or 62 ft. The width of the building will be 8(3)(40 in. + 6 in.)/12 + 2 ft, or 94 ft. The building length, in ft, will be 80(48 in. + 8 in.)/12 + 12.5 ft + 0.45(48), or 407.43 ft. Thus, a building at least 62 ft tall, 94 ft wide, and 407.43 ft long will be required to accommodate the AS/RS. Depending on other functions to be located in the facility and the amount of staging or unit load accumulation space required "in front" of the AS/RS, additional space might be required.

### Rack Cost Calculation

To estimate the acquisition cost of the storage rack for an AS/RS, we multiply the cost per rack opening (CRO) by the number of rack openings. The cost per rack opening can be estimated as follows:

$$CRO = \alpha[0.92484 + 0.025v + 0.0004424w - (w^2/82{,}500{,}000) \\ + 0.23328n - 0.00476n^2] \tag{12.64}$$

## Example 12.16

Suppose a storage rack is to be provided to accommodate 10,000 unit loads, weighing 2500 pounds, and having dimensions of 42 in.$\times$48 in.$\times$46.5 in. Further, suppose the rack will support 10 unit loads vertically, that is, 10 tiers of storage will be accommodated by the rack. If the storage rack cost parameter, $\alpha$, equals \$30, what will be the cost of the storage rack?

From the problem statement, $x = 42$ in., $y = 48$ in., $z = 46.5$ in., $w = 2500$ lb; $n = 10$ tiers; and $\alpha = \$30$. A calculation establishes that $v = 54.25$ ft$^3$. Substituting the parameters in (12.64) yields a value of \$155.0442 per rack opening. Hence, the storage rack will cost approximately \$1,550,442, installed.

### S/R Machine Cost Calculation

The cost of an S/R machine is obtained by summing the cost contribution due to the height of the system, the weight of the unit load, and the type of machine control used. Specifically, CSR, the cost per S/R machine, is given by

$$CSR = A + B + C \tag{12.65}$$

where

A = S/R machine cost based on the height of the system

B = S/R machine cost based on the weight of the unit load

C = S/R machine cost based on the type of machine control used

| H | A | w | B |
|---|---|---|---|
| < 35' | $\beta$ | < 1000# | $\gamma$ |
| [35',50') | $2\beta$ | [1000#,3500#) | $2\gamma$ |
| [50',75') | $3\beta$ | [3500#,6500#) | $3\gamma$ |
| [75',110') | $4\beta$ | ≥6500# | $4\gamma$ |
| ≥110' | $5\beta$ | | |

| Control Logic | C |
|---|---|
| manual | $\phi$ |
| on-board | $2\phi$ |
| end-of-aisle | $3\phi$ |
| central console | $4\phi$ |

# Example 12.17

Suppose six S/R machines are to be used to lift 2500 pound loads to a height of 55 ft and a central console is to be used to control the S/R machines. What will be the cost of the S/R machines? Based on current market prices, the values of all three cost parameters are estimated to equal $25,000.

From the problem statement, H = 55 ft; $w$ = 2500 lb, central console, and $\beta = \gamma = \phi = \$25,000$. Hence,

$$CSR = 3\beta + 2\gamma + 4\phi = 9(\$25,000) = \$225,000 \text{ per S/R machine}$$

## Building Cost Calculation

To estimate the cost of the building, we first compute the size of the footprint of the building in square feet. Then, we multiply the footprint of the building by the product of CF (the conversion factor) and $\delta$ (the cost per square foot to construct a 25-ft-tall building). Hence, the building cost (BC) can be estimated to be

$$BC = (BW)(BL)(CF)\delta \tag{12.66}$$

| BH | CF |
|---|---|
| 25' | 1.00 |
| 40' | 1.25 |
| 55' | 1.50 |
| 70' | 1.90 |
| 85 | 2.50 |

# *Example 12.18*

Suppose a 60-ft-tall building is to be used to house an AS/RS and it costs $30/ft$^2$ to construct a 25-in.-tall building. Further, suppose the footprint of the building to house the AS/RS is 150 ft $\times$ 440 ft. What will be the cost of the building?

Interpolating to obtain the value of CF yields a value of 1.633. Hence,

$$BC = 150(440)(1.633)(30)$$
$$= \$3,234,000$$

# *Example 12.19*

Suppose storage space of 1,540,000 ft$^3$ must be provided. A 55-ft-tall building will yield a footprint of 28,000 ft$^2$ at a cost of 1.5(28,000)$\delta$, or 42,000$\delta$. If a 25-ft-tall building is built, a footprint of 61,600 ft$^2$ is required and the cost will be 61,600$\delta$. Likewise, a 70-ft-tall building will cost approximately 41,800$\delta$ and an 85-ft-tall building will cost approximately 45,294$\delta$. Thus, a 70-ft-tall building would minimize the cost of the building required to store a constant volume of material.

## S/R Machine Cycle Times

To determine the time required to perform an operation (either a storage or a retrieval), expected S/R machine cycle time formulas will be used. Recall, the S/R machine is capable of traveling simultaneously in the aisle both vertically and horizontally. Hence, the time required to travel from the P/D station to a storage or retrieval location is the maximum of the horizontal and vertical travel times, which is also known as Chebyshev travel.

Because of the importance of the cycle time in estimating the throughput capacity of the system, considerable emphasis is placed on the time for the S/R machine to perform a **single command** or a **dual command** cycle. A single command cycle consists of either a storage or a retrieval, but not both; whereas a dual command cycle involves both a storage and a retrieval.

In determining the cycle times, it is assumed that a single command storage cycle begins with the S/R at the P/D station; it picks up a load, travels to the storage location, deposits the load, and returns empty to the P/D station. A single command retrieval cycle is also assumed to begin with the S/R at the P/D station; it travels empty to the retrieval location, picks up the load, travels to the P/D station, and deposits the load. One pickup and one deposit occur during a single command cycle.

A dual command is assumed to begin with the S/R at the P/D station; it picks up a load, travels to the storage location, deposits the load, travels empty to the retrieval location, picks up the load, travels to the P/D station, and deposits the load. A total of two pickups and two deposits are performed during a dual command cycle.

In developing models of expected single command cycle times and expected dual command cycle times, the following notation is used. Let

$E(SC)$ = the expected travel time for a single command cycle

$E(TB)$ = the expected travel time from the storage location to the retrieval location during a dual command cycle

$E(DC)$ = the expected travel time for a dual command cycle

$L$ = the rack length (in feet)

$H$ = the rack height (in feet)

$b_v$ = the horizontal velocity of the S/R machine (in ft/min)

$v_v$ = the vertical velocity of the S/R machine (in ft/min)

$t_b$ = time required to travel horizontally from the P/D station to the furthest location in the aisle, $(t_b = L/b_v$ mins)

$t_v$ = time required to travel vertically from the P/D station to the furthest location in the aisle, $(t_v = H/v_v$ mins)

In order to determine the above expected travel times in a compact form, we first "normalize" the rack by dividing its shorter (in time) side by its longer (in time) side [12]. That is, we let

$$T = \max(t_b, t_v), \tag{12.67}$$

$$Q = \min(t_b/T, t_v/T), \tag{12.68}$$

where $T$ (i.e., the "scaling factor") designates the longer (in time) side of the rack, and $Q$ (i.e., the "shape factor") designates the ratio of the shorter (in time) side to the longer (in time) side of the rack. Note that the normalized rack is $Q$ units long in one direction and one unit long in the other direction, where $0 < Q \le 1$. If $Q = 1$, the rack is said to be "square-in-time" (SIT).

Assuming randomized storage, an end-of-aisle P/D station located at the lower left-hand corner of the rack, constant horizontal and vertical S/R machine velocities (i.e., no acceleration or deceleration is taken into account), and a continuous approximation of the storage rack (which is sufficiently accurate), the expected travel times for the S/R machine are given in [12] as:

$$E(SC) = T\left[1 + \frac{Q^2}{3}\right] \tag{12.69}$$

$$E(TB) = \frac{T}{30}[10 + 5Q^2 - Q^3] \tag{12.70}$$

$$E(DC) = \frac{T}{30}[40 + 15Q^2 - Q^3] \tag{12.71}$$

Note that $E(DC)$ in Equation 12.71 is obtained by observing that $E(DC) = E(SC) + E(TB)$. Furthermore, in [12] it is shown that $E(SC)$ and $E(DC)$ are minimized at $Q = 1$; i.e., given a rack with a fixed area (but variable dimensions), the expected single and dual command travel times are minimized when the rack is SIT.

The expected total cycle time for the S/R machine is obtained by adding the load handling time to the expected travel time. That is, letting

$T_{SC}$ = the expected single command cycle time

$T_{DC}$ = the expected dual command cycle time

$T_{P/D}$ = time required to either pickup or deposit the load we have

$$T_{SC} = E(SC) + 2T_{P/D} \tag{12.72}$$

$$T_{DC} = E(DC) + 4T_{P/D} \tag{12.73}$$

## *Example 12.20*

For the rack system given in Example 12.15 determine the cycle times for single and dual command operations. Assume the S/R requires 0.35 min to perform a P/D operation, travels horizontally at an average speed of 350 fpm, and travels vertically at an average speed of 60 fpm.

The dimensions (inches) of the storage rack are 56 *m* (length) by 58 *n* (height), where *m* is the number of horizontal addresses and *n* is the number of vertical addresses. For the example, *m* = 80 and *n* = 12. Therefore, the rack dimensions are

$$L = 56(80)/12 = 373.33 \text{ ft}$$
$$H = 58(12)/12 = 58 \text{ ft}$$

(*Note:* It is assumed the S/R machine travels a maximum of 373.33 ft horizontally and 58 ft vertically. Actually, the shuttle mechanism must be lifted only to the load support position of the twelfth storage level, rather than to the top of the storage rack. The horizontal travel distance is probably underestimated from 2 to 10 ft, depending on the exact location of the P/D station.)

The maximum horizontal and vertical travel times are determined as follows:

$$t_h = L/h_v = 373.33/350 \text{ fpm} = 1.067 \text{ min}$$
$$t_v = H/v_v = 58/60 \text{ fpm} = 0.967 \text{ min}$$

Thus,

$$T = \max(1.067, 0.967) = 1.067 \text{ min}$$

and

$$Q = \frac{0.967}{1.067} = 0.906$$

Therefore, the single command and dual command cycle times are

$$
\begin{aligned}
T_{SC} &= T\left(1 + \frac{Q^2}{3}\right) + 2T_{P/D} \\
&= 1.067\left[1 + \frac{(0.906)^2}{3}\right] + 2(0.35) \\
&= 2.06 \text{ min per single command cycle}
\end{aligned}
$$

$$
\begin{aligned}
T_{DC} &= \frac{T}{30}(40 + 15Q^2 - Q^3) + 4T_{P/D} \\
&= \frac{1.067}{30}[40 + 15(0.906)^2 - (0.906)^3] + 4(0.35) \\
&= 3.23 \text{ min per dual command cycle}
\end{aligned}
$$

Thus, the average cycle time per operation is 2.06 min with a single command cycle and 1.615 min with a dual command cycle.

## *Example 12.21*

For the situation considered in Examples 12.15 and 12.20, suppose 40% of the storages and 40% of the retrievals are performed as single command operations; the remaining are performed as dual command operations. If 120 storages per hour and 120 retrievals per hour are to be handled by the AS/RS, what will be the percent utilization of the S/R machines?

Assuming no transfer cars and assuming the 8 S/R machines are loaded uniformly, then each S/R must perform 15 storages per hour and 15 retrievals per hour. Since 6 storages per hour and 6 retrievals per hour are performed using a single command operation (i.e., 40%), there will be 9 dual command trips per hour. The workload on the S/R machine is obtained as follows:

$$\text{Workload per S/R} = 2(6)(2.06) + 9(3.23)$$
$$= 53.79 \text{ min/hr}$$

Thus, each S/R will be utilized 53.79/60, or 89.65%. The average cycle time per operation is 53.79/30, or 1.793 min per operation, which implies that the maximum throughput capacity of the system is equal to (60/1.793)8, or 267.71 operations/hr.

Since the times between requests for storages and/or retrievals are often random variables and since the locations of the storage and/or retrieval addresses to be visited on a single or dual command cycle are random variables, waiting line analysis can be used to aid in the AS/RS design. However, due to the mixture of single and dual command operations and the general probability distribution for travel times, a simulation approach is suggested.

Transfer cars are utilized when the activity level in an aisle is not sufficient to justify an S/R machine dedicated to each aisle. In order to determine the cycle time for an AS/RS having transfer cars, the time required to perform the transfer must be included. Additionally, S/R waiting lines can occur when multiple S/R machines are served by a single transfer car. Because of the network of queues involved in the design of an AS/RS with transfer cars, simulation analysis is generally used to aid in the determination of the number of S/R machines, the number of transfer cars, and associated operating disciplines. However, since the transfer car cost might be as large as half the S/R machine cost, most AS/R systems are installed without transfer cars; hence, there is generally an S/R machine for each aisle.

## *12.7*   ORDER PICKING SYSTEMS

With (unit load) storage/retrieval systems, each storage or retrieval operation involves one unit load; that is, the parts are stored as one unit load and retrieved as one unit load. In some storage systems, however, the parts may be stored as a unit load but retrieved in less–than–unit–load quantities. Such systems are generally known as *order picking* (OP) systems, where each order typically calls for less–than–unit–load quantities. For example, in a mail-order music warehouse, many copies of the same title are stored together as one "unit load," while the customers almost never order two copies of the same title (but they may order one copy of several titles). Hence, parts of the same type (in our example, CDs or tapes of a particular title) are stored as one unit load but they are retrieved in less–than–unit–load quantities. Naturally, the system is replenished periodically either by restocking the empty containers or by removing the empty containers and replacing them with full containers.

There are two principal approaches to order picking. The first one is based on the picker traveling to each container to be visited in order to pick the parts on one or more orders. Since the containers are typically stored along aisles, the general term used for such systems is "in–the–aisle OP." When we prepare a shopping list and go to the grocery store, for example, we are, in essence, performing in–the–aisle OP as we walk up and down the aisles and fill our cart with one or two pieces of each grocery item we need.

The second principal approach (which we will discuss later) is known as end–of–aisle OP. With such systems, the containers are brought to the end–of–the aisle, where the picker removes the appropriate number of items. Each container is then stored back in the system until it is needed again. An end–of–aisle OP system may at first resemble a unit load storage/retrieval system in that we store and retrieve unit loads. However, with an end–of–aisle OP system, a container is typically retrieved several times (until it is empty). In a unit load storage/retrieval system, once a container is retrieved, it leaves the system.

## In–the–Aisle Order Picking

Various configurations and material handling equipment are available for performing in–the–aisle OP. The grocery shopping example we gave above applies to many warehouses where the pickers fill their carts with one or more orders as they walk along the aisles. (Such systems are also known as "walk–and–pick" systems.) In some in–the–aisle OP systems, however, each picker is dedicated to a particular aisle and, to reduce fatigue and improve travel times, the picker travels to each container via an automated or semiautomated device. The person–on–board AS/RS (described in Chapter 6), for example, is one such case. With a person–on–board AS/RS, the S/R machine automatically stops in front of the appropriate container and waits for the picker to perform the pick. The parts that are picked are usually taken to a P/D station (which is typically located at the end of the aisle).

To avoid unnecessary trips between the containers in the rack and the P/D station, the picker performs several picks (i.e., he or she visits several containers) between successive stops at the P/D station. (Although the picker may actually pick several items from a particular container, here we define it as one "pick.") In other words, the picker starts at the P/D station, visits the appropriate containers in the rack, returns to the P/D station and discharges the parts before starting the next trip. The throughput capacity of such systems, that is, the number of picks performed per hour, depends on the expected time it takes to complete one trip. The time required per trip, in turn, depends on the travel speed of the S/R machine, the rack size, and rack shape (i.e., the scaling factor and the shape factor as defined by Equations 12.67 and 12.68, respectively), the time it takes to complete one pick (once the S/R machine has stopped), the time it takes to discharge the parts at the P/D station, and the sequence in which the containers are visited. Note that unnecessary S/R travel may occur if the containers to be visited on one trip, that is, the "pick points" in the rack, are not carefully sequenced. (The reader familiar with operations research would readily recognize the above sequencing problem as the traveling salesman problem [50]).

Except for the total S/R machine travel time for the trip, it is relatively straightforward to estimate the above parameters (i.e., the time spent at each pick point and the time spent at the P/D station between two successive trips). In [8] it is empirically shown that the expected time required to travel from the P/D station, make $n$ stops (or picks) in the aisle, and return to the P/D station can be estimated by the following expression:

$$E(P, X_1, X_2, \ldots, X_n, D) \approx$$
$$T\left[\frac{2n}{(n + 1)} + 0.114703n\sqrt{Q} - 0.074257 - 0.041603n + 0.459298Q^2\right]$$

$$(12.74)$$

for $3 \le n \le 16$. The above expression is based on the assumption that the P/D station is located at the lower left-hand corner of the rack and that the pick points, which are randomly and uniformly distributed over the rack, are sequenced in an optimum fashion (i.e., the total S/R machine travel time is minimized). Of course, we remind the reader the S/R machine travels according to the Chebyshev metric and that the travel time is not affected by whether a pick point is on the left-hand or right-hand side of the aisle. Also, no acceleration or deceleration effects have been taken into account.

## Example 12.22

Consider a storage rack that is 360 ft long and 60 ft tall. Suppose that the S/R machine travels at a constant speed of 400 fpm and 80 fpm in the horizontal and vertical direction, respectively. Further suppose that 10 stops are made during each order picking trip, i.e., $n = 10$. From Equation 12.67, the scaling factor, $T$, is equal to max(360/400, 60/80) = 0.90 min. The shape factor, $Q$, is equal to 0.8333 (by Equation 12.68). Hence, from Equation 12.74, the expected travel time for the above trip with 10 picks is estimated to be equal to 2.4245 min. That is, the travel time required to start at the P/D station, pick parts from 10 containers (according to the optimum sequence), and return to the P/D station equals approximately 2.4 min.

## Example 12.23

Suppose in Example 12.22 it takes the picker, on the average, 15 sec to pick one or more items from each container. Further suppose that the picker spends 1.2 min at the P/D station between successive trips (to discharge the parts he or she picked and to prepare for the next trip). The throughput capacity of the system, expressed in number of picks completed per hour, is computed as follows. The total approximate time required to complete one trip is equal to 2.4 + 10(15/60) + 1.2 = 6.1 min. Since 10 picks are performed in about 6.1 min, the maximum throughput capacity is approximately equal to 98 picks/hour. In other words, the picker will visit approximately 98 containers/hour on the average. Given the expected number of items picked from each container, the above figure can also be expressed as the number of items picked per hour.

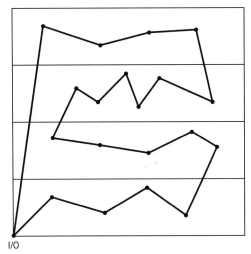

**Figure 12.10** The four-band heuristic. (Due to Chebyshev travel, the solid lines do not necessarily reflect the travel path of the S/R machine.)

Sequencing the pick points in an optimum fashion may impose a heavy computational burden. In fact, although the number of points is quite small (generally no more than 20 to 24 points per trip), obtaining the optimum sequence with the Chebyshev metric can require a large amount of computer time (see [10] for details). Hence, assuming Chebyshev travel, a number of "fast" heuristic procedures (see [10] and [34], among others) have been empirically shown to generate optimum or near-optimum solutions for $3 \leq n \leq 24$, the one most often encountered in practice is a simple heuristic which is known as the "band heuristic."

With the band heuristic, the rack is divided into $k$ equal-sized horizontal bands, where $k$ is an even number. Starting with the first band, the S/R machine travels in a serpentine fashion, picking sequentially along the $x$-axis until all the picks are performed. An example with four bands is shown in Figure 12.10. Note that the S/R machine completes all the picks in the current band before proceeding to the next band. Consequently, depending on the number of bands (i.e., the band width), some "zigzagging" may occur in each band. Since the band heuristic is a heuristic approach, the expected S/R machine travel time per trip will be greater than that obtained using an exact (i.e., optimum seeking) solution procedure. However, the band heuristic performs reasonably well and it is simple to implement.

## Example 12.24

Suppose the 360 ft×60 ft rack used in Example 12.22 is divided into two bands and the coordinate locations of the 10 picks to be made are as given in Table 12.7. Using the band heuristic the items would be visited in the following sequence: 5, 7, 3, 1, 4, 2, 8, 10, 6, and 9. A variation of the band heuristic divides the rack into $k$ equal-sized bands, but determines the band width by dividing $k$ into the largest $y$-coordinate among the pick points. In this example, the division of the two bands would be at $y = 27.5'$ and the same sequence would

**Table 12.7   Pick Locations for Example 12.24**

| Pick # | Coordinates | Pick # | Coordinates |
|--------|-------------|--------|-------------|
| 1 | (200, 20) | 6 | ( 50, 50) |
| 2 | (350, 30) | 7 | (100, 10) |
| 3 | (150, 15) | 8 | (250, 40) |
| 4 | (300, 25) | 9 | ( 25, 55) |
| 5 | ( 75,  5) | 10 | (225, 35) |

be used to pick the 10 items. Note that, with the above variation, the bands are problem specific; that is, they generally vary from one trip to the next.

Using the band heuristic, in [8] it is shown that the expected time required to perform $n$ picks (that are randomly and uniformly distributed over the rack) can be approximated as follows:

$$E(k, n, T, Q) = T\left[2kA + kB(n - 1) + \frac{k - 1}{k} Q\right],$$   (12.75)

where $k$ designates the number of bands, and

$$A = \frac{C}{2} + \frac{1 - (1 - C)^{n+2}}{C(n + 1)(n + 2)},$$   (12.76)

$$B = \frac{C}{3} + \frac{2nC + 6C - 2 + 2(1 - C)^{n+3}}{C^2(n + 1)(n + 2)(n + 3)},$$   (12.77)

$$C = \frac{Q}{k^2}.$$   (12.78)

## Example 12.25

Consider the same data given for Example 12.22; that is, $T = 0.90$ min and $Q = 0.8333$, with $n = 10$ picks per trip. Using Equations 12.75 through 12.78, and assuming that the rack is divided into two bands (i.e., $k = 2$), we obtain $C = 0.2083$, $B = 0.1166$, and $A = 0.1383$, which yields $E(k, n, T, Q) \approx 2.7617$ min. Compared with the optimum expected travel time computed in Example 12.22 (i.e., 2.4245 min), the two-band heuristic results in an increase of approximately 14% in the expected travel time per trip. However, the above increase does not imply that the throughput capacity of the system using the band heuristic will be proportionally less. This is primarily due to the time required per pick and the time required at the P/D station between two successive trips, both of which are generally independent of the sequence in which the picks are made. Using the same data given for Example 12.23 (i.e., 15 sec per pick and 1.2 min at the P/D station between successive trips), for the two-band heuristic we obtain a maximum throughput capacity of about 93 picks/hour, which is 5% less than that obtained with the optimum pick sequence.

## Example 12.26

Continuing with the previous example, suppose the rack is divided into four bands and the number of picks performed per trip is increased to 50. (Note that the maximum number of picks performed per trip is generally dictated by the capacity of the S/R machine relative to

Table 12.8   *"Optimum" Number of Bands for the Band Heuristic*

| No. of Picks | No. of Bands |
|---|---|
| [   1,   24] | 2 |
| [ 25,   72] | 4 |
| [ 73, 145] | 6 |
| [146, 242] | 8 |
| [243, 363] | 10 |

the weight and/or volume of the parts to be picked.) From Equations 12.75 through 12.78 we obtain $C = 0.052$, $B = 0.0269$, and $A = 0.0328$, which yields $E(k, n, T, Q) \approx 5.54$ min. Interestingly, in comparison with Example 12.25, increasing the number of picks by 400% increases the expected travel time by about 100%. This is primarily due to the fact that after more than five picks per trip, the total travel time increases very slowly with the number of picks. We remind the reader, however, that the above increase may not be as (relatively) small when the S/R machine acceleration and deceleration are taken into account.

For a given $n$ value, since it is possible to approximate the expected travel time as a function of $k$, the number of bands that will minimize the expected travel time can be determined. For a SIT rack, Table 12.8 shows the "optimum" number of bands as a function of the number of picks performed per trip [8]. Alternately, given the number of bands, Equation 12.75 can be used to determine the "optimum" configuration of an aisle (i.e., the rack length and height) to minimize the expected distance traveled.

## *Example 12.27*

Rather than continue with the previous data set, assume that a storage area of 18,000 ft² is to be provided within a storage aisle. Since storage racks are provided on both sides of the aisle, 9000 ft² of rack space will be provided on each side. Recall that the P/D station is located at the lower left-hand corner of the rack, randomized storage is used, and the S/R machine follows the Chebyshev metric.

Suppose the S/R machine travels at 300 fpm and 60 fpm in the horizontal and vertical directions, respectively. Further suppose that 30 picks will be performed on each pick trip and the rack is divided into two bands, that is, $n = 30$ and $k = 2$. As shown in Table 12.9, one alternative is to provide a 15 ft tall storage area that is 600 ft long. (The length and height of the storage area are based on the travel region for the S/R machine, not necessarily the storage rack dimensions.) The expected travel for the 15 ft×600 ft rack is equal to 4.3951 min. Given other configurations shown in Table 12.9, in this particular case the optimum rack height is approximately 25 ft 3 in., plus or minus 3 in., and the minimum expected travel time is 3.5634 min. Note that the "optimum" shape factor for the rack is equal to 0.3542, which is not even close to being SIT. We believe this is due to the band heuristic (with $k = 2$) which tends to yield less "zigzagging" if the band width is small.

According to [10], the most desirable rack shape depends on the heuristic used for sequencing the pick points. With certain heuristics, the expected trip length *decreases* as the rack becomes SIT. Some of these heuristics generate sequences that require less travel time than those obtained from the band heuristic; the reader may refer to [10] for details. Also, the results in [26] suggest that, for sufficiently large $n$ values, the exact shape of the region has little or no impact on the expected length of the optimum picking sequence. Furthermore,

Table 12.9   *Alternative Configurations for Example 12.27*

| Alternative | Height | Length | T | Q | E(k, n, T, Q) |
|---|---|---|---|---|---|
| 1 | 15.00 | 600.00 | 2.0000 | 0.1250 | 4.3951 |
| 2 | 20.00 | 450.00 | 1.5000 | 0.2222 | 3.7321 |
| 3 | 25.00 | 360.00 | 1.2000 | 0.3472 | 3.5636 |
| 4 | 30.00 | 300.00 | 1.0000 | 0.5000 | 3.6586 |
| 5 | 40.00 | 225.00 | 0.7500 | 0.8889 | 4.2433 |
| 6 | 24.00 | 375.00 | 1.2500 | 0.3200 | 3.5712 |
| 7 | 27.00 | 333.30 | 1.1111 | 0.4050 | 3.5783 |
| 8 | 25.50 | 352.94 | 1.1765 | 0.3613 | 3.5638 |
| 9 | 24.75 | 363.64 | 1.2121 | 0.3403 | 3.5645 |
| 10 | 25.25 | 356.44 | 1.1881 | 0.3542 | 3.5634* |

we remind the reader that the above "optimum" rack shape is determined only with respect to the expected travel time for the S/R machine. When the rack cost is included, the "optimum" shape factor may be different from the one we computed in Table 12.9.

In deriving the above results, we assumed randomized storage and Chebyshev travel. Many deviations from these assumptions occur in practice. For example, walking along an aisle and picking parts from bins or flow racks (i.e., "walk–and–pick" systems) might be represented as a single dimensional picking problem, with negligible vertical travel. Likewise, instead of a person–on–board S/R machine, if the picker uses, say, an order picker truck or lift truck, our Chebyshev travel assumption would not apply. (Although some order picker/lift trucks are capable of simultaneous travel, for safety reasons simultaneous travel is allowed only if the forks are not fully raised.) Also, assigning the "fast movers" to the "closest" storage positions (i.e., turnover based storage) violates the assumption of uniformly and randomly distributed pick points.

In designing in–the–aisle OP systems, the primary design decisions are: the number, length, and height of storage aisles to provide; and the number of pickers. As a first-cut in designing such systems, it is useful to consider the situation in which *the storage aisles are SIT and each picker is dedicated to one aisle.* Under such an assumption, a single design decision is required: the number of storage aisles. The resulting model is

> Minimize:   Number of storage aisles
> Subject to:   Throughput constraint
> Storage space constraint

The following notation is used to specify the design algorithm [8]:

$S$ = total storage space required, expressed in square footage of storage rack,

$R$ = total throughput required, expressed in picks per hour,

$n$ = average number of stops (i.e., picks) per trip, including the P/D station ($3 \leq n \leq 24$),

$p$ = average pick time per stop, expressed in minutes,

$K$ = time spent at the P/D station between successive pick trips, expressed in minute

$b_v$ = horizontal velocity of picking vehicle or the S/R machine, expressed in ft/min,

$v_v$ = vertical velocity of picking vehicle or the S/R machine, expressed in ft/min,

$t = p + (K/n)$,

$q = [S/(2h_v v_v)]^{0.5}$, and

$u = 0.073331 + (0.385321/n) + 2/(n + 1)$; assuming that the pick points are sequenced in an optimum fashion.

In the following algorithm, $\lceil m \rceil$ denotes the smallest integer greater than or equal to $m$ (i.e., the "ceiling function").

**ALGORITHM 1.**

**STEP 0.** Set $M = \lceil Rt/60 \rceil$.

**STEP 1.** If $M(60M - Rt)^2 \geq (Rqu)^2$, then go to step 3; otherwise, go to step 2.

**STEP 2.** Let $M \leftarrow M + 1$, go to step 1.

**STEP 3.** STOP, $M$ is the minimum number of aisles (or pickers).

## Example 12.28

Suppose $S = 72{,}000$ ft$^2$, $R = 400$ picks/hr, $n = 5$ picks/trip, $p = 0.25$ min/pick, $h_v = 450$ fpm, $v_v = 90$ fpm, and $K = 0.75$ min/trip. Thus, $t = 0.40$, $q = 0.9428$, and $u = 0.4837$. The right-hand side of the inequality in Step 1 has the value 33,274.48. Enumerating over $M$, the left-hand side of the inequality has the following values:

| $M$ | $M(60M - 160)^2$ |
|---|---|
| 3 | $1200 < 33{,}274.48$ |
| 4 | $25{,}600 < 33{,}274.48$ |
| 5 | $98{,}000 > 33{,}274.48$ |

Hence, the minimum number of storage aisles and pickers required is 5; each aisle (or rack) will be 37.95 ft tall and 189.74 ft long. Note that the above solution is based on the rate at which the picks must be made (i.e., $R$); there is no explicit allowance made for replenishing the containers.

### End-of-Aisle Order Picking

As we described earlier, end–of–aisle OP is based on bringing the containers to the end of the aisle where the picker performs the pick and the container is returned to the storage rack. Although there are several end–of–aisle OP systems available, our analysis will be limited to the miniload AS/RS, which is used primarily for small- to medium-sized parts. We stress that the results we show here would apply to other systems as long as the device bringing the containers to the picker follows the Chebyshev metric and the storage area is rectangular.

The same assumptions used for in–the–aisle OP apply to end–of–aisle OP. Specifically, it is assumed that the containers are randomly located in the storage rack (or the S/R machine is equally likely to visit any point in the rack in order to retrieve a container), a single "pick" is made from each container (although one such pick may involve several items), the picker is dedicated to the aisle, a continuous approximation of the storage rack is sufficiently accurate, acceleration and deceleration losses are negligible, and where multiple aisles are required they are identical in activity and rack shape.

A three-aisle miniload AS/RS is shown in Figure 12.11. Note that in front of each aisle there are two *pick positions*; for the first aisle they are designated as pick position A and pick position B. The operation of, say, the first aisle may be described as follows (recall that one picker is assigned to each aisle): Suppose there are two containers—one in each pick position—and the S/R machine is idle at the P/D station. Further suppose that the picker is picking from the container in position A. After completing a pick, the picker presses a "pick complete" button, which signals to the S/R machine that the picker is done with the container in position A. At that point, the picker switches over to pick position B and starts picking from the container there while the S/R machine picks up the container in pick position A and stores it in the rack. The S/R machine then travels directly to another location in the rack and retrieves the next container for pick position A.

If the picker completes the pick at position B before the S/R machine delivers the next container for position A, then the picker becomes idle until the S/R machine arrives at the P/D station. On the other hand, if the S/R machine delivers the new container for position A before the picker completes the pick at position B, then the S/R machine becomes idle at the P/D station until the picker hits the "pick complete" button. In either case (i.e., whether the S/R machine waits on the picker or vice versa), the next (system) cycle begins with the picker picking from the new container in position A and the S/R machine performing another (dual command) trip to retrieve

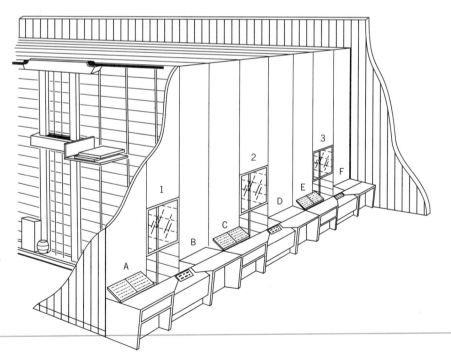

**Figure 12.11** A three-aisle miniload AS/R system. (Reprinted from [89])

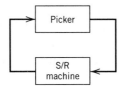

**Figure 12.12**   Two-server cyclic queuing system for the miniload AS/RS.

a new container for position B. Hence, the "system cycle" time is defined to be the maximum of the S/R machine cycle time and the pick time. The S/R machine cycle time is defined by a dual command cycle, which includes two pickups and two deposits.

In [13] it is noted that the above system can be modeled as a closed, two-server cyclic queuing system as shown in Figure 12.12. In such a system, each "customer" circulates in the loop indefinitely, served alternately by the picker and the S/R machine (i.e., as soon as a customer completes service at one of the servers, it joins the queue for the other server). Since there are two pick positions, the (constant) number of customers in the loop is set equal to two. The "mean service time" for the picker is set equal to the mean pick time, while the "mean service time" for the S/R machine is set equal to the expected dual command cycle time. (In the model, the same two customers circulate in the loop; however, in the actual system, a different container is retrieved by the S/R machine each time it stores one. This difference has no impact on the performance of the system from a queuing standpoint as long as the S/R machine service times are independent of the particular containers stored or retrieved on each trip.) A new "system cycle" starts whenever the two servers start serving a new customer.

The performance of the miniload AS/RS depends on the utilization of the picker. Given the expected value of the "system cycle" time, it is straightforward to obtain the expected value of the picker utilization. The determination of the expected value of the "system cycle" time, however, is complicated by either or both the S/R machine cycle time and the pick time being random variables. Specifically, the expected value of the system cycle time is the expected value of the maximum of two (independent) random variables. Even if the pick time is exponential, it is not a simple calculation because of the complex form of the probability distribution for dual command cycles (see [28]). (Note that in Equation 12.71 we presented the mean travel time for a dual command cycle but we did not show its variance.)

By approximating the S/R machine travel time with a uniform distribution, in [13] a design algorithm is presented for deterministic or exponential pick times. This algorithm works with a rack shape factor ($0 < Q \leq 1$) specified by the user a priori. The limits of the above uniform distribution are designated by $\hat{k}_1$ and $\hat{k}_2$ for a "normalized" rack. To specify the algorithm, we introduce additional notation as follows; let:

$E(\hat{DC})$ = expected value of a dual command cycle in a "normalized" ($Q \times 1$) rack,

$S(\hat{DC})$ = standard deviation of dual command cycles in a "normalized" ($Q \times 1$) rack,

$E(PU)$ = expected value of picker utilization,

$E(SRU)$ = expected value of S/R machine utilization,

$E(CT)$ = expected value of system cycle time,

$C$ = total container handling time per dual command cycle ($4T_{P/D}$).

## ALGORITHM 2.

**STEP 0.** Set $E(\widehat{DC}) = (4/3) + (Q^2/2) - (Q^3/30)$ and $S(\widehat{DC}) = (0.3588 - 0.1321Q) E(\widehat{DC})$. Compute the values of $\hat{k}_1$ and $\hat{k}_2$ as follows:

$$\hat{k}_1 = E(\widehat{DC}) - 1.7321S(\widehat{DC})$$

$$\hat{k}_2 = E(\widehat{DC}) + 1.7321S(\widehat{DC})$$

Lastly set $M = \lceil Rp/60 \rceil$ and $q = [S/(2Qb_vv_v)]^{0.5}$.

**STEP 1.** Set $r = q/\sqrt{M}$. Given the value for $r$, set $k_1 = r\hat{k}_1$, $k_2 = r\hat{k}_2$, and $E(DC) = rE(\widehat{DC})$. Let $t_1 = k_1 + C$ and $t_2 = k_2 + C$.

**STEP 2.** If the pick time is deterministic, go to step 2a; if the pick time is exponentially distributed, go to step 2b.

**STEP 2a.** Compute $E(CT)$ as follows:

$$E(CT) = \begin{cases} E(DC) + C & \text{for } 0 < p \le t_1 \\ \dfrac{p^2 - 2pt_1 + t_2^2}{2(t_2 - t_1)} & \text{for } t_1 < p \le t_2 \\ p & \text{for } t_2 < p \end{cases}$$

**STEP 2b.** Compute $E(CT)$ as follows:

$$E(CT) = E(DC) + C + \frac{p^2}{(t^2 - t_1)} [exp(-t_1/p) - exp(-t_2/p)]$$

**STEP 3.** Compute

$$E(PU) = p/E(CT)$$

$$E(SRU) = [E(DC) + C]/E(CT)$$

**STEP 4.** If $E(PU)(60/p)M \ge R$, go to step 6; otherwise go to step 5.

**STEP 5.** Let $M \leftarrow M + 1$ and go to step 1.

**STEP 6.** STOP, $M$ is the "minimum" number of aisles.

Assuming that the vertical side of the rack is shorter in time, the $M$ value obtained from Algorithm 2 can be used to determine the resulting rack dimensions as follows:

rack length,  $L = [Sb_v/(2QMv_v)]^{0.5}$,

rack height,  $H = QLv_v/b_v$.

Note that, unlike in–the–aisle OP systems, the number of containers per order does not play a role in the above design algorithm. This is primarily due to the fact that with in–the–aisle OP systems the number of containers per order affects the picking sequence and the resulting S/R machine travel time. With end–of–aisle OP systems, however, as long as the S/R machine operates on a dual command basis (and the containers for an order are retrieved in no particular sequence), the number of containers per order does not affect the S/R machine cycle time. Of course, we are assuming that the containers for one order are not intermixed with the containers for another order. That is, the S/R machine will not retrieve the containers for the next order until all the containers for the current order have been retrieved. To avoid unnecessary single command trips, the last container of the current order is interleaved with the first container of the next order. In other words, when the S/R machine picks up the last container from, say, position A, and stores it in the rack, it retrieves the first container for the next order. The same holds true for position B. This allows the S/R machine to operate on a dual command basis 100% of the time without intermixing the containers that belong to different orders.

Also note in Algorithm 2 that the expected S/R machine utilization does not play a direct role in determining the throughput capacity of the system. As indicated by the left-hand side of the inequality shown in step 4, the throughput capacity of the system is dictated by the expected picker utilization, $E(PU)$. Last, we remind the reader that no allowance has been made for "replenishment cycles." If replenishment is performed on a second shift, it should not affect the results obtained from Algorithm 2. However, if replenishment cycles are interleaved with picking cycles (i.e., "new" containers are stored in the system while existing ones are being depleted), some allowance must be made for S/R machine time lost due to replenishment cycles. Alternately, the picker may have to replenish a near-empty container when the S/R machine retrieves it. In such cases, an allowance must be made for the picker to replenish such containers.

## Example 12.29

Suppose $S = 72,000$ ft$^2$, $R = 300$ picks/hr, $p = 1.20$ min/pick, $b_v = 450$ fpm, $v_v = 90$ fpm, and $C = 0.60$ min/cycle. Assuming that the user desires a SIT rack, we set $Q = 1$. We further assume that the pick time is constant. Applying Algorithm 2, we obtain the following results:

0. $E(\widehat{DC}) = 1.80$ and $S(\widehat{DC}) = 0.4081$. Therefore, $\hat{k}_1 = 1.0932$ and $\hat{k}_2 = 2.5068$. Also, $M = 6$ and $q = 0.9428$.

1. $r = 0.3849$; therefore, $k_1 = 0.4208$, $k_2 = 0.9649$, and $E(DC) = 0.6928$. Hence, $t_1 = 1.0208$ and $t_2 = 1.5649$.

2a. Since $1.0208 < p < 1.5649$, $E(CT) = 1.3223$.

3. $E(PU) = 0.9075$ and $E(SRU) = 0.9777$.

4. $E(PU)(60/p)M = 272.25 < 300$.

5. $M = 7$.

1. $r = 0.3563$; therefore, $k_1 = 0.3895$, $k_2 = 0.8932$, and $E(DC) = 0.6413$. Hence, $t_1 = 0.9895$ and $t_2 = 1.4932$.

2a.  Since $0.9895 < p < 1.4932$, $E(CT) = 1.2853$.

  3.  $E(PU) = 0.9336$ and $E(SRU) = 0.9658$.

  4.  $E(PU)(60/p)M = 326.76 > 300$.

  6.  STOP; the "minimum" number of storage aisles and pickers required is 7. Each aisle will be 32.07 ft tall and 160.36 ft long.

Note that the throughput *capacity* of the system exceeds the throughput *requirement.* Consequently, the expected system utilization will be approximately 92% ($300/326.76 = 0.9181$). This implies that the "effective" picker and S/R machine utilizations will be approximately equal to 86% ($0.9336 \times 0.9181$) and 89% ($0.9658 \times 0.9181$), respectively. Further note that when $E(PU)$ increases, $E(SRU)$ decreases. In general, $E(PU)$ and $E(SRU)$ have a negative correlation; that is, as one of them decreases, the other one increases (unless, of course, it is already at 100%). This is primarily due to the closed, two-server cyclic queuing system we showed earlier in Figure 12.12. Last, the system throughput capacity (326.76 picks/hr) is not rounded off to an integer value because it reflects an hourly average.

In Algorithm 2, it is assumed that the shape factor for the rack is specified by the user. It is generally understood that the user will run Algorithm 2 for various $Q$ values and generate several alternative configurations to evaluate them in terms of cost, throughput capacity, and space requirements.

## *12.8*  CONVEYOR MODELS

Our interest in conveyors stems from the fact that conveyor systems are one of the most widely used material handling systems to date. The models discussed in this section provide some basic design procedures for towline and trolley-type conveyors and for powered belt and roller conveyors.

### Towline or Trolley Conveyors

In this section a deterministic model of a towline or trolley conveyor is considered. The towline conveyor and trolley conveyor have received considerable attention from researchers [83]. The analysis of such conveyors is referred to as **conveyor analysis**. Conveyor analysis is a term that is used to refer to the analysis of closed-loop, irreversible, or recirculating conveyors with discretely spaced carriers.

The interest in analyzing recirculating conveyors can be traced to two engineers: Kwo, with General Electric, and Mayer, with Western Electric. Kwo [47, 48] developed a deterministic model of materials flow on a conveyor having one loading station and one unloading station. He proposed a number of rules to ensure the input and output rates of materials were compatible with the design of the conveyor. Mayer [55] employed a probabilistic model in analyzing a conveyor with multiple loading stations.

Conveyor analysis was developed because of a number of operating problems that occurred with overhead trolley conveyors. Namely,

1.  At a loading station, no empty carriers were available when a loading operation was to be performed.

2.  At an unloading station, no loaded carriers were available when an unloading operation was to be performed.

Because of the interference problems analytical and simulation studies were undertaken in order to gain insight into the relationships among the design parameters and operating variables.

A conveyor system consists of the combination of the conveyor equipment, loading and unloading stations, and the operating discipline employed. Hence, system design parameters include not only equipment parameters, but also such considerations as waiting space allowances and the number, spacing, and sequencing of loading and unloading stations.

Conveyors provide many of the links among workstations in modern production systems. They are used not only to transport or deliver materials, but also to provide storage capacity for materials. Some conveyor systems are basically handling or transport systems, some are storage or accumulation systems, and some perform both handling and storage functions.

Kwo [48, 48] developed three principles to be used in designing closed-loop, irreversible conveyors:

1.  *Uniformity principle:* Materials should be uniformly distributed over the conveyor.

2.  *Capacity principle:* The carrying capacity of the conveyor must be greater than or equal to the system throughput requirements.

3.  *Speed principle:* The speed of the conveyor (in terms of the number of carriers per unit time) must be within the permissible range, defined by loading and unloading station requirements and the technological capability of the conveyor.

He also developed some numerical relations based on the uniformity principle that provide **sufficient** conditions for steady-state operations. Since the work of Kwo, more general results have been developed by Muth [56–58]. Our presentations will be based on the multistation analysis of Muth [58].

A schematic representation of the recirculating conveyor considered by Muth is given in Figure 12.13. There are $s$ stations located around the conveyor, *numbered in reverse sequence to the rotation of the conveyor.* Each station can perform loading and/or unloading. There are $k$ carriers equally spaced around the conveyor. The passage of a carrier by a workstation establishes the increment of time used to define material loading and unloading sequences.

For convenience, station 1 is used as a reference point in defining time; consequently, carrier $n$ becomes carrier $n + k$ immediately after passing station 1. The sequence of points in time at which a carrier passes station 1 is denoted $\{t_n\}$, where $t_n$ is the time at which carrier $n$ passes station 1. The amount of material loaded on carrier $n$ as it passes station $i$ is given by $f_i(n)$ for $i = 1, 2, \ldots, s$.[4] The amount of material carried by carrier $n$ immediately after passing station $i$ is denoted $H_i(n)$.

In order for the conveyor to operate for long periods of time (achieve steady-

---

[4]Negative values of $f_i(n)$ are permissible and denote unloading.

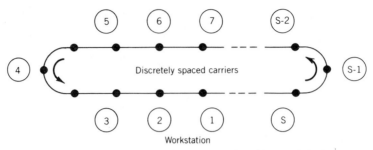

**Figure 12.13**   Conveyor layout considered by Muth [58].

state operations), the total amount of material loaded on the conveyor must equal the total amount of material unloaded. Since it is assumed that the conveyor will be operated over an infinite period of time, the sequences $\{f_i(n)\}$ are assumed to be periodic, with period $p$. Thus,

$$\{f_i(n)\} = \{f_i(n + p)\} \tag{12.79}$$

Additionally,

$$\sum_{n=1}^{p} \sum_{i=1}^{s} f_i(n) = 0 \tag{12.80}$$

because of the requirement that all material loaded on the conveyor must be unloaded and vice versa.

In analyzing the conveyor it is convenient to employ the following relation:

$$F_1(n) = \sum_{i=1}^{s} f_i(n) \tag{12.81}$$

Muth obtains the following results:

1.   $k/p$ cannot be an integer for steady-state operations.
2.   Letting $r = k \bmod p$, $r/p$ must be a proper fraction for general sequences $\{F_1(n)\}$ to be accommodated.[5]
3.   It is desirable for $p$ to be a prime number, as conveyor compatibility results for all admissible values of $k$.

The materials balance equation for carrier $n$ can be given as

$$H_1(n) = H_1(n - r) + F_1(n) \tag{12.82}$$

In order to determine the values of $H_i(n)$ the following approach is taken.

1.   Let $H_1^*(n)$ be a particular solution to Equation 12.82 and use the recursion

$$H_1^*(n) = H_1^*(n - r) + F_1(n) \tag{12.83}$$

by letting $H_1^*(1) = 0$.

---

[5]The operation of $r = k \bmod p$ means $r$ is given by the remainder following the division of $k$ by $p$, for example, 12 mod 7 = 5 and 26 mod 8 = 2.

2. Given $H_i^*(n)$, the value of $H_{i+1}^*(n)$ is obtained from the relation

$$H_{i+1}^*(n) = H_i^*(n) - f_i(n) \tag{12.84}$$

3. Given $\{H_i^*(n)\}$ for $i = 1, \ldots, s$, let

$$c = \min_{i,n} H_i^*(n) \tag{12.85}$$

4. The desired solution is given by

$$H_i(n) = H_i^*(n) - c \tag{12.86}$$

5. The required capacity per carrier is

$$B = \max_{i,n} H_i(n) \tag{12.87}$$

## Example 12.30

In order to illustrate the deterministic solution procedure, consider a situation involving a single loading station and a single unloading station, as depicted in Figure 12.14. The conveyor has nine carriers, equally spaced. The loading and unloading sequences have periods of 7 time units and are given to be

$$\{f_1(n)\} = \{1, 1, 2, 2, 2, 1, 1\}$$

and

$$\{f_2(n)\} = \{0, 0, 0, 0, 0, -5, -5\}$$

Thus,

$$\{F_1(n)\} = \{1, 1, 2, 2, 2, -4, -4\}$$

and

$$r = k \bmod p$$
$$= 9 \bmod 7$$
$$= 2$$

Notice $k/p = 9/7$ is not an integer, $r/p = 2/7$ is a proper fraction, and $p = 7$ is a prime number.

Arbitrarily, we let $H_1^*(1) = 0$ and employ the relation $H_1^*(n) = H_1^*(n - r) + F_1(n)$ to determine $H_1^*(3)$,

$$H_1^*(3) = H_1^*(1) + F_1(3)$$
$$H_1^*(3) = 0 + 2 = 2$$

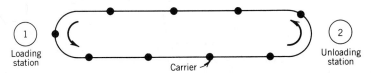

**Figure 12.14**   Conveyor layout for Example 12.30.

Likewise, for $H_1^*(5)$

$$H_1^*(5) = H_1^*(3) + F_1(5)$$
$$= 2 + 2 = 4$$

Since $r = 2$, the next computation is for $H_1^*(7)$,

$$H_1^*(7) = H_1^*(5) + F_1(7)$$
$$= 4 - 4 = 0$$

Next, $H_1^*(9)$ is obtained; however, since $p = 7$, $H_1^*(9) = H_1^*(2)$. Hence,

$$H_1^*(2) = H_1^*(7) + F_1(2)$$
$$= 0 + 1 = 1$$

Likewise, for $H_1^*(4)$,

$$H_1^*(4) = H_1^*(2) + F_1(4)$$
$$= 1 + 2 = 3$$

Similarly,

$$H_1^*(6) = H_1^*(4) + F_1(6)$$
$$= 3 - 4 = -1$$

Hence, $\{H_1^*(n)\} = \{0, 1, 2, 3, 4, -1, 0\}$.

The values of $H_2^*(n)$ are obtained using the relation $H_2^*(n) = H_1^*(n) - f_1(n)$. Therefore,

$$\{H_2^*(n)\} = \{0, 1, 2, 3, 4, -1, 0\} - \{1, 1, 2, 2, 2, 1, 1\}$$
$$= \{-1, 0, 0, 1, 2, -2, -1\}$$

The value of $c$ is found to be

$$c = \min_{i,n} H_i^*(n) = -2$$

for $i = 2$ and $n = 6$. Consequently, the sequences $\{H_i(n)\}$ are given by

$$\{H_1(n)\} = \{2, 3, 4, 5, 6, 1, 2\}$$
$$\{H_2(n)\} = \{1, 2, 2, 3, 4, 0, 1\}$$

and $B = \max_{i,n} H_i(n) = 6$, for $i = 1$ and $n = 5$. Therefore, each carrier must have sufficient capacity to accommodate 6 units of product.

As shown in Table 12.10 the amount of materials on the carriers will change with changing values of $k$. Additionally, the required carrier capacity $B$ is affected by the value of $k$.

The sequences $\{f_1(n)\}$ and $\{f_2(n)\}$ indicate, for example, that 5 units are removed from the sixth carrier at station 2 and, when it arrives at station 1, 1 unit is placed on the sixth carrier. When the sixth carrier passes station 1, its steady-state load $H_1(6)$ equals 1 unit of product; additionally, it becomes labeled the fifteenth carrier if $k = 9$.

When the fifteenth carrier reaches station 2 the action taken is given by $f_2(15) = f_2(8) = f_2(1) = 0$, since $p = 7$. Thus, nothing happens at station 2; however, at station 1, $f_1(1) = 1$ and 1 unit of product is added to the carrier, raising its total to $H_1(1) = 2$.

The carrier is now labeled the twenty-fourth carrier; when it reaches station 2, since $f_2(24) = f_2(3) = 0$, nothing happens at station 2. However, at station 1, $f_1(3) = 2$ and the content of the carrier is increased to 4 units. The carrier with 4 units arrives at station 2 num-

Table 12.10  Values of {$H_i(n)$} for Various Values of k in Example 12.30

| n | k = 8 | | k = 9 | | k = 10 | | k = 11 | | k = 12 | | k = 13 | |
|---|---|---|---|---|---|---|---|---|---|---|---|---|
| | $H_1(n)$ | $H_2(n)$ | $H_1(n)$ | $H_2(n)$ | $H_1(n)$ | $H_2(n)$ | $H_1(n)$ | $H_2(n)$ | $H_1(n)$ | $H_2(n)$ | $H_1(n)$ | $H_2(n)$ |
| 1 | 2 | 1 | 2 | 1 | 5 | 4 | 4 | 3 | 6 | 5 | 9 | 8 |
| 2 | 3 | 2 | 3 | 2 | 2 | 1 | 7 | 6 | 5 | 4 | 8 | 7 |
| 3 | 5 | 3 | 4 | 2 | 5 | 3 | 5 | 3 | 5 | 3 | 7 | 5 |
| 4 | 7 | 5 | 5 | 3 | 7 | 5 | 3 | 1 | 4 | 2 | 5 | 3 |
| 5 | 9 | 7 | 6 | 4 | 4 | 2 | 6 | 4 | 3 | 1 | 3 | 1 |
| 6 | 5 | 4 | 1 | 0 | 1 | 0 | 3 | 2 | 2 | 1 | 1 | 0 |
| 7 | 1 | 0 | 2 | 1 | 3 | 2 | 1 | 0 | 1 | 0 | 5 | 4 |
| | $B = 9$ | | $B = 6$ | | $B = 7$ | | $B = 7$ | | $B = 6$ | | $B = 9$ | |

bered the thirty-third carrier. Since $f_2(33) = f_2(5) = 0$ nothing happens; however, at station 1, $f_1(5) = 2$, and the carrier's load is increased to 6 units.

When the carrier arrives at station 2 it is numbered the forty-second carrier. Since $f_2(42) = f_2(7) = -5$, the content of the carrier is reduced to 1 unit. When it arrives at station 1, $f_1(7) = 1$ and 1 unit is added to bring the total amount to 2 units.

The carrier is numbered the fifty-first carrier when it arrives at station 2. Since $f_2(51) = f_2(2) = 0$, nothing happens. At station 1 the content of the carrier is increased by 1 unit to total 3 units and it becomes the sixtieth carrier.

The sixtieth carrier arrives at station 2 with 3 units; since $f_2(60) = f_2(4) = 0$, nothing happens. At station 1, 2 units are added to the carrier and the number of the carrier is increased to 69. When the carrier arrives at station 2, $f_2(69) = f_2(6) = -5$, and 5 units are removed. Hence, the carrier arrives at station 1 empty, just as it did when it was labeled carrier six.

The above description for carrier 6 can be duplicated for each carrier on the conveyor by considering a period of time equal to $kp = 63$ time periods. The content of the carrier as it left station 1 followed the sequence {1, 2, 4, 6, 2, 3, 5}, which is a permutation of $\{H_1(n)\}$; likewise, the content of the carrier as it left station 2 followed the sequence {0, 1, 2, 4, 1, 2, 3}, which is the same permutation of $\{H_2(n)\}$ that generated the sequence for station 1.

If one tracks the content of any given carrier for $kp$ time periods, it will be found that each carrier as it leaves station $i$ will carry an amount of material given by a permutation of the sequence $\{H_i(n)\}$. Hence, each carrier will, at some time during the repeating cycle $kp$ carry an amount of material equal to each element in the sequence $\{H_i(n)\}$ as it leaves station $i$. Consequently, a conveyor designed using Muth's results will satisfy the uniformity principle.

# Example 12.31

Consider a conveyor system with one loading station and one unloading station. The material flow sequences are

$$f_1(n) = (0, 0, 2, 3, 1)$$
$$f_2(n) = (0, 0, 0, -2, -4)$$

If there are seven carriers on the conveyor, then $r = 2$. Solving for the sequences $\{H_i(n)\}$ for $i = 1, 2$ gives

$$\{H_1(n)\} = \{3, 2, 5, 3, 2\}$$
$$\{H_2(n)\} = \{3, 2, 3, 0, 1\}$$

Thus, $B = 5$.

Suppose the distance from station 2 to station 1 on the conveyor is changed, such that the loading and unloading sequences are given by

$$f_1(n) = (2, 3, 1, 0, 0)$$
$$f_2(n) = (0, 0, 0, -2, -4)$$

Solving for the sequences $\{H_i(n)\}$ gives

$$\{H_1(n)\} = \{3, 3, 4, 1, 0\}$$
$$\{H_2(n)\} = \{1, 0, 3, 1, 0\}$$

and $B = 4$.

## Example 12.32

Suppose a closed-loop, irreversible conveyor with discretely spaced carriers is used to serve three workstations located as shown in Figure 12.15. The material flow sequences are assumed to be

$$f_1(n) = (4, 4, 0)$$
$$f_2(n) = (-3, -1, 0)$$
$$f_3(n) = (0, -2, -2)$$

and $k = 16$, such that $r = 1$. Solving for $\{H_i(n)\}$ gives

$$\{H_1(n)\} = \{4, 5, 3\}$$
$$\{H_2(n)\} = \{0, 1, 3\}$$
$$\{H_3(n)\} = \{3, 2, 3\}$$

with $B = 5$.

If the workstations are rearranged around the conveyor such that

$$f_1(n) = (0, -2, -2)$$
$$f_2(n) = (4, 4, 0)$$
$$f_3(n) = (-3, -1, 0)$$

then the appropriate computations establish that

$$\{H_1(n)\} = \{4, 5, 3\}$$
$$\{H_2(n)\} = \{4, 7, 5\}$$
$$\{H_3(n)\} = \{0, 5, 5\}$$

with $B = 7$.

The last two examples illustrate the effect on $\{H_i(n)\}$ of the locations of the workstations served by the closed-loop, recirculating conveyor. Not only can the capacity requirement for a carrier be affected by changing the number of carriers on

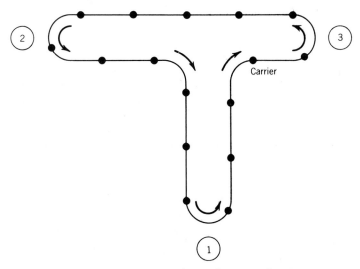

**Figure 12.15**   Conveyor layout for Example 12.32.

the conveyor, but it also can be affected by the locations of the workstations around the conveyor.

Muth's modeling approach is descriptive, rather than prescriptive. Consequently, optimum conveyor designs are not obtained directly from the analysis. However, for a particular situation one can evaluate a number of alternative designs and compare the economic performance of each. As has been shown, the carrier capacity can be affected in a number of ways. Consequently, one would want to consider not only alternative conveyor lengths (measured in the number of carriers), but also the locations of stations around the conveyor.

## Horsepower Calculations

In the previous section we considered a model of a recirculating trolley conveyor. The model allowed us to consider the effect on the carrying capacity of an individual carrier (trolley) of changes in loading and unloading sequences. Further, the model provided us with an indication of the amount of inventory on the conveyor during steady state conditions.

We noted that the cost of the trolley conveyor is affected by the length of the conveyor and the number and size of carriers installed on the conveyor. In addition to the cost of the hardware represented by the trolleys and connecting chain, the installation cost is also a significant factor. Not to be overlooked, the cost of powering the conveyor is a function of its design. In particular, the horsepower requirement depends on the conveyor speed and the weight of the conveyor and the material it transports. The greater horsepower requirement, typically, the greater the cost of the motor(s) used.

In this section, we consider the horsepower requirements for powered belt and roller conveyors used to transport unit loads. Drawing upon models provided in design manuals of various conveyor manufacturers, we present in this section approximate or "quick and dirty" models, rather than extremely precise models, of horsepower requirements.

Our purpose in considering the horsepower requirements that result from design decisions for a particular conveyor is to demonstrate that what might seem to be inconsequential design decisions can significantly impact horsepower requirements. We are not trying to make you an expert in determining the horsepower requirements for a particular conveyor; we realize that the accuracy of the assumptions made regarding material to be conveyed, as well as decisions regarding the placement of a motor on the conveyor, can affect final decisions regarding the size (and cost) of the motor installed.

### *Powered Unit and Package Conveyors*

To determine the horsepower requirements for a belt conveyor or a roller conveyor to be used to convey packages, toteboxes, pallet loads, and so on, the following notation will be used:

HP = horsepower requirement, hp;

S = speed of the conveyor, feet per minute (fpm);

$L$ = load to be carried by the conveyor, in pounds;

$TL$ = total length of the conveyor, in feet;

$RC$ = spacing between center lines of rollers, i.e., roller spacing, in inches;

$WBR$ = width between rails, in inches (approximates belt width and/or roller length; rollers are generally at least 2.5″ longer than the width of the load being transported);

$\alpha$ = angle of incline, in degrees;

$LL_I$ = live load on incline, in pounds (weight of material on the conveyor section that is inclined);

$BV$ = base value;

$FF$ = friction factor; and

$LF$ = length factor.

## Belt Conveyors

To estimate the horsepower requirement for a belt conveyor, BV can be approximated by multiplying WBR by two-thirds. To determine the value of FF, it is important to know how the belt will be supported: if it is roller supported, then FF = 0.05; if it is slider bed supported, then FF = 0.30. The value of LF can be obtained from Table 12.11.

The horsepower requirement can be approximated as follows:

$$HP = [BV + LF(TL) + FF(L) + LL_I(\sin \alpha)](S)/14{,}000 \qquad (12.88)$$

## Example 12.33

Consider a 100-ft-long, roller-supported belt conveyor inclined at an angle of 10°, with a 6 in. spacing between rollers and a WBR value of 27 in. (Use roller beds, not slider beds, for inclines; 35° is the maximum angle of incline for most situations.) Suppose the conveyor is used to transport toteboxes of material, with each totebox being 18 in. long and weighing 35 lb, loaded. (For stability of the load, it is recommended that a minimum of two rollers be under the load at all times; on an incline, three or more rollers should be under the load.) A spacing of 12 in. is to be provided between toteboxes. The conveyor speed is planned to be 90 ft/min. What is the approximate horsepower requirement for the conveyor? What is the maximum loading condition for the conveyor?

**Table 12.11  *LF Values for Belt Conveyors, Given Specific RC and WBR Values***

| RC Value | WBR = 15″ | WBR = 19″ | WBR = 23″ | WBR = 27″ | WBR = 33″ | WBR = 39″ |
|---|---|---|---|---|---|---|
| 4.5″ | 0.41 | 0.50 | 0.61 | 0.72 | 0.85 | 1.00 |
| 6.0″ | 0.34 | 0.42 | 0.51 | 0.61 | 0.72 | 0.85 |
| 9.0″ | 0.29 | 0.35 | 0.43 | 0.52 | 0.62 | 0.73 |
| 12.0″ | 0.25 | 0.31 | 0.38 | 0.46 | 0.54 | 0.65 |
| S. B. | 0.50 | 0.70 | 0.80 | 1.00 | 1.20 | 1.50 |

S. B. denotes a slider bed supported belt conveyor.

To facilitate the determination, let a load-segment be defined as a totebox plus the clearance between toteboxes. In this case the length of the load-segment is equal to 30 in. Hence, exactly 40 load segments will be on the 100-ft-long conveyor at all times. Thus, the weight of the load will be 40(35) or 1400 lb.

In this case, $S = 90$ fpm, $RC = 6''$, $WBR = 27''$, $BV = \frac{2}{3}(27) = 18$, $LF = 0.61$, $TL = 100'$, $FF = 0.05$, $L = LL_I = 1400$ lb, and $\alpha = 10°$. Hence,

$$HP = [18 + 0.61(100) + 0.05(1400) + 1400(0.1737)](90)/14{,}000 = 2.52 \text{ hp}$$

Thus, a 2.75 hp or 3.0 hp motor should meet the demands placed on the belt conveyor.

What if the spacing between consecutive toteboxes had been 8 in., rather than 12 in.? Dividing the conveyor length by the length of the load segment yields a value of 46, plus a fraction. Hence, there would be 46 complete load segments, plus a partial load-segment. The 46 load segments have a length of 46(18'' + 8''), or 1196''. Thus, the maximum weight condition would be 46 full toteboxes and 4 in. of a partial totebox. If the weight of a totebox is uniformly distributed over its length, then the partial totebox contributes (4/18)(35), or 7.78 lb. Therefore, the maximum load would be 46(35) + 7.78, or 1617.78 lb and the horsepower requirement would be 2.83 hp.

## Example 12.34

Continuing the previous example, suppose it is desired to elevate the toteboxes 17.36'. Three alternatives are under consideration, using angles of incline of 10°, 20°, and 30°. Rounding off to the nearest foot, the resulting conveyor lengths would be 100 ft, 51 ft, and 35 ft in length. The horsepower requirement for the first alternative has already been determined, 2.52 hp.

For the case of $\alpha = 20°$ and $TL = 51$ ft, a calculation establishes that 20 complete load segments and 12 in. of a partial load segment will be accommodated on the conveyor. Thus, $L = LL_I = 20(35) + (12/18)(35) = 723.33$ lb. The horsepower requirement will be approximately

$$HP = [18 + 0.61(51) + 0.05(723.33) + 723.33(0.3420)](90)/14{,}000 = 2.14 \text{ hp}$$

For the case of $\alpha = 30°$ and $TL = 35$ ft, a calculation establishes that 14 complete load segments will be accommodated on the conveyor. Hence, $L = LL_I = 14(35) = 490$ lb; hence, the horsepower requirement will be approximately

$$HP = [18 + 0.61(35) + 0.05(490) + 490(0.50)](90)/14{,}000 = 1.99 \text{ hp}$$

Depending on circumstances of the application and the overall layout of the facility, it appears that the steepest feasible angle of incline would be preferred since it requires a smaller horsepower motor and a shorter conveyor.

## Example 12.35

In Example 12.33, suppose the toteboxes are to be transported horizontally over a distance of 100 ft, rather than elevated. Also, suppose a 2.5 hp motor is available and must be used. What is the maximum weight per totebox that should be transported by the conveyor?

From the Example 12.18, $S = 90$ fpm, $RC = 6''$, $WBR = 27''$, $BV = \frac{2}{3}(27) = 18$, $LF = 0.61$, $TL = 100'$, $FF = 0.04$, and $\alpha = 0°$. Hence,

$$HP = [18 + 0.61(100) + 0.05L](90)/14{,}000 = 2.5 \text{ hp}$$

**Table 12.12** *LF Values for Roller Conveyors, Given RC and WBR Values*

| RC Value | WBR = 15 in. | WBR = 19 in. | WBR = 23 in. | WBR = 27 in. | WBR = 33 in. | WBR = 39 in. |
|---|---|---|---|---|---|---|
| 3.0" | 1.3 | 1.5 | 1.7 | 2.0 | 2.4 | 2.6 |
| 4.5" | 0.9 | 1.0 | 1.2 | 1.4 | 1.6 | 1.8 |
| 6.0" | 0.7 | 0.8 | 0.9 | 1.0 | 1.2 | 1.5 |
| 9.0" | 0.5 | 0.6 | 0.7 | 0.8 | 0.9 | 1.0 |

Solving for L gives a value of 6197.78 lb. Since 40 toteboxes will be on the conveyor at the same time (based on a 30 in. load-segment), the maximum weight per totebox is 154.94 lb.

### Roller Conveyors

Using different values of FF, BV, and LF, the same formula can be used to determine the horsepower requirement for roller conveyors. The following values are used for the friction factor: $FF = 0.10$ when a flat belt is used to drive the rollers; $FF = 0.085$ when a zero-pressure accumulating belt is used to power the rollers; $FF = 0.075$ when a v-belt is used to drive the rollers; and $FF = 0.05$ when the rollers are chain driven. The base value for roller conveyors is given by the following relation:

$$BV = 4.60 + 0.445(WBR) \tag{12.89}$$

LF values are provided in Table 12.12 for various RC and WBR values.

## Example 12.36

Consider the conveying requirement given in Example 12.33, albeit without the requirement to incline the load. (Although it is possible to use a roller conveyor to elevate unit loads, belt conveyors are preferred due to the slippage between the load and the rollers on an incline. The friction of the belt prevents slippage of the load on belt conveyors.) Consider a flat belt driven roller conveyor, with the rollers still on 6 in. centers. (For stability of transport, at least three rollers should be under the load at all times; for soft loads, closer spacing is required.) Hence, for an 18-in.-long load, the largest roller spacing would be 6 in.)

For the example, $S = 90$ fpm, $RC = 6$ in., $WBR = 27$ in., $BV = 4.60 + 0.445(27) = 16.62$, $LF = 1.0$, $TL = 100$ ft, $FF = 0.10$, $L = 1400$ lb, and $LL_1 = 0$ lb, since $\alpha = 0°$. Hence,

$$HP = [16.62 + 1.0(100) + 0.10(1400)](90)/14,000 = 1.65 \text{ hp}$$

Thus, a 1.75 hp or 2.0-hp motor should satisfy the conveyance requirements.

What is the upper bound on the weight of a totebox if no more than 2.5 horsepower can be required? A calculation establishes that each totebox must not weigh more than 68 lb.

## Example 12.37

In the previous example, suppose the toteboxes go directly from the roller conveyor operating at a speed of 90 fpm to a 100-ft-long, flat belt driven roller conveyor operating at 150 fpm. What would be the horsepower requirement for the second conveyor?

If 18-in.-toteboxes are moving at a speed of 90 fpm, with a clearance of 12 in., what will be the spacing of toteboxes on the second conveyor? Be careful! Note that the rate at which toteboxes exit the first conveyor equals (90 ft/min)(12 in./ft)/30 in./totebox), or 36 toteboxes/min. Since the rate they exit the first conveyor must equal the rate at which they enter the second conveyor, then the spacing of toteboxes on the second conveyor will be (150 ft/min)(12 in./ft)/(36 toteboxes/min), or 50 in./totebox. Since the toteboxes are 18 in. long, then the clearance on the second conveyor will be 32 in.

(Some might have thought that increasing the speed by two-thirds would cause the clearance to increase by two-thirds, to attain a value of 20 in. Not so! If you have ever walked along an airport concourse and stepped on to a moving sidewalk, you have experienced the spacing differential that occurs with the toteboxes. Since the totebox is neither stretched nor shrunk as it moves from one conveyor to another operating at a different speed, the clearance must accommodate the expansion/contraction for the load segment.)

On the second conveyor there will be exactly 24 complete load segments. Hence, the following parameter values apply to the second conveyor: S = 150 fpm, RC = 6 in., WBR = 27 in., BV = 16.62, LF = 1.0, TL = 100 ft, FF = 0.10, and L = 840 lb. Therefore,

$$HP = [16.62 + 1.0(100) + 0.10(840)](150)/14{,}000 = 2.15 \text{ hp}$$

## *Example 12.38*

From the two previous examples, we found that a 1.65-hp motor and a 2.15-hp motor would be required to transport the toteboxes over a distance of 200 ft by using two 100-ft-long conveyors operating at 90 fpm and 150 fpm. Suppose a single 200 ft long conveyor is used. What would be the horsepower requirement if it is operated at a speed of 120 fpm and conveys the toteboxes at a rate of 36 toteboxes/min?

A calculation establishes that the length of a load segment is 40 in. and 60 complete load segments are accommodated on the conveyor. Thus, for the example, S = 120 fpm, RC = 6 in., WBR = 27 in., BV = 16.62, LF = 1.0, TL = 200 ft, FF = 0.10, and L = 2100 lb. Hence,

$$HP = [16.62 + 1.0(200) + 0.10(2100)](120)/14{,}000 = 3.66 \text{ hp}$$

Depending on the costs of the motors, the requirements of the particular application, and the costs of 100 ft and 200 ft conveyor sections, one might prefer to have a single 200-ft conveyor to having two 100 ft conveyors.

## *12.9*   WAITING LINE MODELS[6]

Among the many analytical approaches that can be used to aid in the design of facilities is that of **waiting line analysis** or **queueing theory**. As the name implies, waiting line analysis involves the study of waiting lines. Examples of waiting lines include the accumulation of parts on a conveyor at a workstation, pallet loads of material at receiving or shipping, in-process inventory accumulation, customers at a tool crib, customers at a checkout station in a grocery store, and lift trucks waiting for maintenance, to mention but a few.

Waiting line problems can be analyzed either mathematically or by using sim-

---

[6]The material presented in this section draws heavily from the text by White, Schmidt, and Bennett [84].

ulation. In this section, we present some *quick*-and-*dirty* methods for analyzing queueing problems mathematically. These mathematical methods are useful in preliminary planning and in many cases will yield results not significantly different from those obtained by simulation.

Because waiting lines occur in a variety of forms and contexts, it is important for the facilities planner to define clearly and accurately the waiting line situation under study. The following four elements are useful in defining a waiting line system: customers, servers, queue discipline, and service discipline.

The term **customer** can be interpreted very broadly. *Trucks* arriving at a receiving dock, *pallet loads* of materials arriving at a stretchwrap machine, *cases* arriving at a case sealing station, *employees* arriving at a tool crib, *patients* arriving at the hospital admissions window, and *airplanes* arriving at an airport are examples of customers. In brief, customers are the entities that arrive and require some form of service.

The term **server** can also be interpreted very broadly. For example, S/R machines, shrinkwrap tunnels, order pickers, industrial trucks, cranes, elevators, machine tools, computer terminals, inspector/packers, postal clerks, and nurses can be considered to be servers in certain contexts. Servers are the entities or combination of entities that provide the service required by customers.

The term **queue discipline** refers to the behavior of customers in the waiting line, as well as the design of the waiting line. In some cases, customers may refuse to wait; some customers may wait for a period of time, become discouraged, and depart before being served. Some waiting line systems have a limited amount of waiting space; some systems employ a single waiting line, while others have separate waiting lines for each server.

The term **service discipline** refers to the manner in which customers are served. For example, some systems have servers that serve customers singly, while others have servers that serve more than one customer at a time. Additionally, some systems serve customers on a first-come, first-serve basis, while others use a priority or random basis.

There exist a large number of different varieties of queueing problems. To facilitate the discussion of queueing problems, we will adopt the following classification scheme.

$$(x/y/z):(u/v/w)$$

where

$x$ = the arrival (or interarrival) distribution

$y$ = the departure (or service time) distribution

$z$ = the number of parallel service channels in the system

$u$ = the service discipline

$v$ = the maximum number allowed in the system (in service, plus waiting)

$w$ = the size of the population

The following codes are commonly used to replace the symbols $x$ and $y$:

$M$ = Poisson arrival or departure distributions (or equivalently exponential interarrival or service time distributions—$M$ refers to the Markov property of the exponential distribution)

$GI$ = General independent distribution of arrivals (or interarrival times)

$G$ = General distribution of departures (or service times)

$D$ = Deterministic interarrival or service times

The symbols $z$, $v$, and $w$ are replaced by the appropriate numerical designations. The symbol $u$ is replaced by a code similar to the following:

FCFS = first come, first served

LCFS = last come, first served

SIRO = service in random order

SPT = shortest processing (service) time

GD = general service discipline

Additionally, a superscript is attached to the first symbol if bulk arrivals exist and to the second symbol if bulk service is used. To illustrate the use of the notation, consider $(M^b/G/c):(FCFS/N/\infty)$. This denotes exponential interarrival times with $b$ customers per arrival, general service times, $c$ parallel servers, "first come, first serve" service discipline, a maximum allowable number of $N$ in the system, and an infinite population.

## Poisson Queues

The simplest waiting lines to study mathematically are those involving Poisson-distributed arrivals and services. In the case of an infinite customer population, if $\lambda$ is the average number of arrivals per unit time, then the probability distribution for the number of arrivals during the time period $T$ is given by

$$p(x) = \frac{e^{-\lambda T}(\lambda T)^x}{x!} \quad x = 0, 1, 2, \ldots \tag{12.90}$$

There exists an interesting, and important, relationship between the Poisson distribution and the exponential distribution. Namely, if arrivals are Poisson distributed, then the time between consecutive arrivals (interarrival time) will be exponentially distributed.

If services are Poisson distributed, then service times are exponentially distributed and can be represented by

$$f(t) = \mu e^{-\mu t} \quad t > 0 \tag{12.91}$$

where $\mu$ is the service rate or expected number of services per unit time by a busy server, and $(1/\mu)$ is the expected service time.

In analyzing waiting line problems, we will let $P_n$ denote the probability of $n$ customers in the system (in service or waiting). Letting $\lambda_n$ denote the arrival rate and $\mu_n$ denote the service rate when there are $n$ customers in the system, it can be shown that

$$P_n = \frac{\lambda_0 \lambda_1 \ldots \lambda_{n-1}}{\mu_1 \mu_2 \ldots \mu_n} P_0 \tag{12.92}$$

where $P_0$ is the probability of 0 customers in the system.

If the system can hold not more than $N$ customers in total, then $\lambda_N = 0$. Additionally, because the sum of the probabilities must be unity, then

$$P_0 + P_1 + \ldots P_N = 1 \tag{12.93}$$

From Equation 12.92 it is possible to express $P_n$ in terms of $P_0$. From Equation 12.93 we then solve for $P_0$ and substitute the value of $P_0$ in Equation 12.92 to obtain the value of $P_n$ for each value of $n$.

## Example 12.39

In order to illustrate the approach suggested for analyzing a waiting line problem, suppose a workstation receives parts automatically from a conveyor. An accumulation line has been provided at the workstation and has a storage capacity for five parts ($N = 6$). Parts arrive randomly at the switching junction for the workstation; if the accumulation line is full, parts are diverted to another workstation. Parts arrive at a Poisson rate of 1/min; service time at the workstation is exponentially distributed with a mean of 45 sec. Hence,

$$\lambda_n = \begin{cases} 1 & n = 0, 1, \ldots, 5 \\ 0 & n = 6 \end{cases}$$

(*Note*: the system has a capacity of 6, the waiting line has a capacity of 5.)

$$\mu_n = \begin{cases} 0 & n = 0 \\ \frac{4}{3} & n = 1, \ldots, 6 \end{cases}$$

From Equation 12.92

$$P_1 = \frac{\lambda_0}{\mu_1} P_0 = \frac{1}{\frac{4}{3}} P_0 = \frac{3}{4} P_0$$

Likewise,

$$P_2 = \frac{\lambda_0 \lambda_1}{\mu_1 \mu_2} P_0 = \frac{(1)(1)}{(\frac{4}{3})(\frac{4}{3})} P_0 = \left(\frac{3}{4}\right)^2 P_0$$

Table 12.13 summarizes the calculations involved in solving for $P_n$. Notice, in terms of $P_0$, the sum of the probabilities equals $(14{,}197/4{,}096)P_0$; thus, $P_0$ equals $4{,}096/14{,}197$. Given the value of $P_0$, the values of $P_n$ are computed for $n = 1, \ldots, 6$.

From Table 12.13 it is seen that $P_0 = 0.2885$ and $P_6 = 0.0513$; hence, the workstation will be idle 28.85% of the time and the accumulation line will be full 5.13% of the time. The average number of parts in the accumulation (waiting) line $L_q$ is given by

$$L_q = 0(P_0 + P_1) + 1P_2 + 2P_3 + 3P_4 + 4P_5 + 5P_6$$

or

$$\begin{aligned} L_q &= 1(0.1623) + 2(0.1217) + 3(0.0913) + 4(0.0685) + 5(0.0513) \\ &= 1.2101 \text{ parts in the accumulation line} \end{aligned}$$

The average number of parts in the system, $L$, is given by

$$\begin{aligned} L &= \sum_{n=0}^{N} nP_n \\ &= 0(0.2885) + 1(0.2164) + 2(0.1623) + 3(0.1217) + 4(0.0913) + 5(0.0685) + 6(0.0513) \\ &= 1.9216 \text{ parts in the system} \end{aligned}$$

**Table 12.13**   *Computation of* $P_n$ *for Example 12.39*

| $n$ | $\lambda_n$ | $\mu_n$ | $P_n$ | $P_n$ |
|---|---|---|---|---|
| 0 | 1 | 0 | $1P_0$ | $\dfrac{4{,}096}{14{,}197} = 0.2885$ |
| 1 | 1 | $\frac{4}{3}$ | $\dfrac{\lambda_0}{\mu_1}P_0 = \dfrac{3}{4}P_0$ | $\dfrac{3{,}072}{14{,}197} = 0.2164$ |
| 2 | 1 | $\frac{4}{3}$ | $\dfrac{\lambda_1}{\mu_2}P_1 = \left(\dfrac{3}{4}\right)^2 P_0$ | $\dfrac{2{,}304}{14{,}197} = 0.1623$ |
| 3 | 1 | $\frac{4}{3}$ | $\dfrac{\lambda_2}{\mu_3}P_2 = \left(\dfrac{3}{4}\right)^3 P_0$ | $\dfrac{1{,}728}{14{,}197} = 0.1217$ |
| 4 | 1 | $\frac{4}{3}$ | $\dfrac{\lambda_3}{\mu_4}P_3 = \left(\dfrac{3}{4}\right)^4 P_0$ | $\dfrac{1{,}296}{14{,}197} = 0.0913$ |
| 5 | 1 | $\frac{4}{3}$ | $\dfrac{\lambda_4}{\mu_5}P_4 = \left(\dfrac{3}{4}\right)^5 P_0$ | $\dfrac{972}{14{,}197} = 0.0685$ |
|  |  |  | $\dfrac{\lambda_5}{\mu_6}P_5 = \left(\dfrac{3}{4}\right)^6 P_0$ | $\dfrac{729}{14{,}197} = \dfrac{0.0513}{1.0000}$ |
| 6 | 0 | $\frac{4}{3}$ | $\dfrac{14{,}197}{4{,}096}P_0 = 1$ |  |

The rate at which parts actually enter the accumulation line is not 1/min because the accumulation line is sometimes full and parts are diverted elsewhere. The effective arrival rate $\overline{\lambda}$ is defined as the rate at which customers *enter* the system and is given by

$$\overline{\lambda} = (L - L_q)\mu \tag{12.94}$$

or

$$\overline{\lambda} = (1.9216 - 1.2101)(\tfrac{4}{3})$$
$$= 0.9487 \text{ parts per minute}$$

Alternately, in this case

$$\overline{\lambda} = (1 - P_N)\lambda$$
$$= 0.9487 \text{ parts per minute}$$

The percentage of arriving parts that do not enter the accumulation line is $P_N = 0.0513$. Suppose the production manager desires that no more than 2% of the arriving parts be diverted, how long should the accumulation line be? We will provide an answer to this question subsequently.

## Example 12.40

The previous example involved a single server ($c = 1$). We consider now an example in which two servers are present ($c = 2$). Suppose in the previous situation the accumulation line supplies parts for two workstations, as depicted in Figure 12.16. Parts are removed from the accumulation line, worked on, and placed on the conveyor which delivers the parts to the packaging department. In this case, we assume $\lambda = 2$ parts per minute and let each server operate at a rate of $\frac{4}{3}$ parts per minute. Hence, $N = 7$ (waiting space for 5 parts, plus 2 being serviced) and

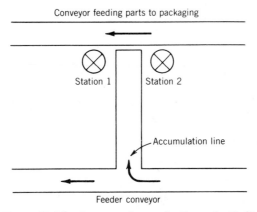

**Figure 12.16** Conveyor layout for Example 12.40.

Table 12.14  *Computation of* $P_n$ *for Example 12.40*

| $n$ | $\lambda_n$ | $\mu_n$ | $P_n$ | $P_n$ |
|---|---|---|---|---|
| 0 | 2 | 0 | $1P_0$ | 0.1613 |
| 1 | 2 | $\frac{4}{3}$ | $\frac{\lambda_0}{\mu_1}P_0 = \frac{6}{4}P_0$ | 0.2420 |
| 2 | 2 | $\frac{8}{3}$ | $\frac{\lambda_1}{\mu_2}P_1 = \frac{6}{4}\left(\frac{6}{8}\right)P_0$ | 0.1815 |
| 3 | 2 | $\frac{8}{3}$ | $\frac{\lambda_2}{\mu_3}P_2 = \frac{6}{4}\left(\frac{6}{8}\right)^2 P_0$ | 0.1361 |
| 4 | 2 | $\frac{8}{3}$ | $\frac{\lambda_3}{\mu_4}P_3 = \frac{6}{4}\left(\frac{6}{8}\right)^3 P_0$ | 0.1021 |
| 5 | 2 | $\frac{8}{3}$ | $\frac{\lambda_4}{\mu_5}P_4 = \frac{6}{4}\left(\frac{6}{8}\right)^4 P_0$ | 0.0766 |
| 6 | 2 | $\frac{8}{3}$ | $\frac{\lambda_5}{\mu_6}P_5 = \frac{6}{4}\left(\frac{6}{8}\right)^5 P_0$ | 0.0574 |
| 7 | 0 | $\frac{8}{3}$ | $\frac{\lambda_6}{\mu_7}P_6 = \frac{6}{4}\left(\frac{6}{8}\right)^6 P_0$ | 0.0430 |
| | | | $(3,250,112/524,288)P_0 = 1$ | 1.0000 |

$$\lambda_n \begin{cases} 2 & n = 0, 1, \ldots, 6 \\ \\ 0 & n = 7 \end{cases} \qquad \mu_n = \begin{cases} 0 & n = 0 \\ \frac{4}{3} & n = 1 \\ \frac{8}{3} & n = 2, 3, \ldots, 7 \end{cases}$$

Notice that if both servers are busy ($n \geq 2$), then the rate at which departures (services) occur is twice the service rate for a single server.

From Table 12.14 it is seen that $P_0 = 0.1613$, $P_1 = 0.2420$, and $P_7 = 0.0430$. Hence, both workstations are idle 16.13% of the time, one workstation is busy 24.20% of the time, and the accumulation line is full 4.30% of the time. The average number of parts in the accumulation line, $L_q$, and the average number of parts in the system, $L$, are found to be

$$L_q = 1P_3 + 2P_4 + 3P_5 + 4P_6 + 5P_7$$
$$= 1.0146 \text{ parts in the accumulation line}$$

and

$$L = \sum_{n=0}^{7} nP_n$$

$$= 2.4501 \text{ parts in the system}$$

In analyzing waiting lines a number of operating characteristics are typically used. Among them are the following:

1.  $L$, expected number of customers in the system
2.  $L_q$, expected number of customers in the waiting line
3.  $W$, expected time spent in the system per customer
4.  $W_q$, expected time spent in the queue per customer
5.  $U$, server utilization
6.  $\bar{\lambda}$, effective arrival rate

The values of $L$, $L_q$, $W$, and $W_q$ are related as follows:

$$L = \sum_{n=0}^{N} nP_n \tag{12.95}$$

$$L_q = \sum_{n=c}^{N} (n - c)P_n \tag{12.96}$$

$$U = (L - L_q)/c \tag{12.97}$$

$$\hat{\lambda} = (L - L_q)\mu \tag{12.98}$$

$$W = L/\hat{\lambda} \tag{12.99}$$

$$W_q = L_q/\hat{\lambda} \tag{12.100}$$

$$L = L_q + \frac{\hat{\lambda}}{\mu} \tag{12.101}$$

$$W = W_q + \frac{1}{\mu} \tag{12.102}$$

Equations 12.98, 12.101, and 12.102 are based on the assumption that an individual server maintains the same service rate ($\mu$) regardless of the number of customers in the system ($n > 0$). If a server has a changeable service rate then a more complicated situation exists and the values of $W$ and $W_q$ are not easily obtained.

## Example 12.41

Consider a tool crib with two attendants ($c = 2$). Time studies indicate customers arrive in a Poisson fashion at a rate of 12 per hour so long as no more than one customer is waiting to be served. If two customers are waiting, the rate at which people enter the tool crib is reduced to 8 per hour. If three customers are waiting, customers enter at a rate of 4 per hour. No additional customers enter if four customers are waiting. Hence, the maximum number of customers in the system (N) equals 6. The time required to fill a customer's order is exponentially distributed with a mean of 10 min.

From Table 12.15 it is seen that $L = 2.7211$ and $L_q = 1.0923$. Because the service rate for an individual server, $\mu$, is constant for $0 < n < 6$, then

**Table 12.15** *Computation of* $P_n$, $L$, *and* $L_q$ *for Example 12.41*

| $n$ | $\lambda_n$ | $\mu_n$ | $P_n$ | $P_n$ | $nP_n$ | $(n-c)P_n$ |
|---|---|---|---|---|---|---|
| 0 | 12 | 0 | $1P_0$ | 0.0928 | — | — |
| 1 | 12 | 6 | $2P_0$ | 0.1856 | 0.1856 | — |
| 2 | 12 | 12 | $2P_0$ | 0.1856 | 0.3712 | — |
| 3 | 12 | 12 | $2P_0$ | 0.1856 | 0.5568 | 0.1856 |
| 4 | 8 | 12 | $2P_0$ | 0.1856 | 0.7424 | 0.3712 |
| 5 | 4 | 12 | $\frac{4}{3}P_0$ | 0.1237 | 0.6185 | 0.3711 |
| 6 | 0 | 12 | $\frac{4}{9}P_0$ | 0.0411 | 0.2466 | 0.1644 |
| | | | $\frac{97}{9}P_0 = 1$ | 1.0000 | $L = 2.7211$ | $L_q = 1.0923$ |

$$U = \frac{(2.7211 - 1.0923)}{2}$$
$$= 0.8144 \text{ or } 81.44\%$$

$$\hat{\lambda} = (L - L_q)\mu$$
$$= (2.7211 - 1.0923)(6)$$
$$= 9.7728 \text{ customers per hour}$$

and

$$W = L/\hat{\lambda}$$
$$= \frac{2.7211}{9.7728}$$
$$= 0.2784 \text{ hr per customer}$$
$$= 16.706 \text{ min per customer}$$

$$W_q = \frac{L_q}{\hat{\lambda}}$$
$$= \frac{1.0923}{9.7728}$$
$$= 0.1118 \text{ hr per customer}$$
$$= 6.706 \text{ min per customer}$$

## $(M/M/c):(GD/N/\infty)$ Results

We consider next the special case of an infinite population with Poisson arrivals and services having arrival rates and service rates defined by

$$\lambda_n = \begin{cases} \lambda & n = 0, \ldots, N-1 \\ 0 & n = N \end{cases} \qquad \mu_n = \begin{cases} n\mu & n = 0, \ldots, c-1 \\ c\mu & n = c, \ldots, N \end{cases}$$

In words, the arrival rate is constant and equal to $\lambda$ until the system is full, at which time $\lambda_N = 0$. The service rate per busy server is $\mu$ for each server; hence, if $n$ customers are present, then the overall service rate for the system is $n\mu$, up to a maximum value of $c\mu$ when all servers are busy. It can be shown that Equation 12.92 reduces to

$$P_n = \begin{cases} \dfrac{(c\rho)^n}{n!} P_0 & n = 0, 1, \ldots, c-1 \\ \dfrac{c^c\rho^n}{c!} P_0 & n = c, \ldots, N \\ 0 & n > N \end{cases} \qquad (12.103)$$

**Table 12.16** *Summary of Operating Characteristics for the (M/M/c) Queue*

| | $(M \mid M \mid c):(GD \mid N \mid \infty)^a$ | $(M \mid M \mid c):(GD \mid \infty \mid \infty)^b$ |
|---|---|---|
| $L_q$ | $\dfrac{\rho(c\rho)^c P_0}{c!(1 - \rho)^2}[1 - \rho^{N-c+1} - (N - c + 1)(1 - \rho)\rho^N]$ | $\dfrac{\rho(c\rho)P_0}{c!(1 - \rho)^2}$ |
| $L$ | $L_q + \dfrac{(c\rho)^c(1 - \rho^{N-c+1})P_0}{(c - 1)!(1 - \rho)} + \sum\limits_{n=0}^{c-1} nP_n$ | $L_q + \dfrac{\lambda}{\mu}$ |
| $W_q$ | $\dfrac{(c\rho)^c[1 - \rho^{N-c+1} - (N - c + 1)(1 - \rho)\rho^N]}{c!c\mu(1 - \rho)^2(1 - P_N)}P_0$ | $\dfrac{(c\rho)P_0}{c!c\mu(1 - \rho)^2}$ |
| $W$ | $W_q + \dfrac{1}{\mu}$ | $W_q + \dfrac{1}{\mu}$ |
| $U$ | $\rho(1 - P_N)$ | $\rho$ |
| $\overline{\lambda}$ | $\lambda(1 - P_N)$ | $\lambda$ |
| $P_0$ | $\left[\dfrac{(c\rho)^c(1 - \rho^{N-c+1})}{c!(1 - \rho)} + \sum\limits_{n=0}^{c-1} \dfrac{(c\rho)^n}{n!}\right]^{-1}$ | $\left[\dfrac{(c\rho)^c}{c!(1 - \rho)} + \sum\limits_{n=0}^{c-1} \dfrac{(c\rho)^n}{n!}\right]^{-1}$ |

$^a\rho \neq 1.$
$^b\rho < 1.$

where $\rho = \dfrac{\lambda}{c\mu}$. Solving for $P_0$ gives

$$P_0 = \begin{cases} \left[\dfrac{(c\rho)^c(1 - \rho^{N-c+1})}{c!(1 - \rho)} + \sum\limits_{n=0}^{c-1} \dfrac{(c\rho)^n}{n!}\right]^{-1} & \rho \neq 1 \\[3ex] \left[\dfrac{c^c}{c!}(N - c + 1) + \sum\limits_{n=0}^{c - 1} \dfrac{c^n}{n!}\right]^{-1} & \rho = 1 \end{cases} \qquad (12.104)$$

The quantity $\rho$ is referred to as the traffic intensity for the system.

Formulas for the operating characteristics are given in Table 12.16. Also, given in Table 12.16 are the values of the operating characteristics when $N$ is infinitely large, that is, there is no limitation on the number the system can accommodate. (Notice that it must be true that $\rho < 1$ for the case of $N$ infinitely large.) Values of the operating characteristics are given in Table 12.17 for the special case of a single server ($c = 1$).

## Example 12.42

Recall, in Example 12.36 it was given that $c = 1$, $\lambda = 1$, and $\mu = \frac{4}{3}$. It was desired to determine the length of an accumulation line ($N$) such that $P_N \leq 0.02$. From Table 12.17 it is seen that

$$P_0 = \frac{1 - \rho}{1 - \rho^{N+1}}$$

where $\rho = \lambda/\mu$ or 0.75. From Equation 12.103 it is seen that the value of $N$ is to be determined such that

$$P_N = \frac{\rho^N(1 - \rho)}{1 - \rho^{N+1}} \leq 0.02$$

**Table 12.17   *Summary of Operating Characteristics for the (M/M/1) Queue***

| | $(M \mid M \mid 1){:}(GD \mid N \mid \infty)^a$ | $(M \mid M \mid 1){:}(GD \mid \infty \mid \infty)^b$ |
|---|---|---|
| $L_q$ | $\dfrac{\rho^2[1 - \rho^N - N\rho^{N-1}(1 - \rho)]}{(1 - \rho)(1 - \rho^{N+1})}$ | $\dfrac{\rho^2}{1 - \rho}$ |
| $L$ | $\dfrac{\rho[1 - \rho^N - N\rho^N(1 - \rho)]}{(1 - \rho)(1 - \rho^{N+1})}$ | $\dfrac{\rho}{1 - \rho}$ |
| $W_q$ | $\dfrac{\rho[1 - \rho^N - N\rho^{N-1}(1 - \rho)]}{\mu(1 - \rho)(1 - \rho^N)}$ | $\dfrac{\rho}{\mu(1 - \rho)}$ |
| $W$ | $\dfrac{1 - \rho^N - N\rho^N(1 - \rho)}{\mu(1 - \rho)(1 - \rho^N)}$ | $\dfrac{1}{\mu(1 - \rho)}$ |
| $U$ | $\dfrac{\rho(1 - \rho^N)}{1 - \rho^{N+1}}$ | $\rho$ |
| $\bar{\lambda}$ | $\dfrac{\lambda(1 - \rho^N)}{1 - \rho^{N+1}}$ | $\lambda$ |
| $P_0$ | $\dfrac{1 - \rho}{1 - \rho^{N+1}}$ | $1 - \rho$ |

[a] $\rho \neq 1$.
[b] $\rho < 1$.

or

$$\rho^N - \rho^{N+1} \le 0.02 - 0.02\rho^{N+1}$$

$$\rho^N(1 - 0.98\rho) \le 0.02$$

$$\rho^N \le \frac{0.02}{1 - 0.98\rho}$$

$$(0.75)^N \le \frac{0.02}{1 - 0.98(0.75)}$$

Taking the logarithm of both sides gives

$$N \log (0.75) \le \log (0.07547)$$

$$N \ge 8.982$$

Thus, $N = 9$, and the accumulation line will have space to accommodate 8 parts. The operating characteristics for the system can be obtained using the results given in Table 12.17. Namely, for $N = 9$, $\lambda = 1$, $\mu = \frac{4}{3}$, $c = 1$, and $\rho = \lambda/c\mu = 0.75$,

$$L = \frac{\rho[1 - \rho^N - N\rho^N(1 - \rho)]}{(1 - \rho)(1 - \rho^{N+1})} = 2.403$$

$$L_q = \frac{\rho^2[1 - \rho^N - N\rho^{N-1}(1 - \rho)]}{(1 - \rho)(1 - \rho^{N+1})} = 1.668$$

$$W = \frac{1 - \rho^N - N\rho^N(1 - \rho)}{\mu(1 - \rho)(1 - \rho^N)} = 2.452$$

$$W_q = \frac{[1 - \rho^N - N\rho^{N-1}(1 - \rho)]}{\mu(1 - \rho)(1 - \rho^N)} = 1.702$$

$$U = L - L_q = 2.403 - 1.668 = 0.735$$

Values of $L$ and $L_q$ can be obtained using Figures 12.17 and 12.18 for the $(M/M/c){:}(GD/\infty/\infty)$ case. Notice, when $N = \infty$, then $U = \rho$. Figures 12.17 and 12.18 indicate an important

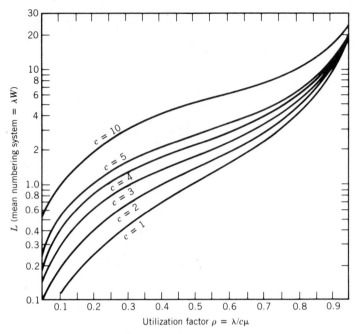

**Figure 12.17**   $L$ (mean number in system $= \lambda W$) for different values of $c$ versus the utilization factor $\rho$. (*From White et al. [84] with permission*)

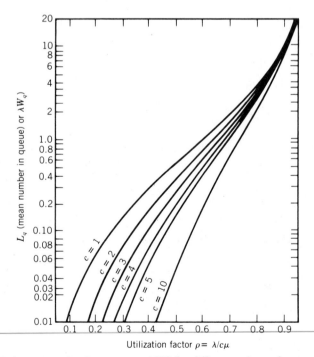

**Figure 12.18**   $L_q$ (mean number in queue, $= \lambda W_q$) for different values of $c$ versus the utilization factor $\rho$. (*From White et al. [84] with permission*)

concept when designing material handling systems. Namely, as equipment utilization ($\rho$) increases, the waiting lines for material to be moved increases exponentially. It is a common tendency for managers to evaluate material handling effectiveness on the basis of equipment utilization; in many cases, that is counterproductive for the overall system.

## Example 12.43

A fleet of lift trucks is used to place pallet loads of materials in storage and to retrieve pallet loads from storage. Job requisitions are received in a Poisson fashion at a rate of 10 per hour. The time required to perform the required material handling for a requisition is exponentially distributed with a mean of 15 min. Management wishes for the lift trucks to be utilized at least 80% of the time; hence, $\rho \geq 0.80$ or $\lambda/c\mu \geq 0.80$, which means $c \leq 10/4(0.80)$ or $c \leq 3.125$. Because an integer number of lift trucks must be assigned, $c \leq 3$.

The operating characteristics for the system having $\lambda = 10$, $\mu = 4$, $c = 3$, and $N = \infty$ are obtained from Table 12.16 as follows

$$P_0 = \left[ \frac{(c\rho)^c}{c!(1 - \rho)} + \sum_{n=0}^{c-1} \frac{(c\rho)^n}{n!} \right]^{-1} = \left[ \frac{(2.5)^3}{6(0.1667)} + 1 + 2.5 + \frac{(2.5)^2}{2} \right]^{-1}$$

$$= 0.0449$$

$$L_q = \frac{\rho(c\rho)^c}{c!(1 - \rho)^2} P_0 = 3.5078 \text{ requisitions}$$

$$L = L_q + \frac{\lambda}{\mu} = 6.0078 \text{ requisitions}$$

$$W_q = \frac{(c\rho)^c P_0}{c!c\mu(1 - \rho)^2} = 0.35078 \text{ hr} = 21.0468 \text{ min per requisition}$$

$$W = W_q + \frac{1}{\mu} = 21.0468 + 15.0 = 36.0468 \text{ minutes per requisition}$$

When the manager realizes that the average time a requisition is delayed before processing begins equals approximately 21 min, it is decided that the objective should be to have orders wait, on the average, less than 10 min. Hence, $W_q \leq 10$ minutes per requisition or

$$L_q = W_q\lambda \leq \frac{10 \text{ min}}{\text{req.}} \left( \frac{10 \text{ req.}}{\text{hour}} \right) \left( \frac{\text{hour}}{60 \text{ min}} \right) = 1.667$$

From Figure 12.18 it is seen that for $c = 4$, $\rho = 0.626$, and $L_q \approx 0.60$. Hence, $c = 4$ should provide the level of service desired, yet the utilization of the lift trucks will be only 62.5% rather than the 80% figure desired initially.

### $(M^b/M/1):(GD/\infty/\infty)$ Results

In a number of situations customers do not arrive singly, but in groups. Such a situation is termed a bulk arrival queueing system. Likewise, customers can often be serviced in groups rather than individually. In this section, we consider the bulk arrival situation under the usual Poisson assumptions; we also assume unlimited waiting space and a single server.

The arrival of $b$ customers per arrival instant results in the following operating characteristics:

$$L = \frac{\rho(1 + b)}{2(1 - p)} \tag{12.105}$$

$$L_q = L - \frac{b\lambda}{\mu} \tag{12.106}$$

$$W = \frac{1 + b}{2\mu(1 - \rho)} \tag{12.107}$$

$$W_q = W - \frac{1}{\mu} \tag{12.108}$$

where $\rho = b\lambda/\mu$ is also the utilization of the single server. The probability the server will already be busy when the arrival occurs is $\rho$.

When the number of customers that arrive at a given arrival instant is a random variable, then the operating characteristics become

$$L = \frac{\lambda[V(b) + E^2(b) + E(b)]}{2[\mu - \lambda E(b)]} \tag{12.109}$$

$$L_q = L - \frac{\lambda E(b)}{\mu} \tag{12.110}$$

$$W = \frac{V(b) + E^2(b) + E(b)}{2E(b)[\mu - \lambda E(b)]} \tag{12.111}$$

$$W_q = W - \frac{1}{\mu} \tag{12.112}$$

where $E(b)$ and $V(b)$ are the expected value of $b$ and the variance of $b$, respectively, and $\rho = \lambda E(b)/\mu$.

# Example 12.44

Truckloads of material arrive at a receiving dock according to a Poisson process at a rate of two per hour. The number of unit loads per truck equals 10. The time required to remove an individual unit load from a truck is exponentially distributed with a mean of 2.4 min. A single lift truck is used in the receiving area for transporting unit loads from the truck to a conveyor that delivers the unit load to storage.

In this situation $b = 10$, $\lambda = 2$, $\mu = 25$, and $c = 1$. Hence, $\rho = 0.80$ and

$$L = \frac{\rho(1 + b)}{2(1 - \rho)} = \frac{0.8(11)}{2(0.2)} = 22 \text{ unit loads}$$

$$L_q = L - \frac{b\lambda}{\mu} = 22 - \frac{10(2)}{25} = 21.2 \text{ unit loads}$$

$$W = \frac{1 + b}{2\mu(1 - \rho)} = \frac{11}{2(25)(0.2)} = 1.1 \text{ hr per unit load}$$

$$W_q = W - \frac{1}{\mu} = 1.1 - 0.04 = 1.06 \text{ hr per unit load}$$

## Example 12.45

Suppose the number of unit loads per truck is a random variable. Given the data from the previous example, consider the situation where $b$ is binomially distributed with a mean of 10 and a variance of 5. Since $E(b) = 10$, $V(b) = 5$, $\lambda = 2$, $\mu = 25$, and $c = 1$, then $\rho = 0.80$ and

$$L = \frac{2[5 + 100 + 10]}{2[25 - 2(10)]} = 23 \text{ unit loads}$$

$$L_q = 23 - 0.80 = 22.20 \text{ unit loads}$$

$$W = \frac{5 + 100 + 10}{2(10)[25 - 2(10)]} = 1.15 \text{ hr per unit load}$$

$$W_q = W - \frac{1}{\mu} = 1.15 - 0.04 = 1.11 \text{ hr per unit load}$$

## (M/M/c):(GD/K/K) Results

The previous analysis of waiting lines was based on an assumption that the population of customers is infinitely large. When the customer population is sufficiently large that the probability of an arrival occurring is not dependent on the number of customers in the waiting line, then the assumption of an infinite population is appropriate.

In many cases the number of customers is sufficiently small that the probability of an arrival is dependent on the number of customers in the queueing system. In such a situation, it is recommended that the arrival rate for an individual customer be determined. In particular, $\lambda$ is defined as the reciprocal of the time between a customer completing service and requiring service the next time. If the number of customers is $K$ and each customer has an identical arrival rate, then the arrival rate at the queueing system is given by

$$\lambda_n = (K - n)\lambda \quad n = 0, 1, \ldots, K \tag{12.113}$$

Furthermore, since we are assuming service times are exponentially distributed, then

$$\mu_n = \begin{cases} n\mu & n = 0, 1, \ldots, c - 1 \\ c\mu & n = c, c + 1, \ldots, K \end{cases} \tag{12.114}$$

Substituting Equations 12.113 and 12.114 into Equation 12.92 gives

$$P_n = \begin{cases} \binom{K}{n} (c\rho)^n P_0 & n = 0, 1, \ldots, c - 1 \\ \binom{K}{n} \frac{n! c^c \rho^n}{c!} P_0 & n = c, c + 1, \ldots, K \\ 0 & n > K \end{cases} \tag{12.115}$$

where, as before, $\rho = \lambda/c\mu$, and

$$P_0 = \left[ \sum_{n=0}^{c-1} \binom{K}{n} (c\rho)^n + \sum_{n=c}^{K} \binom{K}{n} \frac{n! c^c \rho^n}{c!} \right]^{-1} \tag{12.116}$$

Peck and Hazelwood [62] have developed extensive tables of computational results for the $(M/M/c):(GD/K/K)$ queue which can be used to determine the operating characteristics for a given situation. A sample of the results obtained by Peck and

**Table 12.18**  *Results Obtained by Peck and Hazelwood for (M/M/c) : (GD/K/K)*

| K | c | D | F | K | c | D | F | K | c | D | F |
|---|---|---|---|---|---|---|---|---|---|---|---|
|  |  | X = 0.05 |  | 12 | 1 | 0.879 | 0.764 | 10 | 1 | 0.987 | 0.497 |
| 4 | 1 | 0.149 | 0.992 |  | 2 | 0.361 | 0.970 |  | 2 | 0.692 | 0.854 |
| 5 | 1 | 0.198 | 0.989 |  | 3 | 0.098 | 0.996 |  | 3 | 0.300 | 0.968 |
| 6 | 1 | 0.247 | 0.985 |  | 4 | 0.019 | 0.999 |  | 4 | 0.092 | 0.994 |
|  | 2 | 0.023 | 0.999 | 14 | 1 | 0.946 | 0.690 |  | 5 | 0.020 | 0.999 |
| 7 | 1 | 0.296 | 0.981 |  | 2 | 0.469 | 0.954 | 12 | 1 | 0.998 | 0.416 |
|  | 2 | 0.034 | 0.999 |  | 3 | 0.151 | 0.992 |  | 2 | 0.841 | 0.778 |
| 8 | 1 | 0.343 | 0.977 |  | 4 | 0.036 | 0.999 |  | 3 | 0.459 | 0.940 |
|  | 2 | 0.046 | 0.999 | 16 | 1 | 0.980 | 0.618 |  | 4 | 0.180 | 0.986 |
| 9 | 1 | 0.391 | 0.972 |  | 2 | 0.576 | 0.935 |  | 5 | 0.054 | 0.997 |
|  | 2 | 0.061 | 0.998 |  | 3 | 0.214 | 0.988 | 14 | 2 | 0.934 | 0.697 |
| 10 | 1 | 0.437 | 0.967 |  | 4 | 0.060 | 0.998 |  | 3 | 0.619 | 0.902 |
|  | 2 | 0.076 | 0.998 | 18 | 1 | 0.994 | 0.554 |  | 4 | 0.295 | 0.973 |
| 12 | 1 | 0.528 | 0.954 |  | 2 | 0.680 | 0.909 |  | 5 | 0.109 | 0.993 |
|  | 2 | 0.111 | 0.996 |  | 3 | 0.285 | 0.983 |  | 6 | 0.032 | 0.999 |
| 14 | 1 | 0.615 | 0.939 |  | 4 | 0.092 | 0.997 | 16 | 2 | 0.978 | 0.621 |
|  | 2 | 0.151 | 0.995 |  | 5 | 0.024 | 0.999 |  | 3 | 0.760 | 0.854 |
|  | 3 | 0.026 | 0.999 | 20 | 1 | 0.999 | 0.500 |  | 4 | 0.426 | 0.954 |
| 16 | 1 | 0.697 | 0.919 |  | 2 | 0.773 | 0.878 |  | 5 | 0.187 | 0.987 |
|  | 2 | 0.195 | 0.993 |  | 3 | 0.363 | 0.975 |  | 6 | 0.066 | 0.997 |
|  | 3 | 0.039 | 0.999 |  | 4 | 0.131 | 0.995 |  | 7 | 0.019 | 0.999 |
| 18 | 1 | 0.772 | 0.895 |  | 5 | 0.038 | 0.999 | 18 | 2 | 0.994 | 0.555 |
|  | 2 | 0.243 | 0.991 | 25 | 2 | 0.934 | 0.776 |  | 3 | 0.868 | 0.797 |
|  | 3 | 0.054 | 0.999 |  | 3 | 0.572 | 0.947 |  | 4 | 0.563 | 0.928 |
| 20 | 1 | 0.837 | 0.866 |  | 4 | 0.258 | 0.987 |  | 5 | 0.284 | 0.977 |
|  | 2 | 0.293 | 0.988 |  | 5 | 0.096 | 0.997 |  | 6 | 0.118 | 0.993 |
|  | 3 | 0.073 | 0.988 |  | 6 | 0.030 | 0.999 |  | 7 | 0.040 | 0.998 |
| 25 | 1 | 0.950 | 0.771 | 30 | 2 | 0.991 | 0.664 | 20 | 2 | 0.999 | 0.500 |
|  | 2 | 0.429 | 0.978 |  | 3 | 0.771 | 0.899 |  | 3 | 0.938 | 0.736 |
|  | 3 | 0.132 | 0.997 |  | 4 | 0.421 | 0.973 |  | 4 | 0.693 | 0.895 |
|  | 4 | 0.032 | 0.999 |  | 5 | 0.187 | 0.993 |  | 5 | 0.397 | 0.963 |
| 30 | 1 | 0.992 | 0.663 |  | 6 | 0.071 | 0.998 |  | 6 | 0.187 | 0.988 |
|  | 2 | 0.571 | 0.963 |  |  | X = 0.20 |  |  | 7 | 0.074 | 0.997 |
|  | 3 | 0.208 | 0.994 | 4 | 1 | 0.549 | 0.862 |  | 8 | 0.025 | 0.999 |
|  | 4 | 0.060 | 0.999 |  | 2 | 0.108 | 0.988 | 25 | 3 | 0.996 | 0.599 |
|  |  | X = 0.10 |  |  | 3 | 0.008 | 0.999 |  | 4 | 0.920 | 0.783 |
| 4 | 1 | 0.294 | 0.965 | 5 | 1 | 0.689 | 0.801 |  | 5 | 0.693 | 0.905 |
|  | 2 | 0.028 | 0.999 |  | 2 | 0.194 | 0.976 |  | 6 | 0.424 | 0.963 |
| 5 | 1 | 0.386 | 0.950 |  | 3 | 0.028 | 0.998 |  | 7 | 0.221 | 0.987 |
|  | 2 | 0.054 | 0.997 | 6 | 1 | 0.801 | 0.736 |  | 8 | 0.100 | 0.995 |
| 6 | 1 | 0.475 | 0.932 |  | 2 | 0.291 | 0.961 |  | 9 | 0.039 | 0.999 |
|  | 2 | 0.086 | 0.995 |  | 3 | 0.060 | 0.995 | 30 | 4 | 0.991 | 0.665 |
| 7 | 1 | 0.559 | 0.912 | 7 | 1 | 0.883 | 0.669 |  | 5 | 0.905 | 0.814 |
|  | 2 | 0.123 | 0.992 |  | 2 | 0.395 | 0.941 |  | 6 | 0.693 | 0.913 |
|  | 3 | 0.016 | 0.999 |  | 3 | 0.105 | 0.991 |  | 7 | 0.446 | 0.963 |
| 8 | 1 | 0.638 | 0.889 |  | 4 | 0.017 | 0.999 |  | 8 | 0.249 | 0.985 |
|  | 2 | 0.165 | 0.989 | 8 | 1 | 0.937 | 0.606 |  | 9 | 0.123 | 0.995 |
|  | 3 | 0.027 | 0.999 |  | 2 | 0.499 | 0.916 |  | 10 | 0.054 | 0.998 |
| 9 | 1 | 0.711 | 0.862 |  | 3 | 0.162 | 0.985 |  | 11 | 0.021 | 0.999 |
|  | 2 | 0.210 | 0.985 |  | 4 | 0.035 | 0.998 |  |  |  |  |
|  | 3 | 0.040 | 0.998 | 9 | 1 | 0.970 | 0.548 |  |  |  |  |
| 10 | 1 | 0.776 | 0.832 |  | 2 | 0.599 | 0.887 |  |  |  |  |
|  | 2 | 0.258 | 0.981 |  | 3 | 0.227 | 0.978 |  |  |  |  |
|  | 3 | 0.056 | 0.998 |  | 4 | 0.060 | 0.996 |  |  |  |  |

*Source*: From Peck and Hazelwood [62].

Hazelwood is given in Table 12.18. To interpret these data, let the service factor $X$ be defined as

$$X = \frac{\lambda}{\lambda + \mu} \tag{12.117}$$

and let

$$F = \frac{W - W_q + \lambda^{-1}}{W + \lambda^{-1}} \tag{12.118}$$

Peck and Hazelwood provide values of $F$ for various combinations of $X$, $K$, and $c$. Based on the value of $F$, the following values are obtained.

$$L = K[1 - F(1 - X)] \tag{12.119}$$

$$L_q = K(1 - F) \tag{12.120}$$

$$W = \frac{1 - F(1 - X)}{\mu F X} \tag{12.121}$$

$$W_q = \frac{1 - F}{\mu F X} \tag{12.122}$$

## Example 12.46

A distribution center delivers material to the retail stores in its service region. Twelve trucks are used for delivery. The time required to load a truck is exponentially distributed with a mean of 40 min; the time required to deliver a load and return is exponentially distributed with a mean of 6 hr. There are two crews available for loading trucks. The distribution center operates continuously, that is, 24 hr per day and 7 days per week.

Based on the data for the distribution center $K = 12$, $\lambda = \frac{1}{6}$, $\mu = \frac{3}{2}$, and $c = 2$; hence, $X = 0.1$. From Table 12.18, it is found that $F = 0.970$. Therefore,

$$L = K[1 - F(1 - X)] = 12[1 - 0.970(1 - 0.1)]$$
$$= 1.524 \text{ trucks}$$

$$L_q = K(1 - F) = 12(1 - 0.970)$$
$$= 0.360 \text{ trucks}$$

$$W = \frac{1 - F(1 - X)}{\mu F X} = \frac{1 - 0.970(1 - 0.1)}{1.5(0.970)(0.1)}$$
$$= 0.873 \text{ hr per truck}$$

$$W_q = \frac{1 - F}{\mu F X} = \frac{1 - 0.970}{1.5(0.970)(0.1)}$$
$$= 0.206 \text{ hr per truck}$$

The results given above are based on the assumption that either of the two crews can load any truck. It has been suggested that trucks 1, ..., 6 be serviced by crew 1 and trucks 7, ..., 12 be serviced by crew 2. In such a case $K = 6$, $\lambda = \frac{1}{6}$, $\mu = \frac{3}{2}$, $c = 1$, and $X = 0.1$ for the first group of trucks. From Table 12.18, $F = 0.932$. Hence,

$$L = 6[1 - 0.932(1 - 0.1)] = 0.9672 \text{ trucks}$$

$$L_q = 6(1 - 0.932) = 0.408 \text{ trucks}$$

$$W = \frac{1 - 0.932(1 - 0.1)}{1.5(0.932)(0.1)} = 1.153 \text{ hr per truck}$$

$$W_q = \frac{1 - 0.932}{1.5(0.932)(0.1)} = 0.486 \text{ hr per truck}$$

Since there are two crews, then the average number of trucks in the total system is $L = 2(0.9672) = 1.9344$ trucks; likewise, the average number of trucks waiting to be loaded is $L_q = 2(0.408) = 0.816$ trucks. Under the proposed plan a complete delivery cycle by a truck will require $6 + 1.153 = 7.153$ hr; therefore, a truck can be expected to make $7(24)/7.153 = 23.49$ deliveries per week. Under the present plan a truck can be expected to make $7(24)/6.873 = 24.44$ deliveries per week.

## Non-Poisson Queues

In general, when arrivals and/or services are not Poisson distributed, mathematical results are difficult to obtain and simulation is often used. However, there are some simple non-Poisson queues for which the operational characteristics are known. We will consider three non-Poisson queueing systems: $(M/G/1):(GD/\infty/\infty)$, $(D/M/1):(GD/\infty/\infty)$, and $(M/G/c):(GD/c/\infty)$. The first case allows any general service time distribution; the second case is appropriate when the time between consecutive arrivals is deterministic (constant); and the third case allows any general service time distribution when no waiting is provided.

### $(M/G/1):(GD/\infty/\infty)$ Results

Assume arrivals are Poisson distributed and a single server is present. If the average service time is $\mu^{-1}$ and the variance of service time is $\sigma^2$, then the following expressions are available for determining the operating characteristics of the system:

$$L = \rho + \frac{\lambda^2\sigma^2 + \rho^2}{2(1 - \rho)} \tag{12.123}$$

$$L_q = \frac{\lambda^2\sigma^2 + \rho^2}{2(1 - \rho)} \tag{12.124}$$

$$W = \frac{1}{\mu} + \frac{\lambda\left(\sigma^2 + \frac{1}{\mu^2}\right)}{2(1 - \rho)} \tag{12.125}$$

$$W_q = \frac{\lambda\left(\sigma^2 + \frac{1}{\mu^2}\right)}{2(1 - \rho)} \tag{12.126}$$

where $\rho = \lambda/\mu$. Letting $\alpha^2 = \mu^2\sigma^2$, values of $L$ and $L_q$ are depicted in Figures 12.19 and 12.20 for various values of $\alpha^2$. Notice that when service times are constant then $\alpha^2 = 0$ and when service times are exponentially distributed then $\alpha^2 = 1$.

## *Example 12.47*

Unit loads arrive randomly at a shrinkwrap machine. An average of 15 unit loads arrive per hour at a Poisson rate. The time required for a unit load to be processed through the shrink wrapping operation is a constant of 2.5 min. The operating characteristics are obtained using Equations 12.123 through 12.126 by letting $\lambda = 15$, $\mu = 24$, $\sigma^2 = 0$, and $\rho = 0.625$.

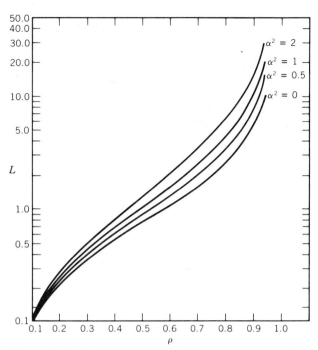

**Figure 12.19** Values of $L$ for the $(M/G/1):(GD/\infty/\infty)$ queue. (*From White et al. with permission*)

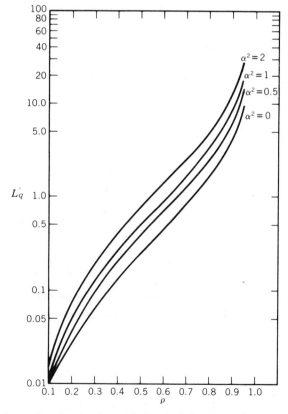

**Figure 12.20** Values of $L_q$ for the $(M/G/1):(GD/\infty/\infty)$ queue. (*From White et al. [84] with permission*)

$$L = 0.625 + \frac{(15)^2(0) + (0.625)^2}{2(1 - 0.625)} = 1.1458 \text{ unit loads}$$

$$L_q = \frac{(15)^2(0) + (0.625)^2}{2(1 - 0.625)} = 0.5208 \text{ unit loads}$$

$$W = \frac{L}{\lambda} = 0.07639 \text{ hr per unit load}$$

$$= 4.583 \text{ min per unit load}$$

$$W_q = W = \frac{1}{\mu} = 2.083 \text{ min per unit load}$$

## (D/M/1):(GD/∞/∞) Results

When the time between consecutive arrivals is deterministic (constant) and services are Poisson, then for the single server queue the following results are available:

$$L = \frac{\theta}{1 - \theta} \tag{12.127}$$

$$L_q = \frac{\theta^2}{1 - \theta} \tag{12.128}$$

$$W = \frac{1}{\mu(1 - \theta)} \tag{12.129}$$

$$W_q = \frac{\theta}{\mu(1 - \theta)} \tag{12.130}$$

where $\theta$ is a fractional-valued parameter $(0 < \theta < 1)$ satisfying the relation

$$\theta = e^{-(1-\theta)/\rho} \tag{12.131}$$

with $\rho = \lambda/\mu$ and the time between consecutive arrivals is $\lambda^{-1}$. Values of $\theta$ are given in Figure 12.21 for various values of $\rho$.

Since a more accurate estimate of $\theta$ might be needed, the following numerical procedure can be used to refine the value obtained from Figure 12.21. Let $\theta_0$ be the value obtained from Figure 12.21. An updated estimate, $\theta_1$, can be obtained from the relation,

$$\theta_1 = e^{-(1-\theta_0)/\rho}$$

More generally, the *k*th estimate of $\theta$ can be obtained from

$$\theta_k = e^{-(1-\theta_{k-1})/\rho} \tag{12.132}$$

The iterative process can be continued until successive estimates of $\theta$ are less than, say, 0.001.

## Example 12.48

Parts are supplied by a conveyor to an inspection station from a numerically controlled milling machine. Because of the insignificant variation in machining time and conveyor speed, parts arrive at the inspection station in a deterministic fashion at a rate of 6 per minute. In-

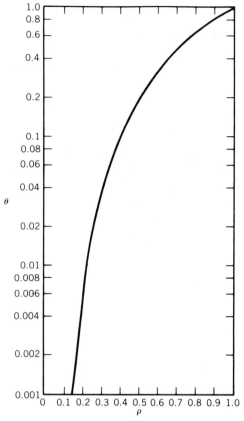

**Figure 12.21**   Relationship between $\theta$ and $\rho$.

spection is performed manually and the inspection times are exponentially distributed with a mean of 7.5 sec. In this situation, $\lambda = 6$, $\mu = 8$, $c = 1$, and $\rho = 0.75$. From Figure 12.21, $\theta = 0.55$. Therefore, if an estimate for $\theta$ is desired such that $|\theta_k - \theta_{k-1}| < 0.001$ then, for the example with $\rho = 0.75$, the following iterative process can be used:

$$\theta_k = e^{-(1-\theta_{k-1})/0.75}$$

Calculations show that if $\theta_0 = 0.55$, then $\theta_1 = 0.5488116$, $\theta_2 = 0.5479427$, and $\theta_3 = 0.5473083$. Therefore, substituting 0.5473083 for $\theta$ in Equations 12.127 through 12.130 yields

$$L = \frac{0.5473083}{0.4526917} = 1.209 \text{ parts}$$

$$L_q = \frac{0.5473083}{0.4526917} = 0.662 \text{ parts}$$

$$W = \frac{1}{8(0.4526917)} = 0.276 \text{ min per part}$$

$$W_q = \frac{0.5473083}{8(0.4526917)} = 0.151 \text{ min per part}$$

## *(M/G/c):(GD/c/∞) Results*

When no waiting space is provided and arrivals are Poisson distributed, the situation gives rise to a formula called *Erlang's loss formula*,

$$P_c = \frac{(\lambda/\mu)^c/c!}{\sum\limits_{k=0}^{c} (\lambda/\mu)^k/k!} \tag{12.133}$$

which gives the probability all servers are busy in a $(M/G/c): (GD/c/\infty)$ queue. Values of $P_c$ are provided in Figure 13.7 for selected values of $c$ and $\lambda/\mu$. (Multiplying both the numerator and denominator of Equation 12.133 by exp-$(\lambda/\mu)$ gives an interesting result. Namely, the numerator is the probability of a Poisson distributed random variable with an expected value of $\lambda/\mu$ equalling c; the denominator is the probability of c or less occurrences for the Poisson distributed random variable. Hence, by consulting tables of the Poisson distribution, one can easily compute the value of $P_c$.)

The operating characteristics of the $(M/G/c):(GD/c/\infty)$ queue are easily obtained since $L_q = 0$, $W_q = 0$, and $W = 1/\mu$. The expected number in the system is given by

$$L = \lambda(1 - P_c)/\mu \tag{12.134}$$

## *Example 12.49*

A conveyor belt is used to feed parts to 10 workstations located along the conveyor belt. For convenience it is assumed that travel along the belt between workstations is instantaneous. Parts arrive at the first workstation at a Poisson rate of 12 per minute. If the first station is busy, then the part moves to the second station, and so forth, until, finally, if all 10 workstations are busy, then the part passes all workstations and is accumulated until the second-shift when they are processed by other workstations.

If the service times at the individual workstations are normally distributed with a mean of 45 sec and a standard deviation of 5 sec, what percentage of the parts will overflow the system? In this case $\lambda = 12$, $\mu = \frac{4}{3}$, $c = 10$, and from Figure 12.22, $P_c = 0.17$ or approximately 17% of the parts will pass all 10 workstations. The average number of busy workstations is equal to $L$, or $12(.83)(0.75) = 7.47$ busy workstations.

Having presented a number of waiting line models for use in analyzing facilities planning requirements for buffer spaces, waiting lines, and in-process storage, we would be negligent if we did not admit that many situations exist that do not fit the assumptions underlying the models we presented. In such a case, what should you do? Three alternatives come to mind: utilize a more advanced waiting line model, one that more closely fits your situation; utilize one of the simple models we presented, recognizing that the solutions obtained are, at best, approximations; and utilize a simulation model, rather than a waiting line model.

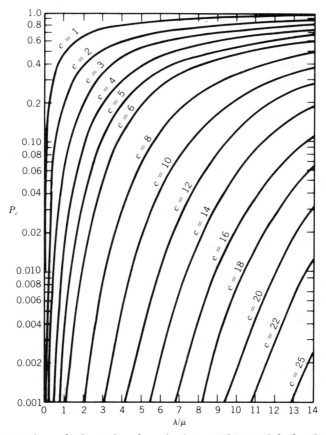

**Figure 12.22**  Values of Erlang's loss formula. (*From White et al. [84] with permission*)

## *12.10*  SIMULATION MODELS

Considerable improvements have occurred in the past decade in developing simulation packages for use in modeling material flow problems. Indeed, simulation technology has evolved to the point that serious facilities planners will use simulation one or more times during the facilities planning cycle. Relatively sophisticated simulations can be performed today using microcomputers; hand-held pocket computers can be used to perform rudimentary simulations; spreadsheet software typically includes random number generators for use in performing simulations. Advances in software graphics allow simulation results to be displayed using animated representations of systems.

Due to the explosion of options available for performing simulations of flow processes, we cannot provide more than a brief introduction to simulation. Because we can only scratch the surface, it was tempting to omit coverage of simulation. However, its power and increasingly important role in facilities planning mandated that we provide at least an introduction for those unfamiliar with simulation. So, for those who are knowledgeable about simulation, our apologies for taking up space in your book!

Simulation involves building a model of a system and experimenting with the model to determine how the system reacts to various conditions. Simulation does not provide an optimum solution, since it is a descriptive model. It simply provides us with a mechanism to use in understanding and, perhaps, predicting the behavior of a system. By asking enough "what if" questions, the configuration of the system that best satisfies some criteria can be chosen.

Some of the major reasons for using simulation are:

1.   When a mathematical solution cannot be obtained easily or at all.
2.   Selling the facilities plan to management.
3.   Explaining to operating personnel how a proposed system will function.
4.   Testing the feasibility of a proposed system.
5.   Developing throughput and storage requirements.
6.   Validating mathematical models.
7.   Predicting the impact of a change in the physical system, the environment, or operating procedures.

Simulation can result in improved understanding of the facilities plan. Some ways in which this can occur are as follows:

1.   The process of creating a simulation model requires a detailed understanding and documentation of the activity being simulated. For example, the flow of information through a warehouse typically seems straightforward at first glance. Once a simulation of this information flow is developed, however, many exceptions and alternative information flows are identified.

2.   The teaching of some concepts is quite difficult because of the complex interrelationship among variables. Simulation can be used in a "gaming" sense to relate these complex interrelationships. For example, determining the number and location of distribution centers to serve a market is based on the customer service that can be provided, the transportation cost, and the cost of carrying inventories.

3.   The orientation of employees on a system can often create significant problems. Simulation can be used to orientate existing employees to new systems or new employees to existing systems. For example, the pattern of loading parts onto a power-and-free conveyor system has a significant impact on the balance of effort in different paint booths. If parts are loaded incorrectly, excessive queues in front of some paint stations stop the conveyor while other paint stations are idle. This imbalance impacts not only the paint booths but also drying and assembly stations. In an effort to gain experience in loading the conveyor prior to actually loading parts, an operator may be trained on a simulator where the only impact of loading errors will be a greater understanding of the system.

It is virtually impossible to plan a facility of any magnitude without conducting some type of simulation. The question is not whether a simulation is to be done, but whether it is to be done formally or informally, and what will be the scope of the simulation. If the simulation is to be done informally, such as a mental simulation of a lift truck picking up a load and traveling down an aisle, a detailed design obviously will not be required. Formal simulation experiments requiring detailed designs are often an important portion of the facilities planning process.

The scope of facilities planning simulation experiments may include an individual workstation, a piece of equipment, an entire facility, or a whole series of facilities. A very brief example of the use of simulation is given in Example 12.50. For one desiring to understand the process of simulation, this example illustrates the process. For one interested in learning to develop simulation models, enough examples cannot be cited to properly present the "art" of simulation. The more serious reader is referred to [2], [49], [65], and [71], among others.

## Example 12.50

An existing warehouse is 50 years old and is to be abandoned as it requires considerable renovation to promote efficient warehousing practice. A warehouse is to be built adjacent to the existing site to handle all existing business. The question that has been asked is, "How many dock bays should be constructed?" The first thought is that waiting line analysis could be used to answer this question. Unfortunately, a surge of vehicles arrives in the middle of the morning and the middle of the afternoon. This prevents defining arrival rates in a manner acceptable for mathematical analysis. The best method of determining the number of dock bays appears to be simulation.

A time study is made of the arrival of trucks to obtain the data given in Tables 12.19 and 12.20. The time required to load and unload vehicles does not vary with the time of day. A time study of these activities results in the data given in Table 12.21.

**Table 12.19** *Truck Arrivals Between 8:00 A.M. and 10:00 A.M., 11:00 A.M. and 2:00 P.M. and 3:00 P.M. and 5:00 P.M.*

| Time Between Arrivals (Hours) | Relative Frequency | Cumulative Frequency |
|---|---|---|
| 0– .25 | 0.02 | 2 |
| 0.251– .50 | 0.07 | 9 |
| 0.501– .75 | 0.19 | 28 |
| 0.751–1.00 | 0.34 | 62 |
| 1.001–1.25 | 0.26 | 88 |
| 1.251–1.50 | 0.08 | 96 |
| 1.501–1.75 | 0.03 | 99 |
| 1.751–2.00 | 0.01 | 100 |

**Table 12.20** *Truck Arrivals Between 10:00 A.M. and 11:00 A.M., and 2:00 P.M. and 3:00 P.M.*

| Time Between Arrivals (Hours) | Relative Frequency | Cumulative Frequency |
|---|---|---|
| 0–0.25 | 0.36 | 36 |
| 0.251–0.50 | 0.41 | 77 |
| 0.501–0.75 | 0.23 | 100 |

**Table 12.21** *Truck Loading and Unloading Times*

| Unloading Time (Hours) | Relative Frequency | Cumulative Frequency |
|---|---|---|
| 0– .5 | 0.01 | 1 |
| 0.51–1.0 | 0.10 | 11 |
| 1.01–1.5 | 0.22 | 33 |
| 1.51–2.0 | 0.20 | 53 |
| 2.01–2.5 | 0.20 | 73 |
| 2.51–3.0 | 0.18 | 91 |
| 3.01–3.5 | 0.06 | 97 |
| 3.51–4.0 | 0.02 | 99 |
| 4.01–4.5 | 0.01 | 100 |

The truck spotting time is observed to be a constant of 0.1 hr. The existing warehouse has three dock bays. Considerable knowledge is available with respect to "typical" operations for three docks. Hence, the simulation model can be validated with three dock bays. The logic underlying the simulation model is as follows:

1. Initialize the model. Begin with three dock bays and add one bay each time it is reinitialized.
2. Generate a series of random numbers from 0 to 99.
3. Transform the random numbers to a series of truck interarrival times by relating random numbers to the cumulative frequencies in Tables 12.19 and 12.20. For example, in Table 12.20, random numbers 0 to 35 would result in an interarrival time of 0.125 hr, random numbers 36 to 76 would result in an interarrival time of 0.375 hr and random numbers 77 to 99 would result in an interarrival time of 0.625 hr.
4. Generate a series of random numbers from 0 to 99.
5. Transform the random numbers to a series of truck loading and unloading times by relating the random number to the cumulative frequencies in Table 12.21.
6. Assign trucks to dock bays, or if unavailable, to a queue. Unload the trucks, dispatch waiting trucks, and assign spotting time. Perform the truck loadings and unloadings for the entire day and maintain statistics.
7. Determine whether steady state is reached. If so, and six dock-bays have been considered, print all statistics and terminate the model.
8. If steady state is reached and less than six dock bays have been considered, return to step 1.
9. If steady state is not reached, return to step 6 and simulate another day's operation.

By running this model, the types of data that are available are

| Factor | Number of Bays | | | |
|---|---|---|---|---|
| | 3 | 4 | 5 | 6 |
| Average truck waiting time (minutes) | 46.3 | 19.6 | 3.2 | 1.6 |
| Longest truck waiting time (minutes) | 60.4 | 26.1 | 4.3 | 2.0 |
| Truck waiting time variance (minutes²) | 4.1 | 1.9 | .2 | .07 |
| Average time truck spent at warehouse (minutes) | 167.4 | 139.7 | 124.2 | 116.3 |
| Average dock bay utilization (percentage) | 82% | 61% | 49% | 41% |

The data the model generated for three dock bays is seen to be consistent with what is experienced with the present operation. Therefore, the model is considered to be a valid representation of the truck being loaded and unloaded for various numbers of dock bays. The data presented for four, five, and six dock bays can be evaluated by management and a decision reached with respect to the number of dock bays to be built for the new facility.

## Simulation Software

Simulation models are often implemented using a specialized simulation language, rather than more general programming languages such as BASIC, C, PASCAL, or FORTRAN. Indeed, there are several simulation languages that have been designed with material flow and facilities planning targeted as the principal application focus. Among the simulation languages that are either designed for or well suited for simulating material handling systems are AutoMod from AutoSimulations, Incorporated [59], [60], FACTOR from Pritsker Corporation [51], GPSS/H™ from Wolverine Software Corporation [73], GPSS/World™ from Minuteman Software [19], [20], MOGUL from High Performance Software, Inc., ProModel for Windows from PROMODEL Corporation [37], SIMAN V® and ARENA from Systems Modeling Corporation [4], [31], [63], SIMSCRIPT II.5® and SIMFACTORY II.5 from CACI Products Company [32], SLAM II and SLAMSYSTEM® from Pritsker Corporation [61], Taylor II from F&H Logistics and Automation B.V. in the Netherlands [40], and WITNESS from AT&T Istel. For a comparison of these software products, see [3].

In recent years, considerable improvements have occurred in the development of animation software to accompany the simulation packages cited above. Indeed, most of the software cited above includes animation capabilities. However, it is also the case that general purpose animation packages exist; PROOF Animation from Wolverine Software Corporation is an example of an animation package that can be driven by any software that can write ASCII data to a file. As noted by Banks [2], you can build a simulation model graphically using PROOF Animation, complete the model using MOGUL, execute the model using GPSS/H, and view an animation using PROOF Animation.

As noted in Chapter 8, several object oriented software packages have been developed to facilitate facilities planning. Likewise, a number of object-oriented simulation models have been developed for material handling applications [5], [15], [22], [27], [41], [44], [45], [78]. Among the material handling technologies commonly targeted for simulation studies are automated guided vehicles (AGV) and automated storage and retrieval systems (AR/RS); however, it is safe to assert that simulation applications have included all material handling technologies in use [6], [33], [36], [42], [64], [67], [76], [80], [82], [86].

## *12.11*  SUMMARY

In this, the longest chapter, we attempted to provide you with a wide range of models that can be used in facilities planning. Some of the models have broad application, whereas others are constrained to very specific applications. Likewise, some are relatively simple representations of complex applications, while others more accurately represent the situation being studied.

Although the range of problems addressed in the chapter is sizeable, we make no claim to having covered all the pertinent models. Indeed, we are unable to assert that we included the most important and the most relevant. Why? Because new models are being developed even as we write the chapter.

Our objective for this chapter was threefold: to expose you to the richness of the world of quantitative modeling in facilities planning; provide you with a large buffet of relatively simple models to launch you on your facilities planning journey and to whet your appetite for more; and present models that can provide you with additional insights regarding tradeoffs, sensitivities, and data requirements in facilities planning. We hope we accomplished our objective *as it relates to you*! If so, then you will enjoy exploring the references provided in the bibliography for the chapter.

# BIBLIOGRAPHY

1.  Askin, R. G., and Standridge, C. R., *Modeling and Analysis of Manufacturing Systems*, John Wiley and Sons, NY, 1993.
2.  Banks, J., Carson, J. S. II, and Nelson, B. L., *Discrete-Event System Simulation*, 2nd ed., Prentice-Hall, Englewood Cliffs, NJ, 1996.
3.  Banks, J., "Software for Simulation," *1994 Winter Simulation Conference Proceedings*, eds. J. D. Tew, S. Manivannan, D. A. Sadowski, and A. F. Seila, Association for Computing Machinery, New York, pp. 26–33.
4.  Banks, J., Burnette, B., Rose, J. D., and Kozloski, H., *SIMAN V and CINEMA V*, John Wiley & Sons, Inc., New York, NY., 1995.
5.  Basnet, C. B., Karacal, S. C., and Beaumariage, T. G., "Experiences in Developing Object-Oriented Modeling Environment for Manufacturing Systems," *1990 Winter Simulation Conference Proceedings*, Association for Computing Machinery, New York, 1990, pp. 477–481.
6.  Benjaafar, S., "Intelligent Simulation for Flexible Manufacturing Systems: An Integrated Approach," *Computers and Industrial Engineering*, vol. 22, no. 3, 1992, pp. 297–311.
7.  Blaisdell, W. E., and Haddock, J., "Simulation Analysis Using SIMSTAT 2.0," *1993 Winter Simulation Conference Proceedings*, eds. G. W. Evans, M. Mollaghasemi, E. C. Russell, and W. E. Biles, Association for Computing Machinery, New York, 1993, pp. 213–217.
8.  Bozer, Y. A., "Optimizing Throughput Performance in Designing Order Picking Systems," Ph.D. Dissertation, Georgia Institute of Technology, Atlanta, 1985.
9.  Bozer, Y. A., "Guided Vehicle Systems: Information/Control System Implications of Alternative Design and Operation Strategies," *Advanced Information Technologies for Industrial Material Flow Systems*, eds., S. Y. Nof and C. L. Moodie, Springer-Verlag, 1989.
10. Bozer, Y. A., Schorn, E. C., and Sharp, G. P., "Geometric Approaches to Solve the Chebyshev Traveling Salesman Problem," *IIE Transactions*, vol. 22, no. 3, 1990, pp. 238–254.
11. Bozer, Y. A., and Srinivasan, M. M., "Tandem Configuration for AGVS and the Analysis of Single Vehicle Loops," *IIE Transactions*, vol. 23, no. 1, pp. 72–82, 1991.
12. Bozer, Y. A. and White, J. A., "Travel-Time Models for Automated Storage/Retrieval Systems," *IIE Transactions*, vol. 16, no. 4, 1984, pp. 329–338.

13. Bozer, Y. A. and White, J. A., "Design and Performance Models for End-of-Aisle Order Picking Systems," *Management Science*, vol. 36, no. 7, 1990, pp. 852–866.

14. Breugnot, D., Gourgand, M., Hill, D., and Kellert, P., "GAME: An Object-Oriented Approach to Computer Animation in Flexible Manufacturing System Modeling," *Proceedings of the 24th Annual Simulation Symposium*, New Orleans, 1991, pp. 217–227.

15. Carrie, A. S., Moore, J. M., Roczniak, R., and Seppanen, J. J., "Graph Theory and Computer Aided Facilities Design," *OMEGA*, the International Journal of Management Science, vol. 6 no. 4, pp. 353–361, 1978.

16. Cho, M., "Design and Performance Analysis of Trip-Based Material Handling Systems in Manufacturing," Ph.D. Dissertation, The University of Michigan, Ann Arbor, MI, 1990.

17. Collins, N., and Watson, C. M., "Introduction to ARENA," *1993 Winter Simulation Conference Proceedings*, eds., G. W. Evans, M. Mollaghasemi, E. C. Russell, and W. E. Biles, Association for Computing Machinery, New York, 1993, pp. 205–212.

18. Conway, R. W., and Maxwell, W. L., "A Note of the Assignment of Facility Locations," *The Journal of Industrial Engineering*, vol. 12, no. 1, pp. 34–36, 1961.

19. Cox, S. W., "GPSS World: A Brief Review," *1991 Winter Simulation Conference Proceedings*, eds. B. L. Nelson, W. D. Kelton, and G. M. Clark, Association for Computing Machinery, New York, 1991, pp. 59–61.

20. Cox, S. W., "Simulation Studio$^{TM}$," *1992 Winter Simulation Conference Proceedings*, eds. J. J. Swain, D. Goldsman, R. C. Crain, and J. R. Wilson, Association for Computing Machinery, New York, 1992, pp. 347–351.

21. DeMars, N. A., Matson, J. O., and White, J. A., "Optimizing Storage System Selection," in *Proceedings of the 4th International Conference on Automation in Warehousing*, Tokyo, Japan, 1982.

22. Drolet, J. R., Moodie, C. L., and Montreuil, B., "Object Oriented Simulation with Smalltalk-80: A Case Study," *1991 Winter Simulation Conference Proceedings*, eds. B. L. Nelson W. D. Kelton, and G. M. Clark, Association for Computing Machinery, New York, 1991, pp. 312–321.

23. Earle, J. J., and Henriksen, J. O., "PROOF Animation: Better Animation for Your Simulation," *1993 Winter Simulation Conference Proceedings*, eds. G. W. Evans, M. Mollaghasemi, E. C. Russell, and W. E. Biles, Association for Computing Machinery, New York, 1993, pp. 172–178.

24. Egbelu, P. J., "The Use of Non-Simulation Approaches in Estimating Vehicle Requirements in an Automated Guided Vehicle Based Transport System," *Material Flow*, vol. 4, 1987, pp. 17–32.

25. Egbelu, P. J., and Tanchoco, J. M. A., "Characterization of Automatic Guided Vehicle Dispatching Rules," *International Journal of Production Research*, vol. 22, no. 3, 1984, pp. 359–374.

26. Eilon, S., Watson-Gandy, C. D. T., and Christofides, N., *Distribution Management: Mathematical Modelling and Practical Analysis*, Hafner Publishing Co., New York, 1971.

27. Eom, J. K., *Selection, Design, Control of Material Handling Storage Systems: An Object-Oriented and Knowledge-Based Approach*, Ph.D. Dissertation, Department of Industrial Engineering, North Carolina State University, Raleigh, NC, 1992.

28. Foley, R. D., and Frazelle, E. H., "Analytical Results for Miniload Throughput and the Distribution of Dual Command Travel Time," *IIE Transactions*, vol. 23, no. 3, 1991, pp. 273–281.

29. Francis, R. L., McGinnis, L. F., and White, J. A., *Facility Layout and Location: An Analytical Approach*, 2nd ed., Prentice-Hall, Englewood Cliffs, NJ, 1992.

30. Francis, R. L., and White, J. A., *Facility Layout and Location: An Analytical Approach*, First Edition, Prentice-Hall, Englewood Cliffs, NJ, 1974.

31. Glavach, M. A., and Sturrock, D. T., "Introduction to SIMAN/Cinema," *1993 Winter Simulation Conference Proceedings*, eds. G. W. Evans, M. Mollaghasemi, E. C. Russell, and W. E. Biles, Association for Computing Machinery, New York, 1993, pp. 190–192.

32.  Goble, J. "Introduction to SIMFACTORY II.5," *1991 Winter Simulation Conference Proceedings*, eds. B. L. Nelson, W. D. Kelton, and G. M. Clark, Association for Computing Machinery, New York, 1991, pp. 77–80.

33.  Gobal, S. L., and Kasilingam, R. G., "A Simulation Model for Estimating Vehicle Requirements in Automated Guided Vehicle Systems," *Computers and Industrial Engineering*, vol. 21, nos. 1–4, 1991, pp. 623–627.

34.  Goetschalckx, M. P. and Ratliff, H. D., "Sequencing Picking Operations in a ManAboard Order Picking System," *Material Flow*, vol. 4, no. 4, 1988, pp. 255–264.

35.  Goldsman, D., Nelson, B. L., and Schmeiser, B., "Methods for Selecting the Best System" *1991 Winter Simulation Conference Proceedings*, eds. B. L. Nelson, W. D. Kelton, and G. M. Clark, Association for Computing Machinery, New York, 1991, pp. 177–186.

36.  Gonzalez, C. J., *A Design Procedure for Microload Automated Storage/Retrieval Systems*, Project Report, Department of Industrial Engineering, North Carolina State University, Raleigh, NC, 1990.

37.  Harrell, C. R., and Leavy, J. J., "ProModel Tutorial," *1993 Winter Simulation Conference Proceedings*, eds. G. W. Evans, M. Mollaghasemi, E. C. Russell, and W. E. Biles, Association for Computing Machinery, New York, 1993, pp. 184–189.

38.  Heragu, S. S., "Recent Models and Techniques for Solving the Layout Problem," *European Journal of Operations Research*, Vol. 57, pp. 136–144, 1992.

39.  Heragu, S. S., and Kusiak, A., "Machine Layout in Flexible Manufacturing Systems," *Operations Research*, vol. 36, no. 2, pp. 258–268, 1988.

40.  Hillen, D. W., and Werner, D., "Taylor II Manufacturing Simulation Software," *1993 Winter Simulation Conference Proceedings*, eds. G. W. Evans, M. Mollaghasemi, E. C. Russell, and W. E. Biles, Association for Computing Machinery, New York, 1993, pp. 276–280.

41.  Hollinger, D., and Bell, G., "An Object-Oriented Approach for CIM Systems Specification and Simulation," *Computer Applications in Production and Engineering*, Elsevier Science Publishers B. V., New York, 1987.

42.  Jeyabalan, V. and Otto, N. C., "Simulation Models of Material Delivery Systems," *1991 Winter Simulation Conference*, eds. B. L. Nelson, W. D. Kelton, and G. M. Clark, Association for Computing Machinery, New York, 1991, pp. 356–364.

43.  Kaspi, M., and Tanchoco, J. M. A., "Optimal Flow Path Design of Unidirectional AGV Systems," *International Journal of Production Research*, vol. 28, no. 6, pp. 1023–1030, 1990.

44.  Kim, K. S., *Object-Oriented Design and Simulation for AGV Systems*, Ph.D. Dissertation, Department of Industrial Engineering, North Carolina State University, Raleigh, NC, 1993.

45.  King, C. U., and Fisher, E. L., "Object-Oriented Shop-Floor Design, Simulation, and Evaluation," *Proceedings of the Fall IIE Conference*, Atlanta, 1986.

46.  Koff, G. A., "Automatic Guided Vehicle Systems: Applications, Controls and Planning," *Material Flow*, vol. 4, 1987, pp. 3–16.

47.  Kwo, T. T., "A Theory of Conveyors," *Management Science*, vol. 5, no. 1, pp. 51–71, 1958.

48.  Kwo, T. T., "A Method for Designing Irreversible Overhead Loop Conveyors," *The Journal of Industrial Engineering*, Vol 11, No. 6, pp. 459–466, 1960.

49.  Law, A. M. and Kelton, W. D., *Simulation Modeling and Analysis*, 2nd ed., McGraw-Hill, New York, 1991.

50.  Lawler, E. L., Lenstra, J. K., Rinnooy K., A. H., Shmoys, D. B., (Eds.), *The Traveling Salesman Problem*, Wiley, 1985.

51.  Lilegdon, W. R., Martin, D. L., and Pritsker, A. A. B., "FACTOR/AIM: A Manufacturing Simulation System," *Simulation*, vol. 62, 1994, pp. 367–372.

52.  Love, R. F., Morris, J. G., and Wesolowsky, G. O., *Facilities Location: Models and Methods*, North-Holland Publishing Company, NY, 1988.

53. *Matlab User's Guide*, The MathWorks, South Natick, MA, 1989.

54. Matson, J. O., "The Analysis of Selected Unit Load Storage Systems," PhD Thesis, School of Industrial and Systems Engineering, Georgia Institute of Technology, 1982.

55. Mayer, H., "Introduction to Conveyor Theory," *Western Electric Engineer*, vol. 4, no. 1, pp. 43–47, 1960.

56. Muth, E. J., "Analysis of Closed-Loop Conveyor Systems," *AIIE Transaction*, vol. 4, no. 2, pp. 134–143, 1972.

57. Muth, E. J., "Analysis of Closed-Loop Conveyor Systems: the Discrete Flow Case," *AIIE Transactions*, vol. 6, no. 1, pp. 73–83, 1974.

58. Muth, E. J., "Modelling and Systems Analysis of Multistation Closed-Loop Conveyors," *International Journal of Production Research*, vol. 13, no. 6, pp. 559–566, 1975.

59. Norman, V. B., "Simulation and Advanced Manufacturing System Design," *1986 Winter Simulation Conference*, Association for Computing Machinery, New York, pp. 554–558.

60. Norman, V. B., and Farnsworth, K. D., "AutoMod," *1993 Winter Simulation Conference Proceedings*, eds. G. W. Evans, M. Mollaghasemi, E. C. Russell, and W. E. Biles, Association for Computing Machinery, New York, 1993, pp. 249–254.

61. O'Reilly, J. J., "Introduction to SLAM II and SLAMSYSTEM," *1993 Winter Simulation Conference Proceedings*, eds. G. W. Evans, M. Mollaghasemi, E. C. Russell, and W. E. Biles, Association for Computing Machinery, New York, 1993, pp. 179–183.

62. Peck, L. G., and Hazelwood, R. N., *Finite Queueing Tables*, Wiley, New York, 1958.

63. Pegden, C. D., Shannon, R. E., and Sadowski, R. P., *Introduction to Simulation using SIMAN*, McGraw-Hill, New York, 1990.

64. Prasad, K., and Rangaswami, M., "Analysis of Different AGV Control Systems in an Integrated IC Manufacturing Facility Using Computer Simulation," *1988 Winter Simulation Conference Proceedings*, Association for Computing Machinery, New York, 1988, pp. 568–574.

65. Pritsker, A. A. B., *Introduction to Simulation and SLAM II*, 3rd ed., John Wiley and Sons, New York, 1986.

66. Pritsker, A. A. B., Signal, E., and Hammesfahr, J., *SLAM II: Network Models for Decision Support*, Prentice-Hall, Englewood Cliffs, NJ, 1989.

67. Quiroz, M. A. *Simulation of a Microload Automated Storage/Retrieval System*, M.S. Thesis, School of Industrial and Systems Engineering, Georgia Institute of Technology, Atlanta, 1986.

68. Roberts, S. D., and Flanigan, M. A., "Simulation Modeling and Analysis with INSIGHT: A Tutorial," *1990 Winter Simulation Conference Proceedings*, Association for Computing Machinery, New York, 1990, pp. 80–88.

69. Russell, E. C., "SIMSCRIPT II.5 and SIMGRAPHICS Tutorial," *1993 Winter Simulation Conference Proceedings*, eds. G. W. Evans, M. Mollaghasemi, E. C. Russell, and W. E. Biles, Association for Computing Machinery, New York, 1993, pp. 223–227.

70. Schmeiser, B., "Modern Simulation Environments: Statistical Issues," *First Industrial Engineering Research Conference Proceedings*, Chicago, 1992, pp. 139–143.

71. Shannon, R. E., *Systems Simulation: The Art and Science*, Prentice-Hall, Englewood Cliffs, NJ, 1975.

72. Sinriech, D., and Tanchoco, J. M. A., "The Central Projection Method for Locating Pickup and Delivery Stations in Single-Loop AGV Systems," *Journal of Manufacturing Systems*, vol. 11, no. 4, pp. 297–307, 1992.

73. Smith, D. S., and Crain, R. C., "Industrial Strength Simulation Using GPSS/H," *1993 Winter Simulation Conference Proceedings*, eds. G. W. Evans, M. Mollaghasemi, E. C. Russell, and W. E. Biles, Association for Computing Machinery, New York, 1993, pp. 223–227.

74. Stein, D. M., "Scheduling Dial-a-Ride Transportation Systems," *Transportation Science*, vol. 12, no. 3, pp. 232–249, 1978.

75. Tanchoco, J. M. A., and Sinriech, D., "OSL—Optimal Single-Loop Guide Paths for AGVS," *International Journal of Production Research*, vol. 30, no. 3, pp. 665–681, 1992.

76. Thomasma, T., and Ulgen, O. M., "Modeling of a Manufacturing Cell Using a Graphical Simulation System Based on Smalltalk-80," *1987 Winter Simulation Conference Proceedings*, Association for Computing Machinery, New York, 1987, pp. 683–691.

77. Thompson, W. B., "A Tutorial for Modeling with the WITNESS Visual Interactive Simulator," *1993 Winter Simulation Conference Proceedings*, eds. G. W. Evans, M. Mollaghasemi, E. C. Russell, and W. E. Biles, Association for Computing Machinery, New York, 1993, pp. 159–164.

78. Ulgen, O. M., "Simulation Modeling in an Object-Oriented Environment using Smalltalk-80," *1986 Winter Simulation Conference Proceedings*, Association for Computing Machinery, New York, 1986, pp. 474–484.

79. Ulgen, O. M., and Thomasma, T., "SmartSim: An Object Oriented Simulation Program Generator for Manufacturing System," *International Journal of Production Research*, vol. 28, no. 9, 1990, pp. 1713–1730.

80. Ulgen, O. M., and Kedia, P., "Using Simulation in Design of a Cellular Assembly Plant with Automatic Guided Vehicles," *1990 Winter Simulation Conference Proceedings*, Association for Computing Machinery, New York, 1990, pp. 683–691.

81. Vincent, S. G., and Law, A. M., "Unifit II: Total Support for Simulation Input Modeling," *1993 Winter Simulation Conference Proceedings*, eds. G. W. Evans, M. Mollaghasemi, E. C. Russell, and W. E. Biles, Association for Computing Machinery, New York, 1993, pp. 199–204.

82. Webster, R. L., and Foster, D. F., "Building Flexible AGV and AS/RS System Models for Facility Design Phase Applications," *1990 Winter Simulation Conference Proceedings*, Association for Computing Machinery, New York, 1990, pp. 692–698.

83. White, J. A., and Muth, E. J., "Conveyor Theory: A Survey," *AIIE Transaction*, vol. 11, no. 41, pp. 270–277, 1979.

84. White, J. A., Schmidt, J. W., and Bennett, G. K., *Analysis of Queueing Systems*, Academic Press, New York, 1975.

85. Wimmert, R. J., "A Mathematical Method of Equipment Location," *The Journal of Industrial Engineering,*" vol. 9, no. 6, pp. 498–505, 1958.

86. Xu, Z., and Pirasteh, R. M., "Airline-Catering Plant Material Handling System Analysis with Simulation and Scaled Automation," *1991 Winter Simulation Conference Proceedings*, eds. B. L. Nelson and G. M. Clark, Association for Computing Machinery, New York, 1991, pp. 402–410.

87. Zollinger, H. A., "Planning, Evaluating, and Estimating Storage Systems," Presented at the Institute of Material Management Education First Annual Winter Seminar Series, Orlando, February, 1982.

88. *Considerations for Planning an Automated Storage/Retrieval System*, The Material Handling Institute, Charlotte, North Carolina, 1982.

89. *Warehouse Modernization and Layout Planning Guide*, Department of the Navy, Naval Supply Systems Command, NAVSUP Publication 529, Washington, D.C., 1978.

# PROBLEMS

12.1. Let four existing facilities by located at $P_1 = (0,10)$, $P_2 = (8,10)$, $P_3 = (5,15)$, and $P_4 = (10,5)$ with $w_1 = 15$, $w_2 = 20$, $w_3 = 5$, and $w_4 = 30$. Determine the optimum location for a single new facility when cost is proportional to rectilinear distance.

**12.2.** The XYZ Company has six retail sales stores in the city of Lafayette. The company needs a new warehouse facility to service its retail stores. The location of the stores and the expected delivery per week from the warehouse to each store are:

| Store | Location (miles) | Expected Deliveries |
|-------|------------------|---------------------|
| 1 | (1,0) | 1 |
| 2 | (2,5) | 7 |
| 3 | (3,8) | 5 |
| 4 | (1,6) | 3 |
| 5 | (−5,−1) | 6 |
| 6 | (−3,−3) | 2 |

Assume that travel distance within the city of Lafayette is rectilinear and that after each delivery the delivery truck must return to the warehouse. If there are no restrictions on the warehouse location, where should it be located?

**12.3.** A new back-up power generator is to be located to serve a total of six precision machines in a manufacturing facility. Separate electrical cables are to be run from the generator to each machine. The locations of the six machines are $P_1 = (0,0)$, $P_2 = (30,90)$, $P_3 = (60,20)$, $P_4 = (20,80)$, $P_5 = (70,70)$ and $P_6 = (90,40)$. Determine the location for the generator that will minimize the total required length of the electrical cable. Assume rectilinear distance.

**12.4.** Six housing subdivisions within a city area are targeted for emergency service by a centralized fire station. Where should the new fire station be located such that the maximum rectilinear travel distance is minimized? The centroid locations (in miles) and total value of the houses in the subdivisions are as follows:

| Subdivision | X-Coordinate | Y-Coordinate | Total Value |
|-------------|--------------|--------------|-------------|
| A | 20 | 15 | 50 mil. |
| B | 25 | 25 | 120 mil. |
| C | 13 | 32 | 100 mil. |
| D | 25 | 14 | 250 mil. |
| E | 4 | 21 | 300 mil. |
| F | 18 | 8 | 75 mil. |

**12.5.** The city council of Greater Lafayette has decided to locate an emergency response unit within the city. This unit is responsible for four housing sectors (A) and three major street intersections (P) as shown in the figure below. Assume that the weights are uniformly distributed over the housing sectors.
   **a.** Determine the minimum location based on the weights given in the table below.
   **b.** Determine the minimax location based on the weights given in the table below.

| Housing Sector | Weight | Intersection | Weight |
|----------------|--------|--------------|--------|
| A1 | 10 | P1 | 20 |
| A2 | 15 | P2 | 10 |
| A3 | 20 | P3 | 5 |
| A4 | 40 | | |

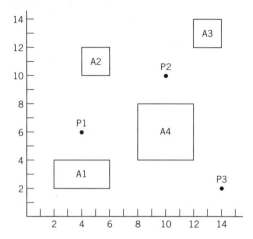

**12.6.** A new elementary school is needed in a suburban area of Denver, Colorado. After extensive research, the school board has narrowed down its choice to three possible sites. The locations for the current residential areas, expected students from each residential area, and possible locations for the school are shown in the tables below. Determine the optimal location for the new elementary school such that the total distance the students have to travel is minimized. It is fair to assume that the construction cost for all three sites are similar and distance is measured rectilinearly.

| Residences | X-Coordinate | Y-Coordinate | Weight |
|---|---|---|---|
| A | 20 | 25 | 100 |
| B | 36 | 18 | 400 |
| C | 62 | 37 | 500 |
| D | 50 | 56 | 300 |
| E | 25 | 0 | 200 |

Possible sites for the new elementary school:

| Possible Sites | X-Coordinate | Y-Coordinate |
|---|---|---|
| 1 | 50 | 50 |
| 2 | 30 | 45 |
| 3 | 65 | 28 |

**12.7.** The Chicago Exponent plans to rent building space for a new print shop within the city limits. The locations for current distribution centers, expected deliveries, and possible locations for the facility are shown in the tables and figures below.
**a.** Determine the optimal location for the new print shop.
**b.** Rank the alternative locations in order of preference using contour lines.

Current Distribution Centers:

| Center | X-Coordinate | Y-Coordinate | Weight |
|---|---|---|---|
| A | 5 | 10 | 200 |
| B | 50 | 15 | 400 |
| C | 25 | 25 | 500 |
| D | 35 | 5 | 100 |
| E | 15 | 20 | 400 |
| F | 30 | 30 | 600 |

Possible locations for the new print shop:

| Building | X-Coordinate | Y-Coordinate |
|---|---|---|
| 1 | 20 | 20 |
| 2 | 40 | 25 |
| 3 | 25 | 35 |

**12.8.** Plot the contour line passing through the point (1,5) for the following formula:

$$f(X, Y) = 5|X - 6| + 3|X - 4| + 2|Y - 1|$$

**12.9.** A small machine shop has five existing machines (M1 through M5) located at coordinate locations $P_1 = (8,25)$, $P_2 = (10,12)$, $P_3 = (16,30)$, $P_4 = (30,10)$, and $P_5 = (40,25)$. Two new machines (N1 and N2) are to be located in the shop. It is anticipated that there will be four trips per day between the new machines. The number of trips per day between each machine and each existing machine is:

| M/C | M1 | M2 | M3 | M4 | M5 |
|---|---|---|---|---|---|
| N1 | 8 | 6 | 5 | 4 | 3 |
| N2 | 2 | 3 | 4 | 6 | 6 |

a. *Formulate* the objective function assuming that rectilinear distance is used.
b. *Formulate* the objective function assuming that Euclidean distance is used.

**12.10.** Three new facilities are to be located relative to six existing facilities. The weighting factors are [$v_{jk}$ = flow intensity between new machines $j$ and $k$, and $w_{ji}$ = flow intensity between new machine $j$ and existing machine $i$];

| | | | | |
|---|---|---|---|---|
| $v_{12} = 6$, | $v_{13} = 1$, | $v_{23} = 4$, | $w_{11} = 4$, | $w_{12} = 0$, |
| $w_{13} = 2$, | $w_{14} = 0$, | $w_{15} = 6$, | $w_{16} = 5$, | $w_{21} = 0$, |
| $w_{22} = 0$, | $w_{23} = 0$, | $w_{24} = 0$, | $w_{25} = 8$, | $w_{26} = 1$, |
| $w_{31} = 0$, | $w_{32} = 0$, | $w_{33} = 8$, | $w_{34} = 10$, | $w_{35} = 2$ |
| $w_{36} = 6$ | | | | |

The coordinate locations of the existing facilities are $P_1 = (0,12)$, $P_2 = (2,1)$, $P_3 = (10,2)$, $P_4 = (6,12)$, $P_5 = (20,10)$, and $P_6 = (5,20)$.
a. *Formulate* the location problem assuming rectilinear distance.
b. *Formulate* the location problem assuming Euclidean distance.

**12.11.** An enterpreneur plans to locate coffee shops in a newly constructed office building. The current office tenants are located at $P_1 = (20,70)$, $P_2 = (30,40)$, $P_3 = (90,30)$, and $P_4 = (50,100)$. 50 persons per day are expected to visit the first office, 30 in the second office, 70 in the third office, and 60 in the last office. Seventy percent of the visitors are expected to drop by the coffee shop. For each additional unit distance a customer has to travel, the coffee shop is expected to lose $0.25 in revenue. The daily operating cost for $n$ coffee shops is $C(n) = 5000n$. Determine the optimal number of coffee shops to construct and specify their locations.

**12.12.** Five special-purpose machines are located in a plant at the point (0,0), (0,10), (30,25), (15,10), and (20,20). The machines require maintenance at expected frequencies of 10, 16, 8, 5, and 12 times per month, respectively. Due to the nature of the maintenance, all machine maintenance must be performed at the maintenance center. The annual cost of owning and operating a maintenance center is $5,000. The cost of moving machine to the maintenance center is estimated to be $10 per unit distance.

a. What are the total cost and location of the maintenance center if *one* maintenance center is the optimum number?

b. What are the total cost and locations of the maintenance centers if *two* maintenance centers is the optimum number?

12.13. Amco Frozen Corp. is planning on locating its food distribution centers in Dallas, Texas, to serve the southwestern part of the country. After extensive research, the company has narrowed down its choice to five existing facilities. The monthly cost of meeting the customer's demands and rental cost for each of these five facilities are summarized in the figure below. Determine which facilities are the optimal site for the distribution center such that the overall cost is minimized.

**Existing Facility Sites**

| Customer | A | B | C | D | E |
|---|---|---|---|---|---|
| 1 | 2,500 | 500 | 3,600 | 10,000 | 8,000 |
| 2 | 1,800 | 6,000 | 5,400 | 1,200 | 7,200 |
| 3 | 5,000 | 5,000 | 4,700 | 1,500 | 6,500 |
| 4 | 2,800 | 2,600 | 4,800 | 6,000 | 7,000 |
| 5 | 6,000 | 1,500 | 9,000 | 8,000 | 6,300 |
| Rental Cost | 5,000 | 7,000 | 6,000 | 2,000 | 8,000 |

12.14. A warehouse company plans to locate its regional warehouse in Miami, Florida, to serve the southeastern part of the country. The company has four sites to choose from. The *monthly* costs of meeting the customers' demands and *annual* rental cost for each of these four facilities are summarized below. Determine the optimal size for the distribution center such that the overall cost is minimized.

**Existing Facility Sites**

| Customer | A | B | C | D |
|---|---|---|---|---|
| 1 | 12,000 | 20,000 | 15,000 | 24,000 |
| 2 | 18,000 | 20,000 | 10,000 | 15,000 |
| 3 | 25,000 | 20,000 | 12,000 | 32,000 |
| 4 | 16,000 | 18,000 | 10,000 | 24,000 |
| 5 | 52,000 | 48,000 | 25,000 | 58,000 |
| 6 | 32,000 | 30,000 | 10,000 | 55,000 |
| Rental Cost | 50,000 | 75,000 | 72,000 | 45,000 |

12.15. A manufacturing plant is to be constructed in the city of Cleveland, Ohio, in order to provide a faster service to its major customers in the surrounding area. There are three possible locations the company is considering. The distances (in miles) from these three locations to the four major customers are listed in the table below. The transportation cost is approximately $1.00/mile. The annualized fixed cost of constructing the plant and the customers' demand (number of trips required per year) are also given in the table. Find the best location for this plant.

**Possible Locations**

| Customer | No. of Trips | A | B | C |
|---|---|---|---|---|
| 1 | 1,000 | 25 | 15 | 20 |
| 2 | 2,400 | 30 | 45 | 40 |
| 3 | 4,200 | 10 | 10 | 50 |
| 4 | 3,200 | 25 | 25 | 35 |
| Construction Cost | | 120,000 | 75,000 | 70,000 |

**12.16.** Five departments are involved in the processing required for the products in a manufacturing company. A summary of the material flow matrix (between departments) and
distance matrix (between sites in feet) can be found in the matrices below. The material
handling cost is directly proportional to the distance traveled.
  **a.** Determine the lower bound of the material handling cost.
  **b.** Determine the material handling cost for the current layout design.

$$V = \begin{bmatrix} 0 & 4 & 8 & 6 & 5 \\ 4 & 0 & 3 & 7 & 9 \\ 8 & 3 & 0 & 2 & 1 \\ 6 & 7 & 2 & 0 & 6 \\ 5 & 9 & 1 & 6 & 0 \end{bmatrix} \quad D = \begin{bmatrix} 0 & 70 & 45 & 62 & 35 \\ 70 & 0 & 57 & 47 & 91 \\ 45 & 57 & 0 & 28 & 71 \\ 62 & 47 & 28 & 0 & 62 \\ 35 & 91 & 71 & 62 & 0 \end{bmatrix}$$

**12.17.** Six cells are to be located in a two by three facility as shown in the figure below. The
material flow matrix is given in a separate table. The material handling cost is directly
proportional to the distance traveled (from centroid to centroid).
  **a.** Determine the lower bound of the material handling cost.
  **b.** Determine the initial cell layout and total material handling cost.
  **c.** Use the improvement procedure to determine cellular layout and calculate the total
  material handling cost.

| A | C | E |
|---|---|---|
| B | D | F |

| Cell | 1 | 2 | 3 | 4 | 5 | 6 |
|------|---|---|---|---|---|---|
| 1 | — | 100 | 200 | 150 | 0 | 0 |
| 2 | | — | 20 | 10 | 150 | 0 |
| 3 | | | — | 30 | 250 | 150 |
| 4 | | | | — | 0 | 50 |
| 5 | | | | | — | 10 |
| 6 | | | | | | — |

**12.18.** Four manufacturing cells are served by an automatic guided vehicle (AGV) on a linear
bidirectional track as shown below. The product routing information and required production rates are given in the table below. Determine a cell arrangement based on an
improvement procedure. (Assuming that the pickup/delivery stations are located along
the AGV track at the midpont of the cell edge.)

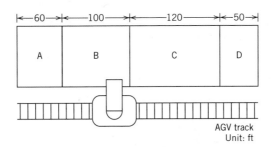

AGV track
Unit: ft

| Product | Processing Sequence | Weekly Production |
|---------|---------------------|-------------------|
| 1 | A B D C D | 1,000 |
| 2 | B A D C A | 700 |
| 3 | B D A | 300 |
| 4 | A B C A | 400 |

**12.19.** Six machines are located on either side of a bidirectional conveyor system as shown in the figure below. For the flow information given in the table, how can the machines be rearranged to minimize the time products spend on the conveyor? Assume the load transfer station for each machine is located in the midpoint of the machine edge facing the conveyor. Distances are given in feet.

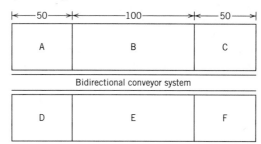

Bidirectional conveyor system

| | A | B | C | D | E | F |
|---|---|---|---|---|---|---|
| A | — | 100 | 200 | 20 | 100 | 100 |
| B | | — | 150 | 250 | 250 | 75 |
| C | | | — | 0 | 50 | 0 |
| D | | | | — | 125 | 225 |
| E | | | | | — | 300 |
| F | | | | | | — |

**12.20.** Consider an array of storage locations as illustrated below. The array represents storage bays. Rectilinear travel is used and is assumed to originate and/or terminate at the centroid of the storage bay. Products are received through one of the I/O points (docks). The storage activity is divided equally between the docks.

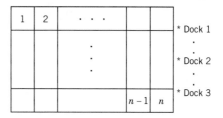

Define the following parameters and variable:

$m$ = number of items

$A_i$ = number of grid squares required by item $i = 1, \ldots, m$.

  Assume $n = \sum A_i$.

$d_{kj}$ = distance between dock $k$ (= 1, . . . , $p$) and the centroid of grid $j$
  ($j$ = 1, . . . , $n$).

$w_{ik}$ = cost/unit distance incurred in transporting item $i$ between dock $k$ and its
  storage region

$X_{ij}$ = if item $i$ is assign to grid square $j$, 0 otherwise.

a. Challenge the average distance item $i$ travels between dock $k$ and its storage region.

b. Calculate the average cost of transporting item $i$ between dock $k$ and its storage region.

c. Formulate an integer programming model which minimizes the total transportation cost.

12.21. Consider a warehouse illustrated by the figure below. Thirty-two bays (25 ft.×25 ft.) are available for storage. Four different types of products (A, B, C, and D) are to be stored. Each type of product cannot share a bay with any other type of product. Products are received from Dock 1 and are shipped out in equal proportion from docks 2 and 3. The area requirement and weekly load rate for each of these four products are shown in the table below. Determine the layout that will minimize the average distance traveled per week.

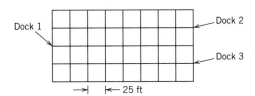

| Product | Area (ft²) | Weekly Load Rate |
|---|---|---|
| A | 4,375 | 500 |
| B | 7,500 | 600 |
| C | 1,500 | 700 |
| D | 6,250 | 200 |

12.22. An existing warehouse will be used for the storage of six product families. The warehouse consists of storage bays of size 20 ft×20 ft. Dock 1 has been designated as the receiving dock, while dock 2 is used as the shipping dock. The area requirement and monthly load rate for each product family are shown in the following table. Determine the layout that will minimize the total expected travel distance.

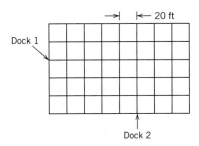

| Product Family | Area (ft$^2$) | Load Rate |
|:---:|:---:|:---:|
| 1 | 2,400 | 500 |
| 2 | 3,200 | 250 |
| 3 | 2,000 | 650 |
| 4 | 2,800 | 450 |
| 5 | 4,000 | 375 |
| 6 | 1,600 | 750 |

**12.23.** Three classes of products (A, B, and C) are to be stored in the warehouse depicted in the figure below. Product storage requirements are 15 bays for A, 5 bays for B, and 16 bays for C. Forty percent of the shipment has to go through dock 1, and the other shipment are evenly distributed between docks 2 and 3. All products require the same number of trips from/to storage (dock) per day. Recommend a layout which maximizes throughput.

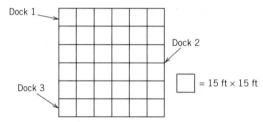

**12.24.** Shown in the figure below is a layout of an existing warehouse. One-way aisles are 8 ft wide, while two-way aisles are 16 ft wide. Each storage bay is approximately 20 ft×20 ft. Three products (X, Y, and Z) are to be stored. Door A and B serve as the receiving and shipping docks, respectively. The area requirements and load rates are shown in the table below. Determine the layout that will minimize the expected travel distance.

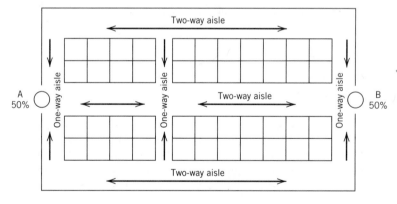

| Product | Area | Load Rate |
|:---:|:---:|:---:|
| X | 6,400 | 300 |
| Y | 8,800 | 400 |
| Z | 2,400 | 500 |

**12.25.** Four equal sized (20 ft×20 ft) machines are grouped into a manufacturing cell in a linear layout. The material handling device is a two-way shuttle cart system. The pickup and delivery station(s) for each type of machine is shown in the figure below. The material flow information is given in the table.

a. Determine the machine location that will minimize the loaded travel distance.
b. Calculate the total loaded travel distance (assuming the clearance between each pair of machines if 8 ft).

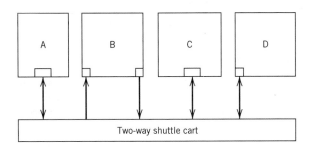

| M/C | A | B | C | D |
|-----|-----|-----|-----|-----|
| A | — | 150 | 100 | 100 |
| B | 150 | — | 300 | 50 |
| C | 100 | 300 | — | 0 |
| D | 100 | 50 | 0 | — |

**12.26.** Block stacking is used for the storage of unit loads of paper products. The dimensions of the unit load are 48 in.×52 in. A clearance of 8 in. is provided between storage rows. The storage aisle is 12 ft wide. The product is stacked four levels high in the warehouse. Forty loads of a particular product have been received.
   a. Determine the depth of the storage rows that will minimize the average amount of floor space required. Assume uniform depletion of product over the life of the forty loads. Use enumeration to obtain the solution.
   b. By using a continuous approximation, what row depth would minimize the average amount of floor space?

**12.27.** Deep lane storage is used for storing a soft drink product. The dimensions of the unit load are 48 in.×42 in. The flue clearance is 14 in., the rack is 6 in. wide, the clearance between a load and the rack is 4 in., and the aisle width is 8 ft. Seven tiers of storage are used. Two hundred pallet loads of the product are to be stored. Determine the following values:
   a. $x$, using continuous approximation.
   b. $x$, when the quantity of pallet loads to be stored is reduced to 35 and using enumeration.

**12.28.** Block stacking is used for the storage of 48 in.×52 in. unit loads. A counterbalanced lift truck is used to store and retrieve the unit loads, which are stacked five-high. The storage aisle is 13 ft. wide. A clearance of 8 in. is used between storage rows.
   a. If 250 loads of a particular product are received and are to be stored, what storage depth will minimize the average floor space required during the length of time the product is stored? Assume the product is withdrawn from storage at a uniform rate. Use continuous approximation and round off the answer to the nearest integer value.
   b. Suppose the lot size in a) had been 50 unit loads. Using enumeration, determine if the optimum row depth is 3, 4, or 5.
   c. Suppose the lot size in a) had been 50 unit loads and the inventory levels for the product were distributed as follows during its life cycle.

| Inventory | Percent | Inventory | Percent |
|-----------|---------|-----------|---------|
| 50 | 20 | 25 | 5 |
| 45 | 20 | 20 | 5 |
| 40 | 15 | 15 | 5 |
| 35 | 10 | 10 | 5 |
| 30 | 10 | 5 | 5 |

Determine the expected number of storage rows required using a storage row depth of 3.

d. In (c), determine the expected amount of floor space required using a storage row depth of four.

12.29. Twenty-five 48 in.×40 in. unit loads of a particular product are to be stored in storage racks. The warehouse manager cannot decide whether to store the product in single deep or double deep storage racks. Seven tiers of storage are used; the storage aisle is 8 ft wide. A clearance of 4 in. is used between a load and the 4 in. upright truss and a clearance of 4 in. side-to-side between loads. Two loads are stored side-by-side on a load beam. The flue space has a width of 12 in. between loads. Withdrawals occurs at a uniform rate over the life cycle of the 25 loads.

a. Determine the expected amount of floor space required to store the product using single deep storage rack.

b. Determine the expected amount of floor space required to store the product using double deep storage rack.

c. Suppose a scissors type reach attachment will allow 3-deep storage in the storage rack. As with double deep storage, 2 loads are stored side-by-side on a load beam. The same clearances apply. Based on assumptions similar to those used for double deep lane storage, determine the expected amount of floor space required to store the product.

12.30 Block stacking is used to store unit loads, 48″×50″; each unit load is 54″ high, including the pallet. A clearance of 10″ is provided between storage rows. The storage aisle is 13 ft. wide; storage is assumed to occur on both sides of the aisle. Unit loads are stacked 3-high.

a. Given $Q = 25$, using enumeration determine $x$ to minimize $S$.

b. Suppose $Q = 250$. Using continuous approximation, how deep should the storage rows be?

c. Suppose $Q = 250$ and $s = 25$ loads. Use continuous approximation to determine $x$.

d. Suppose $Q = 25$ and $s = 10$ loads. Use enumeration to determine $x$.

12.31 An appliance manufacturer stores its finished goods using block stacking and clamp attachments on counterbalanced lift trucks. Leased warehouse space is currently being used to store freezers. The outside dimensions of the freezers are 36″×60″×48″, with the latter being the height of the freezer. The unit can be stored in either the 36″×60″ configuration or the 60″×36″ configuration. Units can be stacked 3-high in storage. If the 36″×60″ configuration is used, then a clearance of 12″ between storage lanes is required; if the 60″×36″ configuration is used, then a clearance of 24″ is required. A 13 ft. storage aisle is used.

a. If the 36 in.×60 in. storage configuration is used, what row depth will minimize the average amount of floor space required over the life of a storage lot of size 20, assuming uniform withdrawals from storage? (Use enumeration to obtain you answer).

b. Given the conditions of part (a), what row depth is indicated when using continuous approximation with a lot of size 300?

c. Given the conditions of part (a), what row depth is indicated when using continuous approximation with a lot of size 200 and a safety stock of 25?

d. For a lot of size 20 and uniform withdrawal, what is the resulting minimum square footage for the best configuration, that is, the minimum value of $S$ for either 36 in. $\times$ 60 in. $\times$ 48 in. or 60 in. $\times$ 36 in. $\times$ 48 in.?

**12.32.** The Most Delicious Ice Cream Company used block stacking to store pallet loads of its finished goods in frozen storage. The dimensions of the pallet load are 36 in.$\times$48 in. Each load consists of 40 half-gallon cartons. Eight cartons to a layer, and five layers to a pallet. Using walkie stacker lift trucks, pallet loads of ice cream are stacked 3-high. The storage aisle is 8 ft wide, and a clearance of 12 in. is to be used between storage rows. A production run of 20-pallet loads of Hot Fudge Delight occurs. It is expected that the product will be withdrawn from storage at a uniform rate of 4/day; however, the first withdrawal will not occur immediately. Specifically, the inventory level for day k is given to be

$I_k = 20$      for      $k = 1, \ldots, 4$

$I_k = 36\text{-}4k$      for      $k = 5, \ldots, 8$.

a. If loads are stacked 2-deep in a row, what will be the average amount of floor space required in the freezer over the life of the production lot?

b. If loads are stacked 3-deep in a row, what will be the average amount of floor space required in the freezer over the life of the production lot?

**12.33.** A storage system is to be designed for storing 60 in. $\times$ 38 in. pallet loads of a wide variety of products. The storage alterantives to be considered are block stacking, deep lane storage, single-deep pallet rack, double-deep pallet rack, and triple-deep pallet rack.

In the case of block stacking, a counterbalanced lift truck will be used. The resulting design parameters are 13-ft storage aisles, 10 in. clearance between storage rows, and 4-high storage of products.

With deep lane storage, an S/R machine, will be used. Also, the following design parameters are to be used: a 6-ft aisle, 12-in. flue space, 3-in. clearance between the load and the rack, 4-in. rack member, and 8-high storage.

With single-deep, double-deep and triple-deep pallet rack, a narrow-aisle lift truck will be used. Two loads are stored side-by-side on a common load beam. The following design parameters are to be used: an 8-ft aisle, 12-in. flue space, 5-in. clearance between loads side-by-side on the load beam, 5-in. clearance between the load and the rack, 5-in. rack member, and 6-high storage.

a. Based on a lot size of 25, determine the average amount of floor space required for block stacking with the row depth that minimizes the average amount of floor space, measured in square feet. Solve for the row depth using enumeration.

b. Solve a) for the case of deep-lane storage, but determine the lane depth using a continuous approximation; round-off the lane depth to the nearest integer value.

c. Based on a lot size of 25, determine the average amount of floor space required to store the product using single-deep pallet rack. Express answer in square feet.

d. Solve part c) using double-deep pallet rack.

e. Solve part c) using triple-deep pallet rack.

**12.34.** Deep-lane storage is used to store 36 in.$\times$48 in. unit loads of Captain Crunchy Goober Butter. The clearance between the load and the upright rack member is 4 in.; the upright rack member is 4 in. wide; a 12-in. flue space is provided at the end of the deep lane, the storage aisle is 5 ft. wide; there are 15 storage levels in the deep lane system. The system consists of 5-deep lanes, 10-deep lanes, and 15-deep lanes.

a. Suppose 225 unit loads are received. Which lane depth is the worst choice for storing the Captain Crunchy Goober Butter in terms of minimizing the average square footage of floor space required? What is the resulting square footage? Assume the 225 units loads are withdrawn uniformly.

b. In part (a), suppose a safety stock of 75 units exists. Which of the three lane depths is the best in terms of minimizing the average square footage of storage space? What will be the resulting square footage?

c. In part b), suppose double-deep storage rack is used for the product. With two loads stored side-by-side on a load beam, 4 in. clearances between the loads and the 4-in. upright trusses and between loads side-by-side, a 12 in. flue space, and an 8-ft. aisle, what would be the average square footage required if the rack contained 6 storage levels?

12.35. 60 in.×48-in. loads are block stacked 5-high using 10-ft. wide aisle. One hundred unit loads of a product are received; a safety stock level of 15 unit loads is desired. In this case, suppose 15 units loads of the the previous lot of the product are in the warehouse at the time of receipt of the current lot. Due to the random nature of demand for the product, the probability distribution for the amount of the particular lot of product in inventory is given as follows:

| Inventory Level | Probability | Inventory Level | Probability |
|---|---|---|---|
| 100 | .12 | 50 | .04 |
| 95 | .02 | 45 | .04 |
| 90 | .02 | 40 | .06 |
| 85 | .02 | 35 | .06 |
| 80 | .02 | 30 | .06 |
| 75 | .02 | 25 | .08 |
| 70 | .02 | 20 | .08 |
| 65 | .02 | 15 | .08 |
| 60 | .02 | 10 | .08 |
| 55 | .04 | 5 | .10 |

Determine the lane depth for block stacking that minimizes the average amount of floor space over the lifetime of a lot of the particular product in question.

a. Use continuous approximation based on uniform demand.

b. Using enumeration, determine a local minimum near the solution obtained in part (a).

12.36. An AS/R system is being designed for 48×42 in. unit loads that are 45 in. high. One of the design alternatives involves 12 aisles, no transfer cars, 1440 openings per aisle, 12 levels and 60 columns of storage on each side of the aisle, and sprinklers. The S/R will travel horizontally at an average speed of 320 fpm; simultaneously, it travels vertically at an average speed of 75 fpm. The time required to pick up or deposit a load equals 0.25 min. Each S/R will have to store and retrieve at a rate of 15 storages per hour and 15 retrievals per hour. It is felt that 70% of the storages and retrievals will be performed on a dual-command cycle. What will be the utilization of each S/R? What will be the cost of the building if a 25-ft-high building costs $20/ft²? What will be the cost of the AS/RS if a central computer console is used and if each unit load weighs 2500 lb?

12.37. Five aisles of storage are provided with an S/R machine dedicated to each aisle. Randomized storage is used. The S/R machine travels simultaneously horizontally and vertically at speeds of 400 fpm and 80 fpm, respectively. Each aisle is 300 ft long and 50 ft high. The time required to pick up or put down a load is 0.30 min. Over an 8-hr period a total of 400 single-command cycles and 400 dual-command cyles are performed. Determine the system utilization.

12.38. An AS/R system is being designed for 42×48 in. unit loads that are 42 in. high. One of the design alternatives involves eight aisles, no transfer cars, 1100 openings per aisle,

11 levels and 50 columns of storage on each side of the aisle, and sprinklers. The S/R will travel horizontally at an average speed of 300 fpm; simultaneously, it travels vertically at an average speed of 70 fpm. The time required for the shuttle to perform either a pickup (P) or a deposit (D) operation is 0.35 min. Each S/R will have to store and retrieve at a rate of 18 storages per hour and 18 retrievels per hour. Due to the uncertainty concerning the proportion of the operations that will be single command versus dual command, you are to determine the maximum percentage of single-command operations so that utilization of the S/R does not exceed 85%. Also, the cost of the building, rack, and S/R machines are to be determined. Assume 2000-lb loads, off-board computer controls (but not a central console), and a $20/ft$^2$ building cost for a 25-ft-high building.

12.39. An AS/RS is being designed; randomized storage is to be used. One alternative being considered is to have each aisle be 70 ft tall and 310 ft long. The S/R has a horizontal speed of 425 fpm and a vertical speed of 80 fpm. It requires 0.40 min to pick up or deposit a load. The activity level for the system varies during each shift. The heaviest operating period is a time during which a large number of retrievals must be performed. During peak activity each S/R must be capable of performing 30 retrievals/per hour and 10 storages per hour. It is expected that 100% of the storages will be performed with dual-command cycles.

a. Can the system handle the required workload? Why or why not?

b. If during the peak period the number of retrievals is 28 per hour and the number of storages is six per hour, what will be the utilization for each S/R?

12.40. An AS/RS is to be designed for a 40×48-in. unit load that is 48 in. high. There are eight aisles; each aisle is 12 loads high and 50 loads long. There are to be sprinklers and no transfer cars. It takes 0.30 min to perform a P/D operation. The S/R travels horizontally at a speed of 300 fpm and travels vertically at a speed of 60 fpm. Thirty percent of the retrievals and 30% of the storages are single-command operations.

a. How many storages per hour and retrievals per hour can each S/R handle without being utilized more than 90%?

b. Estimate the cost of racks, S/R machines, and building. Assume it costs $25/ft$^2$ to build a 25-ft-high building, loads weigh 3000 lb, central console, and conventional structure (not rack supported).

12.41. An AS/R system is to be designed to store approximately 12,000 loads. Due to space limitations in the existing facility, it is known that the system will have eight tiers of storage. The choice has been reduced to having either (a) 10 aisles, 75 openings long, (b) 9 aisles, 84 openings long, or (c) 8 aisles, 94 openings long. The S/R will travel horizontally at an average speed of 450 fpm; simultaneously, it travels vertically at an average speed of 80 fpm. 48 × 40-in. unit loads, 44 in. high and weighing 2500 lb are to be stored. The P/D station is located at the end of the aisle, at floor level. The time required to pick up (P) or to deposit (D) a load is 0.30 min. The thoughout requirement on the system is to perform a total of 180 storages per hour and 180 retrievals per hour. Randomized storage is used. Forty percent single-command operations are anticipated. A maximum utilization of 95% can be planned for an individual S/R. A central computer console is to be used to control the S/Rs. Determine the least-cost alternative that satisfies the throughout constraint. Consider only rack cost and S/R machine cost; do not consider building cost. (Assume sprinklers and no transfer cars are to be used.)

12.42. An AS/RS is to be designed for a 48×54 in. unit load that is 50 in. high. There are 10 aisles; each aisle is 14 loads high and 60 loads long. There are to be sprinklers and no transfer cars. It takes 0.40 min to perform a P/D operation. The S/R travels horizontally at a speed of 425 fpm and travels vertically at a speed of 80 fpm. During peak activity each S/R in the system must be capable of performing 24 retrievals/hr. and 6 storages/

hr. During peak activity it is expected that 100% of the storages will be performed with dual-command cycles. The maximum utilization of an S/R during peak activity must be less than or equal to 98%. Does the system satisfy the utilization constraint?

12.43. If a 42×46-in. pallet load (62 in. high including the pallet) is to be stored in an AS/RS, what building dimensions are required to accommodate six aisles? Assume each aisle has 40 columns and 10 levels of storage along each side of the aisle. Do not allow for transfer cars; do consider sprinklers.

12.44. Based on a peak rate of 100 storages per hr. and 100 retrievals per hr, determine the minimum number of S/R machines required without having greater than 75% utilization of the equipment. Assume 60% of the operations are performed on a dual-command basis. The shuttle on the S/R requires 15 sec to perform a P/D operation. The average horizontal speed of the S/R is 400 fpm; the average vertical speed of the S/R is 80 fpm; horizontal and vertical movement is simultaneous. Each aisle is 300 ft long and 60 ft high.

12.45. For an AS/RS, suppose it has been determined that A sq.ft. of rack space is required on either side of each aisle. Given that A = L H and assuming that $b_v$ and $v_v$ are fixed, determine the rack shape (i.e., the Q value) for which the expected S/R machine travel time for a single command cycle is minimized.

12.46. In deriving the cycle time equations, the P/D station was assumed to be located at the lower-left-hand corner of the rack. Based on the particular type of hardware involved, however, the exact location of the P/D station may not coincide with the lower-left-hand corner of the rack. In fact, suppose the P/D station is located 12 ft from the lower-left-hand corner of the rack as shown below:

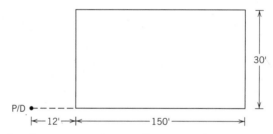

Further assume that the storage rack is small; say, 150 ft long and 30 ft high. The S/R machine travels at 400 fpm and 80 fpm in the horizontal and vertical directions, respectively. Calculate the expected S/R machine travel time for single- and dual-command cycles for the above case. Compare your results with the expected travel times obtained from Equations 12.69 and 12.71 (show the percent difference). Note: In traveling from the P/D station to a location in the rack (and vice versa), the S/R machine is **not** required to pass through the lower-left-hand corner of the rack.

12.47. Consider a normalized storage rack Q units long in the vertical direction. The P/D station is located at the lower-left-hand corner of the rack, and each trip originates and terminates at the P/D station. Given that the rack will be serviced by a straddle truck, it is assumed that the truck will travel according to the *rectilinear* metric.
   a. Derive (in closed-form) the normalized expected travel time for a single-command cycle.
   b. Derive (in closed-form) the normalized expected travel time for a dual-command cycle.

12.48. An AS/RS is to be designed for the storage of 60 in.×48 in. unit loads. The S/R machine will travel horizontally at a speed of 500 fpm; simultaneously, it will travel vertically at a speed of 80 fpm. The storage aisle will be 400 ft long and 60 ft tall. The pickup and deposit station will be located at the end of the aisle and will be elevated 20 ft above floor level. Assume randomized storage will be used and the storage region served by the S/R is a continuous, rectangular region 400 ft×60 ft. The P/D time will be 0.30 min.

a. Determine the single-command cycle time, including P/D times.

b. Determine the dual-command cycle time, including P/D times.

c. Suppose three classes of products are to be stored in the storage aisle. Fast movers will be stored in a rectangular region defined by the following coordinates: (0,0), (0,40), (125,40), and (125,0). Medium movers will be stored in an L-shaped region between the fast and slow movers. Slow movers will be stored in a rectangular region defined by: (250,0), (250,60), (400,60), and (400,0). Fast movers represent 60% of the storage and retrieval operations performed; medium movers account for 30% of the activity; and slow movers represent the balance. Determine the average single command cycle time, including P/D times.

**12.49.** Consider a unit load AS/RS rack that is 400 ft long and 80 ft high. The S/R machine travels at 450 fpm and 90 fpm in the horizontal and vertical directions, respectively. The P/D station is located at the lower-left-hand corner of the rack. Each trip starts and ends at the P/D station. Randomized storage is used. Suppose two types of products, labeled A and B, are stored in the rack. As shown below, product type A is stored in a 40 ft by 200 ft rectangular region located at the center of the rack. The remainder of the rack is used for storing product type B.

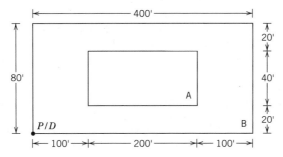

a. Determine the expected *single command* travel time (in minutes) for handling product type A.

b. Determine the expected *single command* travel time (in minutes) for handling product type B.

**12.50.** Consider a unit load AS/RS rack that is 500 ft long and 45 ft high. The S/R machine travels at 400 fpm and 120 fpm in the horizontal and vertical directions, respectively. The P/D station is located at the lower-left-hand corner of the rack. As shown by the shaded area below, the S/R machine will **not** be allowed to use one-third of the rack during the first shift. Compute the expected dual command travel time (in minutes) to perform a dual command trip assuming that the first operation is in region A and the second operation is in region B. (Each trip starts and ends at the P/D station.)

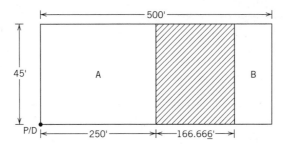

**12.51.** Consider a single-aisle AS/R system where the S/R machine travels at 450 fpm and 70 fpm in the horizontal and vertical directions, respectively. The pickup station is located at the lower-left-hand corner of the rack while the deposit station is located at the lower-right-hand corner. Loads to be stored must be picked up at the pickup station. Loads that are retrieved must be delivered to the deposit station. The rack is 400 ft long and 70 ft high. All the operations are performed on a dual command basis where the storage operation is assumed to be always performed before the retrieval.

Currently the system operates only for 8 hours through the first shift. An IE recently hired by the firm has proposed to use the S/R machine to move loads **within** the rack during the second shift (8 hr). Since no loads enter or leave the system during the second shift, the S/R machine can move loads with virtually no human intervention. Suppose the rack is divided into two equal pieces with an "imaginary" line as shown below:

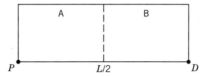

According to the proposal, during the second shift the S/R machine will move each load one-at-a-time from region A to region B. That is, the S/R machine will pick up a load from region A and deposit the load in an empty opening in region B. Subsequently, it will return to region A and pick up the next load to be moved to region B and so on. The above activity will continue for 8 hours as long as region A is not empty and region B is not full. The purpose is to move each load closer to the deposit station while providing more empty openings close to the pickup station for storages made during the *first* shift.

**a.** For the second shift, suppose the point where a load is picked up is randomly located within region A. Likewise, the point of deposit is assumed to be randomly located within region B. Determine the one-way expected S/R machine travel time (in minutes) to move a load from region A to region B. Compare your answer with the expected travel between time if the above two points were randomly located within the entire rack.

**b.** If the above proposal is implemented, the *first* shift operations are estimated to be affected as follows:

| | |
|---|---|
| $Pr$(storage in region A) = 0.75 | $Pr$(retrieval from region A) = 0.20 |
| $Pr$(storage in region B) = 0.25 | $Pr$(retrieval from region B) = 0.80 |

where $Pr(k)$ denotes the probability of event $k$. On the other hand, if the proposal is not implemented, then the storage and retrieval locations for the first shift will remain randomly distributed over the entire rack. If implemented, would the proposal made by the new IE reduce the expected S/R machine travel time during the first shift? Why or why not?

**12.52.** Consider an in-the-aisle order picking system based on the person-on-board AS/RS. Suppose the storage rack is 400 ft long and 80 ft tall. The S/R machine travels at a constant speed of 450 fpm and 90 fpm in the horizontal and vertical direction, respectively. On each trip, the picker performs eight picks on the average.

**a.** Assuming that the pick points are sequenced in an optimum fashion, compute the expected travel time required to complete a trip with eight picks.

**b.** Solve part (a) assuming that the two-band heuristic is used.

**c.** Solve part (a) assuming that the four-band heuristic is used.

**12.53.** Suppose in Problem 12.52 it takes the picker, on the average, 20 sec to complete a pick. Further suppose the picker spends 0.80 min at the P/D station between successive trips.

    a. Assuming that the pick points are sequenced in an optimum fashion, compute the throughput capacity of the system in picks/hr.

    b. Solve part (a) assuming that the two-band heuristic is used.

    c. Solve part (a) assuming that the four-band heuristic is used.

12.54. Suppose the system in Problem 12.53 does not meet the required throughput capacity. Two alternative solutions have been proposed: (a) increase the S/R machine travel speed by 10% both in the horizontal and vertical direction, or (2) reduce the time per pick from 20 sec to 17 sec by improving the packaging of the items. Assuming that the pick points are sequenced in an optimum fashion, which one of the above two alternatives would increase the thoughput capacity of the system more than the other?

12.55. Consider an in-the-aisle order picking system based on the person-on-board AS/RS. The user would like to store 7000 loads in the system, where each load requires 18 ft² of rack surface (including necessary clearances). The system must be capable of performing 215 picks/hr. On each trip, the picker performs eight picks; each pick requires 1.5 min. The S/R machine travels at a speed of 400 fpm and 100 fpm in the horizontal and vertical direction, respectively. Last, the picker spends 1.20 min at the P/D station between successive trips.

    a. Assuming that the pick points are sequenced in an optimum fashion, determine the "minimum" number of pickers and storage aisles required.

    b. Determine the rack dimensions for the "minimum" number of storage aisles.

    c. What fraction of the system's throughput capacity will be utilized?

12.56. Consider an in-the-aisle order picking system based on the person-on-board AS/RS. Total storage space required (expressed in square footage of storage rack) is equal to 300,000 ft². On each trip, the picker performs 10 picks; each pick requires 0.90 min. The S/R machine travels at a speed of 400 fpm and 100 fpm in the horizontal and vertical direction, respectively. The picker spends 1.40 min at the P/D station between successive trips. Assuming that the pick points are sequenced in an optimum fashion:

    a. Determine the throughput capacity of a three-aisle system.

    b. Determine the throughput capacity of a four-aisle system.

    c. Determine the throughput capacity of a five-aisle system.

    d. What can you say about the relationship between the throughout capacity of the system and the number of storage aisles?

12.57. Consider an end-of-aisle order picking system based on the miniload AS/RS. The P/D station is located at the lower-left-hand corner of the rack, and each trip originates and terminates at the P/D station. One aisle is assigned to each picker. Furthermore, the S/R machine travels at 400 fpm and 100 fpm in the horizontal and vertical directions, respectively. It takes 0.10 min to pick up or deposit a container. The user would like to store approximately 7000 loads in the system where each load requires 18 ft² of rack surface (including the necessary clearances). Last, the system must be capable of performing at least 215 picks/hr. The pick time is constant at 1.5 min/pick.

    a. For $b = 0.90$, determine the near minimum number of pickers required to meet the throughput and storage space requirements.

    b. Assuming that the vertical dimension of the rack yields the value of $Q$, determine the resulting rack length and height in feet.

    c. Determine the *effective* values of the expected picker and S/R machine utilizations for the resulting system.

12.58. Consider one aisle of a miniload AS/R system used by a book distributor for order picking. The P/D station is located at the lower-left-hand corner of a rack which is 360 ft long and 63 ft tall. The S/R machine travels at 450 fpm and 90 fpm in the horizontal and vertical directions, respectively. It takes 0.15 minutes to pick up or deposit a tray. On the average, the picker picks 8 books from each tray. Time and motion studies indicate that, including the necessary paperwork, on the average it takes the picker 15 sec to pick one book. It has also been determined that the total time required to make the necessary picks from a tray is exponentially distributed.

a. Compute the expected dual command cycle time for the above system.

b. Suppose the expected dual command cycle time is equal to 2.30 min/cycle. Compute the *number of books picked per hour* by the above picker.

12.59. Consider a miniload AS/RS with two pick positions per aisle. It is assumed that the P/D station is located at the lower left-hand corner of the rack. Randomized storage is used. The S/R machine travels at 400 fpm and 80 fpm in the horizontal and vertical direction, respectively. The *total* load handling time on a dual command trip is equal to 0.42 min. The total storage area required is equal to 80,000 ft². Last, the rack is required to have a shape factor of 0.90. Assuming that the pick time is *constant* (i.e., deterministic) at 1.50 min/pick, determine the minimum number of aisles required to obtain a picker utilization of 100%.

12.60. Consider a miniload AS/RS used for end-of-aisle order picking. It is assumed that the P/D station is located at the lower-left-hand corner of the rack and that there are two pick positions. Consider a single-aisle system with one picker. Under the current mode of operation, the containers *within a given order* are retrieved on a closest-container-first basis. That is, the container closest to the P/D station (in travel time) is retrieved first and so on. Ties are broken randomly. (Recall that no other container can be retrieved before all the containers needed for filling the current order have been retrieved.) Furthermore, the parts handled by the system are small enough so that multiple stock keeping units (SKUs or part numbers) can be stored in each container (or tray).

Currently, the system is unable to meet the required throughput level. Moreover, it has been established that:

(i) It is impossible to further reduce the mean and/or variance of the time required to pick an SKU.

(ii) It is impossible to further increase the travel velocity (or acceleration and deceleration) of the S/R machine (in either direction, horizontal or vertical). Likewise, the container handling time (or pick-up/deposit time) is fixed.

(iii) The space requirement is expected to remain at its present level. Therefore, the rack size may not be reduced.

(iv) Management is unwilling to purchase another miniload aisle. However, they are willing to support reasonable modification(s) to the current system.

Given the above situation, how would you attempt to improve the throughput performance of the system? Explain and discuss *at least three* conceptually different alternatives. For each alternative, you must carefully argue *why, how,* and *under what conditions* you would expect the throughput to show a reasonable improvement. The economic comparison of the alternatives and the *relative amount* of improvement offered by each alternative is **not** within the scope of the question. However, your suggestions must be realistic and reasonable from an implementation standpoint.

12.61. A trolley conveyor serves four workstations. There are 543 carriers equally spaced around the conveyor. The material flow patterns for the work stations are as follows:

$\{f_1(n)\} = (0,0,3,2,0,0,0,3,2,0,0,0,3,2, \ldots)$

$\{f_2(n)\} = (-3,0,-3,0,-3,0,-3,0-3,0-3,0, \ldots)$

$\{f_3(n)\} = (3,0,4,0,2,0,4,0,2,0,4,0, \ldots)$

$\{f_4(n)\} = (0,-3,0,0,-3,0,0,-3,0,0,-3,0,0, \ldots)$

a. Given the patterns, what is the smallest period?

b. Can general sequences be accommodated? If so, why? If not, why not?

c. Can steady-state operations be achieved? If so, why? If not, why not?

d. Is conveyor compatibility guaranteed for all values of $k$ not an integer multiple of the period of the material flow patterns? If so, why? If not, why not?

12.62. A trolley conveyor is used to transport parts among four production stations. There are 31 carriers equally spaced around the conveyor, with 10-ft separation between carriers. The material flow patterns for the work stations are

$$\{f_1(n)\} = (5,0,0); \{f_2(n)\} = (0,-3,0); \{f_3(n)\} = (0,0,4); \{f_4(n)\} = (4,0,0)$$

a. Determine the value of $B$.
b. Suppose the number of carriers on the conveyor can be decreased by one or two without affecting the cost of the conveyor. Consider each of the three options and specify the number of carriers that will minimize the amount of inventory on the conveyor. Use as an estimate of inventory on the conveyor the sum of the $H_i(n)$ values for all $i$ and $n$.
c. Suppose the unloading patterns at stations 1 and 2 can be changed. Consider each of the options shown below and specify the patterns (including the original patterns) that will minimize the amount of inventory on the conveyor. Use as an estimate of inventory on the conveyor the sum of the $H_i(n)$ values for all $i$ and $n$.

(1) $\{f_1(n)\} = (0,0,-4); \{f_2(n)\} = (-4,0,0)$
(2) $\{f_1(n)\} = (-4,0,0); \{f_2(n)\} = (0,-4,0)$

12.63. A trolley conveyor is used to transport parts among four production stations. There are 31 carriers equally spaced around the conveyor, with 10-ft separation between carriers. The material flow patterns for the work stations follows:

$$\{f_1(n)\} = (-2,0,-2,0); \qquad \{f_2(n)\} = (0,-2,0,-2);$$
$$\{f_3(n)\} = (0,0,4,0); \qquad \{f_4(n)\} = (4,0,0,0);$$

a. Can steady-state operations be achieved? If so, why? If not, why not?
b. Can general sequences for $\{F_i(n)\}$ be accommodated? If so, why? If not, why not?
c. Determine the value of $H_3(2)$.
d. Determine the value of $B$.
e. Suppose the number of carriers on the conveyor can be decreased by one or two without affecting the cost of the conveyor. Consider each of the three options and specify the number of carriers that will minimize the amount of inventory on the conveyor. Use as an estimate of inventory on the conveyor the sum of the $H_i(n)$ values for all $i$ and $n$. Specify the value obtained for inventory on the conveyor.
f. Suppose the loading patterns at stations 3 and 4 can be changed. Consider each of the options shown below and specify the patterns (including the original patterns) that will minimize the amount of inventory on the conveyor. Use as an estimate of inventory on the conveyor the sum of the $H_i(n)$ values for all i and n.

(1) $\{f_3(n)\} = (0,0,2,2) \{f_4(n) = (2,2,0,0)$
(2) $\{f_3(n)\} = (2,0,2,0) \{f_4(n) = (0,2,0,2)$

12.64. A trolley conveyor is used to transport parts between three production stations. There are 62 carriers equally spaced around the conveyor, with 10-ft separation between carriers. The material flow patterns for the work stations follow:

$$\{f_1(n)\} = (0,-4,0)$$
$$\{f_2(n)\} = (3,2,3)$$
$$\{f_3(n)\} = (0,-4,0)$$

a. Determine the value of $B$.
b. Suppose the unload sequence for station 3 can be modified by delaying its start time or location. Which of the following permutations of the unloading sequence for station 3 will minimize the value of $B$:

$\{f_3(n)\} = (0,-4,0)$;

$\{f_3(n)\} = (-4,0,0)$; or

$\{f_3(n)\} = (0,0,-4)$?

c. Suppose the unload sequence for station 3 can be modified by delaying its start time or location. What permutation of the unloading sequence for station 3 will minimize the cumulative amount of inventory on the conveyor?

$\{f_3(n)\} = (0,-4,0)$;

$\{f_3(n)\} = (-4,0,0)$; or

$\{f_3(n)\} = (0,0,-4)$?

**12.65.** An overhead trolley conveyor is used to transport assemblies within an assembly department. There are three loading stations in the assembly department and one unloading station. The trolley conveyor has 79 carriers equally spaced around the conveyor, with 20-ft separation between carriers. The material flow patterns for the work stations follow:

$\{f_1(n)\} = (1,0,0)$; $\{f_2(n)\} = (0,2,0)$; $\{f_3(n)\} = (0,0,1)$; and

$\{f_4(n)\} = (0,-4,0)$

a. Determine the value of $H_3(2)$.
b. Determine the value of B and the amount of inventory on the conveyor. Estimate the inventory on the conveyor by summing the $H_i(n)$ values for all i and n.
c. Suppose the unload sequence can be changed to $\{f_4(n)\} = (0,-a,-b)$, where $a + b = 4$. What integer values of a and b will minimize the value of B?
d. Suppose the work performed by the first and third loading stations is combined to yield two loading stations and one unloading station with the following sequences:

$\{f_1(n)\} = (1,0,1)$; $\{f_2(n)\} = (0,2,0)$; and $\{f_3(n)\} = (0,-4,0)$

What, if anything, will be the impact on the value of B?

**12.66.** Cartons are to be conveyed over a distance of 300 ft using a combination of roller and belt conveyors. Movement over the first 200 ft will be via a flat belt driven roller conveyor with WBR dimensions of 33 in. rollers on 9 in. centers, and a speed of 250 fpm. Each tote box is 24 in. long and weighs 35 lb loaded. The spacing between tote boxes on the roller conveyor is 4 in. The tote boxes are transferred automatically from the roller conveyor onto a 100 ft. long roller supported belt conveyor operating at a speed of 300 fpm. The rollers are also on 9-in. centers and the WBR dimension is also 33 in. For the roller conveyor and the belt conveyor, determine the values of LF, BV, and L.

**12.67.** A chain driven roller conveyor is used to move pallet loads 150 ft. from storage to shipping. The pallet is 48 in. × 36 in. For smooth movement over the rollers, the pallet is oriented so that the stringer is aligned with the rollers (i.e., the stringer is perpendicular to the conveyor frames and parallel to the roller length.) The WBR dimension is 39 in. Each pallet load of product weighs 1500 lb. The spacing between pallets is 24 in. Rollers have a 9 in. spacing. The conveyor speed is 60 fpm. As the pallets near shipping, they are transferred directly from the roller conveyor to a roller supported belt conveyor.
a. What are the values of BV, LF, and FF for the roller conveyor?
b. What is the weight of the load on the roller conveyor?
c. If a 36-in. spacing is desired between pallets on the belt conveyor, what must be its speed?

**12.68.** A V-belt driven roller conveyor is used to convey cartons from the order picking area in a warehouse to the order accumulation and packing area in the warehouse. The conveyor is 350 ft long. Rollers are 2.5 in. in diameter and are spaced on 7.5 in. centers. A WBR dimension of 30 in. is used. A variety of cartons are placed on the conveyor. The cartons are equally likely to be 12 in., 18 in., 24 in., 30 in., or 36 in. long; they are equally likely to weigh 5 lb, 10 lb, 15 lb, 20 lb, or 25 lb. The length and the weight are statistically independent. The minimum design clearance between cartons is 6 in.

a. Determine the BV, LF, and FF values.

b. Determine the average load on the conveyor.

c. Determine the "worst case" load on the conveyor.

12.69. A flat belt driven roller conveyor is used to convey tote boxes of material from the receiving area in a warehouse to the storage area. The conveyor consists of two sections, 175 and 125 ft long. Rollers are on 6-in. centers. A WBR dimension of 27 in. is used. The tote boxes are 24 in. long and are equally likely to weigh 10 lb, 20 lb, 30 lb, 40 lb, or 50 lb. The design clearance between tote boxes on the first conveyor (i.e., the 175-ft section is 6 in.

a. The first conveyor operates at a speed of 75 fpm. What should be the speed of the second conveyor if the spacing between tote boxes is to be 8 in.

b. What would be the horsepower requirement for the longest conveyor based on an average load per tote box? Would you recommend this horsepower be used? If not, what loading condition would you use to specify the horsepower requirement?

12.70. In a distribution center, order picking occurs in the following way: Cartons are picked at various points and placed on a belt conveyor for delivery to shipping, where orders are accumulated. The belt conveyor is roller supported with rollers on 8-in. spacing. Each roller weighs 1.0 lb. The length of the conveyor is 200 ft. The WBR dimension is 23 in. A conveyor speed of 120 fpm is used.

The conveyor can be divided into four zones, each 50 ft long, since four order pickers are responsible for placing cartons on the conveyor from different zones in the distribution center.

The weights and lengths of cartons placed on the conveyor from the various zones are as follows:

| Zone | Carton Weight | Carton Length |
|------|---------------|---------------|
| 1 | 10 lb | 18 in. |
| 2 | 20 | 24 |
| 3 | 35 | 36 |
| 4 | 50 | 42 |

For purposes of computing the horsepower, the following number of cartons of each type can be assumed to be on the conveyor in each zone.

| Zone | 10 lb. | 20 lb. | 35 lb. | 50 lb. |
|------|--------|--------|--------|--------|
| 1 | 5 | — | — | — |
| 2 | 5 | 4 | — | — |
| 3 | 5 | 4 | 4 | — |
| 4 | 5 | 4 | 4 | 4 |

a. If a single drive is used to power the conveyor, what load should be used to compute horsepower?

b. If a single drive is used for zones 1 and 2, a single drive is used for zone 3, and a single drive is used for zone 4, what load should be used to compute horsepower for zone 4?

c. If cartons are uniformly distributed over the conveyor, in zone 4 what will be the average clearance between cartons?

**12.71.** Two conveyors having 6-in. RC values are used to transport cartons horizontally 150 ft in a distribution center from receiving to storage. The first conveyor is a 100-ft-long roller supported belt conveyor with a speed of 250 fpm. Cartons are fed directly from the belt conveyor to a belt driven roller conveyor. Cartons weigh 35 lb each, they are 30 in. long. The clearance between cartons on the first conveyor is 36 in.

  **a.** It is desired to establish the speed of the second conveyor that will result in a spacing of 10 in. between cartons on the second conveyor. What must be the speed of the second conveyor to achieve the desired clearance?

  **b.** Suppose the WBR dimensions for the conveyors are 33 in. What is the horsepower requirement for the first conveyor?

  **c.** Given the information in part (b), what is the horsepower requirement for the second conveyor? Assume the second conveyor is a 50-ft-long belt driven roller conveyor with the same WBR and RC values as the first conveyor.

**12.72.** Two roller supported belt conveyors are used to convey tote boxes of material 250 ft from the assembly department to the packing department. The first conveyor is 150 ft long and has a speed of 300 fpm; the second is 100 ft long and has a speed to be determined. The tote boxes are 24 in. in length. Rollers are 2.5 in. in diameter and are spaced on 10-in. centers. The WBR dimension is 27 in. The design clearance between tote boxes on the first conveyor is 12 in. It is reasonable to assume that the weight of consecutive tote boxes on the conveyor is statistically independent. The weight of the loaded tote boxes varies according to the following distribution.

| *Weight* (lb) | *Percentage* |
|---|---|
| 10 | 10 |
| 20 | 20 |
| 30 | 40 |
| 40 | 20 |
| 50 | 10 |

  **a.** Tote boxes are conveyed from the 150-ft-long conveyor directly to the 100-ft-long conveyor. What should be the speed of the second conveyor in order for the spacing between tote boxes to be 6 in.?

  **b.** Determine the lightest, average, and heaviest load on the first conveyor.

**12.73.** Tote boxes are conveyed over a distance of 200 ft using a combination of two 100-ft-long belt conveyors. Both are roller supported using 2.5 in. rollers weighing 5 lbs. each and installed on 6-in. centers. The WBR dimension is 33 in. for both conveyers. A conveyor speed of 150 fpm is used on the first conveyor, which feeds the tote boxes onto the second conveyor operating at a speed of 250 fpm. Each tote box is 20 in. long and weighs 40 lbs loaded. The spacing between tote boxes on the first conveyor is 6 in.

  **a.** Determine the load on the first conveyor.

  **b.** Determine the load on the second conveyor.

  **c.** What should be the maximum roller center for stability of the tote boxes?

**12.74.** Cartons arrive at a workstation via a roller conveyor spur. The arrivals are Poisson distributed with a mean of 0.5/min. The time required to process the carton at the work station is exponentially distributed with a mean of 0.75 min. How long must the accumulation line be such that it will not be full more than 5% of the time? (*Note*: When the accumulation line is full, cartons are diverted to a separate accumulation area.)

**12.75.** A conveyor delivers parts to an inspection station, with a part arriving every 15 sec. The time required to inspect a part is exponentially distributed with a mean of 10 sec. On the average, how many parts will be waiting to be inspected?

**12.76.** Parts arrive for packaging at a Poisson rate of 20/hr. The time required for packaging is normally distributed with a mean of 3 min and a standard deviation of 1 min. If the system operates as an Erlang loss system, how many packaging stations should be provided in order to have no greater than 5% loss?

**12.77.** Unit loads arrive randomly at a quality control (QC) station. An average of 10 unit loads arrive per hour at a Poisson rate. The time required to perform the QC check is normally distributed with a mean of 5 min and a standard deviation of 1 min. On the average, how long will a unit load have to wait at the QC station before the inspection begins?

**12.78.** Cartons arrive at a workstation at a Poisson rate of 10/min. The time required to process a carton is exponentially distributed with a mean of 4 sec. On the average, how many cartons will be waiting for processing at the work station?

**12.79.** Unit loads arrive randomly at a quality control station. Two inspectors are located at the station; the inspection time is exponentially distributed with a mean of 5 min. Loads arrive at a Poisson rate of 20/hr.
   **a.** What is the probability of both inspectors being idle?
   **b.** What is the average number of units loads waiting to be inspected?

**12.80.** An accumulation conveyor is to be provided at a work station. When the conveyor is full, parts are diverted to another area for processing. Parts arrive at a Poisson rate of 1.0/min. The time required to process a part at the work station is exponentially distributed with a mean of 0.80 min per part. It is desired to provide an accumulation line sufficiently long such that less than 5% of the arriving parts will be diverted to another area of processing. What is the minimum number of *waiting* spaces that will satisfy the objective?

**12.81.** A tote box of parts arrives at an inspection station. The inspector selects a random number of parts for inspection; they are inspected individually. The tote boxes arrive at a Poisson rate of 6/min. The number of parts selected for inspection is equally likely to be 1, 2, 3, or 4. The inspection time per part is exponentially distributed with a mean of 2.5 sec. On the average, how many seconds will parts be waiting to be inspected? Recall, $\text{Var}(X) = E(X^2) - [E(X)]^2$.

**12.82.** A tote box of parts arrives at an inspection station. The inspector selects a random number of parts for inspection; they are inspected individually. Suppose tote boxes arrive at a Poisson rate of 6/min. Also, suppose the number of parts selected for inspection is distributed as follows:

| Number of Parts | Probability |
|---|---|
| 1 | 0.2 |
| 2 | 0.4 |
| 3 | 0.3 |
| 4 | 0.1 |

Furthermore, suppose the inspection time per part is exponentially distributed with a mean of 2.5 sec. On the average, how many parts will be waiting to be inspected?

**12.83.** Consider a queueing situation involving a single server. Suppose service time is exponentially distributed with an expected value of 6 min/customer.
   **a.** Suppose arrivals occur in a Poisson fashion and there is unlimited waiting capacity for the system. Determine the maximum arrival rate that will be acceptable if *L* can be no greater than 4.
   **b.** Suppose arrivals occur deterministically. If it is desired that *L* be no greater than 4.0, what is the maximum arrival rate for the customers?
   **c.** Suppose arrivals occur in a Poisson fashion, but with the folllowing pattern.

| #Customers in System | Arrival Rate |
|---|---|
| 0 | 20 |
| 1 | 20 |
| 2 | 20 |
| 3 | 0 |

What will be the value of $L$?

**12.84.** Consider a queueing situation involving a single server. Service time is exponentially distributed with an expected value of: 6 min/customer when there is one customer in the system; 5 min when there are two in the system; 4 min when there are three in the system; 3 min when there are four in the system; and 2 min/customer when there are five in the system.

Suppose arrivals occur in a Poisson fashion at an average rate of 20/hr when no more than one customer is in the system. If one customer is waiting for service, then the arrival rate drops to a level of 15/hr. If two are waiting, then arrivals occur at a rate of 10/hr; if three are waiting, then the arrival rate is 5/hr; if four are waiting, then arrivals no longer occur.

a. Determine the probability of an idle server.

b. If the probability of an idle server equals 0.1, for the arrival and service rates given above, what is the probability of exactly one customer in the system?

**12.85.** Consider a distribution center that uses a fleet of six battery powered automated guided vehicles for material delivery. Due to the wide variety of demands placed on an AGV, the life of a battery, before recharging is required, is a random variable. Suppose the time from recharging a battery for an AGV until it requires recharging again is exponentially distributed with a mean value of 10 hr. Also, suppose the AGV is idle during battery recharging. Finally, suppose the time required to recharge a battery is also exponentially distributed with a mean of 2 hr. There are two battery recharging machines, and the distribution center operates 24 hr/day and 7 days/week.

a. Determine the probability of at least one battery recharging machine being idle.

b. Determine the average number of AGVs *waiting* to be recharged. (Don't include those being recharged.)

**12.86.** Consider a manufacturing situation involving the use of five robots. There is one robot attendant who changes end-of-arm tooling for the robots, programs the robots, and performs setup changes for the robots. The time between requirements for an attendant for an individual robot is exponentially distributed with an average value of 2 hr. The time required to tend to the robot is also exponentially distributed and has an average value of 0.40 hr.

a. What is the probability of an idle attendant?

b. Determine the average number of robots waiting on the attendant to begin performing a service.

**12.87.** Draw a flow chart of a simulation experiment to determine how many receiving docks should be provided at a distribution center. Assume three kinds of receipts occur: small parcel delivery services (e.g., Federal Express, UPS, U.S. Postal Service); commercial carrier (J.B. Hunt, Schneider, Yellow Freight); and private carriers (your firm's and suppliers' trucks). Because of the differences in dock design needed to accommodate small parcel delivery (SPD) trucks, over-the-road (OTR) trailers, and your firm's (OWN) delivery trucks, three different types of customers exist. The simulation is to examine the impact of having general purpose docks versus having three different categories of docks.

**12.88.** Given the situation in Problem 12.87, suppose the time each category of customer occupies a dock space has the following distribution:

| Time at the Dock (mins) | Frequency (%) | | OWN |
|---|---|---|---|
| | SPD | OTR | |
| 0–4 | 60 | 0 | 10 |
| 5–9 | 30 | 10 | 20 |
| 10–14 | 10 | 20 | 35 |
| 15–19 | | 40 | 25 |
| 20–24 | | 20 | 10 |
| 25–29 | | 10 | |

Can theoretical distributions be used in place of the data? If so, what distributions appear to be likely candidates? How should you decide which theoretical distributions to use?

12.89. Consider the situation described in Problem 12.87. Suppose the times (min) between consecutive arrivals of trucks of each category are given by the following distributions:

| Time Between Arrivals (min) | Frequency (%) | | OWN |
|---|---|---|---|
| | SPD | OTR | |
| 0–4 | 22 | 0 | 0 |
| 5–9 | 17 | 0 | 0 |
| 10–14 | 14 | 5 | 0 |
| 15–19 | 12 | 10 | 0 |
| 20–24 | 10 | 20 | 10 |
| 25–29 | 8 | 30 | 20 |
| 30–34 | 6 | 15 | 40 |
| 35–39 | 4 | 10 | 20 |
| 40–44 | 3 | 5 | 10 |
| 45–49 | 2 | 5 | 0 |
| 50–54 | 1 | 0 | 0 |
| 55–59 | 1 | 0 | 0 |
| >59 | 0 | 0 | 0 |

Can theoretical distributions be used in place of the data? If so, what distributions appear to be likely candidates? How should you decide which theoretical distributions to use?

12.90. Given the information in Problems 12.87, 88, and 89, perform a simulation to determine the number of docks required to provide the following level of service: The probability an arriving truck has to wait for a dock is no greater than 0.15. Consider the following design alternatives: designated docks for each category of truck and undesignated docks. In the latter case, any truck can use any dock. What is the difference in the total number of docks required to provide the service level specified? Are there other design alternatives that should be considered?

12.91. Suppose the data provided in Problems 12.88 and 12.89 were obtained by studying for 15 working days the arrival patterns during the 8-hr work shift of arriving trucks and the "dwell times" at the docks for each category of truck. What concerns might you have about the data that were collected? Suppose the distribution center is used to store products that are highly seasonal or have a strong cyclic behavior during the year, during a month, during a week, and during a day.

12.92. A facility has two bridge cranes operating on a single bridge crane runway. To increase the amount of crane capacity, it has been suggested that a third bridge crane be placed on the runway. Given it is structurally feasible, describe the steps to use in developing and using a simulation model to determine the desirability of installing a third bridge crane.

12.93. Consider the facility location models and layout models in Sections 12.2, 12.3, and 12.4. When might simulation be used to aid in facility location and layout decisions?

**12.94.** Consider the storage models in Section 12.5. How might simulation be used in actually designing a warehouse based on the results from the models?

**12.95.** Consider the AS/RS and order picking models in Section 12.6 and 12.7. How might simulation be used in designing an AS/RS or order picking system? What role, if any, would the models presented in Sections 12.6 and 12.7 play if simulation is used?

**12.96.** Consider the conveyor models presented in Section 12.8. What role might simulation play in such analyses? When should simulation be used, rather than the deterministic models?

**12.97.** Perform a literature search (paying particular attention to the Winter Simulation Conference Proceedings); identify and describe simulation software not cited in the chapter that can be used in facilities planning.

**12.98.** Perform a literature search and describe five facilities planning simulation applications for a distribution center or warehouse not referenced in the text. (See the bibliography at the end of Chapter 9 for candidate magazines and journals.)

**12.99.** Perform a literature search and describe five facilities planning simulation applications for a manufacturing facility not referenced in the text. (See the bibliography at the end of Chapter 10 for candidate magazines and journals.)

# Part Five

## EVALUATING, SELECTING, PREPARING, PRESENTING, IMPLEMENTING, AND MAINTAINING

# 13

## *EVALUATING AND SELECTING THE FACILITIES PLAN*

## *13.1* **INTRODUCTION**

The preceding discussion focused on defining facilities requirements (Part 1) and developing alternative facilities plans (Parts 1, 3, and 4). Part V addresses the evaluation, selection, preparation, presentation, implementation, and maintenance of the facilities plan. In this chapter, evaluation and selection of the facilities plan are considered.

As we noted in the early chapters of the book, the process of developing facilities plans contains elements of both art and science. The artist's dependence on *creativity, synthesis,* and *style* combined with the scientist's use of *analysis, reduction,* and *deduction* is the essence of facilities planning.

Analysis is a dissection process; synthesis is a combining or creating process. Through the application of quantitative models, including computer-aided layout models, the facilities planner explores many different solution spaces for the facilities plan; through the application of synthesis, the facilities planner combines the quantitative and qualitative aspects of the plan into a set of alternative facilities plans to be evaluated. Throughout, the facilities planner is engaged in a design process.

Recall that the design process was depicted in Chapter 1 as a six-step process:

1. Define the problem
2. Analyze the problem
3. Generate alternative solutions
4. Evaluate the alternatives
5. Select the preferred solution
6. Implement the preferred solution

It is recommended that the discussion on developing layout alternatives in Section 7.5 be read again before proceeding further. Special attention should be given to the treatise on style in design. In some sense the process of developing alternative facilities plans is best described as a "groping" process. At times it seems to be a search through a maze with few clear directions as to the way out and few, if any, indications whether progress is being made.

This statement is not intended to discourage, but rather as preparation for the realities of facilities planning. In particular, it should be noted that the six-step process is probably a simplistic representation of the facilities planning process for any but the simplest problems. However, we do believe it is very valuable to follow the six steps when possible to do so. At the same time, it is recognized that the "real" design process is often an iterative one, with many steps. Also, it is typically the case that the process is applied to subsets of the design problem.

Perhaps a more accurate representation of the design process would be similar to that given below.

1. Define the design problem and review with management to ensure the scope, objectives, schedule, and budget for the design effort are acceptable.

2. If management accepts the problem definition, then proceed to step 3; otherwise, redefine the problem and repeat the process until acceptance is obtained.

3. Analyze the problem; develop the material flow requirements and the database to be used as the foundation for the design.

4. Review the database with management and operating personnel to ensure acceptance of designs based on the database.

5. If the database review reveals changes are needed in the definition of the problem, adjust the definition; if the database review reveals gaps exist in the database, gather additional data.

6. Break up the problem into subproblems; often, departmental boundaries will serve as a logical basis for subdivision. As an example, if the facility being planned is a warehouse, the problem might be divided into receiving, moving to storage, small parts storage, unit load storage, order picking/retrieval, order accumulation, packing, moving to shipping, and shipping.

7. Generate alternative solutions to each subproblem; macrodesigns or concepts are developed focusing on generic categories of alternatives. Included in the set of alternatives is an improved present method, when an existing system exists.

8. Where appropriate, develop mathematical and/or simulation models of the alternatives to test their feasibility and perform sensitivity tests; if the data collected previously are not sufficient to support the modeling effort, collect additional data.

9. Review the feasible alternatives with operating personnel to verify the reasonability of the concepts; modify the alternatives as appropriate. Often, additional alternatives will be developed at this point; if so, return to step 8.

10. Evaluate the alternatives in terms of the criteria identified in step 1, focusing on tangible and intangible areas of concern. Rank the alternatives.

11. Aggregate the solutions to the subproblem. Pay particular attention to the interfaces to ensure the integration of individual solutions results in a reasonable

overall solution. Where it can be justified, develop a simulation model of the overall system to verify it will meet the throughput requirements for the system and that bottlenecks do not develop at the interfaces.

12. In many cases, alternative solutions will exist for subproblems that are close contenders for selection. If so, incorporate them into the aggregated design to obtain alternative facility plans. (It might occur that the total system will perform best by using one or more solutions to subproblems that are not best when considered alone.)

13. Evaluate alternative aggregate facility plans. Determine the economic impact of each alternative, focusing on capital investment requirements and operating expenses for budget purposes. Perform sensitivity analyses and breakeven analyses, focusing on changes in production volumes, product mix, and technological changes.

14. Rank the alternatives and present the ranking to management for selection of the preferred alternative; obtain management approval to proceed with the detailed design for the preferred alternative.

15. Develop a detailed design for the preferred facility plan. Determine equipment locations; locate columns, doors, aisles, walls; determine floor loadings, facility service requirements, and so on. At this point the architect/engineering/construction plans are initiated. Utilize project management techniques to control the project through implementation.

16. Develop detailed bid specifications for equipment and facilities construction; develop hardware and software specifications for the computer control systems.

17. Send bid specifications to qualified suppliers.

18. Receive and review bid packages; obtain management participation in the selection of the suppliers/contractors; perform any required modifications to the facility plan.

19. Perform the required construction, install the equipment, train personnel, debug the system, and turn over the system to the operating personnel.

20. Perform audits of the facility plan. When changes in condition and/or requirements justify an additional facilities planning effort, return to step 1.

This chapter is concerned with the fourth and fifth steps of the six-step design process: the evaluation of alternatives and selection of a facilities plan. It is assumed that steps 1 through 3 have been completed and that step 6 will follow the completion of the fourth and fifth steps. The input into the fourth step should be a collection of feasible facilities plans. Unfortunately, there exists a tendency to pre-empt the evaluation of alternatives by prejudging feasible plans. Such an approach often eliminates the plan that may have been most promising. To guard against the possibility, a broad cross-section of feasible plans should be included in the initial evaluation and selection steps. For example, the initial evaluation and selection of solutions to a problem involving the transport of parts to an assembly line may include a floor truck alternative, a trolley conveyor alternative, a belt conveyor alternative, a roller conveyor alternative, and an AGVS alternative. To eliminate any of the generic alternatives prior to the initial evaluation and selection might bias the solution. Conversely, to include in the initial evaluation and selection several specific models and brands of floor trucks, trolley conveyors, belt conveyors, roller conveyors, and AGVS would probably be a poor use of time. Detailed data and competitive bidding are not required to perform an initial screening. Once the initial screening is made, specific brands and vendors may be evaluated to determine the most suitable solution.

# *13.2*   EVALUATING FACILITIES PLANS

The evaluation process includes the assessment of each alternative in terms of the criteria identified previously. *Evaluating* is not the same as *selecting!* For example, if the least-cost alternative is to be selected, then the costs for each alternative must be evaluated.

Hopefully, by separating the evaluation of alternatives from selection of the preferred alternative, the selection will be made more objectively. If one is forced to formally evaluate the performance of each alternative, it is less likely that someone's "favorite" (based on subjective factors) will be selected without understanding the impact of the selection.

If the criteria are easily quantified, the process of evaluation is easier to perform. However, it is typically the case in facilities planning that both quantitative and qualitative considerations are employed in evaluating alternatives. Few would disagree with the claim that the evaluation and selection steps are the most difficult steps of the six step design process.

Among the techniques used in evaluating alternative facilities plan are the following:

1.   Listing the positive and negative aspects of each alternative.
2.   Ranking the performance of each alternative against each of several enumerated criteria.
3.   Weighted factor comparison of the alternatives by assigning a numerical weight to each criteria (factor), ranking numerically each alternative against each criteria, and summing the weighted rankings over all criteria to obtain a total weighted factor for each alternative.
4.   Determining the economic performance of each alternative over a specified planning horizon.

## Listing Advantages and Disadvantages

The listing of advantages and disadvantages of each alternative provides a simple way to evaluate facilities plans. Of the evaluation techniques listed, it is probably the easiest to perform. However, it is also difficult to obtain an accurate, balanced, and objective evaluation of alternative facilities plans by simply listing the positive and negative aspects of each.

## Ranking

Ranking the alternatives has the benefit of requiring that all alternatives be compared against a common set of factors. Furthermore, it forces an explicit consideration of the several factors that will influence the selection decision. However, there is no guarantee that all important factors will be considered. Additionally, the ranking process may yield too much information; it may overwhelm the decision makers and confuse the selection process. After using the ranking procedure, the results must be integrated to allow a selection to be made. The integration can be either explicitly or implicitly performed; either way, it occurs!

## Weighted Factor Comparison

Similar to the prioritization matrix described in Chaper 3, the weighted factor comparison method provides an explicit method for integrating the rankings. Numerical values or weights are assigned proportionally to each factor based on their degrees of importance. A numerical score is then assigned to each alternative based on its performance against a particular factor. The scores are multiplied by the weights and the products are summed over all factors to obtain a total weighted score. Although there are some very obvious scaling problems associated with the technique, it is quite popular.

Both the ranking approach and the weighted factor comparison approach involve a comparison of facilities planning alternatives for each factor. In an attempt to make the comparison as objective as possible, a paired comparison is recommended for each factor.

To illustrate the approach, suppose there are four facilities planning alternatives (W, X, Y, and Z). Further, suppose the following preferences are obtained by comparing the alternatives two at a time for a particular factor.

$$W < Y \quad X > Y$$
$$W < X \quad X < Z$$
$$W < Z \quad Y < Z$$

where $W < Y$ means W ranks lower than Y and $X > Y$ means X ranks higher than Y. Combining the paired comparisons yields the following ranking.

$$Z > X > Y > W$$

for the factor considered. For some other factor an entirely different ranking would probably occur.

By performing the paired comparisons, inconsistencies in the rankings will be revealed. For example, suppose in the previous example a ranking of $Y > Z$ had occurred. Obviously, it is inconsistent to state that

$$X > Y \qquad Y > Z \qquad Z > X$$

Even though the ranking procedure indicates preferences, it does not indicate the strength of the preference. For example, is the ranking

$$Z >>> X >> Y > W$$

or

$$Z > X >> Y >>> W$$

Suppose values could be accurately assigned to a factor in direct proportion to its preference. Then $W = 1$, $Y = 3$, $X = 6$, and $Z = 10$ would receive the same ranking $(Z > X > Y > W)$ as would $W = 1$, $Y = 2$, $X = 9$, and $Z = 10$.

In using ranking and weighted-factor comparison approaches, care is required to avoid the halo effect. The halo effect is a phenomenon that occurs when a high ranking on one factor carries over and influences the ranking on other factors. If one alternative clearly dominates all other alternatives on all or almost all factors, then it is likely that the halo effect is present.

Among the factors that are considered in using the ranking procedure and weighted factor comparison are the following:

1. Initial investment required
2. Annual operating costs
3. Return on investment (ROI)
4. Payback period
5. Flexibility or ease of changing or rearranging the installed system
6. Integration with and ability to serve the process operations
7. Versatility and adaptability of the system to accommodate fluctuations in products, quantities, and delivery times (ease of changing or rearranging the installed system)
8. Ease of future expansion
9. Limitations imposed by handling methods on the flexibility and ease of expansion of the layout and/or buildings
10. Space utilization
11. Safety and housekeeping
12. Working conditions and employee satisfaction
13. Ease of supervision and control
14. Availability of trained personnel
15. Frequency and seriousness of potential breakdowns
16. Ease of maintenance and rapidity of repair
17. Volume of spare parts required to stock
18. Interruption or disruption of production and related confusion during the installation period
19. Quality of product and risk of damage to materials
20. Ability to pace, or keep pace with, production requirements
21. Effect on in-process time
22. Personnel problems—available workers with proper skills, training capability, disposition of redundant workers, job description changes, union contracts, or work practices
23. Availability of equipment needed
24. Availability of repair parts
25. Tie-in with scheduling, inventory control, paper work
26. Effect of natural conditions—land, weather, sun temperature
27. Compatibility with the operating organization
28. Potential delays from required synchronization and peak loads
29. Supporting services required
30. Integration with other facilities

31. Tie-in with external transportation
32. Time required to get into operation—installation, training, and debugging
33. Degree of automation
34. Software requirements
35. Promotional or public relations value

A form that is useful in performing a weighted factor comparison is given in Figure 13.1. The factors that are relevant for a particular evaluation and the weights of these factors should be listed in the first two columns. Once the factors and weights have been determined, each of the alternatives should be rated (Rt) and the score

**WEIGHTED FACTOR COMPARISON FORM**

Company _____ Prepared by _____ Date _____
Facility Plan for _____
Sheet _____ Of _____

**Alternatives**

| Factor | Weight | A | | B | | C | | D | | E | |
|--------|--------|-----|-----|-----|-----|-----|-----|-----|-----|-----|-----|
| | | Rt. | Sc. | Rt. | Sc. | Rt. | Sc. | Rt. | Sc. | Rt. | Sc. |
| 1. | | | | | | | | | | | |
| 2. | | | | | | | | | | | |
| 3. | | | | | | | | | | | |
| 4. | | | | | | | | | | | |
| 5. | | | | | | | | | | | |
| 6. | | | | | | | | | | | |
| 7. | | | | | | | | | | | |
| 8. | | | | | | | | | | | |
| 9. | | | | | | | | | | | |
| 10. | | | | | | | | | | | |
| 11. | | | | | | | | | | | |
| 12. | | | | | | | | | | | |
| 13. | | | | | | | | | | | |
| 14. | | | | | | | | | | | |
| 15. | | | | | | | | | | | |
| 16. | | | | | | | | | | | |
| 17. | | | | | | | | | | | |
| 18. | | | | | | | | | | | |
| 19. | | | | | | | | | | | |
| 20. | | | | | | | | | | | |
| Totals | | | | | | | | | | | |

**Figure 13.1** Weighted factor comparison form.

(Sc) determined for each factor by multiplying the rating times the weight. The overall evaluation of each alternative is obtained by adding the scores for all the factors.

## Economic Comparison

Justifying facilities planning investments is of intense interest. Because some have had difficulty justifying their favorite technologies, discounted cash flow (DCF) methods have been criticized and the adoption of new justification approaches have been advocated. Such reactions are akin to shooting the messenger because you don't like the message. It isn't the justification methods that are inadequate, it is the way they have been applied!

In the early 1900s, Henry Ford noted, "If you need a new machine and don't buy it, you pay for it without ever getting it." Cost savings opportunities are often missed due to poorly executed economic justifications.

In performing financial justifications of facilities plans, the following Systematic Economic Analysis Technique (SEAT) can be used to justify investments *that deserve to be justified:*

- specify the feasible alternatives to be compared;
- define the planning horizon to be used;
- estimate the cash flows for each alternative;
- specify the discount rate to be used;
- compare the alternatives using a discounted cash flow (DCF) method;
- perform sensitivity analyses; and
- select the preferred alternative. [22]

### *Specifying the Feasible Alternatives*

Specifying the set of feasible alternatives to be compared is the most important step in justifying a facilities plan. For, if an alternative is not included in the comparison, it will not be selected! Likewise, it is fruitless to include in the comparison those alternatives that will not meet the facilities planning requirements.

A frequently used alternative is the "do-nothing" alternative, since it often serves as the baseline against which the "do-something" alternatives are compared. However, no business situations stand still. Hence, doing nothing is seldom feasible. For while your firm is doing nothing, your competition will be doing something. "Business as usual," if continued, is a failing business strategy. Yet, situations will arise in which "standing pat" is preferred to making a capital investment in a new or improved facility.

An objective assessment of "doing nothing" must include both cost and revenue impacts. Too often, a myopic approach is taken and only those costs that arise *inside* a distribution facility or manufacturing plant are considered. Since investments in facilities often enhance revenues, due to cost reductions, increased quality, and more timely responses to customer demands, marketing representatives should be invited to assess the impact of doing nothing. Many have found facilities planning investments easier to justify with marketing on their side.

In addition to focusing on the do-nothing alternative, a range of alternatives should be considered. Frequently, attempts to justify the acquisition of the ultimate in technological sophistication occur by comparing its economic performance against a very poor current method. No consideration is given to intermediate alternatives, since one might be the better economic choice.

Widespread biasing of justifications toward particular solutions led the treasurer of a Fortune 100 company to admit that he did not trust engineers' justifications. When "important" decisions had to be made, he depended on his intuition and judgment of what was best for the firm, rather than the numbers submitted by the engineers! Care must be taken to ensure that technological preferences and egos do not cloud one's judgment of what is best for the firm over the planning horizon.

A related phenomenon is the engineers' attempts to thwart the use of "size gates" in the approval process, where investments less than a particular size require plant manager approval only, those within another range of sizes require divisional approval, and those above a specified upper limit require corporate approval. When companies delegate approval authority on the basis of the size of the investment, engineers often subdivide the investment into smaller pieces in order to obtain approval at lower levels. Rather than propose an integrated system, several "islands of automation" are defined and one or more are proposed each year, depending on the investment requirements and the size gates used.

Making piecemeal investments can postpone the benefits of an integrated system for years. Also, in trying to "eat the elephant a bite at a time," one or more smaller "bites" might not be sold. Instead of proposing small increments, many firms believe the best strategy is to "design the whole, sell or justify the whole, and then implement the pieces." In particular, the total system is justified and sold to management. When capital is limited or size gates are in use, a phased implementation plan is used and individual components of the total system are implemented over time [9].

## *Defining the Planning Horizon*

The second step in the SEAT approach is the definition of the planning horizon to be used. (For those unfamiliar with the term, the planning horizon is the period of time over which the economic performance of an investment will be measured and evaluated.)

In general, U.S. firms use too small a planning horizon, often less than five years. Japanese firms, on the other hand, typically think in terms of 10 to 20 years. One manifestation of short-sighted management is the use of short planning horizons.

*The planning horizon defines the width of a window through which we will look in evaluating each investment alternative.* An important economic justification principle is that the same window be used in evaluating all alternatives. Hence, a 3-year window should not be used for one alternative and a 10-year window used for another alternative. If a 10-year window is allowed, then each alternative should be evaluated over the 10-year period; if alternatives exist that are not expected to last 10 years, then explicit consideration should be given concerning the intervening investment decisions over the planning horizon.

The planning horizon should be distinguished from the working life of equipment and the depreciable life of equipment, since it might have no relationship to the other two periods of time. The planning horizon is simply the time frame to be used in comparing the alternatives and should realistically represent the period of time over which reasonably accurate cash flow estimates can be provided.

Some firms use standard planning horizons, which, as noted above, are too short. This was the case for a major U.S. textile firm which was losing market share. A major modernization was proposed. However, to realize the full benefits from the investment, the system had to be debugged, all personnel had to be trained, and lost customers had to be regained. It was impossible to accomplish this within the 2-year period required. Faced with the economic consequences of maintaining the status quo, the firm elected to adopt a longer planning horizon.

Planning horizons also can be too long; a balance is needed. Based on our experience in justifying material handling investments, we prefer at least a 10-year planning horizon [10].

## Estimating the Cash Flows

The third step in justifying a facilities plan is estimating the cash flow profiles for each alternative. What is a cash flow profile? Year-by-year estimates of money received (receipts, revenues, income, benefits or savings) and money spent (disbursements, expenses or costs) constitute a cash flow profile. Both the obvious and the not-so-obvious cash flows should be included. For example, investments in new or improved facilities often reduce inventory, increase quality, reduce space, increase flexibility, reduce lead time, increase throughput, reduce scrap, and increase safety. Though real, the economic benefits of each are difficult to quantify. Hence, many call them *intangible benefits* and do not include them in economic justifications.

Robert Kaplan, Harvard Business School accounting professor, noted, "Although intangible benefits are difficult to quantify, there is no reason to value them at zero in a capital expenditure analysis. Zero is, after all, no less arbitrary than any other number. Conservative accountants who assign zero values to many intangible benefits prefer being precisely wrong to being vaguely right. Managers need not follow their example" [2]. In the end a rough estimate of intangible benefits is better than no estimate at all! Intangibles can be effective competitive weapons in increasing revenues and decreasing operating costs.

Since the investment is for the future not the past, estimates of the cash flows for each alternative should reflect *future* conditions not *past* conditions. Best estimates are needed of the annual costs and incomes anticipated over the planning horizon, including a liquidation or salvage value estimate at the end of the planning horizon. Although it is not necessary to include estimates of cash flows that will be the same for all alternatives, it is important to capture the differences in the alternatives. Figure 13.2 provides a cost determination form that may prove useful in accumulating the annual investment and operating costs for each investment alternative.

A number of different approaches can be used to develop cost estimates. Among the estimation techniques commonly used are the following: guesstimates, historical standards, converting methods, ratio methods, pilot plant approaches, and engineered standards.

A **guesstimate** is an educated guess or estimate based on experience. Commonly, when guesstimates are to be used, several individuals are consulted independently and requested to provide estimates; these are screened and pooled to develop a figure to be used.

**COST DETERMINATION FORM**

Company _____ Prepared by _____ Date _____
Handling System for _____
Sheet _____ Of _____

| | Alternative ____ Year ____ | Alternative ____ Year ____ | Alternative ____ Year ____ | Alternative ____ Year ____ | Alternative ____ Year ____ |
|---|---|---|---|---|---|
| Equipment type | | | | | |
| Model | | | | | |
| Vendor | | | | | |
| Date of quote | | | | | |
| Investment | | | | | |
| Invoice-price | | | | | |
| Installation charge | | | | | |
| Maintenance facilities | | | | | |
| Power facilities | | | | | |
| Alterations to facilities | | | | | |
| Freight charges | | | | | |
| Consulting charges | | | | | |
| Supplies | | | | | |
| Other | | | | | |
| Total Investment Cost | | | | | |
| Operating cost | | | | | |
| Interest on investment | | | | | |
| Taxes | | | | | |
| Insurance | | | | | |
| Supervision | | | | | |
| Clerical | | | | | |
| Maintenance labor | | | | | |
| Operating personnel | | | | | |
| Maintenance supplies | | | | | |
| Power costs | | | | | |
| Other | | | | | |
| Total Operating Cost | | | | | |

**Figure 13.2** Cost determination form.

**Historical standards** are developed through years of experience in designing facilities. The more refined and accurate historical standards are those which are expressed in small "chunks" rather than an aggregated standard. As an illustration, equipment installation costs can be expressed as a percentage of acquisition costs. However, if the standard percentage is an aggregate for a wide variety of equipment, then it will not be as accurate an estimate as one that is developed specifically, for, say, cantilever racks.

The **converting method** is closely related to the use of historical standards. Using the converting method to estimate production space requirements, the present space would be converted to square footage per unit produced. Consequently, if the production rate is to be doubled, it would be estimated that twice as much production space is required. Obviously, the method is subject to error and caution should be exercised when it is used.

**Ratio methods** are related to the converting method and involve the development of ratios of, say, number of rack openings to some other factor than can be measured and predicted for the proposed layout. Examples include number of rack openings per pallet load received; square footage of floor space per rack opening; number of receiving employees per ton of materials received; square footage of aisle space per square footage of production space; annual utility cost per square foot of space; annual maintenance cost per dollar invested in equipment; annual operating cost per dollar invested in equipment; annual inventory carrying cost per dollar invested annually in inventory; annual pilferage cost per dollar invested annually in inventory; total investment cost per rack opening in an automated storage system; conveyor investment cost per foot of conveyor; AGV operating cost per foot of guide path; in-process handling cost per dollar spent in production; and material handling cost per dollar spent in production.

The **pilot plant** approach involves the operation of the proposed system on a significantly reduced scale. As a substitute for the pilot plant approach, scale models can be used to develop reasonable estimates for the "real thing."

**Engineered standards** can be developed by systematically analyzing the situation, identifying cause–effect relationships, and developing prediction equations. Using simulation, work measurement, waiting line, inventory control, production planning, and forecasting methods, as well as other analytical models, the engineer can normally develop accurate estimates of many of the important benefit and cost elements.

An example of the use of **engineered standards** in estimating AS/RS equipment costs was described in Chapter 12. Three components were considered: S/R machine, storage racks, and building. The S/R machine cost was represented as a function of the height of the AS/RS, the weight of the unit load, and the type and location of the control logic. The storage rack cost was given as a function of the dimensions of the unit load, the weight of the unit load, and the height of the rack. The building cost was expressed as a function of building height.

"Ballpark" estimates of equipment costs frequently can be obtained from equipment vendors; however, such estimates should not be interpreted as firm quotes or bids. As an example, the **ratio method** is often used to estimate the acquisition cost of storage rack. Based on estimates obtained from four different firms, the costs per 2500-lb load given in Table 13.1 were obtained for various storage methods. (The costs given include the cost of the pallet and storage rack.)

Table 13.1   *Storage Rack Cost Estimates*

| Storage Method | Cost per Pallet Load (dollars) |
|---|---|
| Bulk storage | $ 6–10 |
| Portable stacking rack | 35–75 |
| Drive-in rack | 35–55 |
| Pallet flow rack | 100–125 |
| Selective pallet rack | 25–35 |
| AS/RS rack | 100–125 |

Acquisition cost alone does not provide an accurate estimate of the cost of storage rack. Freight cost ($3 to $4/100 lb of weight) and installation cost should be estimated as well. These, and other, "add-ons" can substantially increase the cost of installation.

Many who evaluate facilities planning investment alternatives fail to recognize the impact such an investment will have on revenues. Consideration of the competition and the market environment are seldom included in the evaluation, but they should be! Often, it is blindly assumed that the only cash flows associated with replacing existing equipment with state-of-the-art equipment are the costs associated with each alternative. The fact that installing the newest equipment will influence customers to select your product rather than the competitor's product is often overlooked. Too frequently, the adverse economic consequences of the "do nothing" alternative are underestimated. Maintaining the status quo can have a substantial impact on both costs and revenues. If the competition is doing "something" while you are doing "nothing", your firm's revenue stream could be reduced significantly [11].

## Specifying the Discount Rate

The next step in justifying facilities planning investments is specifying the discount rate to be used. Variously called a hurdle rate, interest rate, return on investment, and minimum attractive rate of return (MARR), the discount rate is essential to the discounted cash flow (DCF) methods of economic justification.

Underlying all DCF methods is the recognition that *money has time value!* Since money has time value, you would prefer receiving $100,000 today to receiving $100,000 a year from today. By owning the money one year sooner, you can rent (loan) the money to someone who needs it and charge interest for its use. By receiving the money sooner, you can increase its value over time.

If you choose to invest the money in, say, material handling equipment, then you should charge interest against the investment to reflect the loss of income available to you by "loaning the money" or investing it elsewhere and earning a particular return. Using such an approach, the discount rate reflects the foregone opportunity to invest elsewhere and is called an *opportunity cost* discount rate.

The discount rate is used to convert the mix of cash flows occurring throughout the planning horizon to a meaningful economic measure that can be used rationally to compare facilities planning investment alternatives. Due to its impact on the comparative advantages of the investment alternatives, the discount rate used should reflect the real opportunity cost of alternative uses of the investment capital.

An argument given frequently for higher hurdle rates in the United States is the cost of capital. However, as Kaplan [2] pointed out, over the past 60 years the real cost of capital in America has been less than 8%. He believes that the use of an overly large discount rate is a primary reason American industry has underinvested in capital improvements. If a firm is using an after-tax, inflation-free discount rate greater than 10%, a careful examination should be performed to ascertain its negative impact on the long-term viability of the firm.

In our treatment of the economic justification of facilities planning investments, the discount rate will be referred to as the MARR and its value will be based on the opportunity cost concept. Throughout, we will argue that an investment should not be made if it will not earn a return at least as great as the MARR, since the MARR should be based on the opportunities for alternative investment at comparable risk [12].

## Comparing the Investment Alternatives

Comparing the economic performances of facilities planning investment alternatives is the fifth step in the justification process. Because money has time value the following rules apply when comparing investment alternatives that might involve cash flows of different magnitudes occurring at different points in time: Monies can be added and/or subtracted only at the same point(s) in time; to move a cash flow forward in time by one time period, multiply the magnitude of the cash flow by the quantity $(1 + i)$, where $i$ is the time value of money expressed in decimal form; to move a cash flow backward in time by one time period, divide the magnitude of the cash flow by the quantity $(1 + i)$. The process of multiplying and dividing cash flows by $(1 + i)$ is called *compounding* and *discounting,* respectively. The overall process of converting cash flows to single-sum equivalents (called *present worths* and *future worths*) or equivalent annual series (called *annual worth*) is called a discounted cash flow (DCF) process.

Although many discounted cash flow (DCF) methods exist, it is likely that your employer will have a preferred method of comparison. The most popular DCF methods are the present worth and internal rate of return methods. The present worth method involves discounting all future cash flows to a common point in time called the present; the internal rate of return method involves finding the time value of money that yields a present worth of zero. For detailed treatments of alternative DCF methods, see any of a number of economic justification texts [1], [22].

Using the present worth method, a single-sum equivalent is computed for each alternative; the values obtained reflect the sums of the discounted cash flows over the planning horizon. The alternative having the greatest present worth is deemed the most attractive economically.

A simple formula expresses the equivalence between a present amount, $\mathbf{P}$, and a future amount, $\mathbf{F}$, after $\mathbf{n}$ years, when the annual interest rate is $i$ percent per year (expressed as a fraction). Namely,

$$\mathbf{F} = \mathbf{P}\,(1 + i)^{\mathbf{n}} \tag{13.1}$$

Hence, the compound (or future) amount, $\mathbf{F}$, equivalent to the present amount, $\mathbf{P}$, in $\mathbf{n}$ periods is $\mathbf{P}$ times a factor, $(1 + \mathbf{i})^{\mathbf{n}}$, called the ***compound amount factor*** and denoted by $(\mathbf{F}|\mathbf{P}, \mathbf{i}\%, \mathbf{n})$.

The obvious reciprocal relationship exists between **P** and **F**, as denoted by Equation 13.2.

$$\mathbf{P} = \mathbf{F}\,(1 + \mathbf{i})^{-\mathbf{n}} \tag{13.2}$$

Equation 13.2 says that the present value, **P**, equivalent to the future amount, **F**, **n** years in the future, is **F** times a factor, $(1 + \mathbf{i})^{-\mathbf{n}}$. Called the ***present worth factor*** (or "discounting" factor), it is denoted (**P**|**F**, **i**%, **n**).

In addition to determining the present worth of a single sum that occurs at some time in the future and the future worth of some present amount, we can determine the present worth and/or the future worth of a uniform annual series of cash flows.

To determine the present worth, **P**, of a uniform series of **n** equally spaced cash flows, each of magnitude **A**, the following relationship is used:

$$\mathbf{P} = \mathbf{A}[(1 + \mathbf{i})^{\mathbf{n}} - 1]/[\mathbf{i}(1 + \mathbf{i})^{\mathbf{n}}] \tag{13.3}$$

or

$$\mathbf{P} = \mathbf{A}(\mathbf{P}|\mathbf{A}, \mathbf{i}\%, \mathbf{n})$$

*It is important to recognize that the present worth obtained occurs one period before the first uniform series cash flow.*

To determine the uniform series equivalent, **A**, to a present sum, **P**, employ the following reciprocal relation:

$$\mathbf{A} = \mathbf{P}[\mathbf{i}(1 + \mathbf{i})^{\mathbf{n}}]/[(1 + \mathbf{i})^{\mathbf{n}} - 1] \tag{13.4}$$

or

$$\mathbf{A} = \mathbf{P}(\mathbf{A}|\mathbf{P}, \mathbf{i}\%, \mathbf{n})$$

Multiplying the (**F**|**P**, **i**%, **n**) factor by the (**P**|**A**, **i**%, **n**) factor yields the (**F**|**A**, **i**%, **n**) factor; likewise, the reciprocal of the (**F**|**A**, **i**%, **n**) factor is the (**A**|**F**, **i**%, **n**) factor. Values for each of the six factors are tabulated in Appendix 13.A for various values of **n** and for values of **i** equal to 5%, 10%, 15%, 20%, and 25%; for more extensive tables of the discount factors, see [1] and [22], among others.

# Example 13.1

Find the future value after five years of $1,000 when interest is 10 percent annually. From Table A.2 in Appendix 13.A, the value of the (**F**|**P**, 10%, 5) factor is 1.6105. Therefore,

$$\mathbf{F} = \$1,000\ (\mathbf{F}|\mathbf{P},\ 10\%,\ 5)$$
$$= \$1,000\ (1.6105) = \$1,610.50$$

The answer is $1,610.50

# Example 13.2

Find the present value of $1,610.50 occurring five years from now, when interest is 10% annually. From Table A.2, the value of the (**P**|**F**, 10%, 5) factor is 0.6209. Therefore,

$\mathbf{P} = \$1,610.50\ (\mathbf{P}|\mathbf{F},\ 10\%,\ 5)$
$= \$1,610.50\ (0.6209) = \$999.96$

The answer is actually $1,000, with the difference of $0.04 being due to round-off error in the tables.

## *Example 13.3*

To illustrate the DCF process for a more realistic material handling investment, suppose $1,000,000 is to be invested in a distribution center to improve the material handling system. A planning horizon of 10 years is used, along with an inflation-free, after-tax minimum attractive rate of return of 10%. Let the net annual after-tax cash flow resulting from the investment be a positive $200,000 over the 10-year period, plus a salvage value of $250,000. Hence, the cash flow profile for the investment consists of a negative single sum of $1,000,000 occuring at time zero, a uniform series of $200,000 occuring annually at the end of years 1 through 10, and a single sum of $250,000 occuring at the end of year 10.

   To determine the discounted present worth equivalent for the investment, the following equation can be used:

$\mathbf{P} = -\$1,000,000 + \$200,000(\mathbf{P}|\mathbf{A},\ 10\%,\ 10) + \$250,000(\mathbf{P}|\mathbf{F},\ 10\%,\ 10)$

or, using values from Table A.2:

$\mathbf{P} = -\$1,000,000 + \$200,000(6.1446) + \$250,000(0.3855)$
$\mathbf{P} = \$325,295.$

The discounted present worth for the investment is $325,295, which can be interpreted as the excess above a 10% return on the invested capital.

   The internal rate of return is the interest rate that yields a zero present worth for an investment. By trial and error applications of various interest rates, you will find that the internal rate of return for the example is almost 16.58%. The internal rate of return method requires that each increment of investment capital be justified. Although more tedious to use, when applied correctly, the same recommendation will result using the internal rate of return method as using either the present worth or annual worth methods.

   Other DCF methods yield the same recommendations as the present worth method. However, some do not guarantee to provide the same recommendation as the present worth method—chief among them is the payback period method, which is commonly applied using a 2-year payback requirement. It calls for a determination of the length of time required to fully recover the initial investment without regard to the time value of money. In the case of our example, it would take 5 years to recover the initial investment.

Although it is a simple method to use, beware the payback period method! The alternative that is the first to recover the initial investment is the preferred choice; hence, a $1 million investment promising payback of $500,000 per year for two years and $5,000 per year for the next eight years would be preferred to the alternative described earlier.

   The payback period method is short sighted and ignores the financial performance of an investment beyond its payback period; yet it is probably the most popular method in use among U.S. firms. Why? Because it is easy to compute, use, and explain; it provides a rough measure of the liquidity of a project; and it tends to reflect management's attitude under limited capital conditions. However, due its shortcomings, it is recommended that the payback period only be used as a tie-breaker when DCF methods are used [13].

## *Performing Supplementary Analyses*

The sixth step, performing supplementary analysis, though surely the most neglected, often proves to be the most important step. Engineers often resist using formal justification processes due to the absence of accurate data. The notion of garbage-in, garbage-out comes to mind. While the quality of the recommendation can be no greater than the quality of the inputs to the process, it is also true that supplementary analysis can provide valuable insights concerning the economic viability of an investment alternative *in the absence of perfect information.*

Three kinds of supplementary analyses are available: (1) break-even analysis, (2) sensitivity analysis, and (3) risk analysis. Break-even analysis is useful when you have no clue as to the true value of one or more parameters, but you can decide if the value is greater than or less than some break-even value. Break-even analysis, for example, might prove useful when you are unable to quantify exactly the value of reductions in space requirements, cycle times, and inventory levels, as well as increases in quality levels, market share, throughput levels, and flexibility. In each case, the answer to the following question might prove useful: "How much must the reduction (or increase) in _____ be worth for the investment to be justified?." If the answer is $5 million for the value of, say, increased flexibility and you know that such a value is unreasonable, then the investment should not be undertaken. Next, suppose an investment in material handling will reduce from eight to two weeks the time required to fill a customer's order; if at least a 2% increase in customer orders have to occur annually to justify the material handling system, then the investment appears attractive.

Even if your firm does not require supplementary analysis in its economic justification package, give strong consideration to doing it—if for no one else's benefit than yours! By addressing many "what if?" questions, supplementary analysis serves as an "insurance policy" and increases your confidence in your recommendation to management. In summary, do more than *just what is required.* Supplementary analysis provides a means of *going the extra mile* in ensuring that a firm's scarce capital is invested wisely.

In justifying facilities planning investments, remember the axiom: *Don't wade in rivers that on the average are 2 ft deep!* Economic justifications typically use best estimates or "average values" of the parameters embodied in the analysis. Yet, it is seldom the case that the "average condition" will occur. Managers are seldom as concerned with average conditions as they are with rare or unusual conditions! Many people have drowned in rivers that are on the average 2 ft deep; such rivers can have very deep holes and swift currents. The purpose of supplementary analysis is to anticipate the "deep holes and swift currents" and ascertain their impact on the viability of the facilities planning investments under consideration.

The three supplementary analysis techniques differ in the degree of knowledge you must have concerning the "true values" of the "uncertain parameters." Break-even analysis requires knowing if the "true value" of a parameter is above or below the break-even value and sensitivity analysis requires knowledge of the range or possible values of each parameter. Risk analysis, on the other hand, requires knowing both the possible values of the parameters and their probabilities of occurrence.

Typical of the parameters addressed through the use of sensitivity analysis are the discount rate used, the length of the planning horizon, the magnitude of the initial investment, the annual savings realized, and the operating costs required. Sensitivity analysis has as its objective the determination of the sensitivity of the decision to incorrect estimates of the values of one or more parameters. One approach often used is to specify minimum, maximum, and most likely estimates for each parameter in question. By determining the economic performance of each alternative for each combination of values for the "uncertain" parameters, it is anticipated that a better-informed decision will result.

Risk analysis requires the greatest knowledge concerning the future values of the parameters. Simulation is generally used to convert the probability distributions for the parameters to a simulated probability distribution for the measure of economic worth. Based on the results obtained, an estimate can be provided of the probability of the investment yielding a positive outcome. (A more sophisticated version utilizes probability distributions for parameters of probability distributions, but such an approach is not likely to be merited; those who undertake this approach are in danger of succumbing to a terminal case of paralysis of analysis!)

Although it is easy to become enamored of the supplementary analysis techniques, it is important to remember why they are used—to enhance the quality of the ultimate investment decision. One test of this is to ask if you would make the investment if your money were involved. That, indeed, is the bottom line [14]!

## Selecting the Preferred Alternative

At this point, you have available a set of facilities planning investment alternatives, as well as extensive information concerning their financial and operational performance. Additionally, you should have available information concerning the reasons for undertaking the investment, analyses of the impacts of the investment for each alternative, and a recommended preferred alternative. In some instances, only the recommended investment will be submitted for approval; in other cases, you will have to submit the "top candidates."

In preparing economic justification proposals, it is important to *go the extra mile*. Examples include performing sensitivity analyses and performing a competitive benchmarking analysis. The latter should assess the impact of "doing nothing" while the competition is "doing something"; throughout, the impact of the facilities planning investment on the revenue stream should be considered explicitly.

Recognize that the final selection can be quite different from the recommendation you make. Management decisions are based on many considerations, not all of which were included in your analysis. To improve the chances of your recommended action being accepted, consider the following "lessons learned":

- Obtain the support of the users of the recommended system.
- Presell your recommendation!
- Speak the language of the listener.

- Do not oversell the technical aspects of the facilities plan; technical aspects seldom convince management to make the required investment.

- The decision-makers' perspectives are broad; know their priorities and tailor the economic justification package accordingly.

- Relate the proposed investment to the well-being of the firm; show how the investment relates to the firm's strategic plan and the stated corporate objectives.

- Your proposal will be only one of many submitted and many will not be funded.

- Failure to fund your proposal does not mean management is stupid; management's decision to fund your proposal does not necessarily mean they are brilliant.

- Do not confuse unfavorable results with destiny.

- Timing is everything!

- Profit maximization is not always the "name of the game", but "selling" is!

- A firm's ability to finance the proposal is as important as its economic merit.

- You get what you pay for; don't be penny-wise and dollar-foolish.

- Remember the "Golden Rule", those with the gold make the rules!

- Would you fund it if the funds were yours?

- The less you bet the more you stand to lose in case you win [15], [16]!

## *Economic Value Added (EVA®)*[1]

An example of speaking the language of the listener is provided by the emergence of Economic Value Added, or EVA.

Since the mid-1980s, a management tool called *economic value added* has been used by an impressive set of firms to make investment decisions. It focuses management's attention on an important objective: adding value for the shareholders. In fact, it has been a principal tool used by upper management within AT&T, Briggs & Stratton, Coca-Cola, CSX, Eastman Chemical, Quaker Oats, Wal-Mart Stores, and a host of other firms.

EVA is used to facilitate decisions regarding major capital investment, as well as acquisitions and divestitures. It has also been used to analyze products and operating units to determine the economic dogs and cash cows within the firm's portfolio of products and businesses.

What is EVA? Basically, EVA is a management tool that examines the difference between the net operating profit after taxes and the cost of capital, which includes the cost of both debt and equity capital. Hence, the interest charges and bond rates that contribute to the cost of debt capital are combined with the cost to the shareholders of providing the firm with equity capital (by purchasing its stock) [21].

As noted in *Fortune,* the cost of equity capital needs to include the opportunity cost to the shareholders, who can choose where they are going to invest their money. [4]. An analysis of the cost of capital typically reveals that debt capital is far less expensive than equity capital.

---

[1]EVA® is a trademark of Stern Stewart & Co.

Many firms that adopted EVA found that few of their managers knew how much capital was tied up in their business units. Moreover, few managers had a firm grasp on the true cost of capital. EVA seeks to remedy this by focusing attention on adding value for the shareholder through more effective use of capital. The result of an increased emphasis on effective use of capital will result in lower inventories, fewer warehouses, and so on [21].

Stern Stewart argues that there are four ways to create value for the shareholder:

- increase profitability without using additional capital, for example, increase profit margins and increase sales without using additional capital;
- invest in projects that earn more than the cost of capital;
- free-up capital that earns less than the cost of capital; and
- use debt to reduce the cost of capital. [23]

Real estate, equipment, facilities, working capital, and inventories are examples of capital being used within the firm. Other, not so obvious examples of capital are investments in training and in research and development. The investments in training result in increased value of the firm's human capital. While it is not easy to quantify the value of capital in R&D and in training, they should not be completely overlooked in the quest for value.

### Calculating Economic Value Added

The following three examples illustrate the EVA calculation.

# Example 13.4

Two firms (A and B) are being considered for acquisition. The assets of the firms are $100 million and $200 million, respectively. Both are debt free; hence, the equity equals the assets. The annual operating profits for the firms are $40 million and $70 million, respectively. Taxes equal 40% of operating profit. Consequently, the annual net incomes for the firms are $24 million and $42 million, respectively. Dividing the net income by equity yields return on equity of 24% and 21%, respectively. The cost of capital is 12%. Which firm is best from a shareholder value point of view? Some would choose A because it has the greatest ROE (return on equity) or ROA (return on assets). However, B maximizes value for shareholders.

To determine the EVA, subtract the cost of capital from net operating profit after taxes. For A, this yields: $24 million − 0.12 ($100 million) = $12 million. For B, the EVA equals $42 million − 0.12 ($200 million) = $18 million. (As was the case with the IRR method, one cannot choose that alternative that has the greatest return on equity; instead, incremental net operating profits should be compared with the cost of capital.)

In 1924, Donaldson Brown, Chief Financial Officer, General Motors Corporation, recognized that maximizing return on investment was the wrong objective; he noted, "The object of management is not necessarily the highest rate of return on capital, but . . . to assure profit with each increment of volume that will at least equal the economic cost of additional capital required" [23].

# Example 13.5

There are several ways to compute EVA. The method used in the previous example was:

|  | Firm C | Firm D |
|---|---|---|
| Capital Invested | $100 | $200 |
| Operating Profits | $25 | $35 |
| Taxes (40%) | $10 | $14 |
| Net Operating Profit | | |
| After Taxes (NOPAT) | $15 | $21 |
| Cost of Capital (12%) | $12 | $24 |
| EVA | +$3 | −$3 |

Alternatively, EVA can be computed as follows:

|  | | |
|---|---|---|
| Return on Assets (ROA) | 15% | 10.5% |
| Cost of Capital | 12% | 12% |
| Difference | +3% | −1.5% |
| EVA = Difference × Assets | +$3 | −$3 |

# Example 13.6

The difference in EVA and cash flow approaches is illustrated by considering the data for Firm C in Example 13.5. To simplify the analysis, consider a $100 million investment that returns $15 million annually, forever. With a 12% MARR, the net present value (NPV) of the NOPAT can be obtained by dividing the annual NOPAT by the MARR to obtain $125 million. Subtracting the initial investment yields a NPV of $25 million. Alternately, the $100 million investment yields an annual EVA of $3 million. Dividing the EVA by the MARR yields a NPV of $25 million. The NPV for the projected EVA is the same as the NPV of the projected cash flow.

From Example 13.6, since using DCF approaches yields the same recommendation as using discounted EVA, why use discounted EVA? Stern Stewart & Co. note that when using DCF, after an investment is made there is no explicit "accounting" for it since it is a "sunk cost." As a result, "it apparently does not carry an on-going performance obligation." They go on to note, "the incremental profits from the project, on the other hand, make an on-going contribution to earnings growth, thereby stimulating investments that may not make good economic sense. Because cash flow cannot be used for any other purpose—not for expressing goals or appraising performance, for example—the analysis is apparently disconnected from the rest of the financial management system" [25, p. 5].

In support of their argument for using EVA, they note, "the result is the same as discounting cash flow, an important equivalence. The obligation to earn an attractive return on capital is explicit and enduring. The EVA performance targets required to create value fall right out of the analysis. Unlike cash flow, EVA can be used as a performance target and measure. If EVA is used for determining bonuses, managers will see clearly how prospective business decisions will affect their compensation, an important link" [25, p. 5].

## Using Spreadsheets

A variety of software packages are available to facilitate the performance of SEAT's fifth and sixth steps. Of particular significance is the development of spreadsheet software, for example, Excel®, Lotus 1-2-3®, and Quattro® Pro. To show how a spreadsheet can be used to facilitate both financial calculations and the performance of sensitivity analyses, note that a spreadsheet can be developed by using a single formula: $V_0 = V_t(1 + i)^{-t}$, where $V_t$ denotes the value of a cash flow occurring at time $t$, $V_0$ denotes the value of the cash flow at time zero (the present value), and $i$ denotes the time value of money.

Consider the spreadsheet given in Table 13.2. The columns are identified alphabetically and the rows are identified numerically. The interest rate is entered, using decimal format, in cell, B1. The magnitudes of cash flows are entered in column B, beginning with row 4. The value of the discount factor for a particular time period is obtained by dividing the discount factor for the previous period by the sum of one and the interest rate. The present value of an individual cash flow is obtained by multiplying the value of the cash flow and the discount rate. Finally, the net present value (NPV) for the cash flow series is obtained by summing the individual present values, as in cell D10.

Entering data into the spreadsheet, as shown in Table 13.3, and performing the indicated calculations yields a present value of $1,372.36 for the example.

When using Excel 5.0,[2] the net present value of a series of cash flows can be obtained easily. By entering **=NPV(0.10,B5:B9)+B4** in a cell, the net present value of $1,372.36 is obtained. It is important to note that the NPV function computes the present value of a series of cash flows as of one period prior to the first value in the

Table 13.2. *Sample Spreadsheet*

| R\C | A | B | C | D |
|-----|---|---|---|---|
| 1 | Discount Rate = | (enter interest rate) | | |
| 2 | | | | |
| 3 | Time | Cash Flow | Discount Factor | Present Value |
| 4 | 0 | $A_1$ | 1.0000 | D4 = B4*C4 |
| 5 | 1 | $A_2$ | C5 = C4/(1.0 + B1) | D5 = B5*C5 |
| 6 | 2 | $A_3$ | C6 = C5/(1.0 + B1) | D6 = B6*C6 |
| 7 | 3 | $A_4$ | C7 = C6/(1.0 + B1) | D7 = B7*C7 |
| 8 | 4 | $A_5$ | C8 = C7/(1.0 + B1) | D8 = B8*C8 |
| 9 | 5 | $A_6$ | C9 = C8/(1.0 + B1) | D9 = B9*C9 |
| 10 | NPV | | | D10 = SUM(D4:D9) |
| | | | | |

---

[2]Microsoft® Excel, Version 5.0, Microsoft Corporation, 1993–1994.

**Table 13.3** *Example of the Use of a Spreadsheet*

| R\C | A | B | C | D |
|---|---|---|---|---|
| 1 | Discount Rate = | 0.10 | | |
| 2 | | | | |
| 3 | Time | Cash Flow | Discount Factor | Present Value |
| 4 | 0 | ($10,000) | 1.000000 | ($10,000.00) |
| 5 | 1 | $3,000 | 0.909091 | $2,727.27 |
| 6 | 2 | $3,000 | 0.826446 | $2,479.34 |
| 7 | 3 | $3,000 | 0.751315 | $2,253.94 |
| 8 | 4 | $3,000 | 0.683013 | $2,049.04 |
| 9 | 5 | $3,000 | 0.620921 | $1,862.76 |
| 10 | NPV | | | $1,372.36 |
| | | | | |

series. Using Excel's NPV financial function, the net present value is determined for the cash flows entered in cells B5 through B9; hence, to obtain the present value for the entire cash flow series it is necessary to add the entry in cell B4 to the value obtained by using the NPV function over the range of cells from B5 through B9.

After setting up the spreadsheet, it is easy to perform supplementary analyses. For example, if the annual receipts are reduced to $2,500, then, as shown in Table 13.4, the resulting NPV value will equal −$523.03. Likewise, as shown in Table 13.5,

**Table 13.4** *Using a Spreadsheet to Analyze Changes in Cash Flows*

| R\C | A | B | C | D |
|---|---|---|---|---|
| 1 | Discount Rate = | 0.10 | | |
| 2 | | | | |
| 3 | Time | Cash Flow | Discount Factor | Present Value |
| 4 | 0 | ($10,000) | 1.000000 | ($10,000.00) |
| 5 | 1 | $2,500 | 0.909091 | $2,272.73 |
| 6 | 2 | $2,500 | 0.826446 | $2,066.12 |
| 7 | 3 | $2,500 | 0.751315 | $1,878.29 |
| 8 | 4 | $2,500 | 0.683013 | $1,707.53 |
| 9 | 5 | $2,500 | 0.620921 | $1,552.30 |
| 10 | NPV | | | ($523.03) |
| | IRR | 7.9308% | | |
| | | | | |

**Table 13.5   *Using a Spreadsheet to Analyze Changes in the Interest Rate***

| R\C | A | B | C | D |
|---|---|---|---|---|
| 1 | Discount Rate = | 0.16 | | |
| 2 | | | | |
| 3 | Time | Cash Flow | Discount Factor | Present Value |
| 4 | 0 | ($10,000) | 1.000000 | ($10,000.00) |
| 5 | 1 | $3,000 | 0.909091 | $2,586.21 |
| 6 | 2 | $3,000 | 0.826446 | $2,229.49 |
| 7 | 3 | $3,000 | 0.751315 | $1,921.97 |
| 8 | 4 | $3,000 | 0.683013 | $1,656.87 |
| 9 | 5 | $3,000 | 0.620921 | $1,428.34 |
| 10 | NPV | | | ($177.12) |
| | IRR | 15.2382% | | |
| | | | | |

if the interest rate in Table 13.3 is changed to 16%, the net present value changes to −$177.12.

Excel also has a financial function for determining the internal rate of return for an investment. Entering **=IRR(B4:B9)** in any cell in the spreadsheet will yield the value of the discount rate that results in a net present value of zero for the cash flows B4 through B9. As shown in Table 13.4, the internal rate of return for an investment of $10,000 that returns $2,500 a year for 5 years is 7.93%; for the case of $3,000 per year, the internal rate of return is 15.24%.

When faced with choosing the most economic alternative from among several mutually exclusive investment alternatives, rank them on the basis of their present worths and recommend the one having the largest present worth. However, when using the internal rate of return method, it is not necessarily the case that the one with the greatest rate of return is preferred. Hence, it is recommended that present worth analysis be used to determine the preferred alternative and its internal rate of return be computed.

A final note on evaluating facilities planning alternatives is needed. Specifically, a reminder is provided that the evaluation of facilities plans is a separate process from the evaluation of the suppliers and contractors involved in delivering and installing hardware and software. Too often, the evaluation of the design concept is clouded by issues related to individual system suppliers.

Even though we recommend a separation of the evaluation process into *plan* and *supplier* evaluations, the evaluation techniques discussed previously can be applied for either type evaluation. However, the same technique does not have to be used for both evaluations. For example, a weighted factor comparison might be used to evaluate the alternative plans and an economic comparison might be used to evaluate the suppliers.

## *13.3* SELECTING THE FACILITIES PLAN

As mentioned previously, the selection of the facilities plan tends to be a "satisficing" process, as opposed to an optimizing process. The selection decision typically requires a compromise over several important criteria. Furthermore, the decision maker's value system is not necessarily the same as the facilities planner's.

The facilities planner has a responsibility to do a thorough, accurate, and objective job of developing several good facilities plans. The facilities planner also has a responsibility to perform a thorough, accurate, and objective evaluation of the alternative plans. Furthermore, the facilities planner is responsible for presenting professionally the results of the facilities planning effort to management and making a recommendation concerning the preferred plan. However, it is management's responsibility either to approve or disapprove the recommendation or to make the selection decision.

The selection process is critically dependent on the facilities planner selling the facilities plan. Not only must management be sold, but also the end users of the system must be sold.

The sales job for management must address the long-term benefits of the facilities plan. The flexibility, reliability, and adaptability of the system must be addressed. Management may take a strategic, offensive (as opposed to defensive) position and may want to know what is going to be the return on investment, increased services, or other benefits of pursuing the recommended system.

To the contrary, the sales job for users must address the short-term implications of the new system. Noise, safety, and impacts on the user of the system are the key factors. The user often takes a contingency, defensive position, and wants to know how this system will affect the people involved.

Not understanding to whom the system must be justified is a major mistake. Once a survey indicated that 80% of the people who had installed a major material handling system were unhappy with the results. Of these unhappy people, 80% said the reason for their unhappiness was not the equipment aspects of the system, but instead, the people aspects. Systems must be explained and sold to both managers and users and the approach taken to these two groups should be different.

Because of the importance of preparing and selling facilities plans, the subject is treated separately. Chapter 18 is devoted to the preparation of the facilities plan and the development of a sales presentation.

## *13.4* SUMMARY

In summary, the evaluation of alternative facilities plans was considered; four evaluation techniques were described and a seven-step economic justification procedure was presented. The importance of presenting the justification package in a form that "communicates" to management was illustrated by describing the Economic Value Added approach. The use of spreadsheets to facilitate the computational aspects of the systematic approach was also illustrated. A further treatment of the selection of the facilities plan is reserved for Chapter 14, where the preparation and selling aspects of the facilities plan are covered.

# BIBLIOGRAPHY

1.   Canada, J. R., Sullivan, W. G., and White, J. A., *Capital Investment Decision Analysis for Management and Engineering,* 2nd ed., Prentice-Hall, Englewood Cliffs, NJ, 1996.

2.   Kaplan, R. S., "Must CIM be Justified by Faith Alone?," *Harvard Business Review,* **64,** 2, March–April 1986, pp. 87–95.

3.   Sellers, P., "Corporate Reputations," *Fortune,* January 29, 1990, p. 50.

4.   Stern, J. M., "The Mathematics of Corporate Finance—or EVA=$NA[RONA-C]," transcript of a speech given in Atlanta, 1993.

5.   Stewart, G. B., *The Quest for Value,* HarperCollins, NY, 1991.

6.   Tompkins, J. A., and White, J. A., *Facilities Planning,* John Wiley & Sons, New York, 1984.

7.   Tully, S., "The Real Key to Creating Wealth," *Fortune,* September 20, 1993, pp. 38–50.

8.   White, J. A., "How to Justify Equipment Investments," *Modern Materials Handling,* **44,** 1, January 1989, p. 39.

9.   White, J. A., "Specifying the Investment Alternatives," *Modern Materials Handling,* **44,** 2, February 1989, p. 27.

10.  White, J. A., "Defining Your Planning Horizon," *Modern Materials Handling,* **44,** 4, March 1989, p. 29.

11.  White, J. A., "Developing a Cash Flow Profile," *Modern Materials Handling,* **44,** 5, April 1989, p. 27.

12.  White, J. A., "The Cost of Opportunity," *Modern Materials Handling,* **44,** 6, May 1989, p. 29.

13.  White, J. A., "Comparing Investment Options," *Modern Materials Handling,* **44,** 7, June 1989, p. 27.

14.  White, J. A., "Justifying with Break-Even Analysis," *Modern Materials Handling,* **44,** 8, July 1989, p. 29.

15.  White, J. A., "Making Informed Investment Decisions," *Modern Materials Handling,* **44,** 9, August 1989, p. 27.

16.  White, J. A., "Winning Investment Approval," *Modern Materials Handling,* **44,** 10, September 1989, p. 27.

17.  White, J. A., "Leveling the Playing Field," *Modern Materials Handling,* **44,** 12, October 1989, p. 25.

18.  White, J. A., "The Money Value of Time," *Modern Materials Handling,* **44,** 13, November 1989, p. 27.

19.  White, J. A., "Investment, Taxes, and Inflation," *Modern Materials Handling,* **44,** 14, December 1989, p. 27.

20.  White, J. A., "Manage to Better Bottom Line," *Modern Materials Handling,* **45,** 13, November 1990, p. 29.

21.  White, J. A., "Three Ways to Build Economic Value," *Modern Materials Handling,* **49,** 7, June 1994, p. 31.

22.  White, J. A., Agee, M. H., and Case, K. E., *Principles of Engineering Economic Analysis,* 3rd ed., John Wiley & Sons, New York, 1989.

23. *EVA*[TM], Presentation by Stern Stewart & Co., 450 Park Avenue, New York.
24. "Managing," *Fortune,* April 22, 1991, p. 86.
25. *The Quest for Value: How Value is Created and How to Create Value,* Presentation by Stern Stewart & Co., 450 Park Avenue, New York.

# PROBLEMS

13.1 List the advantages and disadvantages of leasing production equipment versus purchasing the equipment.

13.2 List the advantages and disadvantages of providing on-site storage of materials and supplies versus using offsite leased storage.

13.3 List the advantages and disadvantages in building a conventional (25 to 30 ft tall) warehouse versus building a high-rise (55 to 75 ft tall) warehouse.

13.4 List 10 factors to be used in evaluating the following horizontal handling alternatives for moving pallet loads: lift truck, tractor-trailer train, automated guided vehicle, in-floor towline conveyor, roller conveyor for pallets.

13.5 List 10 factors to be used in evaluating site location alternatives for a corporate headquarters building.

13.6 List 10 factors to be used in evaluating site location alternatives for a distribution center.

13.7 Two material handling alternatives are being considered for a new distribution center. The first involves a smaller capital investment, but larger operating costs. Specifically, alternative A involves an initial investment of $200,000 and has annual operating expenses of $500,000. Alternative B involves an initial investment of $1,000,000 and has annual operating expenses of $300,000. Both alternatives are assumed to have negligible salvage values at the end of the 10-year planning horizon. A minimum attractive rate of return of 10% is used for such analyses.
   a. Using present worth analysis, which is preferred?
   b. What is the internal rate of return on the $800,000 incremental investment required for alternative B?
   c. If the payback period method is used, how long would it take to payback the $800,000 incremental investment required for alternative B?

13.8 In Problem 7, suppose there is considerable uncertainty regarding the annual operating expenses for alternative B. What is the break-even value?

13.9 True or False: The present worth method and the internal rate of return method, when applied correctly, always yield the same recommendation.

13.10 True or False: The payback period method will always yield the same recommendation as the present worth method.

13.11 True or False: The use of "size gates" in the authorization of capital expenditures can result in major projects being broken into several seemingly independent investments, some of which might not be able to stand alone in the economic justification process.

13.12 True or False: The "do-nothing" alternative is always the lowest cost alternative and serves as the baseline against which other alternatives are compared.

13.13 True or False: U.S. firms tend to require higher returns on capital investments and tend to use longer planning horizons than their Japanese counterparts.

13.14 Given the following financial data (in millions of dollars) for a firm, determine its EVA.

| | | |
|---|---|---|
| Earnings before Income Taxes | $540 | |
| Taxes | $212 | |
| Net Operating Profit After Taxes | | $328 |
| Investment In: | | |
| Total Assets (Avg) | $5,158 | |
| Payables (Avg) | $1,525 | |
| Net Investment | | $3,633 |
| Cost of Capital | 10% | |

**13.15** Consider two firms, C and D, each with a 12% cost of capital. Based on the financial data for the two firms (in $M's), which has the greatest EVA?

| | Firm C | Firm D |
|---|---|---|
| Equity | $100 | $200 |
| Annual Operating Profit | $25 | $35 |
| Taxes (40%) | $10 | $14 |

**13.16** Three investment alternatives (**A, B,** and **C**) are under consideration. The discounted present worths (**PW**) are $30,000, $24,000, and $18,000, respectively. The three alternatives have very different rankings on time required to fill a customer's order (**T**) and flexibility (**F**). The following weights have been assigned to the three factors (**PW, T,** and **F**): 40, 35, and 25. On a scale from 1 to 10, the following ratings have been assigned to the three alternatives for the three factors:

| | **A** | **B** | **C** |
|---|---|---|---|
| Present worth (**PW**) | 10 | 8 | 6 |
| Fill time (**T**) | 7 | 10 | 8 |
| Flexibility (**F**) | 5 | 7 | 10 |

Using the weighted factor comparison method, which investment alternative would be recommended?

**13.17** Which investment alternative would be recommended in Problem 13.16, if
  a. the weightings for the three factors are 50, 10, and 40?
  b. the rating values for fill time for **A, B,** and **C** had been 5, 7, and 10, respectively?
  c. another factor is introduced, safety (**S**), the weights assigned to the factors (**PW, T, F,** and **S**) and 30, 20, 20, and 70, and the rating values for **S** are 10, 8, and 9 for **A, B,** and **C**, respectively?

**13.18** From solving Problems 13.16 and 13.17, what conclusions do you draw regarding the use of the weighted factor comparison method?

# DISCOUNTED CASH FLOW
# COMPOUND INTEREST TABLES
$i = 5\%, 10\%, 15\%, 20\%, \text{and } 25\%$

Table A.1   *Values of Compound Interest Factors with i = 5%*

| | Single Sums | | Uniform Series | | | |
|---|---|---|---|---|---|---|
| n | Find F<br>Given P<br>F\|P i, n | Find P<br>Given F<br>P\|F i, n | Find F<br>Given A<br>F\|A i, n | Find A<br>Given F<br>A\|F i, n | Find P<br>Given A<br>P\|A i, n | Find A<br>Given P<br>A\|P i, n |
| 1 | 1.0500 | 0.9524 | 1.0000 | 1.0000 | 0.9524 | 1.0500 |
| 2 | 1.1025 | 0.9070 | 2.0500 | 0.4878 | 1.8594 | 0.5378 |
| 3 | 1.1576 | 0.8638 | 3.1525 | 0.3172 | 2.7232 | 0.3672 |
| 4 | 1.2155 | 0.8227 | 4.3101 | 0.2320 | 3.5460 | 0.2820 |
| 5 | 1.2763 | 0.7835 | 5.5256 | 0.1810 | 4.3295 | 0.2310 |
| 6 | 1.3401 | 0.7462 | 6.8019 | 0.1470 | 5.0757 | 0.1970 |
| 7 | 1.4071 | 0.7107 | 8.1420 | 0.1228 | 5.7864 | 0.1728 |
| 8 | 1.4775 | 0.6768 | 9.5491 | 0.1047 | 6.4632 | 0.1547 |
| 9 | 1.5513 | 0.6446 | 11.0266 | 0.0907 | 7.1078 | 0.1407 |
| 10 | 1.6289 | 0.6139 | 12.5779 | 0.0795 | 7.7217 | 0.1295 |
| 11 | 1.7103 | 0.5847 | 14.2068 | 0.0704 | 8.3064 | 0.1204 |
| 12 | 1.7959 | 0.5568 | 15.9171 | 0.0628 | 8.8633 | 0.1128 |
| 13 | 1.8856 | 0.5303 | 17.7130 | 0.0565 | 9.3936 | 0.1065 |
| 14 | 1.9799 | 0.5051 | 19.5986 | 0.0510 | 9.8986 | 0.1010 |
| 15 | 2.0789 | 0.4810 | 21.5786 | 0.0463 | 10.3797 | 0.0963 |
| 16 | 2.1829 | 0.4581 | 23.6575 | 0.0423 | 10.8378 | 0.0923 |
| 17 | 2.2920 | 0.4363 | 25.8404 | 0.0387 | 11.2741 | 0.0887 |
| 18 | 2.4066 | 0.4155 | 28.1324 | 0.0355 | 11.6896 | 0.0855 |
| 19 | 2.5270 | 0.3957 | 30.5390 | 0.0327 | 12.0853 | 0.0827 |
| 20 | 2.6533 | 0.3769 | 33.0660 | 0.0302 | 12.4622 | 0.0802 |
| 21 | 2.7860 | 0.3589 | 35.7193 | 0.0280 | 12.8212 | 0.0780 |
| 22 | 2.9253 | 0.3418 | 38.5052 | 0.0260 | 13.1630 | 0.0760 |
| 23 | 3.0715 | 0.3256 | 41.4305 | 0.0241 | 13.4886 | 0.0741 |
| 24 | 3.2251 | 0.3101 | 44.5020 | 0.0225 | 13.7986 | 0.0725 |
| 25 | 3.3864 | 0.2953 | 47.7271 | 0.0210 | 14.0939 | 0.0710 |
| 26 | 3.5557 | 0.2812 | 51.1135 | 0.0196 | 14.3752 | 0.0696 |
| 27 | 3.7335 | 0.2678 | 54.6691 | 0.0183 | 14.6430 | 0.0683 |
| 28 | 3.9201 | 0.2551 | 58.4026 | 0.0171 | 14.8981 | 0.0671 |
| 29 | 4.1161 | 0.2429 | 62.3227 | 0.0160 | 15.1411 | 0.0660 |
| 30 | 4.3219 | 0.2314 | 66.4388 | 0.0151 | 15.3725 | 0.0651 |
| 32 | 4.7649 | 0.2099 | 75.2988 | 0.0133 | 15.8027 | 0.0633 |
| 34 | 5.2533 | 0.1904 | 85.0670 | 0.0118 | 16.1929 | 0.0618 |
| 36 | 5.7918 | 0.1727 | 95.8363 | 0.0104 | 16.5469 | 0.0604 |
| 38 | 6.3855 | 0.1566 | 107.7095 | 0.0093 | 16.8679 | 0.0593 |
| 40 | 7.0400 | 0.1420 | 120.7998 | 0.0083 | 17.1591 | 0.0583 |
| 42 | 7.7616 | 0.1288 | 135.2318 | 0.0074 | 17.4232 | 0.0574 |
| 44 | 8.5572 | 0.1169 | 151.1430 | 0.0066 | 17.6628 | 0.0566 |
| 46 | 9.4343 | 0.1060 | 168.6852 | 0.0059 | 17.8801 | 0.0559 |
| 48 | 10.4013 | 0.0961 | 188.0254 | 0.0053 | 18.0772 | 0.0553 |
| 50 | 11.4674 | 0.0872 | 209.3480 | 0.0048 | 18.2559 | 0.0548 |
| 55 | 14.6356 | 0.0683 | 272.7126 | 0.0037 | 18.6335 | 0.0537 |
| 60 | 18.6792 | 0.0535 | 353.5837 | 0.0028 | 18.9293 | 0.0528 |
| 65 | 23.8399 | 0.0419 | 456.7980 | 0.0022 | 19.1611 | 0.0522 |
| 70 | 30.4264 | 0.0329 | 588.5285 | 0.0017 | 19.3427 | 0.0517 |
| 75 | 38.8327 | 0.0258 | 756.6537 | 0.0013 | 19.4850 | 0.0513 |
| 80 | 49.5614 | 0.0202 | 971.2288 | 0.0010 | 19.5965 | 0.0510 |
| 85 | 63.2544 | 0.0158 | 1245.0871 | 0.00080 | 19.6838 | 0.05080 |
| 90 | 80.7304 | 0.0124 | 1594.6073 | 0.00063 | 19.7523 | 0.05063 |
| 100 | 131.5013 | 0.0076 | 2610.0252 | 0.00038 | 19.8479 | 0.05038 |
| 120 | 348.9120 | 0.0029 | 6958.2397 | 0.00014 | 19.9427 | 0.05014 |

**Table A.2  Values of Compound Interest Factors with i = 10%**

| | Single Sums | | Uniform Series | | | |
|---|---|---|---|---|---|---|
| n | Find F Given P F\|P i, n | Find P Given F P\|F i, n | Find F Given A F\|A i, n | Find A Given F A\|F i, n | Find P Given A P\|A i, n | Find A Given P A\|P i, n |
| 1 | 1.1000 | 0.9091 | 1.0000 | 1.0000 | 0.9091 | 1.1000 |
| 2 | 1.2100 | 0.8264 | 2.1000 | 0.4762 | 1.7355 | 0.5762 |
| 3 | 1.3310 | 0.7513 | 3.3100 | 0.3021 | 2.4869 | 0.4021 |
| 4 | 1.4641 | 0.6830 | 4.6410 | 0.2155 | 3.1699 | 0.3155 |
| 5 | 1.6105 | 0.6209 | 6.1051 | 0.1638 | 3.7908 | 0.2638 |
| 6 | 1.7716 | 0.5645 | 7.7156 | 0.1296 | 4.3553 | 0.2296 |
| 7 | 1.9487 | 0.5132 | 9.4872 | 0.1054 | 4.8684 | 0.2054 |
| 8 | 2.1436 | 0.4665 | 11.4359 | 0.0874 | 5.3349 | 0.1874 |
| 9 | 2.3579 | 0.4241 | 13.5795 | 0.0736 | 5.7590 | 0.1736 |
| 10 | 2.5937 | 0.3855 | 15.9374 | 0.0627 | 6.1446 | 0.1627 |
| 11 | 2.8531 | 0.3505 | 18.5312 | 0.0540 | 6.4951 | 0.1540 |
| 12 | 3.1384 | 0.3186 | 21.3843 | 0.0468 | 6.8137 | 0.1468 |
| 13 | 3.4523 | 0.2897 | 24.5227 | 0.0408 | 7.1034 | 0.1408 |
| 14 | 3.7975 | 0.2633 | 27.9750 | 0.0357 | 7.3667 | 0.1357 |
| 15 | 4.1772 | 0.2394 | 31.7725 | 0.0315 | 7.6061 | 0.1315 |
| 16 | 4.5950 | 0.2176 | 35.9497 | 0.0278 | 7.8237 | 0.1278 |
| 17 | 5.0545 | 0.1978 | 40.5447 | 0.0247 | 8.0216 | 0.1247 |
| 18 | 5.5599 | 0.1799 | 45.5992 | 0.0219 | 8.2014 | 0.1219 |
| 19 | 6.1159 | 0.1635 | 51.1591 | 0.0195 | 8.3649 | 0.1195 |
| 20 | 6.7275 | 0.1486 | 57.2750 | 0.0175 | 8.5136 | 0.1175 |
| 21 | 7.4002 | 0.1351 | 64.0025 | 0.0156 | 8.6487 | 0.1156 |
| 22 | 8.1403 | 0.1228 | 71.4027 | 0.0140 | 8.7715 | 0.1140 |
| 23 | 8.9543 | 0.1117 | 79.5430 | 0.0126 | 8.8832 | 0.1126 |
| 24 | 9.8497 | 0.1015 | 88.4973 | 0.0113 | 8.9847 | 0.1113 |
| 25 | 10.8347 | 0.0923 | 98.3471 | 0.0102 | 9.0770 | 0.1102 |
| 26 | 11.9182 | 0.0839 | 109.1818 | 0.0092 | 9.1609 | 0.1092 |
| 27 | 13.1100 | 0.0763 | 121.0999 | 0.0083 | 9.2372 | 0.1083 |
| 28 | 14.4210 | 0.0693 | 134.2099 | 0.0075 | 9.3066 | 0.1075 |
| 29 | 15.8631 | 0.0630 | 148.6309 | 0.0067 | 9.3696 | 0.1067 |
| 30 | 17.4494 | 0.0573 | 164.4940 | 0.0061 | 9.4269 | 0.1061 |
| 32 | 21.1138 | 0.0474 | 201.1378 | 0.0050 | 9.5264 | 0.1050 |
| 34 | 25.5477 | 0.0391 | 245.4767 | 0.0041 | 9.6086 | 0.1041 |
| 36 | 30.9127 | 0.0323 | 299.1268 | 0.0033 | 9.6765 | 0.1033 |
| 38 | 37.4043 | 0.0267 | 364.0434 | 0.0027 | 9.7327 | 0.1027 |
| 40 | 45.2593 | 0.0221 | 442.5926 | 0.0023 | 9.7791 | 0.1023 |
| 42 | 54.7637 | 0.0183 | 537.6370 | 0.0019 | 9.8174 | 0.1019 |
| 44 | 66.2641 | 0.0151 | 652.6408 | 0.0015 | 9.8491 | 0.1015 |
| 46 | 80.1795 | 0.0125 | 791.7953 | 0.0013 | 9.8753 | 0.1013 |
| 48 | 97.0172 | 0.0103 | 960.1723 | 0.0010 | 9.8969 | 0.1010 |
| 50 | 117.3909 | 0.0085 | 1163.9085 | 0.00086 | 9.9148 | 0.10086 |
| 55 | 189.0591 | 0.0053 | 1880.5914 | 0.00053 | 9.9471 | 0.10053 |
| 60 | 304.4816 | 0.0033 | 3034.8164 | 0.00033 | 9.9672 | 0.10033 |
| 65 | 490.3707 | 0.0020 | 4893.7073 | 0.00020 | 9.9796 | 0.10020 |
| 70 | 789.7470 | 0.0013 | 7887.4696 | 0.00013 | 9.9873 | 0.10013 |
| 75 | 1271.8954 | 0.00079 | 12708.954 | 0.00008 | 9.9921 | 0.10008 |
| 80 | 2048.4002 | 0.00049 | 20474.002 | 0.00005 | 9.9951 | 0.10005 |
| 85 | 3298.9690 | 0.00030 | 32979.690 | 0.00003 | 9.9970 | 0.10003 |
| 90 | 5313.0226 | 0.00019 | 53120.226 | 0.00002 | 9.9981 | 0.10002 |
| 100 | 13780.612 | 0.00007 | 137796.12 | 0.00001 | 9.9993 | 0.10001 |
| 120 | 92709.069 | 0.00001 | 927080.69 | 0.00000 | 9.99989 | 0.10000 |

**Table A.3** *Values of Compound Interest Factors with i = 15%*

| n | Single Sums | | Uniform Series | | | |
|---|---|---|---|---|---|---|
| | Find F Given P F\|P i, n | Find P Given F P\|F i, n | Find F Given A F\|A i, n | Find A Given F A\|F i, n | Find P Given A P\|A i, n | Find A Given P A\|P i, n |
| 1 | 1.1500 | 0.8696 | 1.0000 | 1.0000 | 0.8696 | 1.1500 |
| 2 | 1.3225 | 0.7561 | 2.1500 | 0.4651 | 1.6257 | 0.6151 |
| 3 | 1.5209 | 0.6575 | 3.4725 | 0.2880 | 2.2832 | 0.4380 |
| 4 | 1.7490 | 0.5718 | 4.9934 | 0.2003 | 2.8550 | 0.3503 |
| 5 | 2.0114 | 0.4972 | 6.7424 | 0.1483 | 3.3522 | 0.2983 |
| 6 | 2.3131 | 0.4323 | 8.7537 | 0.1142 | 3.7845 | 0.2642 |
| 7 | 2.6600 | 0.3759 | 11.0668 | 0.0904 | 4.1604 | 0.2404 |
| 8 | 3.0590 | 0.3269 | 13.7268 | 0.0729 | 4.4873 | 0.2229 |
| 9 | 3.5179 | 0.2843 | 16.7858 | 0.0596 | 4.7716 | 0.2096 |
| 10 | 4.0456 | 0.2472 | 20.3037 | 0.0493 | 5.0188 | 0.1993 |
| 11 | 4.6524 | 0.2149 | 24.3493 | 0.0411 | 5.2337 | 0.1911 |
| 12 | 5.3503 | 0.1869 | 29.0017 | 0.0345 | 5.4206 | 0.1845 |
| 13 | 6.1528 | 0.1625 | 34.3519 | 0.0291 | 5.5831 | 0.1791 |
| 14 | 7.0757 | 0.1413 | 40.5047 | 0.0247 | 5.7245 | 0.1747 |
| 15 | 8.1371 | 0.1229 | 47.5804 | 0.0210 | 5.8474 | 0.1710 |
| 16 | 9.3576 | 0.1069 | 55.7175 | 0.0179 | 5.9542 | 0.1679 |
| 17 | 10.7613 | 0.0929 | 65.0751 | 0.0154 | 6.0472 | 0.1654 |
| 18 | 12.3755 | 0.0808 | 75.8364 | 0.0132 | 6.1280 | 0.1632 |
| 19 | 14.2318 | 0.0703 | 88.2118 | 0.0113 | 6.1982 | 0.1613 |
| 20 | 16.3665 | 0.0611 | 102.4436 | 0.0098 | 6.2593 | 0.1598 |
| 21 | 18.8215 | 0.0531 | 118.8101 | 0.0084 | 6.3125 | 0.1584 |
| 22 | 21.6447 | 0.0462 | 137.6316 | 0.0073 | 6.3587 | 0.1573 |
| 23 | 24.8915 | 0.0402 | 159.2764 | 0.0063 | 6.3988 | 0.1563 |
| 24 | 28.6252 | 0.0349 | 184.1678 | 0.0054 | 6.4338 | 0.1554 |
| 25 | 32.9190 | 0.0304 | 212.7930 | 0.0047 | 6.4641 | 0.1547 |
| 26 | 37.8568 | 0.0264 | 245.7120 | 0.0041 | 6.4906 | 0.1541 |
| 27 | 43.5353 | 0.0230 | 283.5688 | 0.0035 | 6.5135 | 0.1535 |
| 28 | 50.0656 | 0.0200 | 327.1041 | 0.0031 | 6.5335 | 0.1531 |
| 29 | 57.5755 | 0.0174 | 377.1697 | 0.0027 | 6.5509 | 0.1527 |
| 30 | 66.2118 | 0.0151 | 434.7451 | 0.0023 | 6.5660 | 0.1523 |
| 32 | 87.5651 | 0.0114 | 577.1005 | 0.0017 | 6.5905 | 0.1517 |
| 34 | 115.8048 | 0.0086 | 765.3654 | 0.0013 | 6.6091 | 0.1513 |
| 36 | 153.1519 | 0.0065 | 1014.3457 | 0.0010 | 6.6231 | 0.1510 |
| 38 | 202.5433 | 0.0049 | 1343.6222 | 0.00074 | 6.6338 | 0.15074 |
| 40 | 267.8635 | 0.0037 | 1779.0903 | 0.00056 | 6.6418 | 0.15056 |
| 42 | 354.2495 | 0.0028 | 2354.9969 | 0.00042 | 6.6478 | 0.15042 |
| 44 | 468.4950 | 0.0021 | 3116.6334 | 0.00032 | 6.6524 | 0.15032 |
| 46 | 619.5847 | 0.0016 | 4123.8977 | 0.00024 | 6.6559 | 0.15024 |
| 48 | 819.4007 | 0.0012 | 5456.0047 | 0.00018 | 6.6585 | 0.15018 |
| 50 | 1083.6574 | 0.00092 | 7217.7163 | 0.00014 | 6.6605 | 0.15014 |
| 55 | 2179.6222 | 0.00046 | 14524.148 | 0.000069 | 6.6636 | 0.150069 |
| 60 | 4383.9987 | 0.00023 | 29219.992 | 0.000034 | 6.6651 | 0.150034 |
| 65 | 8817.7874 | 0.00011 | 58778.583 | 0.000017 | 6.6659 | 0.150017 |
| 70 | 17735.720 | 0.000056 | 118231.47 | 0.0000085 | 6.6663 | 0.1500085 |
| 75 | 35672.868 | 0.000028 | 237812.45 | 0.0000042 | 6.6665 | 0.1500042 |
| 80 | 71750.879 | 0.000014 | 478332.53 | 0.0000021 | 6.66657 | 0.1500021 |
| 85 | 144316.65 | 0.0000069 | 962104.31 | 0.0000010 | 6.66662 | 0.1500010 |
| 90 | 290272.33 | 0.0000034 | 1935142.2 | 0.00000052 | 6.66664 | 0.15000052 |
| 100 | 1174313.5 | 0.00000085 | 7828749.7 | 0.00000013 | 6.66666 | 0.15000013 |
| 120 | 19219445 | 0.000000052 | 128129627 | 0.000000008 | 6.66667 | 0.150000008 |

**Table A.4**  *Values of Compound Interest Factors with i = 20%*

| | Single Sums | | Uniform Series | | | |
|---|---|---|---|---|---|---|
| n | Find F<br>Given P<br>F\|P i, n | Find P<br>Given F<br>P\|F i, n | Find F<br>Given A<br>F\|A i, n | Find A<br>Given F<br>A\|F i, n | Find P<br>Given A<br>P\|A i, n | Find A<br>Given P<br>A\|P i, n |
| 1 | 1.2000 | 0.8333 | 1.0000 | 1.0000 | 0.8333 | 1.2000 |
| 2 | 1.4400 | 0.6944 | 2.2000 | 0.4545 | 1.5278 | 0.6545 |
| 3 | 1.7280 | 0.5787 | 3.6400 | 0.2747 | 2.1065 | 0.4747 |
| 4 | 2.0736 | 0.4823 | 5.3680 | 0.1863 | 2.5887 | 0.3863 |
| 5 | 2.4883 | 0.4019 | 7.4416 | 0.1344 | 2.9906 | 0.3344 |
| 6 | 2.9860 | 0.3349 | 9.9299 | 0.1007 | 3.3255 | 0.3007 |
| 7 | 3.5832 | 0.2791 | 12.9159 | 0.0774 | 3.6046 | 0.2774 |
| 8 | 4.2998 | 0.2326 | 16.4991 | 0.0606 | 3.8372 | 0.2606 |
| 9 | 5.1598 | 0.1938 | 20.7989 | 0.0481 | 4.0310 | 0.2481 |
| 10 | 6.1917 | 0.1615 | 25.9587 | 0.0385 | 4.1925 | 0.2385 |
| 11 | 7.4301 | 0.1346 | 32.1504 | 0.0311 | 4.3271 | 0.2311 |
| 12 | 8.9161 | 0.1122 | 39.5805 | 0.0253 | 4.4392 | 0.2253 |
| 13 | 10.6993 | 0.0935 | 48.4966 | 0.0206 | 4.5327 | 0.2206 |
| 14 | 12.8392 | 0.0779 | 59.1959 | 0.0169 | 4.6106 | 0.2169 |
| 15 | 15.4070 | 0.0649 | 72.0351 | 0.0139 | 4.6755 | 0.2139 |
| 16 | 18.4884 | 0.0541 | 87.4421 | 0.0114 | 4.7296 | 0.2114 |
| 17 | 22.1861 | 0.0451 | 105.9306 | 0.0094 | 4.7746 | 0.2094 |
| 18 | 26.6233 | 0.0376 | 128.1167 | 0.0078 | 4.8122 | 0.2078 |
| 19 | 31.9480 | 0.0313 | 154.7400 | 0.0065 | 4.8435 | 0.2065 |
| 20 | 38.3376 | 0.0261 | 186.6880 | 0.0054 | 4.8696 | 0.2054 |
| 21 | 46.0051 | 0.0217 | 225.0256 | 0.0044 | 4.8913 | 0.2044 |
| 22 | 55.2061 | 0.0181 | 271.0307 | 0.0037 | 4.9094 | 0.2037 |
| 23 | 66.2474 | 0.0151 | 326.2369 | 0.0031 | 4.9245 | 0.2031 |
| 24 | 79.4968 | 0.0126 | 392.4842 | 0.0025 | 4.9371 | 0.2025 |
| 25 | 95.3962 | 0.0105 | 471.9811 | 0.0021 | 4.9476 | 0.2021 |
| 26 | 114.4755 | 0.0087 | 567.3773 | 0.0018 | 4.9563 | 0.2018 |
| 27 | 137.3706 | 0.0073 | 681.8528 | 0.0015 | 4.9636 | 0.2015 |
| 28 | 164.8447 | 0.0061 | 819.2233 | 0.0012 | 4.9697 | 0.2012 |
| 29 | 197.8136 | 0.0051 | 984.0680 | 0.0010 | 4.9747 | 0.2010 |
| 30 | 237.3763 | 0.0042 | 1181.8816 | 0.00085 | 4.9789 | 0.20085 |
| 32 | 341.8219 | 0.0029 | 1704.1095 | 0.00059 | 4.9854 | 0.20059 |
| 34 | 492.2235 | 0.0020 | 2456.1176 | 0.00041 | 4.9898 | 0.20041 |
| 36 | 708.8019 | 0.0014 | 3539.0094 | 0.00028 | 4.9929 | 0.20028 |
| 38 | 1020.6747 | 0.0010 | 5098.3735 | 0.00020 | 4.9951 | 0.20020 |
| 40 | 1469.7716 | 0.00068 | 7343.8578 | 0.00014 | 4.9966 | 0.20014 |
| 42 | 2116.4711 | 0.00047 | 10577.355 | 0.000095 | 4.9976 | 0.200095 |
| 44 | 3047.7183 | 0.00033 | 15233.592 | 0.000066 | 4.9984 | 0.200066 |
| 46 | 4388.7144 | 0.00023 | 21938.572 | 0.000046 | 4.9989 | 0.200046 |
| 48 | 6319.7487 | 0.00016 | 31593.744 | 0.000032 | 4.9992 | 0.200032 |
| 50 | 9100.4382 | 0.00011 | 45497.191 | 0.000022 | 4.9995 | 0.200022 |
| 55 | 22644.802 | 0.000044 | 113219.01 | 0.0000088 | 4.9998 | 0.2000088 |
| 60 | 56347.514 | 0.000018 | 281732.57 | 0.0000035 | 4.9999 | 0.2000035 |
| 65 | 140210.65 | 0.0000071 | 701048.23 | 0.0000014 | 4.99996 | 0.2000014 |
| 70 | 348888.96 | 0.0000029 | 1744439.8 | 0.00000057 | 4.99999 | 0.20000057 |
| 75 | 868147.37 | 0.0000012 | 4340731.8 | 0.00000023 | 4.999994 | 0.20000023 |
| 80 | 2160228.5 | 0.00000046 | 10801137 | 0.000000093 | 4.999998 | 0.200000093 |
| 85 | 5375339.7 | 0.00000019 | 26876693 | 0.000000037 | 4.999999 | 0.200000037 |
| 90 | 13375565 | 0.000000075 | 66877821 | 0.000000015 | 4.9999996 | 0.200000015 |
| 100 | 82817975 | 0.000000012 | 414089868 | 0.000000002 | 4.99999994 | 0.200000002 |

**Table A.5** *Values of Compound Interest Factors with i = 25%*

| | Single Sums | | Uniform Series | | | |
|---|---|---|---|---|---|---|
| n | Find F<br>Given P<br>F\|P i, n | Find P<br>Given F<br>P\|F i, n | Find F<br>Given A<br>F\|A i, n | Find A<br>Given F<br>A\|F i, n | Find P<br>Given A<br>P\|A i, n | Find A<br>Given P<br>A\|P i, n |
| 1 | 1.2500 | 0.8000 | 1.0000 | 1.0000 | 0.8000 | 1.2500 |
| 2 | 1.5625 | 0.6400 | 2.2500 | 0.4444 | 1.4400 | 0.6944 |
| 3 | 1.9531 | 0.5120 | 3.8125 | 0.2623 | 1.9520 | 0.5123 |
| 4 | 2.4414 | 0.4096 | 5.7656 | 0.1734 | 2.3616 | 0.4234 |
| 5 | 3.0518 | 0.3277 | 8.2070 | 0.1218 | 2.6893 | 0.3718 |
| 6 | 3.8147 | 0.2621 | 11.2588 | 0.0888 | 2.9514 | 0.3388 |
| 7 | 4.7684 | 0.2097 | 15.0735 | 0.0663 | 3.1611 | 0.3163 |
| 8 | 5.9605 | 0.1678 | 19.8419 | 0.0504 | 3.3289 | 0.3004 |
| 9 | 7.4506 | 0.1342 | 25.8023 | 0.0388 | 3.4631 | 0.2888 |
| 10 | 9.3132 | 0.1074 | 33.2529 | 0.0301 | 3.5705 | 0.2801 |
| 11 | 11.6415 | 0.0859 | 42.5661 | 0.0235 | 3.6564 | 0.2735 |
| 12 | 14.5519 | 0.0687 | 54.2077 | 0.0184 | 3.7251 | 0.2684 |
| 13 | 18.1899 | 0.0550 | 68.7596 | 0.0145 | 3.7801 | 0.2645 |
| 14 | 22.7374 | 0.0440 | 86.9495 | 0.0115 | 3.8241 | 0.2615 |
| 15 | 28.4217 | 0.0352 | 109.6868 | 0.0091 | 3.8593 | 0.2591 |
| 16 | 35.5271 | 0.0281 | 138.1085 | 0.0072 | 3.8874 | 0.2572 |
| 17 | 44.4089 | 0.0225 | 173.6357 | 0.0058 | 3.9099 | 0.2558 |
| 18 | 55.5112 | 0.0180 | 218.0446 | 0.0046 | 3.9279 | 0.2546 |
| 19 | 69.3889 | 0.0144 | 273.5558 | 0.0037 | 3.9424 | 0.2537 |
| 20 | 86.7362 | 0.0115 | 342.9447 | 0.0029 | 3.9539 | 0.2529 |
| 21 | 108.4202 | 0.0092 | 429.6809 | 0.0023 | 3.9631 | 0.2523 |
| 22 | 135.5253 | 0.0074 | 538.1011 | 0.0019 | 3.9705 | 0.2519 |
| 23 | 169.4066 | 0.0059 | 673.6264 | 0.0015 | 3.9764 | 0.2515 |
| 24 | 211.7582 | 0.0047 | 843.0329 | 0.0012 | 3.9811 | 0.2512 |
| 25 | 264.6978 | 0.0038 | 1054.7912 | 0.0009 | 3.9849 | 0.2509 |
| 26 | 330.8722 | 0.0030 | 1319.4890 | 0.0008 | 3.9879 | 0.2508 |
| 27 | 413.5903 | 0.0024 | 1650.3612 | 0.0006 | 3.9903 | 0.2506 |
| 28 | 516.9879 | 0.0019 | 2063.9515 | 0.0005 | 3.9923 | 0.2505 |
| 29 | 646.2349 | 0.0015 | 2580.9394 | 0.0004 | 3.9938 | 0.2504 |
| 30 | 807.7936 | 0.0012 | 3227.1743 | 0.0003 | 3.9950 | 0.2503 |
| 32 | 1262.1774 | 0.00079 | 5044.7098 | 0.0002 | 3.9968 | 0.2502 |
| 34 | 1972.1523 | 0.00051 | 7884.6091 | 0.0001268 | 3.9980 | 0.2501268 |
| 36 | 3081.4879 | 0.00032 | 12321.9516 | 0.0000812 | 3.9987 | 0.2500812 |
| 38 | 4814.8249 | 0.00021 | 19255.2994 | 0.0000519 | 3.9992 | 0.2500519 |
| 40 | 7523.1638 | 0.00013 | 30088.6554 | 0.0000332 | 3.9995 | 0.2500332 |
| 42 | 11754.9435 | 0.0000851 | 47015.7740 | 0.0000213 | 3.9997 | 0.2500213 |
| 44 | 18367.0992 | 0.0000544 | 73464.3969 | 0.0000136 | 3.9998 | 0.2500136 |
| 46 | 28698.5925 | 0.0000348 | 114790.3702 | 0.0000087 | 3.9999 | 0.2500087 |
| 48 | 44841.5509 | 0.0000223 | 179362.2034 | 0.0000056 | 3.99991 | 0.2500056 |
| 50 | 70064.9232 | 0.0000143 | 280255.6929 | 0.0000036 | 3.99994 | 0.2500036 |
| 55 | 213821.177 | 0.0000047 | 855280.7072 | 0.0000012 | 3.99998 | 0.2500012 |
| 60 | 652530.447 | 0.0000015 | 2610117.787 | 0.000000383 | 3.999993870 | 0.250000383 |
| 65 | 1991364.89 | 0.000000502 | 7965455.556 | 0.000000126 | 3.999997991 | 0.250000126 |
| 70 | 6077163.36 | 0.000000165 | 24308649.43 | 0.000000041 | 3.999999342 | 0.250000041 |
| 75 | 18546030.8 | 0.000000054 | 74184119.01 | 0.000000013 | 3.999999784 | 0.250000013 |
| 80 | 56597994.2 | 0.000000018 | 226391973.0 | 0.000000004 | 3.999999929 | 0.250000004 |
| 85 | 172723371 | 0.0000000058 | 690893480.4 | 0.000000001 | 3.999999977 | 0.250000001 |
| 90 | 527109897 | 0.0000000019 | 2108439585 | 0.0000000005 | 3.9999999924 | 0.2500000005 |
| 100 | 4909093465 | 0.0000000002 | 19636373857 | 0.0000000001 | 3.9999999992 | 0.2500000001 |

# 14

## PREPARING, PRESENTING, IMPLEMENTING, AND MAINTAINING FACILITIES PLANS

### 14.1  INTRODUCTION

In the previous chapter, we considered the evaluation and selection of the facilities plan; here, we complete our consideration of facilities planning by addressing the physical preparation of the facilities plan, its presentation to management, the implementation of the selected facilities plan, and its maintenance over time. Obviously, the previous 13 chapters have little practical value unless the resultant facilities plan is properly prepared, presented, and sold. However, neither will its value be fully realized if it is not properly implemented and maintained.

### 14.2  PREPARING THE FACILITIES PLAN

If the facilities plan is not sold properly, the probability of acceptance by management is very low no matter how high its quality. As such, it is critical that careful attention be given to the "lessons learned" in justifying capital investments provided in the previous chapter under the seventh step in the SEAT process.

The facilities plan records the results of the entire facilities planning process. It is during the preparation of the facilities plan that decisions must be made concerning several extremely important details. The facilities plans resulting from the analysis of department arrangements yield an overall facility configuration and a department layout. To convert the overall facility configuration into a functioning facility requires the specification of details with respect to the total plot of land. To convert the department layout into a functioning facility, the details of each department must be specified. These plot and layout details may be specified on a plot plan and layout plan, respectively. Section 14.2 of this chapter describes plot plans and Section 14.3 describes layout plans.

There are three components required to successfully sell a facilities plan. The first is a high-quality facilities plan that has been prepared in a clear and accurate manner. The second component is a written report describing the benefits of a facilities plan and documenting why the particular facilities plan selected is best. A description of how to prepare a facilities planning written report is contained in Section 14.4. The third component, an oral presentation, is the focal point of an entire facilities planning project. It is after the oral presentation that a decision will be reached to either implement or not implement a facilities plan. This critical facilities planning component is described in Section 14.5.

In preparing the facilities plan, it is important for both a plot plan and a layout plan to be prepared. Each is described in the following sections.

## Plot Plans

A plot plan (Figure 14.1) is a drawing of the facility, the total site, and the features on the property that support the facility. Features that may be included on a plot plan include:

1. Access roads, driveways, and truck aprons.
2. Railways, waterways, aircraft runways, and heliports.
3. Yards and storage tanks and areas.
4. Parking lots and sidewalks.
5. Landscaping and recreation areas.
6. Utilities, including water, sewage, fire hydrants, gas, oil, electric, and telephone.

A plot plan should be drawn to a scale that is consistent with the size of the plot to be drawn. Common plot scales include 1/16 or 1/32 in. equal to 1 ft or 1 in. equal to 50 or 100 ft for larger areas.

Factors that must be considered when developing a plot plan include:

1. *Expansion:* A plot plan should be constructed while projecting space requirements 10 years into the future. Facility expansion should be planned in at least two directions and particular attention should be given to the expansion of functions that will be difficult to relocate. Sites should be planned so that plant services, utilities, docks, and railways need not be altered when expansion occurs.

2. *Flow patterns:* Separate traffic patterns for materials, employees, and visitors. Plan all flow patterns while considering security, ease of access, and integration of internal flow patterns with external flow patterns.

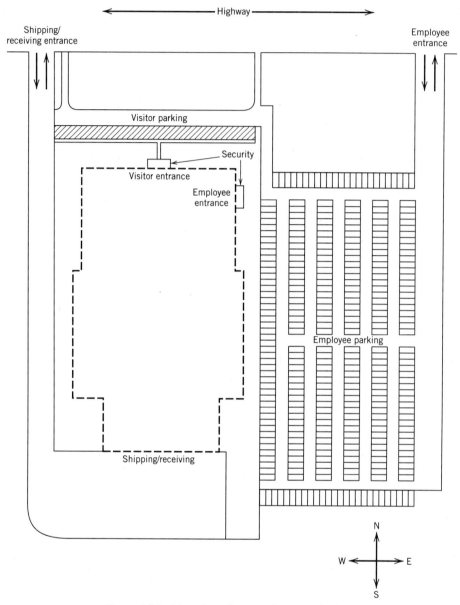

**Figure 14.1**   Plot plan of a manufacturing facility.

3.   *Energy:* Whenever possible, have the building face south and have the shorter dimension running north to south and the longer east to west. Consider the use of underground structures or at least consider the use of large amounts of back-fill on northern and western walls. Avoid asphalt or concrete areas around the building as much as possible, and use deciduous trees for shade during the summer while still allowing sunlight to penetrate in the winter.

4. *Aesthetics:* Locate the facility on the site to obscure unsightly activities. Plan landscaping and land use to blend into the environment and to promote an attractive setting.

## Layout Plans

Layout plans are detailed representations of facilities. The representations may be either two-dimensional and drawn by hand, constructed with templates, or drawn by computer graphics, or three-dimensional and consist of modular building blocks or detailed models. No matter what type of representation is utilized, the same procedure should be followed to create the layout plan. The following sections describe this procedure and illustrate the use of various types of representations.

### Constructing a Layout Plan

Prior to beginning the construction of the layout plan, the following data must be gathered.

1. A department layout that has been developed using appropriate layout techniques.
2. The department and area requirement sheets for each department.
3. The material handling planning charts for each product to be manufactured.

Once the above data are collected, the layout plan should be constructed using a systematic procedure. The systematic procedure for a manufacturing facility is as follows:

1. *Select the scale:* A scale should be selected so that the overall facility configuration, the department layout, and the available layout planning equipment fit appropriately. Whenever feasible, a scale of ¼ in. equals 1 ft should be utilized. If possible, use the same scale used by the architect, construction engineer, or other professionals working on the facility.
2. *Decide on the method of representation:* In general, the selection of the method of representation for layout plans should be based on a combination of clarity and economics. Three-dimensional representations should be used when the third dimension plays an important role in design decisions. Each situation should be treated individually and a decision made as to which approach will result in the most effective layout plan while minimizing the life cycle cost for the layout plan.
3. *Obtain layout plan supplies and/or equipment:* Depending on the method selected in step 2, possibly only paper and pencil are needed or possibly a microcomputer or minicomputer and printer are needed. Whatever supplies and/or equipment are needed, they should be obtained prior to proceeding with the construction of the layout plan.
4. *(For an existing facility) Locate all permanent facilities on the layout plans:* All columns, windows, doors, walls, ramps, stairs, elevators, sewers, cranes, and other permanent facilities should be the first facilities placed on the layout plan for an existing facility. Also when appropriate, floor loading and ceiling heights should be recorded on layout plans for existing facilities.

5. *Locate the exterior wall that includes the receiving function:* It is necessary to begin making decisions concerning the locations of functions in the facility. One approach that works well is to locate first the receiving function and then locate functions along the primary material flow paths until all manufacturing departments are included within the layout plan. Clearly, other starting points may also be selected but as long as the overall department layout, the department and area requirement sheets, and the material handling planning charts are followed, no significant difference should exist in the resulting layout plan.

6. *(For nonexisting facilities) Locate all columns:* The size, span, and location of columns must be among the initial decisions for a facility to be constructed. This must be done to avoid the interference of columns with material flow or with manufacturing equipment. Architectural or construction engineers should be consulted for the column requirements for the type of facility to be constructed. The column spacing decisions must often be made by analyzing the economic trade-off between facility construction cost and material flow cost.

7. *Locate all manufacturing departments and equipment:* Beginning with the receiving department, each department should be tentatively located in the layout plan based on the department layout. Aisles between departments should be included as appropriate. The details of each department should be developed based on the data recorded on the department area and requirements sheet. When necessary, department boundaries and aisles should be altered to allow department details to be shaped into the required department areas.

8. *Locate all personnel and plant services:* Alterations should be made to the manufacturing layout to include all personnel and plant services. Efforts should be made to have uniform aisle spacing and to integrate the service functions into the layout to maximize the use of production space.

9. *Audit the layout plan:* Prior to finalizing the layout plan, it should be audited from both the material perspective and the personnel perspective. The material perspective audit should involve tracing all material flows through the facility. The material handling planning charts should be used as guides and each material move, storage, operation, and inspection should be traced through the facility. The personnel perspective audit should involve tracing the tasks to be performed by every person to be employed within the facility. For example, a production worker's efforts should be traced through at least the following activities:

   a. Park automobile.

   b. Walk into facility.

   c. Hang up coat and store lunch.

   d. Sign in.

   e. Talk with supervisor.

   f. Report to work station.

   g. Begin production.

   h. Handle machine breakdown.

   i. Handle machine setup.

j.  Go to restroom.

k.  Get a drink of water.

l.  Take a break.

m.  Go to lunch.

n.  Receive first aid.

o.  Attend production meeting.

p.  Meet with human resources representative

q.  Perform housekeeping activity at work station.

r.  Interact with quality control.

s.  Interact with production control.

t.  Interact with material handling personnel.

u.  Report production difficulties.

v.  Report production milestones.

w.  Wash hands.

x.  Retrieve coat and lunch box.

y.  Sign out.

z.  Return to automobile.

These types of audits will often result in a change in work station orientation and/or spacing.

10. *Finalize the layout plan:* Once the layout plan is fully audited, it should be finalized by permanently locating everything on the layout plan and recording appropriate headings and clarifying notations. A detailed legend should be included and all unclear areas should be clarified by labeling and, when appropriate, by color coding.

Although a detailed procedure has been given for developing a layout plan, it should not be inferred that development is an easy or a totally straightforward task. To the contrary, while developing the layout plan the majority of data used and assumptions made will be questioned and requestioned. Many iterations will be necessary and many trial-and-error attempts will be made prior to obtaining an acceptable layout plan.

## *Alternative Methods of Representing Layout Plans*

As was previously stated, layout plans may be two- or three-dimensional representations. The alternative methods of representing two-dimensional layout plans are:

1. *Drawings:* Figure 14.2 illustrates a handmade drawing which serves as a rough sketch of an area. Drawings are quickly and conveniently made to illustrate an alternative layout plan. In addition, drawings may be the best approach for layout plans of small areas. However, handmade drawings are much too expensive to produce and to alter for use as final layout plans for large areas. Therefore, handmade drawings should be used as either rough sketches or as final layout plans for small areas.

2. *Templates and tapes:* A common method of creating layout plans for large facilities is to use templates and tapes. Templates may be homemade or pur-

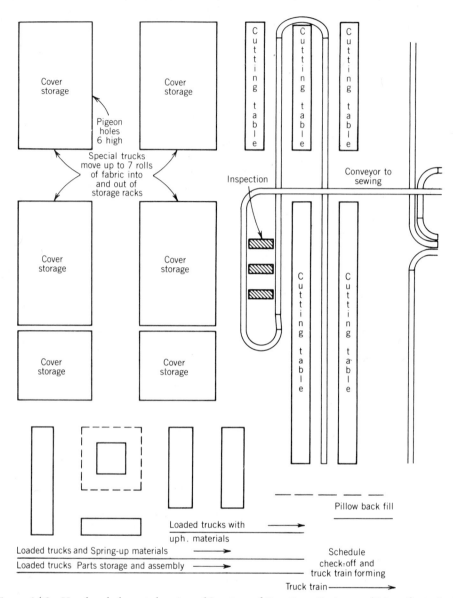

**Figure 14.2**  Handmade layout drawing. *(Courtesy of Furniture Design and Manufacturing)*

chased; typically, they are made from cardboard, mylar, adhesive-backed mylar or magnetized plastic. Templates may be either block templates or contour templates. A block template is simply a labeled rectangle representing the maximum length and width of equipment. A contour template illustrates the contour and the clearances for the movable portions of the machine. A broad range of tapes are available for representing walls, partitions, belt conveyors, roller conveyors, windows, overhead monorails, and aisles. In addition, different colored

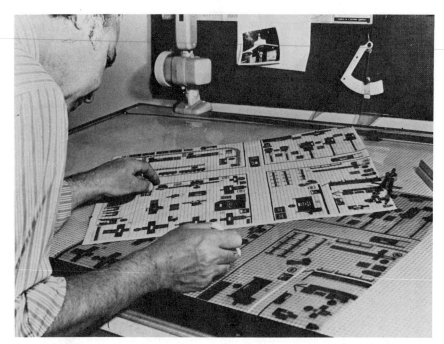

**Figure 14.3** A contour adhesive-backed mylar template and tape layout. *(Courtesy of A.D.S. Company)*

tapes and arrows are available for a wide range of uses. The flexibility and low cost of producing template and tape layouts makes it a popular approach. Figure 14.3 illustrates a contour adhesive-backed mylar template and tape layout. Figure 14.4 illustrates a magnetic template layout.

3. *Computer graphics:* A convenient approach for combining the ease of hand-drawn layouts with the quality of a template and tape layout is computer graphics. A computer graphics system consists of an input terminal, a central processor, and output peripherals. The input terminal typically consists of a keyboard, a light pen, a mouse, and a cathode ray tube. The keyboard, mouse, and light pen are used by the layout planner to provide instructions to the computer graphics system. The cathode ray tube is used to graphically display the layout being developed. The central processor translates the instructions from the layout planner into graphics layouts and stores the templates and symbols to be used in constructing the layout. The output peripherals can include line printers, impact and laser printers, and/or plotters. Figure 14.5 illustrates the use of a computer graphics system for layout planning. A layout may be developed by calling for specific templates and symbols from the central processor. Using the mouse or light pen, the templates and symbols are positioned to obtain a layout. Once a layout is complete, the layout planner may have the layout reproduced using any one of many output peripherals. In addition to the ease and quality of the resulting layouts, computer graphics have the following useful features:

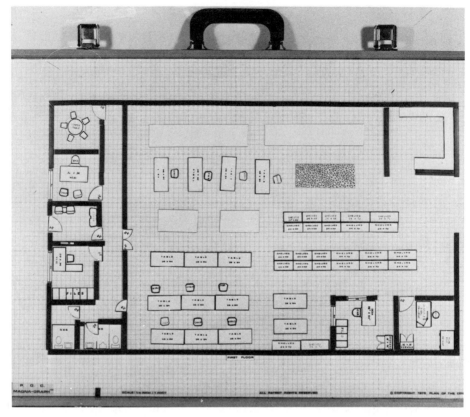

**Figure 14.4**   A magnetic template layout. *(Courtesy of Magna-Graph)*

a. The capability to view a layout from various perspectives. An entire facility, a department, a workstation, or a portion of a workstation may be viewed simply by requesting a different perspective.

b. The scale of layouts can be quickly and easily changed. This allows greater consistency between different users of layout plans.

c. Animated simulations can be performed to test the layout and demonstrate the operation of the system. Congestion points can be tested and evaluated by subjecting the design to peak and changing requirements. Animated 2-D and 3-D representations of the flow of people, material, equipment, information, and paper through the facility can be depicted, providing a degree of realism not possible otherwise. Transforming a static representation of the facilities plan into a dynamic representation, simulation is useful in testing and evaluating a design, as well as selling the design to management and other individuals who will be affected.

An addition to two-dimensional layout plans that often add considerable clarity is the use of overlays. Overlays are rarely used with drawings but are very common

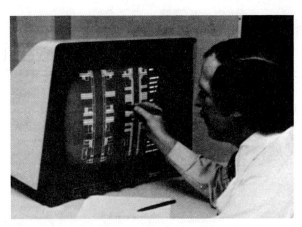

**Figure 14.5**   The use of a computer graphics system for layout planning. *(Courtesy of Computer Graphics, Inc.)*

with either template and tape layout plans or computer graphics layout plans. Overlays for template and tape layouts are typically made of transparent sheets of mylar or acetate and are placed over the layout plan to illustrate certain features. Computer graphics overlays are easily added and deleted. Plumbing, electrical, heating, and lighting overlays can provide considerable insight into the complexities involved with integrating plant services in a layout. An additional overlay that is extremely useful in a manufacturing facility is a material flow overlay. Material flow overlays facilitate evaluating, improving, and presenting the layout plan. All material flows within the facility should be recorded on the overlay. The color of the flow lines may indicate the product being moved or the type of material. Material types may include raw materials, in-process materials, or finished products. In addition, by either changing the width of the flow lines or by using various symbols on the flow lines, the volume of movement and the method of movement may be specified.

Three-dimensional models are the clearest and most easily understood method of developing layout plans. However, because of the difficulty of duplicating a three-dimensional model, a two-dimensional model is still needed. Often, the costs of three-dimensional models cannot be justified. Nevertheless, whenever the third dimension plays a critical role in the layout plan, or whenever a detailed presentation is to be made to a nontechnical group, three-dimensional models should be considered. There are three methods of establishing three-dimensional models.

1.  *Modular block models:* The use of modular block models to develop a three-dimensional model is illustrated in Figure 14.6. The advantage of modular block models is that special models of each machine need not be constructed. The block models may be made of modular building blocks. This significantly reduces the cost and ease of making three-dimensional models without significantly reducing the quality of the layout plan.

2.  *Scale models:* Scale models, like templates, may be either block models or contour models. Contour models show the actual machine contour rather than the

**Figure 14.6**  The use of modular block models to develop a three-dimensional model. *(Courtesy of Model Display Studios)*

machine travel. Figure 14.7 illustrates a contour model three-dimensional layout plan. Although these models are the most easily understood, their cost and difficulty to duplicate and transport limit their use.

3.  *Computer models:* Computerized 3-D representations of facilities can be produced, allowing the designer to rotate the facility, "enter the door," "walk through the facility," and make changes as needed.

In addition to using modular block models, scale models, and 3-D computer-generated representations of a facility, photographs may be used to provide a 3-D representation of facilities in 2-D space. The quality of the 3-D representation has improved in recent years to the point that photographs can be used to portray 3-D models when it is not feasible to take the physical model to the client.

Given the level of research activity underway in holography and "virtual reality," the computer will play a greater role in facilities planning in the future, in design, evaluation, and sales. However, traditional methods will be the "method of choice" for most applications for many years. Regardless of the method of representing the facilities plan, it must have a professional appearance. Color, overlays, tapes, templates, photographs, simulations, multimedia presentations, 2-D and 3-D models—all possibilities should be considered. The final choice will depend on the audience, as well as the costs (and benefits) associated with the alternative methods of presenting the facilities plan.

**Figure 14.7** Contour model three-dimensional layout plans. (*a*) An office layout. (*b*) An automated storage/retrieval system (AS/RS) layout. (*Courtesy of Visual Industrial Products, Inc.*) (*c*) A layout of the waste treatment system for a wool combing plant. (*d*) A two-level air-conditioning system layout. (*Courtesy of Burlington Industries, Inc.*)

**Figure 14.7**   (continued)

## *14.3*   PRESENTING THE FACILITIES PLAN

In addition to providing a 2-D or 3-D depiction of the recommended facilities plan, it is generally accompanied by a written report. Additionally, an oral presentation is typically used to communicate the plan to management in order to provide an opportunity for dialogue regarding the plan and any alternatives or modifications that might be of interest to management. Both the written report and the oral presentation are considered in the following sections.

## The Written Report

The objective of most facilities planning written reports is to describe the benefits of a facilities plan and document why the particular facilities plan selected is the best. It is not the objective of most written reports to describe the detailed procedures required to establish a facilities plan. Hence, it is surprising and disappointing to read facilities planning reports that dwell on the process of establishing the facilities plan and do little to sell it.

The process followed in developing the written report should include the following steps.

1. *Define the objective of the report:* The objective should be written and should state what conclusions should be reached by anyone reading the report. These objectives should be frequently referred to while writing the report.
2. *Establish a table of contents:* The table of contents serves as an outline of a written report. It forces one to organize the material to be placed in the report and to verify that a logical train of thought exists from the introduction to the conclusions.
3. *Determine who is going to read the report:* The style, level of detail, and length of the report will be dictated by the person(s) to whom the report is written. Clearly, a report written to the chief facilities planner would differ from a report that is to be written for a chief executive officer.
4. *Write the report:* The report should be brief but should accurately communicate the reasons for implementing the facilities plan. It should be assumed that the reader will believe the results presented. Therefore, the reader should not be forced to read through detailed justifications or explanations. Justifications and explanations should be included in tables or figures. Full documentation of tables and figures should be placed in appendixes.
5. *Document the report:* Prepare appendixes that define the sources of data and assumptions included in the report. Include within them data and procedural details describing exactly how the conclusions reached in the report were determined. Provide sufficient information in the appendixes so that the reader can re-create the conclusions reached in the body of the report.
6. *Edit the report:* Prior to finalizing the report, allow a few days to pass and then critically review it. Focus your attention on the objective of the report, and need for brevity and clarity.

It is difficult to describe specifically what should be contained in a facilities planning written report. If possible, successful facilities planning reports that have been prepared previously for an organization should be reviewed. A format that succeeded previously within an organization is certainly a format meriting consideration. A format that has been successful is as follows:

- Letter of transmittal
- Cover page

- Executive summary
- Table of contents
- Introduction
- Body
- Implementation plan
- Conclusions
- Appendixes

The **letter of transmittal** is the document that introduces the reader to the report. It sets the stage for the reader by stating what is in the report and why it is being given to the reader. In addition, the letter of transmittal may be used to recognize persons who contributed to the report and to describe future reports or activities that pertain to the report being transmitted.

The **cover page** includes the title of the report, who the report was prepared for, who prepared the report, and the report date.

The **executive summary** lists the major findings of the report. It should not exceed a page and should describe all conclusions reached in the report.

The **table of contents** for a report is like the table of contents for a textbook.

The **introduction** describes the background of the report. The objectives, scope, and boundaries of the report are given in the introduction.

The **body** of the report is broken down into major sections. It leads the reader through the justification and benefits of the facilities plan.

Calculations and listings of data appear in the **appendixes,** not in the body. The appendixes should be referenced in the body.

The **implementation plan** is the most important section in many reports. It describes the specific activities and the dates for completing the activities required to fully implement the facility plan described in the report.

The **conclusions** reiterate the information present in the introduction, body, and implementation plan. The conclusions differ from the executive summary in that the conclusions present a narrative summary of the entire report, whereas the executive summary is simply a listing of the major findings. As has been previously stated, the appendixes contain all supporting data and all backup information. If the appendixes are very long, it may be appropriate to bind them under separate cover.

At least 2 days prior to the oral presentation, the written report should be given to all persons who will be attending the oral presentation. After distributing the report, it may be wise to check with a few key persons who will attend the oral presentation and learn their reactions. Their reactions may be very helpful in preparing the oral presentation.

## The Oral Presentation

The first step in preparing an oral presentation is to determine to whom the presentation is to be made. The perspective of each person attending the presentation should be analyzed and it should be determined what each person expects to hear. Fully understanding the audience and viewing the presentation from their perspective allows the presenter to predict what questions will be asked and what benefits to emphasize. Once an understanding is obtained of who is to attend the presentation, a decision needs to be made concerning what should be presented. The two critical factors to be remembered while deciding what to present are brevity and the need for a logical progression from what to why to how.

The oral presentation should be brief. For most projects, a 1-hr planned oral presentation is all that is necessary. If an oral presentation can be done effectively in 15 min, then only plan for a 15-min presentation.

Oral presentations should begin by stating *what* should be done, describing *why* these things should be done, then *how* to do the things recommended. One should not attempt in an oral presentation of a facilities plan to train the attendee in *how* to establish a facilities plan. To the contrary, an oral presentation should dwell on *what* facilities plans should be implemented, *who* will be impacted, *why* the facilities plans should be implemented, and *how* to implement the facilities plans. Helpful guidelines that may be useful in preparing and presenting an oral presentation include the following:

1. Do not oversell. Be realistic and make certain that all statements may be proven.

2. Refer to other systems that are similar to the one being recommended. Be certain the audience knows that what is being recommended has been successful elsewhere.

3. Just prior to concluding the oral presentation, describe the implementation plan. Explain how to accomplish what is being recommended.

4. Conclude the presentation with the tabulation of the results. Show cost savings, improvements in space utilization, and so on.

5. Rehearse the oral presentation. This breeds self-confidence and makes the presentation easier to accept.

6. Prepare quality visuals. Do not try to place too much information on any one visual. Do not use reproductions from the report.

7. Be appropriately dressed for the presentation. Look successful and your presentation will be much more believable.

8. Begin the presentation on time. Have visual aid equipment ready and focused.

9. Address all objections and questions. Be aware and communicate to the audience that objections and questions are good to get out in the open. Listen carefully and allow objections and questions to be fully voiced. Be open-minded.

10. If you do not know the answer to a question, respond: "I do not know, but I will find out," and do find out!

## *14.4*  IMPLEMENTING THE FACILITIES PLAN

Many excellent systems designs have failed to deliver the promised benefits because of a failure to implement the system effectively. Implementation failures tend either to be technical problems or people problems. Experience suggests that the more frequent and more formidable problem is the people problem.

Invariably a new or improved facilities plan will require that changes be made, and people tend to resist change. The process of managing change and dealing with resistance to change has been studied by many others. Hence, we will not dwell on the subject. Rather, it is recommended that [1], [3], [5], and [6] be consulted.

In the final analysis, the ability to sell the facilities plan is dependent on one's ability to deal effectively with resistance to change. Krick [4] lists a number of causes of resistance to change, including:

1. Inertia
2. Uncertainty
3. Failure to see the need for the proposed change
4. Fear of obsolescence
5. Loss of job content
6. Personality conflict
7. Resentment of implied criticism
8. Poor approach by the facilities planner
9. Inopportune timing
10. Resentment of outside help

Some methods of minimizing resistance to change are also recommended by Krick [4], including:

1. Explaining the need for the change thoroughly and convincingly
2. Involving many people in the development of the facilities plan
3. Tactfully introducing the proposal, being careful of the timing
4. Introducing major changes in stages or steps
5. Emphasizing those aspects of the change most beneficial to the individuals involved
6. Letting others "think of the answer," and letting others receive the credit
7. Being genuinely interested in the feelings and reactions of others.
8. Having changes announced and introduced by the immediate supervisors of those affected.

In implementing the facilities plan it is important for the facilities planner to be sensitive to the sensitivities of all parties involved. Selling the facilities plan does not end with top management; each individual employee must believe the new system will work. Managers may be the decision *makers,* but operating personnel are generally the decision *wreckers.* Hence, it is critical that employees be prepared for the new plan.

As important as are the steps of defining, analyzing, generating, evaluating, and selecting in developing facilities plans, the real payoff comes from implementing. Improvement occurs *if and only if* the plan is implemented properly [2].

Implementation is a tedious, often frustrating "fire-fighting" activity that must be pursued with vigor. Implementation must be carried out by one who has been intimately involved with the selection of the solution and who has the capability and time to coordinate and manage diverse activities. The implementation process includes the following phases [8].

1. *Document the solution:* Develop a series of manuals describing the systems operation, standard operating procedures, systems staffing, maintenance policies, and managerial responsibilities.

2. *Plan for solution implementation:* Develop a Gantt chart or a critical-path network of the activities required for the system to become operational.

3. *Select vendors:* Prepare functional specifications, mail specifications to qualified vendors, and select the vendors to participate in the systems implementation.

4. *Purchase equipment:* Place an order for all components, supports, controls, attachments, and containers required for the system to become operational.

5. *Supervise installation:* Maintain control of the implementation plan and exercise judgment when modifications become necessary.

6. *Make ready:* Establish and install a training program and oversee employee acceptance.

7. *Start up:* Begin operation.

8. *Debug:* Work with employees to ensure the system meets *their* expectations. Check with all shifts; identify and rectify difficulties.

9. *Follow up:* Perform a post-installation audit and verify that the system is operating as *you* expected. If not functioning as expected, determine the causes and rectify.

10. *Maintenance:* Routinely oversee system operation and maintenance to verify the system's continual success.

In discussing the importance of implementation in the context of automated material handling and storage systems, Searls [7] claimed it was the key to a successful system. In defending his claim, he noted that such systems typically involve many different interacting elements. Specifically, fire protection; mechanical and electrical components associated with guided vehicles, conveyors, and storage/retrieval machines; storage racks; and computers must be integrated. If any one component is not installed correctly, it will not perform correctly. Even though "forgiveness" may have been designed into the system, if the components do not function correctly, then neither will the system function correctly. Furthermore, if the system does not function properly, then the planned (and promised) productivity improvements will not be realized.

Searls [7] addresses the implementation problem from the perspective of a storage rack supplier; however, his remarks are pertinent to the generic issue of implementing facilities plans. For this reason the following is excerpted from his presentation.

> Today, most automatic stacker crane systems utilize a control scheme based on XYZ coordinates, and require relative precision in the location of rack elements. The rack must be plumbed to a tolerance that the crane manufacturer's control method and his equipment can tolerate. A prerequisite then to a successful rack installation is a complete and accurate understanding of the overall equipment operating requirements.
>
> It is the responsibility of the system supplier to interpret these overall requirements; and if there is a separate contract for rack, transmit them to his rack supplier, who in turn, must convey them to the installer.

Before the installation contractor can offer a meaningful quotation for installation work, it will be necessary for him to visit the site to determine what constraints affect his ability to perform efficiently. He must determine if there are access roads to the site, if there is sufficient area to store his material near the work area, and he needs to determine the availability of water, power, and other required utilities. In the case of free-standing installations, he needs to check for overhead obstructions and interferences. Most important of all, he must have a clear understanding of the conditions of the floor on which he is to set his rack.

The high spot and low spot is found which determines how much shimming is required. A poor floor can mean that tons of shims will be required at $1,000/ton.

Assuming the scope of work is clear, site conditions understood, and an acceptable proposal made, a signed contract serves to initiate the work.

The construction contract should establish responsibility for insurance to be carried by the several subcontractors on site, and identify those areas where the general contractor has responsibility. Some states allow contractors to self-insure; others require third party insurers. Some construction sites require that all contractors use the same insurance carrier, so that disputes as to responsibility are avoided. It is important to establish when title passes to the owner, as this will establish responsibility for insurance, security, and liability. The construction contract will recognize permits that are required and establish who has responsibility to obtain them—it will also provide for access to the work by the owner or his representative—for inspecting progress and the quality of work.

Contractors should recognize the importance of maintaining an agreeable working relationship with one another on the construction site. The construction site has confusions that arise with the intermingling of different contractors and trades, and a certain amount of give and take is required to get the job done. Those areas where there is a direct interface require particular attention—these would include the fire protection contractor, who will be hanging piping from the rack, and the electrical contractors, who may be attaching conduits and conductor bars to the different parts of the working system.

A good contract contains a clear statement of the work to be done, along with responsibilities of each contractor and of the owner. It will include detailed schedules with easily identifiable milestones so that progress can be monitored by the interested parties. It is essential that there is a clear understanding of responsibilities and reporting lines. Generally, the systems supplier bears responsibility of conveying all the requirements of the owner, as well as his own, to the subcontractors. On occasion, although it is not recommended for the inexperienced, the owner will contract for rack, crane, etc. separately, and act as his own systems coordinator. He may also elect to retain a consultant to monitor progress or, perhaps, even to assume overall responsibility.

Such arrangements generally increase the risk of the individual sub-system suppliers. If the owner, or his representative, is not totally competent and experienced in design and implementation, system integration errors are likely to result, and attempts will be made to pass responsibility through to the individual suppliers. In such instances, the equipment suppliers' management has to be strong enough and knowledgeable enough to withstand efforts to expand his scope of work to cover the integration errors.

Even with great care in overall planning, changes to the scope of work are likely. It is important that the mechanism for handling changes to the scope of work, and the conditions under which added or revised work will be done, be clearly spelled out in the contract. Quite often, there can be tremendous pressure on the installation supervisor to make adjustments, just to keep the work moving. He may have a site crew of 50 or 100 people, with a payroll as high as $20,000 a day, and he is pressed to keep the work moving. Work that is essential, but is not anticipated by contractual arrangements can force expensive delays. Therefore, it is essential that the installation contractor be cognizant not only of the work to be accomplished today, but he must look ahead to tomorrow's and next week's activity, and identify the entire work scope prior to site activities.

The contract should provide for the possibilities of the parties reaching an impasse on a disagreement. Generally, this is provided for by the contract calling for the establishment of an arbitration procedure.

Once a proposal has been made and accepted, and a contract executed, "Project Management"—the business of getting on with the work in a manner that will accomplish time and cost objectives—requires a very careful and well executed plan. Receiving and storing material, the flow of that material through any required on-site fabrication, and through the erection process itself is an industrial engineering feat every bit as complex as that of a manufacturing plant. The magnitude of the feat can be appreciated when it is recognized that the site installation management must organize and execute his work plan with a force that is temporary and of varying levels of skill, and which is seldom motivated by loyalty to the employer. The conditions at the construction site, the independent nature of the work force, and the pressures to complete schedules on time generate problems for installation site management, that are similar but often more intense than those that are faced by their counterparts in manufacturing.

Needless to say, strong personalities emerge at site management, which in itself can create still additional problems. Compromise has to be the order of the day, and disputes must be minimized and resolved quickly. Laborers who may have failed grade school math can figure their paycheck, with all of the additions and deletions required in our bureaucratic society, quicker and more accurately than a computer, and the site clerk must be quick to respond to errors in order to maintain tranquility.

Control of the job site is essential to productivity, and once lost is almost impossible to regain. Therefore, foremen and site superintendents must be quick to discipline, and quick to remove nonproductive or disruptive personnel.

The decision to use union or non-union labor depends on many factors. In some areas, this option is not open. In other areas, particularly right to work states, non-union construction is possible and feasible. If you are a union-installer, it may be possible to work in a non-union area. If you are a non-union installer, you cannot work on job sites where other contractors are union, and where union trades are on site. In areas where the unions are particularly powerful, it may be impossible to even ship material on to the site if the material was not fabricated with union labor. These conditions vary from location to location and from time to time, and help to contribute to the uncertainties that are part of the installation business.

There are advantages to the installer in using union construction. A well-run, well-disciplined union can provide a well-trained and disciplined work force. Unions will also enforce work rules with consistency, which allow the knowledgeable installation contractor to estimate his costs accurately. A pre-job meeting with the appropriate business agents will help prepare the way for a trouble free operation. Problems with local business agents can be referred to the national level, where national agreements are in operation, which is a further assurance of a predictable working situation.

Most rack installations are handled with a composite crew of millwrights and ironworkers, with representation worked out by the two trades. Perhaps the worst experience that the installer can have is to be involved in a jurisdictional dispute between two trades.

The effectiveness of labor on the job site can be enhanced by the careful selection and application of erection equipment. Job trailers should be in good condition and kept neat and organized to set the tone for the performance expected on the job site. Cranes, fork trucks, welders, and handling equipments should be provided with the objective of utilizing manpower as effectively as possible. Lock boxes and cribs must be provided along with procedures for checking out tooling so that pilferage is discouraged.

The site superintendent on an installation project must work very closely with the project engineer to obtain interpretation of drawings and resolutions of questions so as to keep the project moving. Even a well designed system requires input from the project engineer. The worst possible situation is design or redesign at the job site, with engineering details being resolved while project crews wait.

Quality control and inspection are essential to avoiding costly rework and repair situations. Mistakes found in the field are rectified with cutting torches and sledge hammers. They are costly.

Administration at the site must provide for timely and accurate records to meet the legal requirements of the construction business. A laborer cannot be terminated without being paid—so there must be a high degree of flexibility in administrative systems being utilized on the construction site. Any subcontracts based on time and material need to be tracked very carefully to avoid excessive costs. Local, state and national safety requirements must be observed, and records kept of all injuries, so as to avoid fines and unplanned shutdowns.

Relations between the system supplier's project team and the user need to be close and cooperative. A good understanding of the overall system requirements and a close knowledge of daily operations can enhance relationships. Specifications can sometimes be relaxed in certain areas to accommodate construction problems if working customer/contractor relationships are fully developed.

It is essential that installation site management track their progress against planned schedule, and report that progress to their own management, as well as to the customers. Problems that are recognized early are most easily dealt with. Re-planning is a continuing activity on site. On large projects, critical path scheduling may be employed, and is essential when the project is too complex to yield to intuition.

Some projects which are well-designed and carefully executed still face problems toward the end of the installation contract. If workers fear no new work is available, productivity can fall off markedly as the job nears completion.

Rack installation projects (large ones) obtain a high degree of efficiency of certain specified movements, such as the standing of frames. As this work is completed, crews are cut back and the less glamorous functions, such as plumbing and leveling, can become disorganized. This last 5% of the job can eat up a disproportionate share of the resources unless the site superintendent plans the completion of the project just as carefully as he planned the rest of the work. He must be very familiar with the contractual requirements, so there are no unpleasant surprises at the end of a project. Correcting a 100,000 opening rack for any sort of systemic error can use a lot of manhours.

Even assuming that all rack specifications have been met, the installation is not complete until after the system startup. There will be instances of interference between the stacker cranes, shuttle forks, and sprinkler heads, or support arms.

Finally, the installation contractor should be very careful about leaving the site in good condition. Long months of quality workmanship can be blemished by the impressions left by an inadequate cleanup job. People tend to think in exceptions, and the customer's last impression should be a good one.

Installation is a risky business. Problems on the job site are expensive and costs escalate quickly. In automatic storage retrieval systems, the most common problem is dimensional out-of-tolerance. Recognizing that tolerance is a frequent problem, we should anticipate it and approach the design and installation so as to minimize it.

The advent of the rack-supported building presents new requirements on design and installation. An exposed rack structure is more susceptible to wind loading than the completed building by a factor of 2-½. This means that rack-supported building structures must be well-guyed during erection to prevent damage that would be no problem once the building is enclosed. The same is true of paint finishes. The exposure of the rack to sun, wind and rain for extended periods of time, places a more severe requirement on the finish than it will ever see in use. The system supplier and ultimate customer must be made aware of the results of this exposure so that controversy over oxidized paint or rust spots can be avoided. Once the rack structure is enclosed, it is doubtful that any paint is needed if warehouse temperatures are maintained above the dew point.

Strikes invariably increase administrative expenses, and attention should be given to scheduling rack installation projects to avoid periods when contracts will be under negotiation.

Picket lines thrown up by unions other than the construction trades may nonetheless shut down a project. Generally, these problems can be averted by opening up a different "construction gate," and obtaining an injunction against that gate being picketed. Expect at least two or three days of production to be lost if you are faced with this problem. Jurisdictional problems cannot be avoided completely, but it pays to carefully monitor previous practices in the construction area. When problems do arise, it is essential to get in contact with the appropriate business agents, and force the problem to the highest level in the union organization as quickly as possible.

Theft can be a serious problem on a construction site. The high turnover of personnel with a great freedom of access suggests that small tools and equipment be locked up in secure cribs and major equipment bunched together in a protected and lighted area. Common sense precautions are probably the best protection against loss in this area.

Fire can be a problem if trash is allowed to accumulate at the site, or if construction is going on in an existing or adjacent to an existing building. On-site motor fuels, of course, demand special care.

Government regulations can be a problem if not adhered to, and fines and shutdowns can result. Best solution to this problem is communication and education.

## *14.5.*  MAINTAINING THE FACILITIES PLAN

Once the facilities plan has been implemented, it would be nice to conclude that the planning effort is finished. However, as we noted in Chapter 1, the facilities planning cycle is continuous. Changing conditions and requirements necessitate that the facilities plan be audited periodically to ascertain the need for modifications and revisions.

The facilities planning audit involves a comprehensive, systematic examination of the location of the facility, the layout and material handling system, and the building design. The utilization of space, equipment, personnel, energy, and financial resources; the performance of the facility under current conditions; and the capacity of the facility to meet future requirements are to be assessed.

In performing the facilities planning audit, the following areas should be examined [10]:

1.   Layouts, to verify the locations of equipment, storage areas, offices, doors, aisles, and service areas are documented accurately.

2.   Installation, to verify the hardware and software systems installed were, in fact, what were designed and/or purchased.

3.   System performance, to verify satisfactory performance is being provided, focusing on throughput, personnel required, software, uptime, return on investment, costs, and productivity.

4.   Requirements, to verify sufficient capacity exist to meet requirements for space, equipment, and personnel.

5.   Management of the system, to verify the policies, procedures, controls, and reporting channels support effective management of the system.

The audit should be performed in order to learn from the past, to assess the need for change, to develop a database, and to establish or improve credibility within the organization. Despite the fact many benefits can be obtained from formal system audits, very few firms appear to perform such audits.

Some of the more common reasons for firms failing to conduct audits are [10]:

1.   *Fear:* Being afraid promises weren't kept, benefits weren't realized, and resources were wasted.

2.   *Mobility:* The lead times for planning, designing, and installing the system are such that the persons responsible have either been promoted, transferred, or left the company.

3.   *Lack of accountability:* Typically the persons responsible for planning, designing, and selling the system aren't responsible for installing, operating, and maintaining the system.

4.   *Time pressures:* Other needs exist and a "There's never time to do it right" attitude exists.

5.   *Changing requirements:* Due to internal and external factors, by the time the system is installed the need has changed.

6.   *Lack of baseline:* The current status of the system cannot be compared with anything; neither relative nor absolute comparisons can be made.

7.   *Hindsight attitude:* With the belief that "Anyone can look back, but the real pros in the business are those who only look ahead," such an attitude discourages audits.

As with many situations, success has many parents, but failure is an orphan. Unfortunately, audits are associated with the binary judgments of failure and success. Yet, the terms success and failure are generally related to *outcomes,* not *decisions.* It is important to recognize the differences in decisions and outcomes. Good decisions can have good or bad outcomes; likewise bad decisions can have good or bad outcomes. The merit of the decision is often related to the extent to which it was objective, analytical, comprehensive, repeatable, and explainable.

In terms of where the audit should be performed, it is essential that the audit be performed on site. Too often, quick judgments are made concerning systems;

BOLD (based on little data) conclusions are drawn from comments made by persons lacking objectivity. Disgruntled suppliers, jealous users, and politically motivated employees through innuendo and rumor can cast doubts on a system that are difficult to overcome.

The audit must be performed at a level in the organization sufficient to allow it to be performed objectively. The report should be submitted at least at the funding approval level.

The audit should be designed when the system is designed. Too often, audits are designed *after* the system is installed. Such an approach is similar to playing a game and *then* deciding how the scoring will be performed, or submitting bid packages, receiving vendor responses, and *then* deciding what factors will be used to select the supplier. One should know what things are important from an audit point of view when the system is being designed.

The initial audit is best performed approximately 6 months after it has been accepted by the user. To do it sooner is to invite the natural bias of "buyer's remorse." Immediately following startup and during debugging, the user typically undergoes a period of discouragement that can bias attempts to assess the system.

Not only can the audit be performed too soon, but it can also be performed too late. If the initial audit is delayed, then bad habits will have been developed, firm judgments will have been made, and opportunities for improvements will have been foregone.

After the initial audit, annual audits are appropriate for some installations. For others, periodic, but indefinite timing, is preferred.

To facilitate the audit, a checklist should be prepared. It is important for such a checklist to be developed when the audit is being designed.

Performance measures and/or productivity ratios can provide useful information on a continuing basis concerning the performance of the system. A number of such ratios are given in [9]. As examples, consider the following ratios:

$$\text{Aisle space percentage} = \frac{\text{Cube space occupied by aisles}}{\text{Total cube space}}$$

$$\text{Damaged loads ratio} = \frac{\text{Number of damaged loads}}{\text{Number of loads}}$$

$$\text{Energy utilization index} = \frac{\text{BTUs consumed per day}}{\text{Cubic space}}$$

$$\text{Manufacturing cycle efficiency} = \frac{\text{Total time spent on machining}}{\text{Total time spent in productive system}}$$

$$\text{Order picking productivity ratio} = \frac{\text{Equivalent lines or orders picked per day}}{\text{Labor hours required per day}}$$

$$\text{Slot occupancy ratio} = \frac{\text{Number of occupied storage slots}}{\text{Total number of storage slots}}$$

$$\text{Storage space utilization} = \frac{\text{Cube space of material in storage}}{\text{Total cube space required}}$$

Care must be taken to ensure that the performance measures used are reasonable and do reflect positive performance. Additionally, the data required for calculating the performance measure must be readily available.

To be effective, the audit must be comprehensive. Hence, managers, operators, maintenance engineers, designers, system suppliers, and functions being served by the system should be included in the audit. System suppliers can frequently provide recommendations for change or modification.

It is important that the audit be performed formally and that comparisons be made of:

1. Promises versus deliveries
2. Expectations versus realizations
3. Forecasts versus actual occurrences
4. Assumptions versus outcomes
5. Concepts versus installations

Where differences exist, they must be explained.

The results of the audit should be reported formally; they should be reviewed with those affected before being submitted. It is important that feedback occur to ensure that opinions or perceptions are minimized and facts are maximized. Actions for change or continuation must be recommended and the impact of the recommendations must be assessed.

Management must provide the right climate in order for audits to be productive. An adversary relationship must not exist; otherwise the audit process will be doomed to failure. Differences in decisions and outcomes must be recognized.

Management must be patient. Complex systems must be debugged. The period of time following startup will tend to be a frustrating period; management support during this period is crucial.

Engineers need to be more accountable for their designs; they must avoid "pride of authorship" and recognize that changes in the design might be needed. Assumptions, objectives, alternatives considered, and rationale used must be documented throughout the design process for the audit process to be effective. Operating personnel must be involved in the design process. To the extent it is possible to do so, the design must be based on "facts" rather than guesses; it must be rooted in analysis, not hypotheses.

Operating personnel also need to be more accountable. It is important that they follow through on commitments. If reductions in personnel, equipment, and/or space were promised, then they should be made. Operating personnel need to be a part of the solution, not a part of the problem. Resistance to change, lack of confidence in the new system, and a lack of patience with the system do not create the proper climate for an effective audit.

## *14.6* SUMMARY

The preparation of a facilities plan is an important step in developing a workable facility. Several important details must be resolved and considerable fine-tuning must take place during the development of the plan. A facilities plan consists of a plot plan and a layout plan. The critical factors to be considered in developing a plot plan are expansion, flow patterns, energy, and aesthetics. The establishment of a layout plan should follow a systematic procedure. Different approaches are available for presenting the plan, with the computer playing an increasingly important role in both the development of the plan and in presenting visually the plan, itself.

Written and oral reports are important portions of a successful facilities planning project. As much care should be spent on selling a facilities plan using written and oral reports as is spent on developing it. The facilities planner's mission is not to make plans for facilities, but rather to implement facilities plans. A well-written report and a good oral presentation are the initial steps in bringing about implementation.

The implementation and maintenance of facilities plans are also important steps in the facilities planning cycle presented in Chapter 1. It is during the implementation phase that the results of design efforts become evident. By maintaining the plan there is an assurance that the benefits gained will be lasting, rather than transitory.

Not only does this chapter provide a coverage of the final steps of facilities planning (at least until the process is repeated), but it also brings to an end our opportunity to share with you our interest in and commitments to the subject of facilities planning. We have found the subject to be challenging and exciting from our perspectives of educators, researchers, consultants, engineers, and managers. Hopefully, you will find that the coverage of the subject presented herein better prepares you to meet successfully the challenges of facilities planning and infects you with the excitement we feel. Best wishes to you in designing new facilities and in improving existing facilities.

# BIBLIOGRAPHY

1.  Apple, J. M., *Plant Layout and Material Handling,* 3rd ed., John Wiley & Sons, New York, 1977.
2.  Ballinger, R. A., *Layout and Graphic Design,* Van Nostrand Reinhold, New York, 1970.
3.  Blair, E. L., and Miller, S., "Interactive Approach to Facilities Design Using Microcomputers," *Computers and Industrial Engineering,* vol. 9, no. 1, January 1985, pp. 91–102.
4.  Brickner, W. H., *Managing Change,* Lansford Publishing, San Jose, CA, 1974.
5.  Francis, R. L., McGinnis, L. F., and White, J. A., *Facility Layout and Location: An Analytical Approach,* Second Edition, Prentice-Hall, Englewood Cliffs, NJ, 1992.
6.  Gupta, R. M., "Flexibility in Layouts: A Simulation Approach," *Material Flow,* vol. 3, no. 4, June 1986, pp. 243–250.
7.  Hales, H. L., *Computer-Aided Facilities Planning,* Marcel Dekker, New York, 1984.
8.  Irwin, P. H., and Langham, F. W. Jr., "The Change Seekers," *Harvard Business Review,* vol. 44, no. 1, pp. 81–92, January–February 1966.
9.  Konz, S., *Facility Design,* John Wiley & Sons, New York, 1985.
10. Krick, E. V., *Methods Engineering,* Wiley, New York, 1962.
11. Kusiak, A., and Heragu, S. S., "The Facility Layout Problem," *European Journal of Operational Research,* vol. 29, 1987, pp. 229–251.
12. Lawrence, P. R., "How to Deal with Resistance to Change," *Harvard Business Review,* vol. 47, no. 1, pp. 4–12, January–February 1969.
13. Lefferts, R., *Elements of Graphics: How to Prepare Charts and Graphs for Effective Reports,* Harper & Row, New York, 1981.

14. Malouf, L. G., "Initiating Change Within a Systems Framework—How Do You Obtain Employer and Employee Commitment and Improve Results?," presentation to the National Material Handling Forum, Material Handling Institute, Pittsburgh, PA, 1975.

15. Mullins, C. J., *The Complete Writing Guide to Preparing Reports, Proposals, etc.*, Prentice-Hall, Englewood Cliffs, NJ, 1980.

16. Muther, R., *Practical Plant Layout,* McGraw-Hill, New York, 1955.

17. Muther, R., *Systematic Layout Planning,* Industrial Education Institute, Boston, 1961.

18. Muther, R., and Hales, H. L., *Systematic Planning of Industrial Facilities,* vol. I, Management & Industrial Research Publications, Kansas City, MO, 1979.

19. Muther, R., and Hales, H. L., *Systematic Planning of Industrial Facilities,* vol. II, Management & Industrial Research Publications, Kansas City, MO, 1980.

20. O'Brien, C., and Abdel Barr, S. E. Z., "An Interactive Approach to Computer Aided Facility Layout," *International Journal of Production Research,* vol. 18, no. 2, 1980, pp. 201–211.

21. Reed, R., *Plant Layout: Factors, Principles, and Techniques,* Richard D. Irwin, Homewood, IL, 1961.

22. Reed, R., *Plant Location, Layout, and Maintenance,* Richard D. Irwin, Homewood, IL, 1967.

23. Searls, D. B., "Installation: The Key to a Successful Operation," presentation to the 1980 Automated Material Handling and Storage Systems Conference, April 1980.

24. Sims, R. E., Jr., *"Selling Your Solution to Management,"* presentation to the 28th Annual Material Handling Management Course, American Institute of Industrial Engineers; Airlie, VA, June 1981.

25. Souther, J. W., *Technical Report Writing,* 2nd ed., Wiley, New York, 1977.

26. Sule, D. R., *Manufacturing Facilities: Location, Planning, and Design,* 2nd ed., PWS-Kent Publishing Company, Boston, 1994.

27. Tompkins, J. A., "Justification and Implementation: The Critical Link," presentation to the 1981 Automated Material Handling and Storage Systems Conference, Philadelphia, PA, September 1981.

28. White, J. A., *Yale Management Guide to Productivity,* Eaton Corp., Philadelphia, PA, 1979.

29. White, J. A., "Auditing the System After Installation," presentation to the 1980 Automated Material Handling and Storage Systems Conference, April 1980. Subsequently published in *Industrial Engineering,* vol. 12, no. 6, pp. 20–21, June 1980.

# PROBLEMS

14.1 Draw a plot plan of the building housing the classroom for this course. How should the classroom building be expanded if the demand for classrooms doubles? Could the facility have been designed to accommodate expansion better? Explain.

14.2 Why should a facilities planner be concerned with aesthetics?

14.3 How could computer-aided layout techniques be used to study flow patterns outside a facility?

14.4 Describe specific situations where two-dimensional layouts would be superior to three-dimensional layouts and vice versa.

14.5 Describe specific situations where different two-dimensional approaches would be utilized.

14.6 Describe specific situations where different three-dimensional approaches would be utilized.

**14.7**  How could computer-aided layout techniques be integrated with computer graphics? Would this be useful?

**14.8**  Describe specific situations where block templates or models should be used instead of contour templates and models and vice versa?

**14.9**  List five reasons for people resisting change, other than those cited in the chapter.

**14.10**  List five additional ways resistance to change can be overcome.

**14.11**  Visit a manufacturing facility and prepare a written audit report documenting your findings.

**14.12**  Visit a nonmanufacturing facility and prepare a written audit report documenting your findings.

**14.13**  Perform an audit of the university, focusing on the location of facilities and personnel flow.

**14.14**  How might multimedia presentations be used to sell a facilities plan?

**14.15**  Consider the year 2020. What changes in technology do you believe will impact significantly the preparation and presentation of facilities plans?

# INDEX